Electrical Codes, Standards, Recommended Practices and Regulations

T0349013

Electrical Codes, Standards, Recommended Practices and Regulations

An Examination of Relevant Safety Considerations

Robert J. Alonzo P.E.

AMSTERDAM • BOSTON • HEIDELBERG • LONDON • NEW YORK • OXFORD • PARIS
SAN DIEGO • SAN FRANCISCO • SINGAPORE • SYDNEY • TOKYO

William Andrew is an imprint of Elsevier

William Andrew is an imprint of Elsevier
The Boulevard, Langford Lane, Kidlington, Oxford, OX5 1GB, UK
30 Corporate Drive, Suite 400, Burlington, MA 01803, USA

First edition 2010

Notices

Knowledge and best practice in this field are constantly changing. As new research and experience broaden our understanding, changes in research methods, professional practices, or medical treatment may become necessary.

Practitioners and researchers must always rely on their own experience and knowledge in evaluating and using any information, methods, compounds, or experiments described herein. In using such information or methods they should be mindful of their own safety and the safety of others, including parties for whom they have a professional responsibility.

To the fullest extent of the law, neither the Publisher nor the authors, contributors, or editors, assume any liability for any injury and/or damage to persons or property as a matter of products liability, negligence or otherwise, or from any use or operation of any methods, products, instructions, or ideas contained in the material herein.

British Library Cataloguing in Publication Data
Alonzo, Robert J.
   Electrical codes, standards, recommended practices and regulations: an examination of
   relevant safety considerations.
   1. Electrical engineering–Safety measures.
   2. Electrical engineering–Safety regulations.
   3. Electrical engineering–Standards.
   I. Title.
   621.3'0289-dc22

Library of Congress Control Number: 2009938942

ISBN: 978-0-8155-2045-0

For information on all William Andrew publications
visit our website at elsevierdirect.com

Printed and bound by CPI Group (UK) Ltd, Croydon, CR0 4YY

Transferred to Digital Print 2011

Working together to grow
libraries in developing countries

www.elsevier.com | www.bookaid.org | www.sabre.org

ELSEVIER   BOOK AID International   Sabre Foundation

Contents

Preface

It is intention of this text to provide the reader assistance in developing a basic understanding of the complex issue of codes, standards, recommended practices, and regulations for electrical power generation, transmission, and distribution in residential, commercial, industrial, and utility applications. General information is provided on the Canadian and American Standards Development Organizations (SDOs) responsible for the development of codes, standards and recommended practices. Basic outlines are also provided for standards development procedures; code enforcement areas; general code categories; and exposure of some titles for electrical engineering standards for power generation, transmission, and distribution in North America and internationally. Regional SDOs in Europe, Central and South America, and the Pacific Rim area were not examined in detail. The International Electrotechnical Commission (IEC) and International Organization for Standardization (ISO) SDOs are examined in some detail.

Information on the SDOs discussed was developed in part from the Internet websites for those organizations. Should additional information on those organizations be desired, the reader is referred to the SDO websites listed for those organizations in Chapter 1. The listed websites may also be used to research and purchase specific SDO standards documents.

Electrical generation, transmission, and distribution codes, standards, and recommended practices encompasses a large area of information. Some general codes, standards, and recommended practices information was utilized in some discussions; however, the reader should refer to those documents in total when attempting to develop information. SDO websites can be used to provide standards outline information, indexes, and normative references. That information can be extremely helpful in conducting general research in the selection and use of equipment and installation standards.

Codes, standards, and recommended practices are documents that are continually being created, revised, reaffirmed, or withdrawn. It is incumbent upon the reader to verify the latest code, standard, or recommended practice document number, title, validity, and effective issuance date from the appropriate Standards Development Organization sources. The standards titles presented in this book include many of the most commonly used documents involving the generation, transmission, and distribution of electrical energy in 2009. The standards titles were developed using Internet search engines, references in documents, and other means. The titles presented in this book do not represent all of the codes, standards, or recommended practice titles presently available on a specific topic, nor do they necessarily represent the latest effective title information. *The standards titles presented here should not be used as a sole source for standards document information and are presented only for general information, as well as the general education and use of the reader. Any discussion of specific information on standards presented in this book should not be construed as an official explanation of that document. The reader should always review the latest edition of any standard document in its entirety.*

The standards development procedures presented here should only be considered as general outlines of those SDO procedures as of the information available in August, 2009 and should only be used for general information purposes. SDOs have written procedures governing the submittal of new standard proposals and the revisions to or the withdrawal to existing standards. Copies of those procedures should be available from the SDOs or their websites.

SDO websites were used in obtaining some information in this book. The websites listed as references were operational as of November, 2009. Neither the author nor the publisher can be responsible for changes to available information on those websites.

The impetus behind the preparation of this book was to assist individuals assigned with the task of developing electrical power generation, transmission, and distribution equipment and materials specifications. An important segment of that work normally would include listing of *Referenced Standards*. Specifications are sometimes prepared using the term *All Applicable Codes and Standards* in its *Referenced Standards* section. That statement could be interpreted by individuals differently, according to their experience and education. Listing of specific applicable equipment or materials standards titles in the preparation of codes, standards, and recommended practices will provide a better definition of equipment or materials specified.

Information contained in this work has been obtained by the author from sources believed to be reliable. However, neither the publisher nor the author guarantee the accuracy or completeness of any information contained herein. The author and publisher shall not be responsible for any errors or omissions in this publication. They will not be responsible for any damages arising out of the use of the information contained herein for any purpose. The author and publisher are supplying general information only in this work, and are not rendering any engineering or professional services or opinions. Any standards interpretations questions should be referred to the Standards Development Organization responsible for that standard.

Acknowledgments

I would like to thank the following individuals for their help in the proof reading of chapters in this book.

Forrest M. Lotz, Jr, P.E

Charles A. Darnell, P.E

CHAPTER 1

Who, What, Where, When, Why, and How?

My initial idea in writing this book was to limit its coverage to the electrical engineering *codes, standards, recommended practices and regulations* used in the United States. However, research, quickly revealed how the international harmonization of codes, standards, and recommended practices throughout the world has impacted international trade. With the increased globalization of trade and worldwide electrical product development competition, the importance of the development of flexible, cooperative, consensus-driven harmonized American standards has become evident. To assure the continued sale of American exports throughout the world, the United States government must continue to use their influence and cooperation with foreign governments to assure the continued development of voluntary, consensus, internationally accepted American standards.

In the past, the international recognition and use of US codes and standards readily allowed the worldwide sale of American goods and services. However, with the emergence of the European Union and other regional trade organizations, the necessary compliance of American goods and services with local standards requirements and certifications has become challenging. Many emerging economies have adopted ISO and IEC standards. The growth of the use of ISO and IEC standards has challenged US competitiveness.

The electrical engineering codes, standards, and recommended practices examined in this book will include those generally involved with voluntary, consensus standards in electrical power generation, transmission, and distribution in both utilities and residential/commercial/industrial facilities. It will also examine the codes and standards used for the wire and cable aspects for power transmission and distribution. Specific communications, instrumentation, data processing, aviation, marine, automotive/trucking, mining, and railroad equipment aspects will not be examined. Limited shipboard, communications, and instrumentation cabling codes and standards will also be examined.

Anyone associated with electrical design or construction projects has been exposed to the terms *codes*, *standards,* and *recommended practices*. Exactly what do those terms imply? They may imply different things to different individuals, depending upon the individual's experience, training, and responsibility. To anyone in a governmental capacity, the terms may convey compliance with the legislatively mandated regulations and requirements before a Certificate of Occupancy can be issued on a project. To a design engineer involved with a project, it may mean fulfilling their professional responsibility to assure compliance with public safety requirements. To a state fire marshal, they may mean assuring life safety aspects during a structure fire development. Depending upon the occupancy type, it might also mean assurance that specified safety systems will provide automatic notification of the appropriate governmental agencies of a fire or other catastrophic event.

The development of codes, standards, and recommended practices is often necessitated because of substantial loss of life and property or severe personal injuries related to a problematic faulty design or construction/fabrication practices. The development of codes, standards, and recommended practices can be promulgated by industry/manufacturing groups; engineering or professional societies or organizations; or governmental agencies. The promulgating entity will establish a committee comprised of representatives of companies, professionals in the field, academia, and other interested parties, to establish the minimum criteria that will be considered to assure public safety. Implementation of those criteria can be by industry-accepted, voluntary agreements or by Authorities Having Jurisdiction.

On November 28, 1942 a fire occurred at the Cocoanut Grove Night Club in Boston. A total of 492 fatalities resulted from that fire, which was attributed to ignition of combustible decorations by a busboy lighting a match. It was reported that approximately 1000 occupants were in the building at the time of ignition. At that time there were no maximum occupancy requirements as exist today. Most fatalities were the result of inadequate operable exterior exits. The nightclub had only one operable exterior exit, the main entrance revolving door. All other doors were previously bolted shut or bricked over during prohibition. This fire was a motivating force, leading to the development and enforcement of building codes, not only in Boston, but in other cities throughout the United States.

Before proceeding with the examination of some specific codes, standards, and recommended practices, some time must be taken to examine the significance of those terms.

Codes

The Merriam–Webster On-Line Dictionary [1] defines a *code* as *"a systematic statement of a body of law; especially: one given statutory force; a system of principles or rules."* The most recognized Electrical Engineering *Code* in the United States is the National Fire Protection Association's NFPA 70®, *National Electrical Code®* (NEC®). Although it is generally accepted as a nationally accepted consensus code in electrical engineering, it must still be adopted by individual legislative bodies, mandating its acceptance and use by law to the *Authorities Having Jurisdiction (AHJ).*

The *Authority Having Jurisdiction* could be a governmental entity, which through legislative enactment, can mandate by law, adherence to specific engineering practices or *codes*. For example, a municipal government may mandate that a certain edition of the *National Electrical Code* be adhered to in the design of structures. Before a building *Certificate of Occupancy* can be granted, the municipal code enforcement agency must assure that the applicable portions of the approved edition of the NEC® were followed in the design and construction of the structure. Under that scenario, a governmental *regulation* or law mandated the use of a *code*.

The *National Electrical Code* will be reviewed; however, a more detailed examination of that Code will be pursued in Chapter 5. The NEC® is considered an *open-consensus document*. Anyone can promulgate a change to that document or submit a public comment. All such proposals and comments are subject to intensive review. A review and amendment process of the Code is conducted automatically over a three-year cycle. Proposals for change are submitted, reviewed, debated, and voted upon by members of the Code Committee, with final approval by the NFPA Standards Council. Any approved changes are included in the next edition of the document.

The National Fire Protection Association, the organization responsible for the publication of the NEC, indicates:

> The National Electrical Code has become the most widely adopted code in the United States – it is the standard used in all 50 states and all U.S. territories. Moreover, it has grown well beyond the borders of the United States and is now used in numerous other countries. Because the code is a living document, constantly changing to reflect changes in technology, its use continues to grow. [2]

Standards

A very good definition of *standard* is presented in the beginning of ANSI/IEEE Standard 80, *IEEE Guide for Safety in AC Substation Grounding*. That document was produced by the Institute of Electrical and Electronic Engineers (IEEE) and approved as a national consensus standard by the American National Standards Institute (ANSI). The following statement was written by the Secretary, IEEE Standards Board in the "Foreword" to that document. It stated:

> IEEE Standards documents are developed within the Technical Committees of the IEEE Societies and the Standards Coordinating Committees of the IEEE Standards Board. Members of the committees serve voluntarily and without compensation. They are not necessarily members of the Institute. The standards developed within IEEE represent a consensus of the broad expertise on the subject within the Institute as well as those activities outside of the IEEE which have expressed an interest in participating in the development of the standard.

> Use of an IEEE Standard is wholly voluntary. The existence of an IEEE Standard does not imply that there are no other ways to produce, test, measure, purchase, market, or provide other

goods and services related to the scope of the IEEE Standard. Furthermore, the viewpoint expressed at the time a standard is approved and issued is subject to change brought about through developments in the state of the art and comments received from users of the standard. Every IEEE Standard is subject to review at least once every five years for revision or re-affirmation. When a document is more than five years old, and has not been reaffirmed, it is reasonable to conclude that its contents, although still of some value, do not wholly reflect the present state of the art. Users are cautioned to check to determine that they have the latest edition of any IEEE Standard. [3]

The above explanation clearly illustrates the difference between a *code* and a *standard*. Use of a *standard* is wholly voluntary, whereas the use of a *code* may be voluntary or mandated by law. Also, a *standard* does not mandate that there is only one way that a product or procedure can be engineered. That may not necessarily be the case with a *code*. The IEEE Standards Committee allows any proposed change request to be submitted by any interested party, either by IEEE members or others. Also, proposed changes to a *code* involve public notification of the proposed changes for rigorous review. That process may or may not be employed to that extent for changes to a *standard* promulgated by other standard making organizations, particularly organizations associated with special interest groups such as manufacturers. Standards adopted by the American National Standards Institute would require a rigorous public input and organization review process.

Recommended Practices

The purpose of an electrical *recommended practice* is to identify electrical features of systems, products or procedures, which may be important. Recommended practices are electrical design and installation practices, which have been generally accepted in the electrical industry as safe, reliable, efficient, and maintainable. Recommended practices are not considered to be a fixed rule, a *code*, or a *standard*. It is anticipated that sound *engineering judgment* will be utilized when implementing a recommended practice. It is also not the intent that recommended practices should supersede federal, state, or local regulations in their implementation. In summary, recommended practices are generally universally accepted industry rule(s) or practice(s) regarding design, operation, or maintenance of equipment, facilities, installations or procedures.

To better understand this term, we will also examine the term *engineering judgment*, which was used in its definition. A good definition for that term was developed by Alonzo as follows:

> Engineering judgment is the scientific process by which a design, installation, operation/ maintenance or safety problem is systematically evaluated. It utilizes knowledge and experience gained on the subject and applies the scientific method of analysis. It includes gathering all necessary information about the project or problem and systematically sorting the information, to make an informed decision or take action. Part of the evaluation process would include some

sort of hazard or risk analysis, if applicable, and a review of applicable codes, standards, and recommended practices. A thorough knowledge of the process, equipment, or situation is essential in making an engineering judgment. Alternative solutions must be analyzed as well as a critical analysis of any final conclusions or recommendations.

Systematic documentation of the evaluation process is essential in engineering judgment. This would include any calculations, risk or hazard analysis, cause and effect diagrams, list of applicable codes, standards, or recommended practices, etc. It is essential to document the process for both historical and liability reasons. [4]

Who, What, Where, When, and How

To aid in the examination of the terms codes, standards, and recommended practices it is essential to begin by listing and examining the *standards development organizations* and *governmental agencies* that may have been involved in their development or implementation for electrical engineering purposes. They include many of the electrical engineering professional societies, as well as electrical manufacturing industry groups; electrical generating and transmission/distribution industry groups; national standards organizations; independent testing organizations; and governmental agencies. Several of the promulgating agencies and organizations, both in the United States and internationally, will be examined. Each review is designed to provide general information regarding the organization's background, membership, relationships to other standards making organizations and standards making procedures in that organization. Should the reader desire more detailed information, it is recommended to either contact the organization or research its website or other available materials.

The review of standards organizations should begin by examining the role of the major standards organizations both in the United States and internationally. The American National Standards Institute (ANSI) has been designated as the American National Standards approval organization in the United States. It has also been designated as the official United States representative on international standards organizations. Its membership is composed of representatives of several major engineering, professional, and manufacturing standards organizations. Although it is not responsible for the issuance of all American standards, it does jointly issue all documents that have been designated as an *American National Standard*. For example, the Institute of Electrical and Electronic Engineers' (IEEE) standard *IEEE Guide for Safety in AC Substation Grounding*, ANSI/IEEE Standard 80, is jointly issued as an American National Standard. IEEE's Substation Committee of the IEEE Power Engineering Society sponsored the document. The IEEE Standards Board approved the document. It was also approved by the American National Standards Institute as an American National Standard.

American National Standards Institute (ANSI)

American National Standards Institute (ANSI)
1819 L Street, NW
Washington, DC 20036
Phone: (202) 293-8020
Internet: http://www/ansi.org/

The American National Standards Institute (ANSI) was established in 1918 with its mission

> To enhance both the global competitiveness of US business and the US quality of life by promoting and facilitating voluntary consensus standards and conformity assessment systems, and safeguarding their integrity. [5]

Its members are

> Comprised of Government agencies, Organizations, Companies, Academic and International bodies, and individuals. The American National Standards Institute (ANSI) represents the interests of more than 125,000 companies and 3.5 million professionals. [5].

Internationally, ANSI is affiliated as the official US representative to the International Organization for Standardization (ISO). It is also associated with the International Electrotechnical Commission (IEC) via the US National Committee (USNC) and holds membership in the International Accreditation Forum (IAF). Regionally, ANSI maintains membership in standards organization in the Pacific Area (Pacific Area Standards Congress [PASC] and the Pacific Accreditation Cooperation [PAC]), as well as in North and South America (Pan American Standards Commission [COPANT] and Inter American Accreditation Cooperation [IAAC].

Examining the block diagrams [6] in Figures 1.1 and 1.2 below illustrates the complexity and interaction of standards development in today's globalized economy.

ANSI has been associated with the coordination of voluntary standards and conformity assessment in the United States since its inception. It is composed of a diverse membership, including industry standards organizations, academia, professional and technical societies, trade commissions, labor and consumer representatives, et al.

General Information

In 1918 five major existing professional technical societies cooperated to establish an impartial national organization to coordinate standards development. They also sought to develop and approve national consensus standards, thus eliminating confusion with the ultimate standards users. The five initiating professional societies included:

1. American Institute of Electrical Engineers (AIEE)

2. American Society of Mechanical Engineers (ASME)

3. American Society of Civil Engineers (ASCE)

4. American Institute of Mining and Metallurgical Engineers ((AIMME)

5. American Society for Testing Materials (ASTM)

These organizations then invited and granted membership to US Departments of War, Navy, and Commerce as co-founders. It original founding name was the American Engineering Standards Committee (AESC).

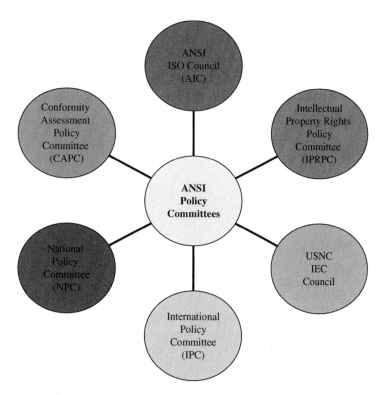

Figure 1.1: ANSI Policy Committees' memberships
Source: ANSI: http://www.ansi.org/about_ansi/organization_chart/chart.aspx?menuid=1

Figure 1.2: ANSI International Regional Committees
Source: ANSI: http://www.ansi.org/about_ansi/organization_chart/chart.aspx?menuid=1

ANSI has the responsibility for issuing national standards for accident prevention and the coordination of national safety codes. The organization has today issued some 1200 safety standards which are designed to protect consumers and the workforce. Since its duties include many engineering discipline areas, the organization has approved approximately 10,500 national standards in mining, electrical engineering, mechanical engineering, construction, and highway traffic safety [7].

ANSI has also cooperated with governmental agencies on safety issues. In 1976 the organization established a joint coordinating committee with the US Department of Labor, Occupational Safety and Health Administration (OSHA). The committee's role was to establish private-public sector communications for voluntary standards which affect safety and health in the workplace. Based on the success of that endeavor, a second joint coordinating committee was established in 1982 with the US Department of Commerce, Consumer Product Safety Commission (CPSC). Its role was to improve standards activities related to consumer products.

ANSI has developed an Internet search engine for standards. It is http://www.nssn.org/. This ANSI Web resource allows information searches in nine different databases [7], including:

1. ANSI Standards

2. Other US Standards

3. ISO/IEC/ITU [8] Approved Standards

4. Non-US National and Regional Standards

5. US DoD [9] Approved Standards

6. ANS Under Development

7. ISO/IEC Development Project

8. DoD Development Project

9. Code of Federal Regulations (CFR) References

This Internet search engine allows the user to check those databases for available standards. It also allows the selection of specific organizations such as ANSI, ASME, IEEE, etc. in which to search. Search results provide the data by document title, number, and scope. Procurement sources are also provided.

ANSI has developed *the United States Standard Strategy* [10]. That strategy provides focus on individual sectors of standards development supported by a dynamic infrastructure. This recognizes that those individuals, groups, governmental entities, etc. involved in a specific sector of standards development are best equipped and most efficient to address the issues and working methods in that area. It recognizes that no single standardization system can satisfy the needs for all standards development. However, the infrastructure provided by ANSI allows them to facilitate and mediate between groups [11]:

1. when cross-sectional issues arise;

2. when sector definitions change; or

3. in venues where a single national voice is required.

To aid standards groups which their development of jointly approved standards, ANSI developed *Essential Requirements: Due Process Requirements for American National Standards*. Section 1.0, January 2009 Edition, Pages 4 and 5 of that document defines the term due process, which was used in the title and is reflected throughout the document's requirements.

> Due process means that any person (organization, company, government agency, individual, etc.) with a direct and material interest has a right to participate by: a) expressing a position and its basis, b) having that position considered, and c) having the right to appeal. Due process allows for equity and fair play. The following constitute the minimum acceptable due process requirements for the development of consensus:
>
> 1.1 Openness
> 1.2 Lack of dominance
> 1.3 Balance
> 1.4 Coordination and harmonization
> 1.5 Notification of standards development
> 1.6 Consideration of views and objections
> 1.7 Consensus vote
> 1.8 Appeals
> 1.9 Written procedures
> 1.10 Compliance with normative American National Standards policies and administrative procedures

These procedures apply to any standard making organization desiring approval as an American National Standard.

International Electrotechnical Commission (IEC)

> International Electrotechnical Commission (IEC)
> 3, rue de Varembé
> P.O. 131, CH-1211
> Geneva 20, Switzerland
> Phone: (+41) 22 919 02 11
> Internet: http://www.iec.org/

The International Electrotechnical Commission (IEC) is an international organization that prepares and publishes standards. It was founded in June, 1906 in London. In 1948 its offices moved to

Geneva, Switzerland. Although its membership was originally primarily European, today it encompasses some 136 countries of which 67 are members and 69 have *Affiliate Country Programme* status. Although its headquarters is in Geneva, it operates regional centers in Singapore; San Paulo, Brazil, and Boston, Massachusetts. The United States is represented by the American National Standards Institute's (ANSI) United States National Committee/IEC (USNC/IEC).

The IEC members are composed of national committees, each representing its nation's electrotechnical interests. The committees may consist of representatives from manufacturing, distribution and sales, consumers, professional societies, trade unions, academia, governmental agencies, national standards bodies, and other interests. The IEC is responsible for issuing electrotechnical standards.

Standards are utilized as the technical basis or references in international contracts, tenders, and trade. All electrotechnical categories are included in the IEC charter. Those technologies include:

> electronics, magnetics, electromagnetics, electroacoustics, multimedia, telecommunication, and energy production and distribution, as well as associated general disciplines such as terminology and symbols, electromagnetic compatibility, measurement and performance, dependability, design and development, safety and the environment. [12]

IEC products [13] include the following two categories of publications:

International Consensus Products:

- International Standards (full consensus) (IS)

- Technical Specification [full consensus not (yet) reached] (TS)

- Technical Reports (information different from an IS or TS)

- Publicly Available Specifications

- Guides (non-normative publications)

Limited Consensus Products

- Industry Technical Agreement

- Technology Trend Assessment

An *International Standard* (IS) is defined as:

> a document, established by consensus and approved by a recognized body, that provides, for common and repeated use, rules, guidelines or characteristics for activities or their results, aimed at the achievement of the optimum degree of order in a given context. An international standard is a standard adopted by an international standardizing/standards organization and made available to the public. [14]

It is further defined by the IEC as:

> a normative document, developed according to consensus procedures, which has been approved by the IEC National Committee members of the responsible committee in accordance with Part 1 of the ISO/IEC Directives as a committee draft for vote and as a final draft International Standard and which has been published by the IEC Central Office. [15]

The word *"consensus"* is very important in the above definition. It is also used in describing codes and standards developed by ANSI, NEMA, and other standards organizations in the United States. "Consensus" indicates that there is a common viewpoint among the standards committee, including representatives from professional technical organizations, academia, manufacturers' representatives, governmental representatives, and others on the committee. The IEC International Standards (IS) require consensus approval within its membership. It should also be noted here that in order for those International Standards (IS) to become effective in any country, it must be adopted by whatever legal mechanism has been established by that country. The International Standards are voluntary and implementation by the IES does not mean that they must be universally accepted and implemented by any sovereign government.

We will briefly examine the remaining IEC consensus document products. The next is a *Technical Specification* (TS). It is also a consensus product and normative in nature; however, it is one that has not received sufficient committee support to be approved as an IS. It should be noted here that the TS might still become an IS at some future time and may be under technical development. A TS requires a two-thirds approval of the initiating technical committee or subcommittee participating members.

A *Technical Report* (TR) would be considered more of a descriptive document than normative like the IS or TS. This document may simply be a collection of data and must only be approved by a simple majority of IEC technical committee or subcommittee participating members.

A *Publicly Available Specification* (PAS) is considered a normative document, like the Technical Report, and is a consensus among experts. The PAS is developed as an urgent market-driven normative industry consortia document under the authority of the IEC.

We will now examine the two IEC *Limited Consensus Products*. They include *Industry Technical Agreements* (ITA) and *Technology Trend Assessments* (TTA). IEC defines an ITA as:

> a normative or informative document that specifies the parameters of a new product or service. It is developed outside the technical structures of the IEC ... [16]

It is an industry new product or market-enabling document; however, it does not deal with all health, safety or environmental aspects. The ITA relies on the *"intrinsic seal of approval"* [17] by the IEC to achieve market acceptance of a new technology.

A *Technology Trend Assessment* is typically considered by the IEC to be *"the result of pre-standardization work or research"* [18]. It may become a standard in the near future and may be issued during the early stages of the technology development.

Similarly to American standards organizations, the IEC publications are subject to *"maintenance cycles"*. It will typically be issued as valid for an established period of time. At the end of that period, it may be subject to amendment or revision. Should the publication be considered obsolete or of no further technical or commercial use, it can be withdrawn.

On November 14, 2002 a *joint agreement of cooperation* was announced between the IEC and the Institute of Electrical and Electronic Engineers (IEEE). That cooperation agreement involved a *dual-logo* arrangement in which some IEEE standards will be accepted and adopted by the IEC and will carry *logos* of both organizations. The IEEE Standards will not be universally accepted; but will be jointly reviewed for IEC's standardization procedures. The selected standards will be processed by the appropriate IEC technical committees. Upon completion of the process and acceptance procedure, they will be published as *IEC/IEEE Dual Logo International Standards* and will require adoption by IEC member countries before they can become national standards. The agreement allows the IEC to develop the final versions of those standards published in the official languages of the IEC.

International Organization for Standardization

> International Organization for Standardization
> 1, ch. de la Voie-Creuse,
> Case postale 56,
> CH-1211 Genève 20, Switzerland
> Phone: (+41) 22 749 01 11
> Internet: http://www.iso.org

The International Organization for Standardization (ISO) was established in February, 1947 in Geneva, Switzerland. It consisted of delegates from 25 countries. As of 2009 it has evolved into "a network of national standards institutes from 162 countries, one member per country" [19]. It is a non-governmental agency, with members from both governmental and private sectors. A block diagram on the ISO website, presented in Figure 1.3, provides an overview of the organization's structure.

The ISO is structured with three different membership levels, including *member bodies*, *correspondent members*, and *subscriber members*. *Member bodies* are full members in the organization and have one vote when approving standards. The United States *member body*

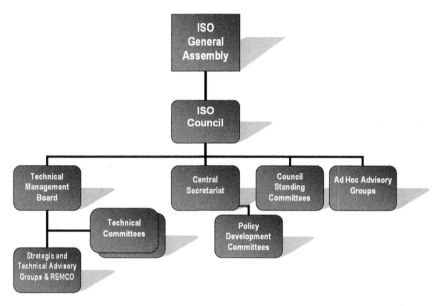

Figure 1.3: ISO Organization Block Diagram [20]

representative is the American National Standards Institute's (ANSI) United States National Committee (USNC). *Correspondent members* participate in policy or technical bodies as observers, without voting rights. Afghanistan is an example of this type of membership. *Subscriber members* are standards institutes from countries with small economies, who wish to maintain a presence in the standardization process. Antigua and Barbuda are examples of this type of membership.

The ISO offers individuals or other entities the ability to participate in standards development on a non-voting basis. This may be done as *Experts on National Delegations*, who are individuals chosen by national member institutes that participate in a technical capacity on ISO committees. *National Mirror Committees* may also be established by national member institutes, composed of individuals or others to aid in the establishment of a national consensus for each national delegation. *Liaison status* can also be granted to international organizations or associations to ISO technical committees. Their purpose is to participate through debates in the development of committee consensus.

Standards Development

A new standard proposal can be made to an ISO member by any business or industry group. The member has the responsibility to present the proposal to the full ISO membership. If the proposal is deemed worthy of acceptance and further study, it will be assigned to an existing technical committee or a new committee(s) established for activity scopes that are not already covered by any of the existing committees.

ISO also has three *policy development committees*, who may assist in standards development activities in cross-sector situations. They include:

- CASCO (conformity assessment)

- COPOLCO (consumer policy), and

- DEVCO (developing country matters)

Actual standards development work is the responsibility of the Technical Committees (TC). They are comprised of business, industry, and technical experts that have requested the standards development. Assistance during the development process may also be provided by representatives of governmental agencies and academia, as well as testing laboratories, consumer advocates or groups, and other interested parties.

Once a technical committee or subcommittee is established, a secretariat is appointed from the member body. In order to increase developing countries' participation in international standards development, a developing country's representative can be assigned as a *"twin"* with a *member body*. A twin committee secretariat may also be appointed. Experts from developing countries may also be appointed as vice-chairpersons on technical committees or subcommittees.

The International Standards process consists of six stages, which are presented in Figure 1.4:

Stage 1: Proposal stage provides confirmation that an International Standard is required. The new work proposal (NP) is submitted to the appropriate technical committees (TC)

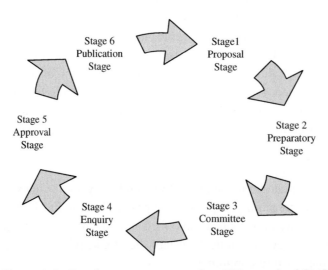

Figure 1.4: Development process of an ISO standard [21]

or subcommittees (SC) for a vote. In order for the NP to be accepted, it must receive a majority vote of the TC/SC members in favor and have at least five members to agree to participate in the process. A project leader will be appointed upon acceptance of the work.

If the proposal submitted is an existing standard developed by another organization and is accepted for review, it then comes under the *Fast-Track Procedure* process and would skip Stages 2 and 3. It would then be submitted to the ISO member bodies as a draft International Standard (DIS). If the proposal has been developed by an ISO recognized international standards body, such as ANSI, it would proceed to Stage 5 as a final draft International Standard (FDIS).

Stage 2: Preparatory stage includes development of a working draft by the technical committee/subcommittee (TC/SC). The working group must arrive at an agreement that the draft is the best technical solution for the proposal. This may require several draft versions, before it is submitted back to the working group parent committee for consensus development.

The Committee stage (*Stage 3*) involves registering the draft by the ISO Central Secretariat and distribution for comments to the participating (P)-members of the TC/SC. Consensus must be reached on the technical content and may require development of several committee drafts. Upon reaching final consensus, the finalized text will be submitted as a draft International Standard (DIS).

Stage 4: Enquiry stage consists of submittal of the DIS to all ISO member bodies. Voting and comments on that proposal must be completed within five months. A two-thirds majority of the TC/SC P-members must approve the submittal for it to be considered a Final Draft International Standard (FDIS). It is also a requirement for passage that not more than 25% of the total cast votes were negative. Should the proposed standard not meet the above criteria, it would be returned to the originating TC/SC for additional study. When a revised document is produced, it would be circulated as a draft International Standard (DIS) for voting and comments.

The Approval Stage (*Stage 5*) involves circulation of the FDIS to all ISO member bodies. A final *Yes/No* vote must be taken within two months. The proposal will be accepted as an International Standard with a two-thirds majority approval of the TC/SC P-members, with not more than 25% negative votes of the total votes cast. Should the proposal not meet the approval criteria, it will be sent back to the originating TC/SC for reconsideration. Any technical reasons submitted with the negative votes will also be submitted to the originating TC/SC committee.

The Publication Stage (*Stage 6*) occurs once approval of a final draft of the International Standard has been approved. No technical comments will be accepted at this stage, only minor editorial changes will be considered. The final text will then be sent to the ISO Central Secretariat for publication.

It is required that all International Standards must be reviewed within three years of publication or within five years after initial review by all ISO member bodies. During that review process, the standard must be confirmed, revised, or withdrawn. That process requires a majority vote of the P-members of the TC/SC.

In 1987, the ISO and the IEC (International Electrotechnical Commission) began joint cooperation in the release of international standards. The ISO/IEC Joint Technical Committees (JTC) and Directives were established. In June, 1991 ISO and the European Committee for Standardization (CEN) reached an agreement in Vienna on an Agreement on Technical Cooperation [22]. An agreement for cooperation in developing international standards was reached in April, 2008 between the ISO and the Institute of Electrical and Electronic Engineers (IEEE) [23]. That agreement will facilitate the process for joint standards development and the adoption of each other's standards. The IEEE is the largest professional technical society in the world and is headquartered in Piscataway, New York.

Association of Edison Illuminating Companies (AEIC)

> Association of Edison Illuminating Companies (AEIC)
> P.O. Box 2641
> Birmingham, Alabama 35291
> Phone: (205) 257-2530
> Internet: http://www.aeic.org/

The Association of Edison Illuminating Companies (AEIC) was established in 1885 by Thomas A. Edison *"to license the illuminating companies to use Edison's inventions and patents"* [24]. It later developed into an association of electric utility companies. Its founding intention was to share utility company views and experiences. It fostered that goal by establishing committees to study utility system operation.

> AEIC's members are electric utilities, generating companies, transmitting companies, and distributing companies – including investor-owned, federal, state, cooperative and municipal systems – from within and outside the United States. Associate members include organizations responsible for technical research and for the promoting, coordinating and ensuring the reliability and efficient operation of the bulk power supply system. [25]

The organization has established six technical committees to assist its members, including:

- *Cable Engineering* – "Provides technical data relating to the quality, physical design, operating conditions and new developments of high-voltage underground and cable accessories that are used for electric utility power delivery systems. The committee publishes specifications and guides in the interest of promoting safe, economical and reliable power cable and accessories." [26]

- *Load Research* – "Promotes responsible load research and analysis in the electric utility industry. The *Committee* develops and disseminates *source material* on the conduct of load research and its appropriate application through annual reports, workshops and seminars, as well as through bi-annual conferences." [27]

- *Meter and Service* – "Provides direction for the industry by studying new technology and reporting operating experience of electric metering equipment and the introduction of service entrance conductors into customer facilities. The Committee maintains representation on ANSI, EPRI, UL, and industry committees to promote metering standardization and research. AEIC's Meter and Service Committee conducts a Joint National Metering Conference with EEI Metering Committee twice a year." [28]

- *Power Apparatus* –"Provides communication between electric utilities and manufacturers of major electric utility power apparatus to encourage the availability of the highest quality and most economical products, consistent with utility needs. The *Committee* also meets with those organizations responsible for research, specifications, standards and safety." [29]

- *Power Delivery* – "Identifies and assesses technological, economical, political, and regulatory issues that will affect the planning, design, construction, maintenance and/or operation of electric utility power delivery systems. It provides a forum for exchanging ideas and exploring changes to improve the delivery of electric power from the generating station to the customer." [30]

- *Power Generation* – "Promotes technological advances in the power generation field by providing a forum for dialogue between manufacturers and users to exchange industry needs and technology developments. The Committee addresses or causes industry resources to be applied to areas of concern." [31]

The AEIC has established a list [32] of specific approved industry specifications, *standards*, and references from the following organizations for use by its member utilities:

- American National Standards Institute (ANSI)

- Electric Power Research Institute (EPRI)

- International Electrotechnical Commission (IEC)

- Institute of Electrical and Electronic Engineers (IEEE)

- Insulated Cable Engineers Association (ICEA)

- National Electrical Manufacturers Association (NEMA)

In addition, the AEIC has developed a series of specifications for utility industry cables, their installation, testing, and use.

American Institute of Chemical Engineers (AIChE)

American Institute of Chemical Engineers (AIChE)
345 East 47th Street
New York, NY 10017
Phone: (212) 705-7338
Internet: http://www.aiche.org/

The first meeting of the American Institute of Chemical Engineers (AIChE) occurred in Philadelphia in 1908. It established restrictive membership requirements and stressed practical chemical engineering over academics. Its membership requirements necessitated a minimum practical manufacturing experience, proficiency in chemistry, minimum age requirements, and gave manufacturing experience credit for an academic degree. The AIChE also established an accreditation system for chemical engineering curricula in colleges and universities.

Although it might seem unlikely that the AIChE would have any codes, standards, recommended practices, or regulations that might be of importance or use in electrical engineering, some of their guidelines deal with static electricity generation mitigation for chemical process flow in piping and vessels. Chemical flammability data is also published and is of significant importance in electrical hazardous area design. Guidelines for hazard evaluation procedures are also published by the AIChE and can be useful in assisting an electrical engineer's design safety evaluation process.

American Petroleum Institute (API)

American Petroleum Institute (API)
1220 L. Street, Northwest
Washington, DC 20005
Phone: (202) 682-8000
Internet: http://www/api.org/

The American Petroleum Institute (API) [33] is a national trade association representing America's oil and natural gas industry. It was established on March 20, 1919 and is composed of approximately 400 corporate members, representing exploration, drilling, production, pipeline, marine transportation, refining, service and supply, and engineering interests in the industry. Because of shortages with drilling equipment during World War I, it was recognized that the oil industry did not have uniformity of pipe sizes, threads, and couplings. That resulted in the organization's development of industry-wide standards starting in 1924.

API produces and maintains more than 500 codes, standards, and recommended practices which are recognized and implemented in both the United States and internationally. Those codes, standards, and recommended practices have been adopted by governmental organizations, such as the United States Department of the Interior, Minerals Management Service (MMS). The International Organization of Standardization (ISO) has also adopted some of API's codes, standards, and recommended practices.

API has developed consensus standards. That process began in 1924:

> API is an American National Standards Institute (ANSI) accredited standards developing organization, operating with approved standards development procedures and undergoing regular audits of its process. [34]

Its codes, standards, and recommended practices cover every segment of the industry. Of electrical engineering interest are those involving electrical hazardous classification for onshore, offshore, and marine drilling, production, processing, refining, and transportation facilities. Also of interest are those involving lightning and static electricity protection, motors, electrical installations, safety systems, etc.

API has established several major standards committees and secretariats, each concerned with specific areas, equipment, and processes. They include [35]:

- *Executive Committee on Standardization of Oilfield Equipment and Materials (ECS) –* "… provides leadership in the efficient development and maintenance of standards that meet the priority needs of the domestic and global oil and gas exploration and production industry by minimizing needs for individual company standards, promoting broad availability of safe, interchangeable oilfield equipment and materials, and, promoting broad availability of proven engineering and operating practices."

- *Committee on Refinery Equipment (CRE) –* "… promotes safe and proven engineering practices in the design, fabrication, installation, inspection, and use of materials and equipment in refineries and related processing facilities."

- *Pipeline Standards Committees –* "… are dedicated to developing, revising, and approving consensus standards for the pipeline industry. These committees are comprised of technical experts, operating companies, vendors, consultants, academia, and regulators to create standards that facilitate safe operation and maintenance of pipelines."

- *Secretariat to ISO/TC 67 Materials, Equipment and Offshore Structures for Petroleum, Petrochemical and Natural Gas Industries –* "… has been delegated to API by ANSI. The scope of ISO/TC 67 is: Standardization of the materials, equipment and offshore structures used in the drilling, production, transport by pipelines and processing of liquid and gaseous hydrocarbons within the petroleum, petrochemical and natural gas industries. Excluded: aspects of offshore structures subject to IMO requirements

(ISO/TC 8)". API is the administrator of the United States National Committee Technical Advisory Group (USNC TAG) participating in ISO/TC 67 (International Organization for Standardization/Technical Committee 67).

- *Safety and Fire Protection Committee (SFPS)* – "... provides proactive safety and occupational health leadership and expertise to the industry, API committees and member companies. The SFPS seeks to advance and improve the industry's overall safety and occupational health performance by combining resources to identify and address important public, employee and company issues."

- *Committee On Petroleum Measurement (COPM)* – "... provides leadership in developing and maintaining cost effective, state of the art, hydrocarbon measurement standards and programs based on sound technical principles consistent with current measurement technology, recognized business accounting and engineering practices, and industry consensus. This is accomplished through the committee's and API's leadership role in the national and international standardization community in the development, publication, promotion, and revision of petroleum measurement standards, through its subcommittee structure, and through elimination of duplicative efforts."

- *The Secretariat to ISO/TC 28 Petroleum Products and Lubricants* – "... has been delegated to API by ANSI. The scope of this group is: Standardization of terminology, classification, specifications, methods of sampling, measurement, analysis and testing for: petroleum; petroleum products; petroleum based lubricants and hydraulic fluids; non-petroleum based liquid fuels; and non-petroleum based lubricants and hydraulic fluids." API is the USNC TAG administrator for ISO/TC 28 and ISO/TC 193 "Natural gas".

- *The Petroleum Industry Data Exchange (PIDX)* – "... API's standards committee on electronic business ... has reengineered entire business processes and operations for greater efficiency and profitability through the implementation of Electronic Data Interchange (EDI) and emerging electronic business technologies such as the Internet and eXtensible Markup Language (XML)."

Any API committee that desires to develop a standard jointly with ANSI must follow the procedures outlined in ANSI's *ANSI Essential Requirements: Due Process Requirements for American National Standards*. API is an ANSI-Accredited Standards Developer.

ASTM International

ASTM International
100 Barr Harbor Drive
West Conshohocken, Pennsylvania 19428-2959
Phone: (610) 832-9585
Internet: http://www.astm.org/

ASTM International, originally known as the American Society of Testing and Materials, was established in 1898 in response to steel rail breaks in the railroad industry. Their work led to standardization [36] for manufacturing of the steel used in rails. ASTM International is a standards organization with a global membership of approximately 30,000. It produces voluntary, consensus-based standards.

American National Standards Institute (ANSI) has approved some of ASTM's standards; however, it has not approved all of ASTM's standards. Those jointly approved by ANSI carry ANSI/ASTM numbers. Additionally, ASTM has jointly approved some International Electrotechnical Commission (IEC) standards. Those standards carry the ANSI/ASTM/IEC designation on their standard number. ASTM has also jointly approved some International Organization for Standardization (ISO) standards. Those standards have the ANSI/ASTM/ISO designation in their standard number. The ISO and IEC standards approved by ASTM may contain some national differences.

To facilitate its place as a world leader in testing and materials standards, ASTM International initiated a *Memoranda of Understanding* (MOU) program in 1991. It was designed to communicate between ASTM International and national standards bodies worldwide, fostering awareness of the standardization systems of all parties involved. The program also facilitates the development of national standards that will aid each country's health, safety, environmental, and economic conditions. These agreements help avoid duplication of effort where possible and mutually promote the standards development activities of ASTM International and the national standards bodies participating in the program [37]. Approximately 57 MOUs have been initiated to date.

> MOUs are designed to encourage, increase, and facilitate the participation of technical experts from around the world in the ASTM standards development process and broaden the global acceptance and use of ASTM International standards. As a benefit of the MOU program, technical experts from any of the countries where MOUs have been signed can participate freely as full voting members in the ASTM standards development process … [38]

Standards Process

Under the standards process [39], any standards proposal submitted to ASTM International is first researched to determine if there is an existing standard in the identified area. That research includes contacting trade associations, governmental agencies, or/and other standards-producing organizations. Once it has been determined that there is no existing standard covering the topic area, key ASTM stakeholders are identified and contacted to determine if market relevance exists. Stakeholders are also asked to commit to participation in the standard review process. Once this process is complete, a formal request will be submitted to an appropriate ASTM Technical Committee task group or subcommittee. Officers will be elected from the stakeholders and documentation procedures are setup and implemented. A liaison representative will be established between other committees with mutual interest or possible conflicts.

ASTM employs specific standards development tools including Draft Standards Templates and Form & Style Manuals that assure that pertinent required information is developed in ASTM's required format. Revisions can be proposed at any time during the process, but must be approved in a ballot. Editorial changes can be made without ballot approval, provided they do not change the technical content. These include:

> (1) those which introduce no change in technical content, but correct typographical errors, modify editorial style, change non-technical information, or reduce ambiguity, and (2) those which correct typographical errors in substance (essential information that could be misused). In the latter case, the year designation of the standard is changed. [40]

Once completed, the draft standard must be submitted to three levels of peer review, including subcommittee, main committee, and Society. Sixty percent of the stakeholder's ballots must be received by the closing date on the ballot for it to be approved. An affirmative vote requires at least two-thirds of the combined affirmative and negative votes cast for approval. Any statements submitted by the voting members are forwarded to the entire technical committee. Any negative vote, without an accompanying statement, is considered an abstention. Negative votes with written statements will be acted upon by the subcommittee or committee through a balloting process and will be resolved at a meeting. Any negative voter, whose negative was found not persuasive by both subcommittee and committee balloting, can submit an appeal to ASTM Headquarters.

Canadian Standards Association (CSA)

> Canadian Standards Association (CSA)
> 178 Rexdale Boulevard
> Rexdale, Ontario M9W IR3, Canada
> Phone: (416) 747-4000
> Internet: http://www.csa.ca/

The Canadian Standards Association (CSA) is part of the CSA Group, a not-for-profit membership association, established in 1919. CSA's responsibility includes *"standards development, information products, sale of publications, training, and membership services"* [41]. CSA International is also a part of the CSA Group and is responsible *"for product testing and certification"* [42]. The CSA Group also has a consumer product evaluation organization called OnSpeX. It is headquartered in Cleveland, Ohio and is involved with *"consumer product evaluation, data management and consulting services"* [43].

Standards Development

CSA utilizes a volunteer committee to develop standards [44], drawn from groups that will be affected by the standard. These volunteers are selected to assure a balanced matrix of expertise, one that is not weighted towards any specific view point. The process consists of eight stages [45]:

- *Preliminary Stage:* On receipt of a request for the development of a standard, an evaluation is conducted and the project is submitted for authorization.
- *Proposal Stage:* Public notice of intent to proceed is published and a Technical Committee is formed – or the project is assigned to an existing Technical Committee.
- *Preparatory Stage:* A working draft is prepared and a project schedule is established.
- *Committee Stage:* The Technical Committee or Technical Subcommittee – facilitated by CSA staff – develops the draft through an iterative process that typically involves a number of committee meetings.
- *Enquiry Stage:* The draft is offered to the public for review and comment, the Technical Committee reaches consensus, CSA staff conduct a quality review and a pre-approval edit is completed.
- *Approval Stage:* The Technical Committee approves the technical content by letter ballot or recorded vote. A second level review verifies that standards development procedures were followed.
- *Publication Stage:* CSA staff conducts a final edit to verify conformity with the applicable editorial and procedural requirements and then publishes and disseminates the standard.
- *Maintenance Stage:* The standard is maintained with the objective of keeping it up to date and technically valid. This may include the publication of amendments, the interpretation of a standard or clause, and the systematic (five-year) review of all standards.

CSA can also issue *Endorsed Standards* [46], which are non-Canadian produced standards. The process in issuing these standards involves their review by an appropriate Technical Committee, and can be approved without modification or issued with national interest changes. A substantial number of those standards have been adopted from International Electrotechnical Commission (IEC) standards. A list of the Canadian *Endorsed Standards* can be found at: http://www.csa.ca/standards/Endorsed_Standards_March_2008.pdf.

Council for Harmonization of Electrotechnical Standards of the Nations of the Americas (CANENA)

Council for Harmonization of Electrotechnical Standards of the
Nations of the Americas (CANENA)
Secretariat, NEMA
1300 North 17th Street, Suite 1752
Rosslyn, Virginia 22209
Phone: (703) 841-3244
Internet: www.canena.org

The Council for Harmonization of Electrotechnical Standards of the Nations of the Americas (CANENA) was founded in 1992 as a standards harmonization organization. Its role is

> to foster the harmonization of electrotechnical product standards, conformity assessment test requirements, and electrical codes between all democracies in the Western Hemisphere. [47]

CANENA's standardization scope involves:

> electrotechnical codes and standards and conformity assessment test methods utilized in North America. Further, CANENA Standardization Activities are not limited to the harmonization or development of standards – conformity assessment, compliance issues, compatibility, interchangeability, interoperability, installation codes, intellectual property and other issues in the broadest definition of standardization may be part of CANENA Standardization Activities. However, CANENA is not a standards developer … and will not hold copyrights or intellectual property rights on the resulting documents. Nothing contained herein shall preclude CANENA Standardization Activities with any country or regional entities such as MERCOSUR, PASC, or the IEC. [48]

CANENA utilizes *Technical Harmonization Committees* (THC). It may organize Technical Harmonization Subcommittees (THSC) or Working Groups "to address specific standards, portions of standards, or any specific or general issue within the scope of the THC" [49]. CANENA may utilize Special Technical Committees (STC) "for unique standardization activities that do not fit into the THC structure or normal operation of a THC" [50]. Another vehicle utilized by CANENA is CANENA Advisory Groups (CAG). They "address special subjects such as intellectual property rights policy. In addition to a scope, each newly created CAG must have an organization, a defined mode of operation, and a stated duration or termination date" [51].

THC and STC Operation

Technical Harmonization Committees (THC) and Special Technical Committees (STC) work for the harmonization or initiation of new or existing standards. Before any work can begin, the committee must receive permission from the copyright holder(s) of the standard. Copyright issues involving *"publishing, distribution, sales rights"* [52] and use of the holder's logo must be resolved. The standard copyright holder has the authority to allow or forbid the use of CANENA's logo on the harmonized standard. CANENA insists that the THCs and STCs utilize the International Electromechanical Committee's (IEC) format when possible. It also encourages harmonization with relevant IEC standards.

> The formal approval of any standard is accomplished outside of CANENA, within and according to the procedures of the organizations involved. Accordingly, there is no formal voting on the standards within the THC or STC. Consensus as determined by the THC or STC Chair will govern the work conduct and the completion of the activities using the definition

of consensus provided in ISO/IEC Guide 2. THCs or STCs shall not be dominated by any single member company or organization. Membership on a THC or STC or any Subcommittee or Working Group of the THC or STC shall be open to all interested CANENA members on an equal basis. [53]

CANENA has developed *Procedures for Harmonizing ANCE/CSA/UL Standards* [54]. The intent of the document is to "produce a harmonized set of requirements to enable manufacturers to build products that can be certified in all the countries involved, meeting fundamental needs in each of them. Extensive national differences do not support this intent" [55].

CANENA realizes that in the harmonization process, there are situations where national differences will be required to be recognized in the standard body. It notes in the above *Procedures* five categories of National Differences (ND) which will be noted on separate lines in the harmonized Co-Published Standard [56]:

- *DR* – These are National Differences based on the *national regulatory requirements.*
- *D1* – These are National Differences which are based on *basic safety principles and requirements*, elimination of which would compromise safety for consumers and users of products.
- *D2* – These are National Differences based on *safety practices.* These are differences for IEC requirements that may be acceptable, but adopting the IEC requirements would require considerable retesting or redesign on the manufacturer's part.
- *DC* – These are National Differences based on the *component standards* and will not be deleted until a particular component standard is harmonized with the IEC component standard.
- *DE* – These are National Differences based on *editorial comments or corrections.*

Illuminating Engineering Society of North America

Illuminating Engineering Society of North America
120 Wall Street, Floor 17
New York, NY 10005
Phone: (212) 248-5000
Internet: http://www.iesna.org

The Illumination Engineering Society was founded in 1906. It is a recognized authority on illumination engineering. IESNA [57] has a membership of approximately 10,000. Members included engineers, architects, lighting designers, academia, manufacturers, interior decorators, contractors, distributors, and others from Canada, Mexico, and the United States. Its membership also consists of similar professionals and manufacturers from throughout the world.

IESNA has issued more than 100 standards and recommended practices, with some jointly approved by American National Standards Institute. Documents published under joint approval would be required to meet the ANSI requirements for development. IESNA offers a multitude of standards, recommended practices, and technical memorandum on all aspects of lighting design, classification, maintenance, research, lighting definitions, and symbols, etc. Applications include marine lighting, aviation lighting, roadway and roadway signs, tunnels, parking structures and lots, sports and recreation facilities, medical and hospital structures, offices, residential lighting, laboratories, industrial, retail and merchandising, light fixture types, lamps, test and measurement, calculations, recommended lighting levels, etc. IESNA has also published several joint American National Standards with the National Electrical Contractors Association (NECA) on the installation procedures of several lighting system types. IESNA has also jointly published an American National Standard with the American Society of Heating, Refrigeration, and Air-Conditioning Engineers, Inc. (ASHRAE) on energy standards for buildings.

The Instrumentation, Systems, and Automation Society

The Instrumentation, Systems, and Automation Society
67 Alexander Drive
P.O. Box 12277
Research Triangle Park, North Carolina 27709
Phone: (919) 549-8411
Internet: http://www.isa.org/

The Instrumentation, System, and Automation Society, formally known as the Instrument Society of America (ISA), was founded on April 28, 1945 by Richard Rimbach. It was established from an amalgamation of some 18 local instrument societies. It issued its first standard, RP 5.1, *Instrument Flow Plan Symbols,* in 1949. The society's membership today is over 28,000 professionals from some 100 countries.

The ISA is divided into two basic groups or divisions, with a variety of technical area divisions in those main groups. They include the following [58]:

Automation and Technology Divisions

- Analysis Division (AD)

- Automatic Control Systems Division (ACOS)

- Computer Technology Division (COMPUTEC)

- Management Division (MAN)

- Process Measurement and Control Division (PMCD)

- Robotics and Expert Systems Division (ROBEXS)

- Safety Division (SAFE)

- Telemetry and Communication Division (TELECOM)

- Test Measurement Division (TMD)

The A&T (Automation and Technology) Department is the administrative "home" for Society Technical Divisions that address areas of automation and technology. The Department serves to stimulate, coordinate and advance Division objectives, and encourages Divisions to draw technical knowledge from, and to transport technology among, all pertinent disciplinary sources. [59]

Industries and Sciences Divisions

- Aerospace Industries Division (ASD)

- Chemical and Petroleum Industries Division (CHEMPID)

- Construction and Design Division (CONDES)

- Food and Pharmaceutical Industries Division (FPID)

- Mining and Metals Industries Division (M&M)

- Power Industry Division (POWID)

- Pulp and Paper Industry Division (PUPID)

- Water and Wastewater Industries Division (WWID)

The I&S (Industries and Sciences) Department is the administrative "home" for Society Technical Divisions that address specific industries and areas of science. The Department serves to stimulate, coordinate and advance Division objectives, and encourages Divisions to draw technical knowledge from, and to transport technology among, all pertinent disciplinary sources. [60]

ISA has established 15 technical committees, including [61]:

- Batch Manufacturing

- Data Processing and Management

- Environmental

- Instruments

- Maintenance and Operations
- Manufacturing Automation
- Measurement
- Motion Systems and Control
- Networks
- Process Automation and Control
- Productivity, Management and Marketing
- Safety
- Security
- Sensors
- Systems Integration

Each technical community is responsible for the development of new standards, as well as the maintenance or withdrawal of existing standards. If a proposed new standard is out of the area of expertise of any of the existing technical committees, an entirely new technical committee may be required.

Standards Preparation [62]

Developing a new standard project begins by submittal of a request to the Manager of ISA Standards Services. That submittal can be from an individual, ISA Divisions or Sections, or ISA Standards Committees [63]. A New Standards Project Proposal (NSP) form is used for that purpose. The proposal is then submitted to the Standards and Practices (S&P) Board Executive Committee. The proposal should meet the requirements outlined in the ISA Standards and Practices Department Procedures Manual [64]. The proposal is reviewed by one or more managing directors. It may then be assigned to an existing committee. It the proposal was generated by an existing Committee and the proposal scope falls under their jurisdiction, then the proposal must be approved by a majority vote of that Committee.

The next step involves the S&P Board Executive Committee establishing a Survey committee. That Committee does not have authority either to write or approve standards. Their function involves the following areas [65]:

(a) define the issue(s) to be addressed by the proposed project
(b) determine whether development of ISA Standard(s) can address the issue(s)
(c) identify the purpose and scope of the proposed ISA Standard(s)
(d) determine priorities for the development of proposed ISA Standard(s)

(e) determine if active volunteers are available and interested in staffing the proposed project

(f) determine whether standards projects are already under way that address the scope of the proposed ISA Standard(s)

(g) develop a schedule, if possible, for the development of the proposed ISA Standard(s)

(h) determine the status, if any, of any equivalent international standards activity.

The Committee can recommend either approving the project or abandoning any future activity. If it recommends approval, it is required to "demonstrate need and the economic impact to undertake the project with a proposed purpose and scope ..." [66]. The Committee is also required to determine a proposal's "relationship with relevant national and international standards and the relationship to other standards committees ..." [67]. The ISA Technical Services Department would also submit their impact assessment on the proposal to the S&P Board. The S&P Board must record a two-thirds approval vote for a new Committee to develop the proposed standard project.

Should the project "develop or revise an ISA Standard as an American National Standard or Draft Standard for Trial Use (DSTU), ISA shall notify ANSI by submitting an ANSI Project Initiation Notification System (PINS) Form for listing in 'ANSI Standards Action'" [68].

The Committee organized to develop the standard should

> be sufficiently diverse to ensure reasonable balance without dominance by a single interest category. The minimum number of voting members to have a viable Committee shall be five ... The scope and purpose of a new Committee, and any changes thereafter, shall be approved by a majority vote of the Board. [69]

In its auditing role, ANSI pays particular interests to the membership composition of the committee to assure that it is not skewed toward any one interest, but represents industry producers or vendors, consultants/educators/others, and users [70]. Subcommittees may be utilized to assist the Committee. Their formation and disbandment requires a majority vote by the Committee. Any drafts or revisions developed by the Subcommittee must be approved by the Committee. All Committee action requires a majority for a quorum. Work may be approved without a quorum by use of a letter or e-mail ballot, or voice vote by conference call.

Standards Committees are composed of two types of members, *voting members* and *information members*. Voting members actively participate in the Committee work and should attend every Committee meeting. Information members may not actively participate, but monitor the work, communicate with voting members, and review drafts. The Committee must be familiar with and strictly follow the ISA's *Standards and Practices Department Procedures*. If the standard is related to any international standards writing group, then it would be necessary that the US Technical Advisory Group (TAG) to that international group or an Expert for that group be a voting or information member on the Committee [71].

Committee action can require either a simple majority vote or two-thirds majority vote, depending upon the action. Administrative action generally falls under the simple majority rule, while the following require a majority of voting members approving *and* two-thirds majority approval vote, excluding abstentions, through a letter ballot or equivalent vote: adopting a new ISA Standard; reaffirming, withdrawing, revising, or addendums to existing ISA Standards; changes in Committee scope; or terminating a Committee.

Should the voting process result in views and objections from Committee members, those must be addressed in writing to attempt to develop a consensus opinion. Substantive changes necessitated to meet objections must be reported to all committee members in writing. New ballots are taken on any revisions and the process repeats itself until a two-thirds majority affirmative vote is obtained and written objections have been addressed.

If the standard is a proposed ANSI American National Standard, certain procedures must be followed. In its auditing function, ANSI must be informed of "notice of final actions on new ANSI/ISA Standards and reaffirmations, revisions, or withdrawals of existing ANSI/ISA Standards" [72]. Notices of these actions are sent to ANSI to be listed in ANSI Standards Action, allowing public review comment. Notification is also required in ISA publications or other means of advertisement. Notification of US National Technical Advisory Group(s) (TAGS) administrators might also be deemed appropriate by the Committee Chair. Any substitutive changes to the *ANSI Public Review* draft of the proposed standard must be resubmitted to the Committee for review. A letter ballot containing the changes and any unresolved objections, including what attempts were made at resolution, would be sent to Committee voting members. This procedure affords the Committee members the right to reaffirm or change their original vote.

Any Standard will be processed after final approval [73], which has received no negative comments from either the Committee or ANSI Public Review and has no outstanding negative votes or comments from the voters. Documents that have been approved by a two-thirds majority of the Committee and have demonstrated proper procedures have been followed will be sent to the Board to attempt to resolve any negative comments received by the Committee. Objecting members with unresolved comments must be informed in writing that an appeals process exists. The document will then be forwarded to the ISA Standards and Procedures Board on a 10-day default ballot. An ANSI/ISA certified document, along with other required accompanying substantiating material, would also be forwarded to ANSI. A two-thirds affirmative vote by the Board is required. Upon approval by the ISA Board and ANSI, the Standard can be published. Should ANSI not approve the standard, then it would require to be resubmitted to the Committee for resolution and restarting of the approval process.

FM Global

> FM Global
> 1301 Atwood Avenue
> P.O. Box 7500
> Johnston, Rhode Island 02919,
> Phone: (401) 275-3000
> Internet: http://www.fmglobal.com

FM Global was started in 1878 as Factory Mutual and served as the loss control and inspection entity of the Associated Factory Mutual Fire Insurance Companies or Factory Mutual. In 1998 a merger of three of the Factory Mutual insurance companies was begun which created FM Global. FM Global provides "state-of-the-art property loss prevention research and engineering and comprehensive insurance products" [74]. To aid in this process, FM Global has developed a number of *Approval Standards* for testing specific manufactured items. FM Global's approval process is divided into five steps [75]:

Step 1: Manufacturer Request

The manufacturer submits a letter, fax or e-mail to FM Approvals requesting Approval for a product or assembly and provides location, scope of work, model numbers, specifications and applicable sales literature. For customers seeking Approval of products designed for use in Hazardous Locations, fill out our application form and request a quote.

Step 2: Proposal Issue and Manufacturer Authorization

A proposal letter is sent by FM Approvals with scope of work, cost estimates, schedule, required tests and sample needs to customer. For new customers, a one-time contractual agreement is also mailed for signature (Master Agreement). The manufacturer then authorizes proposal in writing and submits all requested material and information identified in the proposal.

Step 3: Review, Testing and First Audit

Drawing or specification to product comparisons are made by FM Approvals. If all necessary items are received, testing is scheduled and conducted. The investigator visits the client's facility (if first-time client or new manufacturing location) to review quality control procedures prior to product approval by FM Approvals.

Step 4: Report, FM Approved Mark and Listing

Once testing has been completed successfully, a report is prepared and reviewed for technical accuracy and quality. Samples are retained and archived as necessary, returned to the client or disposed of per client's instructions. FM Approvals sends the final report to the manufacturer. Approval is effective as of date of report. The manufacturer may then label the product as FM Approved and the product is listed in the Approval Guide – a publication of FM Approvals. FM Approved roofing assemblies are entered into RoofNav, our Web-based software.

Step 5: Follow-Up Audits

Follow-up audits of manufacturing facilities are required in order to maintain FM Approved status. The frequency of audits is determined in accordance with authorities having jurisdiction (AHJs) over the installed product.

Institute of Electrical and Electronic Engineers (IEEE)

Institute of Electrical and Electronic Engineers (IEEE)
3 Park Avenue, 17th Floor
New York, N.Y. 10016-5997
Phone: 732 981-0060
Internet: http://www.ieee.org/

The Institute of Electrical and Electronic Engineers (IEEE) is an international Electrical Engineering professional society with more than 375,000 members from 160 countries as of November 2009. It was originally established 1884 as the American Institute of Electrical Engineers (AIEE) in New York. In 1889 it established a *Committee on Standardization*. On January 1, 1963 it merged with the Institute of Radio Engineers (IRE) and became known as the Institute of Electrical and Electronic Engineers (IEEE).

The IEEE is organized into 329 local sections in ten geographic regions [76]. There are membership chapters, 38 societies, and seven technical councils. It has some 1789 student branches in universities and colleges in 80 countries. IEEE has "over 900 completed standards, recommended practices, and guides ... and more than 400 projects in development" [77]. Its standards development process is now handled under its *IEEE Standards Association* (IEEE-SA). The *IEEE Standards Board* has the responsibility under that organization to approve or disapprove a proposed standard, recommended practice or guide.

The IEEE-SA standardization process [78] begins with the submittal of a *Project Authorization Request (PAR)*. It can be submitted by individuals or entity/corporate activity. "*Entities are participants such as academic institutions, corporations, government bodies, partnerships, consortia, standards-development organizations, etc.*" [79]. Each PAR must have a sponsor, usually one or more IEEE Societies or a *Corporate Advisory Group* (CAG). The sponsor has responsibility for oversight of the development of the standard. PARs can also involve revisions or amendments to existing standards. IEEE Standards can either be released as a full status functioning standard or can be issued on a *trial-use* basis. A standard will have a five-year life cycle, after which it can be reaffirmed without revision, revised or amended, or withdrawn. *Trial-Use Standards* will have a two-year life, after which it may be considered for full status, revised, or withdrawn.

The PAR is then submitted to the IEEE staff for review, before forwarding it to the *New Standards Committee (NesCom)* for evaluation. NesCom makes recommendations to the IEEE-SA Standards Board. Once the Board approves the PAR a *Sponsor* is sought. Normally, the Sponsor "will assign a working group to prepare and develop the document" [80]. IEEE-SA has established specific voting rights for those members of a *Working Group*. There are specific procedures that must be strictly followed in the group's work. IEEE templates [81] are utilized in developing the Working Group's standard draft. The *IEEE Standards Style Manual* [82] is one tool that is available in assisting the draft preparation. Graphics, annexes, references and bibliography, copyrights, permissions, patents and trademarks must all be addressed, if applicable, in the draft preparation. Before the draft is sent out for balloting, it must be reviewed by the "IEEE Editorial staff to perform *Mandatory Editorial Coordination (MEC)*" [83].

Once compiled, the draft standard will be distributed to the working group, subcommittee, or technical committee for review and comments by a letter ballot.

> In projects of broad interest, it is sometimes useful to collect a broader spectrum of comments than that available within the working entity involved in the development of the draft. Although the practice is deprecated by the IEEE-SA Standards Board, a small number of IEEE committees publish such drafts for distribution either as separate documents or in Society Transactions. Publication, including electronic, hard copy, or other forms of distribution, shall be carefully controlled to avoid misunderstandings regarding the status of and legal responsibility for such documents (N.B. these documents must not be mistakenly regarded as IEEE standards). The following conditions shall be met for such publication:
>
> (a) The document shall be marked according to IEEE Standards Department directions (see sub clause 4.2 of the IEEE Standards Style Manual).
>
> (b) The draft can be authorized for publication only by the IEEE Standards Department. Committees wishing to have their drafts published and distributed shall have their Sponsor contact the IEEE Standards Department. [84]

Another method of providing a larger review of the draft standard is through its release as a *Trial-Use Standard*.

The standards balloting process includes review of the vote and any comments that may be submitted. Any negative comments received during balloting must be resolved.

> It is up to the ballot resolution committee of the working group to decide whether or not a negative comment is new or an iteration of a previous comment ... during a recirculation ballot, balloters can only vote on the changed portion of the document and/or on any unresolved negative comments. When the comment resolution process is complete, the Sponsor or Chair must determine if a Recirculation Ballot is in order. [85]

There are two reasons why a recirculation ballot may be required. First, if there are still unresolved comments on the first ballot. A second reason would involve any technical or substantive changes to the draft. Voting participants can change their votes during the recirculation balloting process. If votes are received with new negative comments, then the resolution process with additional recirculation ballots must be implemented.

Once the technical issues raised in the balloting are resolved, the draft and accompanying documentation are submitted to the *Review Committee. Standards Board Working Guide for Submittal of Proposed Standards* [86] would be used by the Working Group to submit the draft. That committee issues their recommendation to the IEEE-SA Standards Board [87].

Once submitted to the Board, IEEE has established an appeals process to safeguard participants' rights.

> Persons who have directly and materially affected interests and who have been, or could reasonably be expected to be, adversely affected by a standard within the IEEE's jurisdiction, or by the lack of action in any part of the IEEE standardization process, shall have the right to appeal procedural actions or inactions, provided that the appellant shall have first exhausted the appeals procedures of any relevant subordinate committee or body before filing an appeal with the IEEE-SA Standards Board. [88]

Once any appeals have been resolved, the Standards Board can release the standard for publication, but there are several other steps involved before that can be done. If the standard is a joint certified standard, such as with ANSI, then their final approval and their involvement throughout the entire process must be assured. Secondly, it must receive a thorough review by an *IEEE Standards Editor.* Their job is to ensure that the document is both grammatically and syntactically correct. "The editor can ... make rewordings, editorial changes, and formatting changes to assist in publication of the standard" [89]. Their role does not involve making technical changes to the document. The Editor must work closely with an appointed Working Group representative during this process. If technical errors are found during the editing process, then the Editor must present those to the Review Committee (Rev Com) for their review and opinion.

Insulated Cable Engineers Association, Inc. (ICEA)

Insulated Cable Engineers Association, Inc. (ICEA)
P.O. Box 1568
5 Deerfield Road (30116), Carrollton, Georgia 30112
Phone: (770) 830-0369
Internet: http://www.icea.net/

The Insulated Cable Engineers Association (ICEA) is a professional organization established in 1925 by cable manufacturers to develop electrical cable for use in electric utility transmission and distribution; control and instrumentation; portable equipment; and communications. It is "a 'Not-For-Profit' association whose members are sponsored by over thirty of North America's leading cable manufacturers" [90]. ICEA's technical development work is done under four *semi-autonomous* sections. They include:

Power Cable

Control and Instrumentation Cable

Portable Cable

Communications Cable

The *Power Cable Section* is responsible for "standards for all cables with extruded or laminar insulations and used for the transmission and distribution of electrical energy" [91]. A *Technical Advisory Committee (TAC)* was established for this section and is called the *Utility Power Cable Standards Technical Advisory Committee (UPCS TAC)*. Its membership comes from ICEA members, utilities and cable manufacturing entities, and other specialized interested parties.

The *Communications Cable Section* provides "cable standards, test procedures, and guides for the telecommunications industry" [92]. Its membership is composed of representatives from telecommunications cable manufacturers from North America. A *Technical Advisory Committee* was established for this section and is the *Telecommunications Wire and Cable Standards Technical Advisory Committee (TWCS TAC)*. That committee has a variety of participants, including industry representatives, ICEA members, telecommunication engineers, and other parties with interest in this field.

The TWCS TAC was operational until 2002 [93], when it was disbanded because of a lack of participation and corporate mergers. Its standards work was transferred to the ICEA Communications Division.

"The *Control & Instrumentation Cable Section* is responsible for providing standards, test procedures, and guides for all insulated cables used to control or monitor equipment or power systems, transportation signal systems, and alarms. The *Portable Cable Section* was formed to provide standards for insulated electrical cables for all portable or movable equipment, especially for use in mines or other similar applications, and by the military." [94]

The ICEA has cooperated with many standards organizations in the development of cable standards, including the National Electrical Manufacturers Association (NEMA), American National Standards Institute (ANSI), and Association of Edison Illuminating Companies (AEIC). Standards jointly published with ANSI would require the development process to follow ANSI established guidelines.

NACE International

NACE International
1440 South Creek Drive
Houston, Texas 77084-4906
Phone: (800) 797-6223
Internet: http://www.nace.org/

Formally known as the National Association of Corrosion Engineers, NACE International [95] is charged with developing corrosion prevention and control standards.

NACE was originally established in 1943 by 11 pipeline corrosion engineers. The organization now has some 20,000 members in 100 countries as of November, 2009.

NACE International's technical committee activities [96] are under the responsibility of the *Technical Coordination Committee (TCC)*. The Technical Committees are further organized by *Specific Technology Groups (STGs)*. The STGs are further divided into three administrative areas: *Industry-Specific Technology (N)*; *Cross-Industry Technology (C)*; and *Science (S)*. For example, STG 30 is called *Oil and Gas Production – Cathodic Protection*. Its scope of responsibility is "application and evaluation of cathodic protection of all types of equipment used for oil and gas production" [97].

STG30 is the administrative group that has several *Task Groups (TGs)* and *Technology Exchange Groups (TEGs)* with the following scopes [98]:

1. *TEG 166X* – Cathodic Protection in Seawater

2. *TG 168* – Cathodic Protection Systems, Retrofit, for Offshore Platforms

3. *TG 169* – Cathodic Protection of Pipelines in Seawater

4. *TG 269* – Cathodic Protection Design for Deep Water

STG30 also sponsors several other TGs and TEGs. An example of the responsibility of a Task Group might be TG269's assignment to "Review and/or revise NACE Publication 7L192, Cathodic Protection Design Considerations for Deep Water Structures" [99]. Task Groups are responsible for specific assignments. Those may include development of Technical Committee reports, standards, etc. The Technology Exchange Group is responsible for the exchange of technical information through symposia or other methods.

Task Groups and Technical Exchange Groups have one *Administrative Specific Technology Group (STG)*. "Administrative STG refers to the sponsoring Specific Technology Group that is responsible for supervising a specific Task Group" [100].

NACE International standards are written by "industry professionals, instructors, professors, government officials, and experts from regulatory and governing bodies" [101]. The following is an outline of the NACE International standards procedures [102].

Standards development begins with the preparation of a draft standard by a Task Group. That group must reach a consensus opinion during a meeting, by e-mail or other means. A letter ballot can be conducted, but there must be a two-thirds affirmative vote to pass. The Task Group must forward the draft to NACE International where it is edited to assure correct grammar, spelling, etc. No technical editing can occur during this process; so to assure that, the edited text is returned to the Task Group chairman for review and approval. The edited draft is also forwarded to TCC *Reference Publications Committee (RPC)* to verify adherence with the NACE, International *Publications Style Manual.* It is then retuned to NACE Headquarters. Any comments or questions received from the RPC will be sent to the Task Group and Administrative STG Chairpersons for review.

NACE Headquarters prepares a message and abstract of the draft standard. It is simultaneously sent to members of the Administrative and sponsoring STGs requesting their approval of a letter ballot. A six-week deadline is imposed on receipt of responses. With Chair approval, the ballot is sent out to all committee members and other interested parties. A 50% response from the STG(s) must be received to close out balloting. Upon determination of ballot closeout, the Chairpersons of the sponsoring STG(s) and Task Group are sent all copies of votes with comments along with a close-out letter and report.

A two-thirds affirmative vote, excluding abstentions, must be recorded for approval. After ballot close-out, all negative comments received must be resolved by the Task Group. Voters with descending comments must be invited to discuss the issues. An Open Review will be held during the Administrative STG meeting to assess the attempts to resolve the negative issues. If the issues are resolved, the negative voters must sign the resolution documents for the process to proceed. If the negative comments cannot be resolved, then a re-ballot is necessary. The ballots are distributed to all original voting members with a four-week deadline for response. An affirmative vote of at least 90%, excluding abstentions, must be received for the override ballot to pass. Negative voters must receive written notification of the disposition of their votes and an explanation for the right to appeal.

> The draft is sent to the sponsoring STG Chair(s), Technology Coordinator, and the TCC Chair for approval for publication. Any unresolved negatives are forwarded with the draft standard, along with a statement from the Task Group. [103]

The draft standard is then sent to the RPC for final editorial review. The unresolved negative comments with a statement from the Task Group must accompany the draft standard. The NACE Board of Directors must ratify the voting procedures and standard. Also, if the document is being published as an American National Standard, the approved draft and the confirmation process must also be approved by ANSI. ANSI must receive all backup documentation, along with the approved standard.

National Electrical Manufacturers Association (NEMA)

National Electrical Manufacturers Association (NEMA)
1300 North 17th Street Suite 1752
Rosslyn, Virginia 22209
Phone: (703) 841-3200
Internet: http://www.nema.org/

The National Electrical Manufacturers Association (NEMA) [104] was founded in 1926 and is a trade association representing approximately 450 member companies. Those members are responsible for manufacturing generation, transmission and distribution, control, and end-use electrical products. NEMA is a federation of more than 50 product sections which are grouped into eight divisions. Technical standards are developed within these sections.

NEMA has published approximately 250 product standards [105]. Those include some 52 NEMA/ANSI publications, jointly approved by both organizations. NEMA also participates in some 66 standards harmonization projects [106]. Those projects attempt to harmonize US standards with their international standards counterparts. Twenty-two of those projects involve CANENA (Council for Harmonization of Electrotechnical Standardization of the Nations of the Americas). That organization does not have any responsibility for developing new standards, but solely exists to create commonality between national standards in the Americas. It "provides a forum for harmonization discussions and, upon agreement, the draft harmonized standards are then processed by the respective standards developer in each country" [107].

NEMA also has representatives serving on 59 of the International Electrotechnical Commission's (IEC) Technical Committees (TCs) and Subcommittees (SCs). It also has representatives on one International Organization for Standardization (ISO) technical committee (TCs). NEMA has a substantial investment of resources in *Conformity Assessment*. The ISO/IEC Guide 2: 1996 defines Conformity Assessment as *"any activity concerned with determining directly or indirectly that relevant requirements are fulfilled"* [108]. That process assures that products or services conform to the standards or specifications that governed their creation. Although that process is important in assuring product safety, consumer health, and environmental protection, it does not come under the objective of this book, to review the standards development and application process.

There are eight NEMA *Divisions* which are responsible for the development of *Standards, Recommended Practices, and Guides* for the *Product Groups* in those Divisions. The Divisions [109] include:

Industrial Automation

Lighting Systems

Electronics

Building Equipment

Insulating Materials

Wire and Cable

Power Equipment

Diagnostic Imaging and Therapy Systems

Codes and Standards Development

The Codes and Standards Committee has the responsibility within NEMA for the development of new documents and the maintenance of existing documents for the organization. Members for this committee are appointed yearly by the association's NEMA Standards and Conformity Assessment Policy Committee (SCAPC).

> After a NEMA subdivision has approved a standard by letter ballot, the proposed standard is reviewed by the committee to ensure that it had been coordinated with other NEMA standards and that it conforms with NEMA policies and applicable laws. [110]

There are five steps to the creation of a new NEMA standard [111]:

Project initiation

Developing the draft

Balloting (gathering comments)

Codes and Standards Committee approval

Editing and publication

The process begins with the NEMA section or committee, developing the standard, initiating a voting process regarding their desire to develop the standard. A simple majority in the affirmative is needed to initiate the project. A Project Initiation Request (PIR) form is completed and the process will be tracked by NEMA's engineering department. A technical committee, subcommittee or task force is then formed to develop the standard. The actual work may only be done by a few members of the committee. When a draft of the standard is completed, it is submitted back to the initiating members, for review and comment. The comments are addressed by the initiating members. A revised draft is then prepared and redistributed to the members. This process is repeated until consensus is obtained.

The completed draft is then submitted to the entire technical committee for review and comments. When those comments have been addressed, its membership will determine if the draft will be submitted to the Section (voting classification) for balloting.

To assure conformance to NEMA standards requirements, the draft standard must be presented in a standard format as is defined in NEMA NS-1. It must also be reviewed by legal counsel before being submitted by a letter ballot. The ballot must be acted upon within 30 days and must be returned with an *"affirmative"*, *"negative"* or *"not-voting"* indicated. Members of the technical committee are eligible to participate in the written ballot, as can any member manufacturer who produces the product. The total number of "affirmative" or "negative" votes cast are used as the basis for the vote. A total of two-thirds majority of the total "affirmative" or "negative" votes cast is needed to approve the standard. The "not-voting" ballots are not considered in the total written ballots submitted for voting. The written ballot can be circumvented, if it is presented for approval at a section or voting class meeting with 100% of the section or voting class present.

Should any ballots be received with written comments, the comments are referred to the technical committee for resolution within 30 days of the completion of balloting. If resolution cannot be reached, then the technical committee submits the proposal to the Codes and Standards Committee (C&S). C&S may request written or oral responses from all sides to consider in its review. Voting members may change their ballot to an "affirmative" vote, but it must be submitted in writing. Any written comments may also be withdrawn.

The Codes and Standards Committee must approve all actions by sections and voting classes involving standards approval, revision, reaffirmation, or rescindment. C&S action must be by simple majority action by the committee members by either letter ballot or approval at a meeting.

In its review of a proposed standard, C&S determines whether:

- the standard is in harmony with the policies of NEMA standardization activities and has been developed according to the procedures contained in the NEMA constitution and bylaws;
- the interests of all affected NEMA subdivisions have been considered;
- the standard is technically sound and accurately drawn;
- any recommendations should be made to NEMA Counsel concerning compliance of the standard with NEMA's policies and procedures; and
- the record presented by the subdivision proposing the standard shows that adequate consideration has been give to both safety and user needs. [112]

C&S's review might determine the proposed standard may pose a conflict with other sections or voting groups. If that is the situation, the proposed standard would be submitted to that group or groups for approval. A time limit of 45 days is established for a reply. If one is not received, then C&S may elect to proceed. Should a referral not be needed, then it must be determined that the standard meets all NEMA criteria. When that is established, the standard will be returned to a staff member for final review and approval by the appropriate section representative. Should there be any changes requested by the editorial committee, they would be implemented by the Communications Department. The standard would then be sent to NEMA's publisher for printing and release.

It should be noted here that NEMA does allow downloading of electronic copies of some of their standards, from their website, without a fee. There is a requirement that the recipient must agree to specified terms and conditions regarding their use.

NEMA does issue joint standards with other standard-making organizations. It has jointly issued cable standards with the Insulated Cable Engineers Association (ICEA). NEMA has also done the same with the American National Standards Institute (ANSI).

National Fire Protection Association (NFPA)

National Fire Protection Association (NFPA)
One Batterymarch Park
P.O. Box 9101
Quincy, Massachusetts 02269-9101
Internet: http://www.nfpa.org/

The National Fire Protection Association was founded in 1896 by a group of 18 men who represented a variety of 20 stockholder fire insurance organizations [113]. That formation was in response to an attempt to standardize the installation of fire sprinkler systems. The organization's membership was initially limited to stock fire insurance organizations and their representatives. However, it soon established an Associate Membership category for non insurance-related members, such as the railroad industry, fire fighters, and professional engineers and architects.

With the establishment of the first commercial electrical distribution companies, the installation of electrical lighting sometimes had severe consequences. During that time Factory Mutual Insurance Companies of New England [114] reported that some 23 fires were reported in a six-month time frame in 65 mills where electrical lighting had been installed. By 1895 there were some five electrical codes which had been established in the United States.

In an effort to develop a standardized national electrical code amongst the chaos of the multiple available codes, several national organizations came together in 1896 to establish a committee to study all of the available American, German, and English electrical codes. It established a "National Code" from the best ideas in those codes and in 1896 the National Board of Fire Underwriters adopted "National Electrical Code of 1897". The administration of that code was undertaken by its founding organization the "National Conference on Standard Electrical Rules". In 1911, that organization elected to dissolve itself and the responsibility for the administration of the National Electrical Code was transferred to the National Fire Protection Association.

Today the NFPA has some 75,000 members from over 70 nations [115]. The organization is responsible for some 300 codes and standards covering life safety, building, and fire-related issues.

Standards Development

The NFPA has established a *Standards Council* [116], consisting of 13 members appointed by the Board of Directors, whose purpose is to oversee the codes and standards development for the NFPA. Their duties also include administration of the association's rules and regulations and acting as an appeals body in code development disputes. The Standards Council has more than 250 *Code-Making Panels* and *Technical Committees* [117] acting as consensus bodies with responsibility for revising existing codes and standards and developing new ones. These groups can act on their own submitted technical changes to those documents or they will be responsible for reviewing and acting on changes submitted by any interested party or parties.

A new code proposal submitted to the NFPA will be reviewed by the Standards Council and a public notice will be published in the NFPA's newsletter, website, and the *NFPA News*. Other possible vehicles for a public notice, if appropriate, include the US Federal Register and the American National Standards Institute's *Standards Action*.

NFPA News is a free newsletter providing detailed information on NFPA Codes and Standards activities. *NFPA News* typically includes special announcements, notification of proposal and comment closing dates, requests for comments on NFPA documents, publication of Formal Interpretations (FIs), Tentative Interim Amendments (TIAs), errata, and notice of the availability of Standards Council minutes. [118]

The Standards Council would be responsible for reviewing any input received in response to the advertisements and make a decision on proceeding with the project. The proposal might then be assigned to one or more existing *Technical Committees* and/or *Code Making Panels* or to a new panel or committee created specifically for the project. The number of panels needed to implement the code review process would depend on the areas of expertise needed. The NFPA code review process is based on a consensus opinion from the review panels or technical committees and the membership.

The Codes and Standards Committee Document development process consists of five steps [119]:

Step 1: Call for Proposals

- Proposed new Document or new edition of an existing Document is entered into one of two yearly revision cycles, and a Call for Proposals is published.

Step 2: Report on Proposals (ROP)

- Committee or Panel meets to act on Proposals, to develop its own Proposals, and to prepare its Report.

- Committee votes by written ballot to approve its actions on the Proposals. If approval is not obtained, the Report returns to Committee.

- If approved, the Report on Proposals (ROP) is published for public review and comment.

Step 3: Report on Comments (ROC)

- Committee or Panel meets to act on Public Comments, to develop its own Comments, and to prepare its Report.

- Committee votes by written ballot to approve its actions on the Comments. If approval is not obtained, the Report returns to Committee.

- If approved, the Report on Comments (ROC) is published for public review.

Step 4: Association Technical Meeting

- "Notices of intent to make a motion" are filed, are reviewed, and valid motions are certified for presentation at the Association Technical Meeting. ("Consent Documents" bypass the Association Technical Meeting and proceed directly to the Standards Council for issuance.)

- NFPA membership meets each June at the Association Technical Meeting and acts on Technical Committee Reports (ROP and ROC) for Documents with "certified amending motions."

- Technical Committee(s) and Panel(s) vote on any amendments to the Technical Committee Reports made by the NFPA membership at the Association Technical Meeting.

Step 5: Standards Council Issuance

- Notification of intent to file an appeal to the Standards Council on Association action must be filed within 20 days of the Association Technical Meeting.

- Standards Council decides, based on all evidence, whether or not to issue the Document or to take other action.

The NFPA membership is afforded a chance to amend the Technical Committee Reports (Reports on Proposals and Reports on Comments). This can be done at the Association Technical Meeting which is held during the NFPA Annual Meeting. The motions can include those to accept or reject the Proposal or Comments in whole or part; or return the whole or a portion of the Report to the Technical Committee for further study. The maker of the intended motion must submit in advance a Notice of Intent to Make a Motion (NITMAM). This is necessary to allow them to submit a Certified Amending Motion and certain Follow-Up Motions at the Association Technical meeting for consideration. Other procedural rules must also be followed. NFPA rules sometimes limit those who are authorized to submit an amending motion to a proposed code or standard to the original submitter or their representative. Motions to reject an accepted comment or Return a Technical Committee Report or a portion of that report for Further Study can be made by anyone.

Underwriters Laboratories, Inc. (UL)

Underwriters Laboratories, Inc. (UL)
333 Pfingsten Road
Northbrook, Illinois 60062-2096
Phone: (800) 704-4050
Internet: http://www.ul.com/

Underwriters' Laboratories (UL®) was founded in Chicago in 1894 by William H. Merrill as the Underwriters' Electrical Bureau. It functioned as the Electrical Bureau of the National Board of Fire Underwriters. Merrill was an electrical engineer and had been sent to Chicago to investigate the Palace of Electricity at the Chicago World's Fair.

UL is an independent not-for-profit, non-governmental, product safety certification, Nationally Recognized Testing Laboratory (NRTL). Its goal is *"to promote safe living and working environments by the application of safety science and hazard-based safety engineering"* [120]. It accomplishes that task for more than 72,000 manufacturers in 98 countries. UL evaluates more than 19,000 types of products annually, and more than 20 billion UL Marks appear on products each year. UL's family of companies and its network of service providers include 64 laboratories and certification facilities. [121]

To accomplish product certification, UL has developed more than "1000 Standards for Safety" [122]. In an attempt to harmonize UL standards with others internationally, UL participates in national and international standards technical committees. Many of the UL standards have been approved as an American National Standard by the American National Standards Institute (ANSI). Those are identified by the ANSI/UL designation. ANSI/UL and UL standards utilize slightly different procedures as is noted in [123] (Table 1.1). A list of ANSI/UL standards can be found on the website: http://ulstandardsinfonet.ul.com/catalog/stdstd.html.

UL also issues joint standards with the Canadian Standards Association (CSA) bearing the UL/CSA designation. UL also has adopted some International Electrotechnical Commission (IEC) standards, which may contain some national differences. UL also has established procedures for harmonizing North American standards between UL, CSA, and Asociación de Normalización y Certificación (ANCE). ANCE is the Mexican standards organization. A copy of those procedures can be found at the following UL web location: http://ulstandardsinfonet.ul.com/harm/ANCECSAULprocedures.pdf. UL has also signed harmonization agreements with standards organizations in several other nations.

Also of interest are the UL *Product Directories*. Those documents describe [124]:

- the names of companies whose product samples have been found to comply with the applicable safety requirements and are subject to UL's Follow-Up Services;

- information pertaining to the form and nature of the appropriate Listing Mark, Classification or Recognition Marking to be used;

- limitations or special conditions applying to a product; and

- the title of the Standard used to investigate the product (if there is a published UL Standard or for international directories an appropriate international or regional standard) for the specific product category.

TABLE 1.1   ANSI and UL standards procedures development differences

ANSI/UL standards	UL standards
Proposed standards or changes to standards are announced in ANSI's Standards Action and submitted to public review	Proposed standards or changes to standards are not announced in ANSI's Standards Action. There is no public review. Proposals are circulated to the Standards Technical Panel (STP) and to subscribers to UL standards. Further circulation is prohibited
All public review comments are addressed and circulated. If there is a continuing objection, commenters are given the right to appeal	Comments from subscribers and STP members serve to advise UL staff. UL staff members respond to comments, but are not required to attempt to resolve them to the satisfaction of the commenter
The consensus body is the STP, the members of which are selected to reflect a balance of the interest groups affected by the standard	The consensus body is the UL staff

The Product Directories include [125]:

- Building Materials, Roofing Materials and Systems, Fire Protection Equipment, Fire Resistance (Codes A, O, B and C)

- Recognized Component, Plastics Recognized Component (Codes D and S)

- Electrical Equipment, Hazardous Locations Equipment (Codes E, F, G and R)

- Marine (Code H)

- Heating, Cooling, Ventilating, Cooking Equipment and Food Safety Equipment, Plumbing and Associated Products and Flammable and Combustible Liquids and Gases Equipment (Codes V, P and K)

Occupational Health and Safety Administration (OSHA)

United States Department of Labor
Occupational Health and Safety Administration (OSHA)
200 Constitution Avenue, NW
Washington, DC 20210
Phone: (866) 4USA-DOL
Internet: http://www.osha.gov

Because of concerns for safety in the workplace, the United States Congress promulgated the establishment of the *Occupational Safety and Health Act (OSH Act)* of 1970. It was established:

> To assure safe and healthful working conditions for working men and women; by authorizing enforcement of the standards developed under the Act; by assisting and encouraging the States in their efforts to assure safe and healthful working conditions; by providing for research, information, education, and training in the field of occupational safety and health; and for other purposes. [126]

There are several avenues by which an OSHA standard can be developed. A standard may be proposed by OSHA or it may be proposed

> in response to petitions from other parties, including the Secretary of Health and Human Services (HHS); the National Institute for Occupational Safety and Health (NISOH); state and local governments; any nationally-recognized standards-producing organization; employer or labor representatives; or any other interested person. [127]

That Act established that the

> term "occupational safety and health standard" means a standard which requires conditions, or the adoption or use of one or more practices, means, methods, operations, or processes,

reasonably necessary or appropriate to provide safe or healthful employment and places of employment. [128]

The Act also established the areas in the United States in which its enforcement would be mandated. It noted:

> This Act shall apply with respect to employment performed in a workplace in a State, the District of Columbia, the Commonwealth of Puerto Rico, the Virgin Islands, American Samoa, Guam, the Trust Territory of the Pacific Islands, Wake Island, Outer Continental Shelf Lands defined in the Outer Continental Shelf Lands Act, Johnston Island, and the Canal Zone. The Secretary of the Interior shall, by regulation, provide for judicial enforcement of this Act by the courts established for areas in which there are no United States District Courts having jurisdiction. [129]

The OSH Act of 1970 authorized the Secretary of Labor to establish the Occupational Safety and Health Administration. It was established:

> "to set mandatory occupational safety and health standards applicable to businesses affecting interstate commerce" through public rulemaking.

> OSHA safety standards are designed to reduce on-the-job injuries; health standards to limit workers' risk of developing occupational disease. Most OSHA standards are horizontal – they cover hazards which exist in a wide variety of industries. These are compiled as the OSHA General Industry Standards. Vertical standards apply solely to one industry. OSHA has promulgated vertical standards for the construction, agriculture, and maritime sectors.

> Some general industry standards apply to construction, agriculture, and maritime as well. [130]

OSHA utilizes several advisory committees for the development of recommendations for new standards that they deem needed. There are two standing and ad hoc committees at their disposal. It is mandated that all committees utilized by OSHA shall have members from labor, management, and state agencies. Additionally, one or more designees is to be assigned by the Secretary of Health and Human Services (HHS). The two standing committees consist of [131]:

- the *National Advisory Committee on Occupational Safety and Health (NACOSH)*, which advises, consults with, and makes recommendations to the Secretary of HHS and to the Secretary of Labor on matters regarding administration of the Act; and

- the *Advisory Committee on Construction Safety and Health*, which advises the Secretary of Labor on formulation of construction safety and health standards and other regulations.

Standards recommendations may also come from another agency established under the OSH Act of 1970. That agency is the *National Institute of Occupational Safety and Health (NIOSH)*, which is under the United States Department of Health and Human Services.

NIOSH conducts research on various safety and health problems, provides technical assistance to OSHA and recommends standards for OSHA's adoption. While conducting its research, NIOSH may make workplace investigations, gather testimony from employers and employees and require that employers measure and report employee exposure to potentially hazardous materials. NIOSH also may require employers to provide medical examinations and tests to determine the incidence of occupational illness among employees. When such examinations and tests are required by NIOSH for research purposes, they may be paid for by NIOSH rather than the employer. [132]

OSHA can create, amend or revoke a standard using procedures established in the 1946 *Administrative Procedure Act (APA)*. The APA requires every federal agency to first publish any proposed new regulations, or modification or removal of existing regulations in the *Federal Register* at least 30 days prior to their going into effect [133]. The OSH Act of 1970 provides procedures allowing interested parties comments on a proposed rule:

> The Secretary shall publish a proposed rule promulgating, modifying, or revoking an occupational safety or health standard in the Federal Register and shall afford interested persons a period of thirty days after publication to submit written data or comments. Where an advisory committee is appointed and the Secretary determines that a rule should be issued, he shall publish the proposed rule within sixty days after the submission of the advisory committee's recommendations or the expiration of the period prescribed by the Secretary for such submission.

> On or before the last day of the period provided for the submission of written data or comments under paragraph (2), any interested person may file with the Secretary written objections to the proposed rule, stating the grounds therefore and requesting a public hearing on such objections. Within thirty days after the last day for filing such objections, the Secretary shall publish in the Federal Register a notice specifying the occupational safety or health standard to which objections have been filed and a hearing requested, and specifying a time and place for such hearing. [134]

OSHA is mandated to consult with small businesses and other governmental agencies when proposed rules may significantly affect them. The *Small Business Regulatory Enforcement and Fairness Act* (SBREFA) of 1996 requires OSHA to consult with the *Small Business Administration* and the *Office of Management and Budget* before issuing rules impacting small business.

OSHA will often identify proposed new standards and rules, "Notice of Proposed Rulemaking" or "Advance Notice of Proposed Rulemaking", in the *Federal Register*. The public response time is normally 60 days after publication of the notice. The OSH Act recognizes an individual's right to object to new, amended or withdrawn rules and standards. Any written objection and verbal objections at hearings must be considered in the decision-making process.

Once public hearings, if mandated by legislation, are completed, OSHA is required to publish the full and final text of any proposed new or amended standard, with its effective date and detailed explanations of the document in the *Federal Register*. If the OSHA rule or standard option is implemented over some public objections, individuals in the public have the right to file a petition for judicial review of that rule or standard within 59 days after its promulgation. That review is adjudicated in the United States Court of Appeals in the district in which the objector either lives or maintains a business.

OSHA can issue *Emergency Temporary Standards* under certain conditions. Those standards can be implemented immediately if OSHA determines the possibility exists for grave danger of exposure to toxic substances or agents or physical harm. The *Federal Register* must be utilized to publish notice of that standard. That publication also acts as the first step notification in the procedure to make the Emergency Temporary Standard permanent. The validity of the Emergency Temporary Standard can be challenged in the United States Court of Appeals.

An individual or business may apply to OSHA for a variance from a new or posed standard or regulation. That variance must be based on an inability to comply by the effective date because of one of several legitimate reasons. They may include verifiable shortages of materials or equipment; shortages of professional or technical personnel to implement the rule; existing facilities or methods of operation already in place are "at least as effective" as those required by OSHA. Two types of variances can be issued. They are *Temporary Variances* and *Permanent Variances.*

A *Temporary Variance* for compliance by the mandated affective date of the standard or rule must be based on the unavailability of materials or equipment or the unavailability of professional or technical personnel required for the implementation; or the amount of time required to implement the required facility alternation or new construction would not meet the implementation date. There are procedural, verification, notification of employees, and documentation requirements that the employer must meet.

A *Permanent Variance* may be granted to employers "who prove their conditions, practices, means, methods, operations, or processes provide a safe and healthful workplace as effectively as would compliance with the standard" [135].

An *Experimental Variance* can be granted by OSHA to employers that are participating in an experiment approved by either the Secretaries of Labor or HHS. Also, certain other variances may be granted when it can be proved that implementation of a standard or regulation may impair national defense.

Application for a variance and OSHA's granting of that variance may require a considerable amount of time. An employer may apply for an Interim Order to continue to operate under existing conditions until a decision is made. The terms of an Interim Order must be published in the Federal Register and the employer is required to notify employees or their representative of that action.

References

1. http://www.merriam-webster.com/.
2. National Fire Protection Association, *National Electrical Code® Handbook,* 2002 Edition, page vii. NFPA; Quincy, MA.
3. Institute of Electrical and Electronic Engineers, ANSI/IEEE Std 80-1986, *IEEE Guide for Safety in AC Substation Grounding*; 1986, page 5. New York, NY.
4. Alonzo, Robert J. *Electrical Safety for Petroleum Facilities*; 1997, page 2. ASSE; Des Plaines, IL.
5. ANSI website: http://www.ansi.org/about_ansi/overview/overview.aspx?menuid=1.
6. ANSI website: http://www.ansi.org/about_ansi/organization_chart/chart.aspx?menuid=1.
7. NSSN, ANSI search engine for standards: www.nssn.org/.
8. ITU = International Telecommunication Union.
9. US DoD = United States Department of Defense.
10. http://publicaa.ansi.org/sites/apdl/Documents/Standards%20Activities/NSSC/USSS-2005%20-%20FINAL.pdf.
11. Ibid., Section V – Moving Forward.
12. IEC website: http://www.iec.ch/about/mission-e.htm; Mission.
13. IEC website: http://www.iec.ch/ourwork/iecpub-e.htm; International Consensus Products.
14. Ibid., International Standard.
15. Ibid., International Standard.
16. Ibid., Industrial Technical Agreement.
17. Ibid., Industrial Technical Agreement.
18. Ibid., Technology Trend Assessment.
19. ISO website: http://www.iso.org/iso/about.htm.
20. ISO website: http://www.iso.org/iso/structure.
21. ISO website: http://www.iso.org/iso/standards_development/processes_and_procedures/stages_description.htm.
22. ISO website: http://www.iso.org/iso/standards_development/processes_and_procedures/cooperation_with_cen.htm.
23. ISO website: http://www.iso.org/iso/pressrelease.htm?refid=Ref1125.
24. AEIC website: http://aeic.org/news/expand.html.
25. AEIC website: http://www.aeic.org/.
26. AEIC website: http://www.aeic.org/cable_eng/index.html.
27. AEIC website: http://www.aeic.org/load_research/index.html.
28. AEIC website: http://www.aeic.org/meter_service/index.html.
29. AEIC website: http://www.aeic.org/power_apparatus/index.html.
30. AEIC website: http://www.aeic.org/power_delivery/index.html.

31. AEIC website: http://www.aeic.org/power_generation/index.html.

32. AEIC website: http://www.aeic.org/cable_eng/IndustryReferencesonWebsite3.pdf.

33. API website: http://www.api.org/aboutapi/.

34. API website: http://www.api.org/Standards/.

35. API website: http://committees.api.org/standards/index.html.

36. ASTI International website: http://www.astm.org/ABOUT/aboutASTM.html.

37. ASTM International website: http://www.astm.org/GLOBAL/mou.html.

38. Ibid.

39. ASTM International website: http://www.astm.org/MEMBERSHIP/standardsdevelop.html.

40. ASTN International website: http://www.astm.org/COMMIT/Regs.pdf; Paragraph 10.5.4.1.

41. CSA website: http://www.csagroup.org/Default.asp?language=English.

42. CSA website: http://www.csa-international.org/.

43. CSA website: http://www.onspex.com/about/index.htm.

44. CSA website: ttp://www.csa-international.org/consumers/faq/Default.
asp?articleID=7091.

45. CSA website: http://www.csa.ca/standards/default.
asp?load=development&language=English; Development Process.

46. CSA website: http://www.csa.ca/standards/default.
asp?load=endorsed&language=English.

47. ANENA website: http://www.canena.org/canena/about.html.

48. CANENA website: http://www.canena.org/standards/standardization.aspx; Section 1.1.

49. Ibid., Section 2.2.

50. Ibid., Section 2.3.

51. Ibid., Section 3.4.

52. Ibid., Section 6.2.2.

53. Ibid., Section 6.5.

54. CANENA website: http://www.canena.org/papers/ANCE-CSA-UL-procedures.pdf.

55. Procedures for Harmonizing ANCE/CSA/UL Standards, March 1, 2008, page 4, Para. 1.1.

56. Ibid., page D5.

57. IESNA website: http://iesna.org/about/iesna_about_profile.cfm.

58. ISA website: http://www.isa.org/Graphics/membership/divisions-brochure.pdf.

59. ISA website: http://www.isa.org/MSTemplate.cfm?MicrositeID=9&CommitteeID=4466.

60. ISA website: http://www.isa.org/MSTemplate.cfm?MicrositeID=53&CommitteeID=4510.

61. ISA website: http://www.isa.org/Template.cfm?Section=Technical_Information_and_
Communities&Template=/Taggedpage/CommunityList.cfm.

62. ISA website: http://www.isa.org/Content/NavigationMenu/Products_and_Services/
Standards2/Development/Development_References1/Accredited_Procedures1/2006_
Accredited_Procedures.doc.

63. ISA Standards and Practices Committee Guide, Draft 7, March 2000, Section 2.1.

64. ISA Standards and Practices Department Procedures – 2006 Revision; Section 2.1.

65. Ibid., Section 2.2 (a thru h).

66. Ibid., Section 2.2.

67. Ibid.

68. Ibid., Section 2.4.

69. Ibid., Section 3.

70. Ibid., Section 3.2.5.

71. ISA Standards and Practices Committee Guide, Draft 7, March 2000; Section 4.2.7 (k), page 10.

72. ISA Standards and Practices Department Procedures, 2006 Revision; Section 4.2.7.

73. Ibid., Section 4.2.8.

74. FM Global website: http://www.fmglobal.com/page.aspx?id=01070000.

75. FM Global website: http://www.fmglobal.com/page.aspx?id=50020000.

76. IEEE website: http://www.ieee.org/web/aboutus/home/index.html.

77. IEEE website: http://standards.ieee.org/announcements.bkgnd_stdprocess.html.

78. IEEE website: http://standards.ieee.org/resources/development/initiate/index.html.

79. IEEE website: http://standards.ieee.org/corpforum/participation/faq.html.

80. IEEE website: http://standards.ieee.org/resources/development/wg_dev/index.html.

81. IEEE website: http://standards.ieee.org/resources/development/writing/index.html.

82. IEEE website: http://standards.ieee.org/resources/development/writing/writinginfo.html.

83. IEEE website: http://standards.ieee.org/resources/development/balloting/balloting.html.

84. IEEE website: http://standards.ieee.org/guides/opman/sect8.html#8.2.

85. IEEE website: http://standards.ieee.org/resources/development/balloting/recircballot.html.

86. IEEE website: http://standards.ieee.org/resources/development/final/finalmoreinfo.html.

87. IEEE website: http://standards.ieee.org/resources/development/final/index.html.

88. IEEE website: http://standards.ieee.org/guides/bylaws/sect5.html#5.4.

89. IEEE website: http://standards.ieee.org/guides/companion/part2.html#appeal.

90. ICEA website: http://www.icea.net/index.html.

91. ICEA website: http://www.icea.net/Public_Pages/Organization/Sections.html.

92. Ibid.

93. ICEA website: http://www.icea.net/Public_Pages/Tech/TWCS_TAC.htm.

94. ICEA website: http://www.icea.net/Public_Pages/Organization/Sections.html.

95. NACE website: http://www.nace.org/content.cfm?parentid=1005&currentID=1005.

96. NACE website: http://www.nace.org/content.cfm?parentid=1013&currentID=1013.

97. NACE website: http://web.nace.org/Departments/Technical/Directory/Committee.aspx?id=e7dc32a6-5fef-db11-9194-0017a4466950.

98. Ibid.

99. NACE website: http://web.nace.org/Departments/Technical/Directory/Committee.aspx?id=9257413b-60ef-db11-9194-0017a4466950.

100. NACE, International, *Technical Committee Publications Manual – March 2008*, Section 2.1.1, page 2.

101. NACE website: http://www.nace.org/content.cfm?currentID=1018&parentID=1018.

102. NACE website: http://www.nace.org/content.cfm?parentid=1013&currentID=1343.

103. NACE website: http://events.nace.org/technical/s-r/documentdevelopmentguide.asp.

104. NEMA website: http://www.nema.org/about/.

105. NEMA website: http://www.nema.org/stds/conformity/index.cfm.

106. Ibid.

107. NEMA website: http://www.nema.org/stds/international/canena.cfm.

108. Ibid.

109. NEMA website: http://www.nema.org/about/upload/NEMACorpBrochure05.pdf.

110. NEMA website: http://www.nema.org/stds/codes/.

111. NEMA website: http://www.nema.org/stds/aboutstds/develop.cfm.

112. Ibid.

113. NFPA website: http://www.nfpa.org/itemDetail.asp?categoryID=500&itemID=18020&URL=About%20Us/History.

114. Ibid.

115. NFPA website: http://www.nfpa.org/itemDetail.asp?categoryID=589&itemID=18478&URL=About%20Us/Code%20development%20and%20adoption%20partner.

116. NPFA website: http://www.nfpa.org/categoryList.asp?categoryID=834&URL=Codes%20and%20Standards/Code%20development%20process/Standards%20Council&cookie%5Ftest=1.

117. NFPA website: http://www.nfpa.org/categoryList.asp?categoryID=162&URL=Codes%20and%20Standards/Code%20development%20process/How%20the%20code%20process%20works.

118. NFPA website: http://www.nfpa.org/itemDetail.asp?categoryID=136&itemID=19181&URL=Codes%20and%20Standards/NFPA%20News.

119. NFPA website: http://www.nfpa.org/categoryList.asp?categoryID=162&URL=Codes%20and%20Standards/Code%20development%20process/How%20the%20code%20process%20works.

120. UL website: http://www.ul.com/consumers/mark.html.

121. UL website: http://www.ul.com/global/eng/pages/corporate/aboutUL

122. Ibid.

123. ANSI website: http://ansi.org/news_publications/other_documents/halo_effect.aspx?menuid=7.

124. UL website: www.ul.com/global/eng/pages/corporate/contactus/orderdirectories.

125. Ibid.

126. Public Law 91-596; 84 STAT. 1590; 91st Congress, S.2193; December 29, 1970; as amended through January 1, 2004; Section 1 Introduction.

127. OSHA website: http://www.osha.gov/OCIS/stand_dev.html.

128. Public Law 91-596; 84 STAT. 1590; 91st Congress, S.2193; December 29, 1970; as amended through January 1, 2004; Section 3 Definitions.

129. Ibid. Section 4. Applicability of This Act.

130. OSHA website: ttp://www.osha.gov/pls/oshaweb/owadisp.show_document? p_table=FACT_SHEETS&;p_id=134.

131. OSHA website: http://www.osha.gov/OCIS/stand_dev.html.

132. Ibid.

133. United States Code - Government Organizations and Employees - Chapter 5, Administration Procedure, Subchapter II - Administration, Section 553, Rule Making.

134. Public Law 91-596; 84 STAT. 1590; 91st Congress, S.2193; December 29, 1970; as amended through January 1, 2004; Section 6 Occupational Safety and Health Standards, Section (b) (2) and (3).

135. OSHA website: http://www.osha.gov/OCIS/stand_dev.html.

CHAPTER 2

American versus Global

International harmonization is the goal of American *Standards Development Organizations* (SDOs) for some existing and future codes, standards, and recommended practices. *Harmonization* is a process by which internationally recognized SDOs cooperate to produce *standards* and *conformity assessment procedures* for manufactured goods and services which are accepted by all nations participating in that process. The only accepted exceptions to harmonized standards involve those for specific local normative practices. Harmonization allows the easy movement of goods and services throughout the world, without any questions of product quality, personal safety, fire safety, or trade barriers.

There are three major international standards organizations that deal with electrotechnical issues. They include the *International Organization for Standardization (ISO)*, the *International Electrotechnical Committee (IEC)*, and the *International Telecommunication Union (ITU)*.

As of 2008, a total of 162 nations held memberships of some kind in the ISO. Those membership categories included *Member Body*, *Correspondent Member*, and *Subscriber Member*. Member Body participants have one vote in the International Organization for Standardization's General Assembly. Correspondent Members have no voting privileges in the General Assembly; however, they can participate in any policy or technical discussions. Subscriber Members have no voting rights in the General Assembly or standards committees' discussion participation privileges. They do maintain contacts through the organization. Table 2.1 contains a list of those member nations.

As of November, 2009 IEC was comprised of 76 member nations. Fifty-six with *Full Member* status and 20 maintain *Associate Member* status. Table 2.2 contains a list of those nations. The IEC develops standards and conformity assessments in the fields of electronics, magnetism and electromagnetism, electroacoustics, multimedia, telecommunications, energy production and distribution, and associated fields.

The ITU is based in Geneva, Switzerland, with 191 *Member States* and more than 700 *Sector Members* and *Associates* [1]. This organization coordinates with international governments and the private communications sector to establish worldwide standards for interconnection of communications systems. Those standards deal with Internet and wireless technologies,

Electrical Codes, Standards, Recommended Practices and Regulations; ISBN: 9780815520450

TABLE 2.1 International Organization for Standardization (ISO) member nations

Member Body	Member Body	Member Body	Correspondent Member	Subscriber Member
Algeria	Ecuador	Lithuania	Afghanistan	Antigua and Barbuda
Armenia	Egypt	Luxembourg	Albania	Burundi
		Macedonia, former Yugoslav Republic of		
Argentina	Ethiopia	Malaysia	Angola	Cambodia
Australia	Fiji	Malta	Benin	Dominica
Austria	Finland	Mauritius	Bhutan	Eritrea
Azerbaijan	France	Mexico	Bolivia	Guyana
Bahrain	Germany	Mongolia	Brunei Darussalam	Honduras
Bangladesh	Ghana	Morocco	Burkina Faso	Lao People's Democratic Republic
Barbados	Greece	Netherlands, The	Congo, Republic of the	Lesotho
Belarus	Hungary	New Zealand	Dominican Republic	Saint Vincent and the Grenadines
Belgium	Iceland	Nigeria	El Salvador	Suriname
Bosnia and Herzegovina	India	Norway	Estonia	
Botswana	Indonesia	Oman	Gabon	
Brazil	Iran, Islamic Republic of	Pakistan	Gambia, The	
Bulgaria	Iraq	Panama	Georgia	
Cameroon	Ireland	Peru	Guatemala	
Canada	Israel	Philippines	Guinea	
Chile	Italy	Poland	Hong Kong, China	
China	Jamaica	Portugal	Kyrgyzstan	
Colombia	Japan	Qatar	Latvia	
Congo, Democratic Republic of the	Jordan	Romania	Liberia	
Costa Rica	Kazakhstan	Russian Federation	Macau, China	
Croatia	Kenya	Saint Lucia	Madagascar	
Cuba	Korea, Democratic People's Republic of	Saudi Arabia	Malawi	
Cyprus	Korea, Republic of	Serbia	Mauritania	
Czech Republic	Kuwait	Singapore	Moldova, Republic of	

TABLE 2.1 International Organization for Standardization (ISO) member nations—cont'd

Member Body	Member Body	Member Body	Correspondent Member	Subscriber Member
Côte-d'Ivoire	Lebanon	Slovakia	Montenegro	
Denmark	Libyan Arab Jamahiriya	Slovenia	Mozambique	
		South Africa	Myanmar	
		Spain	Namibia	
		Sri Lanka	Nepal	
		Sudan	Palestine	
		Sweden	Papua New Guinea	
		Switzerland	Paraguay	
		Syrian Arab Republic	Rwanda	
		Tanzania, United Republic of	Senegal	
		Thailand	Seychelles	
		Trinidad and Tobago	Sierra Leone	
		Tunisia	Swaziland	
		Turkey	Tajikistan	
		Ukraine	Turkmenistan	
		United Arab Emirates	Uganda	
		United Kingdom	Yemen	
		United States of America	Zambia	
		Uruguay	Zimbabwe	
		Uzbekistan		
		Venezuela		
		Vietnam		

aeronautical and maritime navigation, data and voice communications, television, radio astronomy, satellite-based meteorology, and next generation networks.

Tables 2.1 and 2.2 provide the international memberships of both ISO and IEC. Reviewing the memberships in those tables indicates the advantage of standards harmonization with the ISO and IEC. Although the United States is a participating nation in both organizations; it has harmonized only a small number of its standards with those organizations.

Standards play a major role in assuring operability, personal safety, environmental safeguards, and fire prevention with manufactured products; however, they also play

TABLE 2.2   International Electrotechnical Committee (IEC) member nations

Full Member	Full Member	Full Member	Associate Member
Algeria	Iraq	Singapore	Albania
Argentina	Ireland	Slovakia	Bahrain
Australia	Israel	Slovenia	Bosnia and Herzegovina
Austria	Italy	South Africa	Colombia
Belarus	Japan	Spain	Cuba
Belgium	Korea, Republic of	Sweden	Cyprus
Croatia	Libyan Arab Jamahiriya	Switzerland	Estonia
Czech Republic	Luxembourg	Thailand	Iceland
Denmark	Malaysia	Turkey	Kazakhstan
Egypt	Mexico	Ukraine	Kenya
Finland	Netherlands, The	United Kingdom	Korea. Democratic People's Republic of
France	New Zealand	United States of America	Latvia
Germany	Norway		Lithuania
Greece	Pakistan		Macedonia, former Yugoslav Republic of
Hungary	Philippines, Republic of the		Malta
India	Poland		Montenegro
Indonesia	Portugal		Nigeria
Iran	Qatar		Sri Lanka
	Romania		Tunisia
	Russian Federation		Vietnam
	Saudi Arabia		
	Serbia		

a major role in international commerce. Countries that have adopted the IEC and ISO standards could place impediments to or restrictions on the importation of goods that do not meet their harmonized standards. That situation emphasizes the need for international harmonization of standards, allowing the unimpeded flow of goods around the world. The challenge for the United States Standards Development Organizations is to develop a balance between the uses of world harmonized standards and American core standards. For instance, IEC 60364, (all parts) and NFPA 70®, *National Electrical Code*® (NEC®) are the respective IEC and ANSI electrical codes. However, there is no public consideration by ANSI that the NEC should either be harmonized with or replaced by IEC 60364. In fact, both electrical building codes reflect somewhat different wiring philosophies.

TABLE 2.3 UL/CSA/ANCE harmonized standards

Developer	Standard No.	Title
UL/CSA/ANCE	UL 6 Ed 14	Electrical Rigid Metal Conduit – Steel
UL/CSA/ANCE	UL 6A Ed 2	Electrical Rigid Metal Conduit – Aluminum, Red Brass, and Stainless Steel
UL/CSA/ANCE	UL 44 Ed 16	Thermoset-Insulated Wires and Cables
UL/CSA/ANCE	UL 50 Ed 12	Enclosures for Electrical Equipment, Non-Environmental Considerations
UL/CSA/ANCE	UL 50E Ed 1	Enclosures for Electrical Equipment, Environmental Considerations
UL/CSA/ANCE	UL 62 Ed 17	Standard for Flexible Cords and Cables
UL/CSA/ANCE	UL 98 Ed 13	Enclosed and Dead-Front Switches
UL/CSA/ANCE	UL 248-1 Ed 2	Low-Voltage Fuses – Part 1: General Requirements
UL/CSA/ANCE	UL 248-2 Ed 2	Low-Voltage Fuses – Part 2: Class C Fuses
UL/CSA/ANCE	UL 248-3 Ed 2	Low-Voltage Fuses – Part 3: Class CA and CB Fuses
UL/CSA/ANCE	UL 248-4 Ed 2	Low-Voltage Fuses – Part 4: Class CC Fuses
UL/CSA/ANCE	UL 248-5 Ed 2	Low-Voltage Fuses – Part 5: Class G Fuses
UL/CSA/ANCE	UL 248-6 Ed 2	Low-Voltage Fuses – Part 6: Class H Non-Renewable Fuses
UL/CSA/ANCE	UL 248-7 Ed 2	Low-Voltage Fuses – Part 7: Class H Renewable Fuses
UL/CSA/ANCE	UL 248-8 Ed 2	Low-Voltage Fuses – Part 8: Class J Fuses
UL/CSA/ANCE	UL 248-9 Ed 2	Low-Voltage Fuses – Part 9: Class K Fuses
UL/CSA/ANCE	UL 248-10 Ed 2	Low-Voltage Fuses – Part 10: Class L Fuses
UL/CSA/ANCE	UL 248-11 Ed 2	Low-Voltage Fuses – Part 11: Plug Fuses
UL/CSA/ANCE	UL 248-12 Ed 2	Low-Voltage Fuses – Part 12: Class R Fuses
UL/CSA/ANCE	UL 248-13 Ed 2	Low-Voltage Fuses – Part 13: Semiconductor Fuses
UL/CSA/ANCE	UL 248-14 Ed 2	Low-Voltage Fuses – Part 14: Supplemental Fuses
UL/CSA/ANCE	UL 248-15 Ed 2	Low-Voltage Fuses – Part 15: Class T Fuses
UL/CSA/ANCE	UL 248-16 Ed 2	Low-Voltage Fuses – Part 16: Test Limiters
UL/CSA/ANCE	UL 486A-486B Ed 1	Wire Connectors
UL/CSA/ANCE	UL 486C Ed 5	Splicing Wire Connectors
UL/CSA/ANCE	UL 486D Ed 5	Sealed Wire Connector Systems
UL/CSA/ANCE	UL 489 Ed 10	Molded-Case Circuit Breakers, Molded-Case Switches, and Circuit-Breaker Enclosures
UL/CSA/ANCE	UL 514A Ed 10	Metallic Outlet Boxes
UL/CSA/ANCE	UL 514B Ed 5	Conduit, Tubing, and Cable Fittings
UL/CSA/ANCE	UL 797 Ed 9	Electrical Metallic Tubing – Steel
UL/CSA/ANCE	UL 845 Ed 5	Motor Control Centers
UL/CSA/ANCE	UL 857 Ed 12	BUSWAYS
UL/CSA/ANCE	UL 891 Ed 11	Switchboards

(Continued)

TABLE 2.3   UL/CSA/ANCE harmonized standards—cont'd

Developer	Standard No.	Title
UL/CSA/ANCE	UL 943 Ed 4	Ground-Fault Circuit-Interrupters
UL/CSA/ANCE	UL 1598 Ed 3	Luminaries
UL/CSA/ANCE	UL 2556 Ed 2	Wire and Cable Test Methods
UL/CSA/ANCE	UL 4248-1 Ed 1	Fuseholders – Part 1: General Requirements
UL/CSA/ANCE	UL 4248-4 Ed 1	Fuseholders – Part 4: Class CC
UL/CSA/ANCE	UL 4248-5 Ed 1	Fuseholders – Part 5: Class G
UL/CSA/ANCE	UL 4248-6 Ed 1	Fuseholders – Part 6: Class H
UL/CSA/ANCE	UL 4248-8 Ed 1	Fuseholders – Part 8: Class J
UL/CSA/ANCE	UL 4248-9 Ed 1	Fuseholders – Part 9: Class K
UL/CSA/ANCE	UL 4248-11 Ed 1	Fuseholders – Part 11: Type C (Edison Base) and Type S Plug Fuse
UL/CSA/ANCE	UL 4248-12 Ed 1	Fuseholders – Part 12: Class R
UL/CSA/ANCE	UL 4248-15 Ed 1	Fuseholders – Part 15: Class T

There are regional international standards harmonization efforts in Europe, the Pacific rim, Africa, the Middle East, and the Americas. One exists in North America between the United States, Canada, and Mexico. It is called *CANENA*. CANENA was founded in 1992 as the Council for Harmonization of Electrotechnical Standards of the Nations of the Americas. Its scope of work was three-fold. First, it was assigned the task to harmonize the electrotechnical products standards. Next, it was given the assignment for the development of conformity assessment test requirements. Third, it was to harmonize the electrical codes between all democracies in the Western Hemisphere.

North American National Standards Development Committees are currently engaged in the process of adopting some of the IEC standards as national standards through CANENA. Those standards may include some national deviations to the IEC standards. These are allowed to facilitate specific local requirements which may be unique to a locality.

The renewal of a previous cooperative agreement between CANENA and the IEC was signed on January 11, 2007. It established objectives, information exchange, cooperation, cross representation in both organizations, and the responsibility to implement the agreement.

An example of that North American cooperation can be seen in Table 2.3, which represents some of the harmonized standards between Underwriters Laboratories (UL), the Canadian Standards Association (CSA), and the National Association of Normalization and Certification of the Electrical Sector (ANCE) [Mexico]. Any product certified under these standards can be marketed and sold throughout North America.

The United States Department of Commerce has designated the American National Standards Institute (ANSI) as the main coordination organization for voluntary, peer reviewed consensus standards development in the United States. ANSI is the United States' voting representative in both the IEC and ISO organizations. ANSI was given the mandate to develop *the United States Standards Strategy.* ANSI works in conjunction with all standards development organizations in the United States which have been recognized and accredited by that organization.

ANSI's *National Standards Strategy for the United States* noted the following in Section V – Moving Forward:

> A sectoral approach recognizes that there is no simple prescription that can be handed down to fit all needs. Sectors must develop their own plans; the purpose of this strategy is to provide guidance, coherence and inspiration constraining creativity or effectiveness. The *U.S. National Standards Strategy* therefore consists of a set of strategic initiatives having broad applicability that will be applied according to their relevance and importance to particular sectors. Stakeholders are encouraged to develop their own initiatives where needed and this strategy suggests some that have widespread applicability. [2]

ANSI's recommendations to continue the sectoral approach for standards harmonization included the following general recommendations: [2]

1 – Strengthen participation by government in development and use of voluntary consensus standards through public/private partnerships

2 – Continue to address the environment, health, and safety in the development of voluntary consensus standards

3 – Improve the responsiveness of the standards system to the views and needs of consumers

4 – Actively promote the consistent worldwide application of internationally recognized principles in the development of standards

5 – Encourage common governmental approaches to the use of voluntary consensus standards as tools for meeting regulatory needs

6 – Work to prevent standards and their application from becoming technical trade barriers to U.S. products and services

7 – Strengthen international outreach programs to promote understanding of how voluntary, consensus-based, market-driven sectoral standards can benefit businesses, consumers and society as a whole

8 – Continue to improve the process and tools for the efficient and timely development and distribution of voluntary consensus standards

9 – Promote cooperation and coherence within the U.S. standards system

10 – Establish standards education as a high priority within the United States private, public and academic sectors

11 – Maintain stable funding models for the U.S. standardization system

12 – Address the need for standards in support of emerging national priorities

ANSI notes in *Section II – Imperatives for Action* [3] in the standards strategy that:

The global economy has raised the stakes in standards development. Competition for the advantages that accompany a widespread adoption of technology has reached a new level, and the impetus to develop globally accepted standards is greater now than ever before.

Globally

- Global standardization goals are achieved in the United States through sector-specific activities and through alliances and processes provided by companies, associations, standards developing organizations, consortia, and collaborative projects.

- This market-driven, private sector-led approach to global standardization is substantially different from the top-down approach favored in many other countries.

- Emerging economies understand that standards are synonymous with development and request standards-related technical assistance programs from donor countries. Increasingly our trading partners utilize such programs to influence the selection of standards by these economies and create favorable trade alliances.

- Policies that protect patents, trademarks, and other intellectual property are not universally or rigorously applied. The standardization process must respect the rights of intellectual property owners while ensuring users have access to the intellectual property rights (IPR) incorporated in standards.

- When standards are utilized as non-tariff barriers to trade, the ability of U.S.-based companies and technologies to compete in the international marketplace is adversely affected.

- Standardization and the manner in which agreements are reached between suppliers and customers continue to evolve and are influenced by advances in technology. Stakeholders are no longer willing in all cases to operate within the boundaries of the formal standards system and they continue to explore new modalities of standards development. Organizations such as consortia and Internet-based processes that enable worldwide participation of stakeholders are creating an innovative environment that is becoming increasingly important in the global marketplace.

- The service industry sector has a significant and rapidly growing presence in the global economy and workforce. The United States must devote more attention to understanding the needs of the service industry sector and establishing service standards initiatives to meet those needs.

Standards Harmonization

In pursuit of the objectives of the *United States Standards Strategy*, ANSI and the ANSI-certified American standards organizations have developed professional cooperation with and memberships in both the ISO and IEC. An example of that cooperation is the Instituted of Electrical and Electronic Engineers (IEEE). IEC and IEEE signed a joint agreement on November 14, 2002 in which IEC agreed to review IEEE standards in electronics, telecommunications, power generation, and other electrotechnical standards for recognition for international status. Those standards chosen by IEC will be processed by IEC technical committees. They will be published as IEC/IEEE Dual Logo International Standards and will be available to IEC member countries for adoption as national standards. Table 2.4 represents the IEEE standards that IEC has jointly adopted through 2007.

Underwriters Laboratories, Inc. has published an investigative report UL 508E, IEC Type "2" Coordination Short Circuit Tests of Electromechanical Motor Controllers in Accordance with IEC Publication 60947-4-1. This UL report is only applicable to some manufacturer-specific selected motor controllers and short circuit protective devices (SCPD). It is an example of one manufacturer's effort to determine the compliance of their products to IEC standards.

Table 2.5 represents the standards that have been harmonized between NEMA/ANSI and IEC. Table 2.6 represents the IEC standards that have been adopted by Underwriters Laboratories. Table 2.6 represents a partial list of UL harmonized standards. For a complete list of UL harmonized standards, refer to the UL website.

Standards Comparison

The United States power distribution equipment manufacturers have pursued the establishment of equipment standardization in the United States through the National Electrical Manufacturers Association (NEMA). As an example, electromechanical motor starter contactors are produced in a range of NEMA standard sizes, with each NEMA size rating serving a range of motor horsepowers. The IEC design philosophy for motor starters took a different approach. An explanation for the differences between NEMA standard motor contactors and IEC motor contactors can be found in NEMA Standard ICS 2.4, *NEMA and IEC Devices for Motor Services – A Guide for Understanding the Differences.* Section 1.5 of that document, Design Philosophies, provides a characterization of both standards organization's design of motor starter contactors [4]. It notes:

1.5.1: Traditional NEMA Contactors

A NEMA contactor is designed to meet the size rating specified in NEMA standards. A philosophy of the NEMA standards is to provide electrical interchangeability among manufacturers for a given NEMA size. Since the installer often orders a controller by the motor

TABLE 2.4 IEC adopted IEEE standards

Developer	Standard No.	Title
IEEE	IEC 61523-3 Ed.1 (2004-09) (IEEE Std 1497™-2001)	Delay and Power Calculation Standards – Part 3: Standard Delay Format (SDF) for the Electronic Design Process
IEEE	IEC 61691-1-1 Ed.1 (2004-10) (IEEE Std 1076™-2002)	Behavioural Languages – Part 1-1: VHDL Language Reference Manual
IEEE	IEC 61691-4 Ed.1 (2004-10) (IEEE Std 1364™-2001)	Behavioural Languages – Part 4: Verilog® Hardware Description Language
IEEE	IEC 61691-5 Ed.1 (2004-10) (IEEE Std 1076.4™-2000)	Behavioural Languages – Part 5: Standard VITAL ASIC (Application Specific Integrated Circuit) Modeling Specification
IEEE	IEC 62050 Ed. 1 (2005-07) (IEEE Std 1076.6™-2004)	IEEE Standard for VHDL Register Transfer Level (RTL) Synthesis
IEEE	IEC 62142 Ed. 1 (2005-06) (IEEE Std 1364.1™-2002)	Standard for Verilog® Register Transfer Level Synthesis
IEEE	IEC 62265 Ed. 1 (2005-07) (IEEE Std 1603™-2003)	Standard for an Advanced Library Format (ALF) Describing Integrated Circuit (IC) Technology, Cells, and Blocks
IEEE	IEC 62530 Ed. 1 (2007-11) (IEEE Std 1800™-2005)	Standard for System Verilog - Unified Hardware Design, Specification, and Verification Language
IEEE	IEC 62531 Ed. 1 (2007-11) (IEEE Std 1850™-2005)	Standard for Property Specification Language (PSL)
IEEE	IEC 60488-1 Ed.1 (2004-07) (IEEE Std 488.1™-2003)	Higher Performance Protocol for the Standard Digital Interface for Programmable Instrumentation – Part 1: General
IEEE	IEC 60488-2 Ed.1 (2004-05) (IEEE Std 488.2™-1992)	Standard Digital Interface for Programmable Instrumentation – Part 2: Codes, formats, protocols and common commands
IEEE	IEC 61588 Ed.1 (2004-09) (IEEE Std 1588™-2002)	Precision Clock Synchronization Protocol for Networked Measurement and Control Systems
IEEE	62243 Ed. 1 (2005-07) (IEEE Std 1232™-2002)	Standard for Artificial Intelligence Exchange and Service Tie to All Test Environments (AI-ESTATE)
IEEE	IEC 62271-111 Ed.1 (2005-11) (IEEE Std C37.60™-2003-Compilation)	High-Voltage Switchgear and Controlgear – Part 111: Overhead, Pad-Mounted, Dry Vault, and Submersible Automatic Circuit Reclosers and Fault Interrupters for Alternating Current Systems Up To 38 kV
IEEE	IEC 62525 Ed. 1(2007-11) (IEEE Std 1450™-1999)	Standard Test Interface Language (STIL) for Digital Test Vector Data
IEEE	IEC 62526 Ed. 1 (2007-11) (IEEE Std 1450.1™-2005)	Standard for Extensions to Standard Test Interface Language (STIL) for Semiconductor Design Environments
IEEE	IEC 62527 Ed. 1 (2007-11) (IEEE Std 1450.2™-2002)	Standard for Extensions to Standard Test Interface Language (STIL) for DC Level Specification
IEEE	IEC 62528 Ed. 1 (2007-11) (IEEE Std 1500™-2005)	Standard Testability Method for Embedded Core-based Integrated Circuits
IEEE	IEC 62032 Ed.1 (2005-03) (IEEE Std C57.135™-2001)	Guide for the Application, Specification and Testing of Phase-Shifting Transformers

TABLE 2.5 NEMA/ ANSI/IEC harmonized standards

Developer	Standard No.	Title
NEMA	ANSI/IEC 60529-2004	Degrees of protection provided by enclosures (IP Code)
NEMA	NEMA ANSI/IEC C78.1195:2001	Electric lamps – double-capped fluorescent lamps – safety specifications
NEMA	NEMA ANSI/IEC C78.1199:2002	Electric lamps – single-capped fluorescent lamps – safety specification
NEMA	NEMA ANSI/IEC C78.60360:2002	For electric lamps – standard method of measurement of lamp cap temperature rise
NEMA	NEMA ANSI/IEC C78.60360:2002	For electric lamps – standard method of measurement of lamp cap temperature rise
NEMA	NEMA ANSI/IEC C78.60432-1:2007	Electric lamps – incandescent lamps – safety specifications – tungsten filament lamps for domestic and similar general lighting purposes – part 1
NEMA	NEMA ANSI/IEC C78.60432-2:2007	Incandescent lamps – safety specifications – part 2: tungsten halogen lamps for domestic and similar general lighting purposes
NEMA	NEMA ANSI/IEC C78.60432-3:2007	Electric lamps – incandescent lamps – safety specifications – part 3: tungsten halogen lamps (non-vehicle)
NEMA	NEMA ANSI/IEC C78.62035:2004	Electric lamps – discharge lamps (excluding fluorescent lamps) – safety specifications
NEMA	NEMA ANSI/IEC C78.901:2005	For electric lamps single base fluorescent lamps – dimensional and electrical characteristics
NEMA	NEMA ANSI/IEC C78.MR11-2:1997	Electric lamps: 1.375 inch (35mm) integral reflector lamps with front covers and gu4 or gz4 bases
NEMA	NEMA ANSI/IEC C81.64:2005	Guidelines and general information for electric lamp bases, lampholders, and gauges

horsepower and voltage rating, and may not know the application or duty cycle planned for the motor and its controller, the NEMA contactor is designed by convention with sufficient reserve capacity to assure performance over a broad band of applications without the need for an assessment of life requirements. Other conventions are that the contacts for most NEMA contactors are replaceable when inspection shows the need and that molded (encapsulated) coils are common on most NEMA devices.

1.5.2: Traditional IEC Contactors

IEC Standards do not define standard sizes. An IEC rating, therefore, indicates that a contactor has been evaluated by the manufacturer or a laboratory to meet the requirements of a number of defined applications (utilization categories).

The goal of the IEC design philosophy is to match a contactor to the load, expressed in terms of both rating and life. Usually, the user or original equipment manufacturer, who requires motors

TABLE 2.6   Partial List of Underwriters Laboratories Harmonized Standards

Developer	Standard No.	Title
UL/ISA/IEC	ANSI/UL 60079-0	Electrical Apparatus for Explosive Gas Atmospheres – Part 0: General Requirements
UL/ISA	ANSI/UL 60079-1	Electrical Apparatus for Explosive Gas Atmospheres – Part 1: Flameproof Enclosures "d"
UL/ISA/IEC	ANSI/UL 60079-5	Electrical Apparatus for Explosive Gas Atmospheres – Part 5: Powder Filling "q"
UL/ISA/IEC	ANSI/UL 60079-6	Electrical Apparatus for Explosive Gas Atmospheres – Part 6: Oil-Immersion "o"
UL/ISA	ANSI/UL 60079-7	Electrical Apparatus for Explosive Gas Atmospheres – Part 7: Increased Safety "e"
UL/ISA/IEC	ANSI/UL 60079-11	Electrical Apparatus for Explosive Gas Atmospheres – Part 11: Intrinsic Safety "i"
UL/ISA/IEC	ANSI/UL 60079-15	Electrical Apparatus for Explosive Gas Atmospheres – Part 15: Electrical Apparatus with Type of Protection "n"
UL/ISA/IEC	ANSI/UL 60079-18	Electrical Apparatus for Explosive Gas Atmospheres – Part 18: Encapsulation "m"
UL/CSA/ ANCE/IEC	ANSI/IL 60947-1	Low-Voltage Switchgear and Controlgear – Part 1: General rules
UL/CSA/ ANCE/IEC	ANSI/IL 60947-4-1A	Low-Voltage Switchgear and Controlgear – Part 4-1: Contactors and motor-starters – Electromechanical contactors and motor-starters
UL/IEC	ANSI/UL 60947-5-2	Standard for Low-Voltage Switchgear and Controlgear – Part 5-2: Control circuit devices and switching elements – Proximity switches
UL/IEC	ANSI/UL 60947-7-1	Standard for Low-Voltage Switchgear And Controlgear – Part 7-1: Ancillary equipment – Terminal blocks for copper conductors
UL/IEC	ANSI/UL 60947-7-2	Standard for Low-Voltage Switchgear and Controlgear – Part 7-2: Ancillary Equipment – Protective Conductor Terminal Blocks for Copper Conductors
UL/IEC	ANSI/UL 60947-7-3	Standard for Low-Voltage Switchgear and Controlgear – Part 7-3: Ancillary equipment – Safety requirements for fuse terminal blocks
UL/IEC	ANSI/UL 61131-2	Standard for Programmable Controllers – Part 2: Equipment Requirements and Tests

and controllers for their specific application, are in the best position to make this match. Typically, the contacts for larger horsepower-rated IEC contactors are replaceable. Most smaller horsepower-rated contactors do not have replaceable or inspectable contacts and are intended to be replaced when their contacts weld or are worn beyond further use. Most IEC contactors are supplied with tape-wound coils.

Some small (below 100 amps) NEMA and IEC devices are designed to comply with the fingersafe and back of hand safe requirements found in IEC60204-1 [Safety of Machinery – Electrical Equipment of Machines – Part 1: General Requirements] [4].

TABLE 2.7 Continuous current rating of NEMA contactors

NEMA contactor size	$I_{Continuous}$ (amps)
00	0 9
0	0 18
1	0 27
2	0 45
3	0 90
4	135
5	270
6	540
7	810
8	1215
9	2250

NEMA contactor sizes range from NEMA 00 to 9, with each contactor size capable of handling a range of motor horsepowers at different frequencies and voltages. Motor plugging and jogging service will affect the size of the NEMA contactor for the service. Table 2.7 references the NEMA contactor size to its continuous current rating, without exceeding the temperature rises permitted in NEMA ICS 1, *Industrial Control and Systems General Requirements*, Section 8.3 Temperature Rise. The NEMA contactor standard is NEMA Standards Publication ICS 2, *Industrial Control and Systems Controllers, Contactors and Overload Relays Rated 600 Volts*.

The IEC uses *Contactor Utilization Categories* (AC1 to AC4) to describe a specific application or use for a contactor. Those categories include those shown in Table 2.8 [5].

In addition to the utilization categories, IEC contactors also rated by motor horsepower (HP) and kilowatt (kW) rating, thermal current (I_{th}), rated operational current (I_e), and rated operational voltage (U_e).

TABLE 2.8 IEC contactor utilization categories

Utilization category	Typical applications
AC-1	Non-inductive or slightly inductive loads, e.g., resistive furnaces
AC-2	(Not covered in NEMA ICS 2.4, Table 2-1)
AC-3	Squirrel cage motors, starting and switching off while running at rated speed. Make locked rotor current and break full load current. Occasionally jog
AC-4	Squirrel cage motors, starting and switching off, while running at less than rated speed, jogging (inching) and plugging (reversing direction of rotation from other than an off condition). Make and break locked-rotor current

NEMA type motor controllers are factory wired, typically consisting of one or more (if required) contactors on a common base plate, overload relays, control transformer, line and load conductor terminals, complete control wiring with field wiring terminals. All of this equipment would be mounted inside an enclosure. IEC motor controller components are usually assembled in the field or by third part contractors. Some IEC components can be mounted on a DIN rail. The components may or may not be installed in an enclosure, depending on the ingress protection (IP) of the component equipment. IEC motor controllers are specifically designed for each motor application. This concept requires detailed information on the motor controller usage, i.e. number of motor starts per hour, lifetime rating, etc., which may not be required when selecting equivalent NEMA components.

Thermal Overload Relays

The NEMA design for electromechanical thermal overload relays includes the capability to field install the thermal overload heater elements. These elements are either bi-metallic type or eutectic alloy heat-sensing devices. They are indirectly heated by the motor line current and are not part of the current path from the contactor to the motor. The NEMA design allows a single overload relay to provide a variety of current ranges. The overload relays are divided into the following three trip classes:

Trip class	Response	Max. trip time
Class 10	Fast trip	10 sec. @ 600% $I_{Full\ Load}$
Class 20	Standard trip	20 sec. @ 600% $I_{Full\ Load}$
Class 30	Slow trip	30 sec. @ 600% $I_{Full\ Load}$

Class 10 overloads are used for hermetic refrigeration compressor motors, submersible pumps, etc. Class 20 would be provided for general purpose motors. Class 30 relays would be applicable for reciprocating pumps, loaded conveyors, etc.

IEC electromechanical thermal overload relays are typically direct-heated, bi-metallic elements. The heater and bi-metallic element are in the line current path and are normally Class 10. They may be equipped with an adjustment dial for an adjustment range of 1.3 to 1.7 times full load current. If a different range is required, then the entire overload relay must be changed.

Section 4.5 of NEMA Standard ICS 2.4 compares the construction of NEMA contactors and motor starters versus their IEC counterparts [6]. It indicates:

> NEMA motor starters and contactors typically have the coil-holding circuit auxiliary contact located on the viewer's left-hand side. IEC contactors typically locate this auxiliary contact on the right.

NEMA magnetic motor starters and contactors typically use more coil power, larger magnets, larger contacts, and stronger contact springs, and have higher short-circuit withstand capability. IEC devices, generally being smaller, consume less coil power.

IEC devices up to 20HP can be mounted on an IEC Standard (DIN) 35 mm rail. This DIN rail mounting permits snap-on interchangeability of one brand of IEC device with another, without additional drilling. IEC devices up to 10 HP, by convention are generally the same width for the same rating.

NEMA has no conventions relating to standard mounting rail, standard widths, nor standard mounting dimension.

Electrical Classified Area Equipment

One example of America's response to the harmonization movement was by the National Fire Protection Association, UL, and the American Petroleum Institute. All three organizations adopted Zone classification systems as an alternative to the Division hazardous area classification system. NFPA added Article 505 Class I, Zone 0, 1, and 2 Locations to NFPA 70, *National Electrical Code.* Eight of the UL 60079, Electrical Apparatus for Explosive Gas Atmospheres series standards, were harmonized with ISA and/or IEC. API added API RP 505, *Recommended Practice for Classification of Locations for Electrical Installations at Petroleum Facilities Classified as Class I, Zone 0, Zone 1, and Zone 2* and API RP 14FZ, *Recommended Practice for Design and Installation of Electrical Systems for Fixed and Floating Offshore Petroleum Facilities for Unclassified and Class I, Zone 0, Zone 1, and Zone 2 Locations.*

In adopting the Zone classification system, NFPA added IEC's eight equipment protection techniques to the National Electrical Code® including:

a. Encapsulation "m"

b. Flameproof "d"

c. Increased Safety "e"

d. Intrinsic Safety "i"

e. Oil Immersion "o"

f. Powder Filling "q"

g. Pressurization "p"

h. Type of Protection "n".

Of those systems, NFPA already recognized and allowed intrinsic safety, oil immersion, and pressurization as acceptable equipment protection methods for hazardous classified areas.

NEMA's "hermetically" sealed method is functionally similar to the IEC "encapsulation" method, although the testing requirements are not the same. It should be mentioned that Underwriters Laboratories, Inc. (UL) has issued with ISA of the standard series IEC based 60079 *Electrical Apparatus for Explosive Gas Atmospheres.* That was done in their adoption of the following standards:

> UL 60079-0, Electrical Apparatus for Explosive Gas Atmospheres – Part 0: General Requirements

> UL 60079-1, Electrical Apparatus for Explosive Gas Atmospheres – Part 1: Flameproof Enclosures "d"

> UL 60079-5, Electrical Apparatus for Explosive Gas Atmospheres – Part 5: Powder Filling "q"

> UL 60079-6, Electrical Apparatus for Explosive Gas Atmospheres – Part 6: Oil-Immersion "o"

> UL 60079-7, Electrical Apparatus for Explosive Gas Atmospheres – Part 7: Increased Safety "e"

> UL 60079-11, Electrical Apparatus for Explosive Gas Atmospheres – Part 11: Intrinsic Safety "i"

> UL 60079-15, Electrical Apparatus for Explosive Gas Atmospheres – Part 15: Electrical Apparatus with Type of Protection "n"

> UL 60079-18, Electrical Apparatus for Explosive Gas Atmospheres – Part 18: Construction, Test and Marking of Type of Protection Encapsulation "m" Electrical Apparatus

Some additional equipment protection techniques are not included in the IEC's repertoire in their IEC 60079 standard series, but are included in NFPA 70 Article 500.7 and 506.8. Among them are:

> Explosionproof Apparatus

> Dust Ignitionproof

> Nonincendive Circuit

> Nonincendive Equipment

> Nonincendive Component

> Nonincendive Field Wiring

NEMA's explosionproof apparatus is similar to IEC's flameproof in gas cooling concept; however, there are major differences. Flameproof enclosures are subjected to routine testing at

1.5 times the enclosure design pressure before leaving the factory. Explosionproof enclosures are tested to a maximum of up to 4 times design pressure and have a substantially thicker enclosure wall design.

Equipment Enclosure Differences

ANSI/NEMA 250, *Enclosures for Electrical Equipment (1000 Volts Maximum)*, is the definitive standard for electrical equipment enclosures in the United States. IEC 60529, *Degrees of Protection Provided by Enclosures (IP Code)*, is a similar standard for nations that have adopted the IEC. Although both standards deal with equipment enclosures, their methods are quite different. An excellent document published by the National Electrical Manufacturers Association provides a comparison of the two standards. It is entitled *A Brief Comparison of NEMA 250 and IEC 60529*. It was published in 2002, by NEMA in Rosslyn, VA.

The term "IP Code" in IEC60529 Standard is defined as *Ingress Protection Code*. It characterizes electrical enclosures or equipment using the letters IP with two *Character Numbers*. The first character number can vary between 0 and 6. As the first character number increases, the degree of ingress protection increases. The Code establishes the level of protection in the first character number (IP _X) with two categories, protection with respect to individuals (persons) and protection with respect to solid foreign object entry. Those designations are defined in Table 2.9.

The IP Code second character number can vary between 0 and 8, representing the degree of protection from water intrusion. The degree of protection against water intrusion increases when the second character number (IP X_) increases while approaching the highest number, 8. This designation involves water ingress only. The standard does not include ingress protection against other fluids. Table 2.10 provides an explanation for the IP Code second character number.

TABLE 2.9 IEC 60529 Ingress Protection Code first character number explanation [7, 9]

IP first character number	Personal protection description	Foreign object protection	Force applied to foreign object [8]
0	No test required	No test required	No test required
1	Back of hand	Objects $\geq$ 50 mm diameter	50 Newton
2	A finger	Objects $\geq$ 12.5 mm diameter	10 Newton
3	A tool	Objects $\geq$ 2.5 mm diameter	3 Newton
4	A wire	Objects $\geq$ 1 mm diameter	1 Newton
5	A wire	Dust-protected	1 Newton
6	A wire	Dusttight	1 Newton

TABLE 2.10   IEC 60529 Ingress Protection Code second character number explanation [9]

IP second character number	Degree of protection from water ingress	Test requirements	Time (min) [10]
0	Non-protected	No test required	N/A
1	Vertically falling water drops	"Drip Box" w/spouts spaced on 20 mm pattern. Rainfall rate 1 mm/min	10
2	Vertically falling water drops with the enclosure elevated up 150	Same "Drip Box" as IP_1. Rainfall 3 mm/min. Enclosure placed in 4-fixed tilt positions @ 15°	2.5/tilt position
3	Spraying water	Water sprayed over 60° arc from vertical w/oscillating tube sprayer w/holes 50 mm apart and water flow rate of 0.07 liters/min/hole	10
		Water sprayed over 60° arc from vertical w/hand held nozzle and water flow rate of 10 liters/min	1 min/m^2 of enclosure surface area, 5 min minimum
4	Splashing water	Same procedure as IP _3, except w/spray arc of 180° from vertical	1 min/m^2 of enclosure surface area, 5 min minimum
5	Water jets	Enclosure sprayed from "all practice directions". Water stream of 12.5 liters/min. Spray from 6.3 mm nozzle @ 2.5–3 meter distance	1 min/m^2 of enclosure surface area, 3 min minimum
6	Powerful water jets	Enclosure sprayed from "all practice directions". Water stream of 100 liters/min. Spray from 12.5 mm nozzle @ 2.5–3 meter distance	1 min/m^2 of enclosure surface area, 3 min minimum
7	The effects of immersion temporarily under water	Lowest point of enclosure $\leq$ 850 mm high immersed in water 1000 mm below the surface	30
		Highest point of enclosure > 850 mm high immersed in water 150 mm below the surface	30
8	The effects of immersion continuously under water	Procedures subject to agreement between manufacturer and user with minimum testing as severe as IP_7	

IEC 60529 Standard does not define or establish tests for any of the following physical requirements regarding enclosures [11, 12]:

Environmental testing (other than water entry)

Door and cover securement

Corrosion resistance testing

External icing testing

Gasket aging and oil resistance

Coolant effects

Protection against risk of explosion

Environmental protection (e.g. against humidity, corrosive atmospheres or fluids, fungus or the ingress of vermin)

The NEMA enclosure designations do have specific requirements for the above noted physical requirements; making it impossible to directly convert IP Code designations to NEMA Type designations. Although IEC 60529 does not have a specific IP Code having an explosion rating, there are other IEC enclosure standards dealing with that possible event. IEC 60079, *Electrical Apparatus for Explosive Gas Atmospheres*, is a series of standards which define acceptable protection techniques for electrical equipment in hazardous (classified) areas. Underwriters Laboratories has harmonized some of those same standards.

NEMA Standard 250, *Enclosures for Electrical Equipment (1000 Volts Maximum)* is the American standard used for selecting enclosures for electrical equipment. See Table 2.11 for an explanation of NEMA enclosure designations. That standard, along with other NEMA product standards or third party certification standards, contains the necessary information on enclosure testing and performance requirements. Examples of third party certification standards which would be used in conjunction with the NEMA 250 Standard are shown in Table 2.12.

NEMA developed a document entitled *Electrical Installation Requirements – A Global Perspective* by Underwriters Laboratories, Inc. Principal Investigator Paul Duks in April 1999. It was an extensive study comparing IEC 60364-5-51 and NFPA 70, *National Electrical Code*. Although both standards have changed somewhat since the time of the article publication, it is a valuable document in understanding the philosophy behind the IEC *Common rules for electrical installations in buildings*.

IEC 60364-5-51, *Electrical Installations of Buildings – Part 5-51: Selection and erection of electrical equipment – Common rules*, was developed in 1969 as a result of attempting to harmonize the European electrical codes. That effort was not successful because of the tremendous differences between the European national wiring standards. Any nation adopting

TABLE 2.11   NEMA 250 enclosure type designations [14]

NEMA 250 type designation	Use	Degree of protection
1	Indoor	Limited amount of falling dirt
2	Indoor	Limited amounts of falling water and dirt
3	Outdoor	Rain, sleet, windblown dust, and damage from external ice formation
3R	Outdoor	Rain, sleet, and damage from external ice formation
3S	Outdoor	Rain, sleet, windblown dust, and provide for external mechanisms when ice laden
4	Indoor or Outdoor	Windblown dust and rain, splashing water, hose-directed water, and damage from external ice formation
4X	Indoor or Outdoor	Corrosion, windblown dust and rain, splashing water, hose-directed water, and damage from external ice formation
5	Indoor	Settling airborne dust, falling dirt, and dripping noncorrosive liquids
6	Indoor or Outdoor	Hose-directed water, entry of water during occasional temporary submersion at a limited depth, and damage from external ice formation
6P	Indoor or Outdoor	Hose-directed water, entry of water during prolonged submersion at a limited depth, and damage from external ice formation
7	Indoor	Locations classified as Class I, groups A, B, C, or D as defined in the National Electrical Code®
8	Indoor or Outdoor	Locations classified as Class I, groups A, B, C, or D as defined in the National Electrical Code®
9	Indoor	Locations classified as Class II, groups E, F, or G as defined in the National Electrical Code®
10		Constructed to meet the applicable requirements of the Mine Safety and Health Administration
12	Indoor	Circulating dust, falling dirt, and dripping noncorrosive liquids
12K	Indoor	Circulating dust, falling dirt, and dripping noncorrosive liquids
13	Indoor	Dust, spraying of water, oil, and noncorrosive liquids

IEC 60364-5-51 would also have to adopt supplemental wiring requirements. IEC 60364-5-51 was not intended to be used by engineers, electricians, or electrical inspectors. It is a broad performance-based document, intended for use as a guide for the development of national wiring standards. IEC 60364-5-51 covers the principals needed to protect against electrical hazards. It does not contain recommendations for installations in electrical hazardous area locations. That is covered in IEC 60079, *Electrical Apparatus for Explosive Gas Atmospheres* [13].

The scope of IEC 60364-5-51 is limited to voltages up to 1000 Volts. The NEC® does not have that restriction dealing with voltages over 600 Volts on a limited basis; however, IEEE C2, *National Electrical Safety Code*, governs utility type installations and transmission and distribution systems with voltages up to 800 kV.

TABLE 2.12 **Enclosure third party certification standards**

Developer	Standard No.	Title
UL	UL 50	Enclosures for Electrical Equipment, Non-Environmental Considerations
UL	UL 50E	Enclosures for Electrical Equipment, Environmental Considerations
UL	UL 489	Molded-Case Circuit Breakers, Molded-Case Switches, and Circuit-Breaker Enclosures
UL	UL 877	Standard for Circuit Breakers and Circuit-Breaker Enclosures for Use in Hazardous (Classified) Locations
UL	UL 2062	Enclosures for Use in Hazardous (Classified) Locations
UL	UL 60079-0	Electrical Apparatus for Explosive Gas Atmospheres – Part 0: General Requirements
UL	UL 60079-1	Electrical Apparatus for Explosive Gas Atmospheres – Part 1: Flameproof Enclosures "d"
UL	UL 60079-5	Electrical Apparatus for Explosive Gas Atmospheres – Part 5: Powder Filling "q"
UL	UL 60079-6	Electrical Apparatus for Explosive Gas Atmospheres – Part 6: Oil-Immersion 'o'
UL	UL 60079-7	Electrical Apparatus for Explosive Gas Atmospheres – Part 7: Increased Safety 'e'
UL	UL 60079-11	Electrical Apparatus for Explosive Gas Atmospheres – Part 11: Intrinsic Safety "i"
UL	UL 60079-15	Electrical Apparatus for Explosive Gas Atmospheres – Part 15: Electrical Apparatus with Type of Protection "n"
UL	UL 60079-18	Electrical Apparatus for Explosive Gas Atmospheres – Part 18: Construction, Test and Marking of Type of Protection Encapsulation 'm' Electrical Apparatus

Conclusions

There are other differences between IEC and ANSI accredited Standards Development Organizations' (SDO) codes, standards, and recommended practices than were examined in this chapter. The examples reviewed here show that the differences between IEC and American standards can generally be related to design philosophies. With the number of nations adopting the IEC standards as the basis for their own national standards it should be obvious for the necessity for some planned form of harmonization of American standards to compete in the global trade economy.

Support from American manufacturers competing in an international marketplace and American SDOs with ANSI's *U.S. Standards Strategy* is essential to remain competitive internationally. Manufacturers can design products certified with international certifications. An example of this can be seen in more American-produced equipment carrying labels identifying certification by American Nationally Recognized Testing Laboratories, CSA, ATEX, CE Mark, and IECEx, with multi-voltage-(120 V/230 V) multi-frequency (50 Hz/60 Hz) listing.

The Institute of Electrical and Electronic Engineers is an international electrical engineering technical organization offering conference technical papers and standards development. Its membership includes engineers from over 150 nations. That organization is influential in presenting internationally the design principals and technical development trends in electrical engineering. Participation in technical papers and studies by its international memberships can influence the use of IEEE standards in international standards adoption. That objective is also one part of ANSI's United States Standards Strategy's objectives. That plan encourages SDOs to

> ... sector management recognize the value of standardization of national and global levels and provide adequate resources and stable funding mechanisms to support such efforts.

> ... respond ... quickly and responsibly to provide standards that address national and international needs. [15]

IEEE's agreement with IEC to have its standards reviewed for possible joint adoption and use by IEC has already proved successful. That strategy is also being pursued by other ANSI accredited SDOs, such as NEMA, UL, and ASTM to list a few.

A simplified analysis of the general difference between IEC and ANSI standards philosophy for some industrial/commercial low-voltage power distribution and control equipment can be presented. The ANSI power distribution and motor control standards design utilize standardized sizes of equipment with sufficient design capacity to be suitable for a range of load sizes, types, and operating requirements. That equipment is normally produced with field interchangeable parts for easy maintenance or modification should the design requirements change slightly. The IEC concept does not support standard, completely factory assembled equipment motor control centers, panelboards, etc., but provides a greater number of various components to allow specific customized designs specifically tailored for each application. IEC parts may not always allow for field replacement of subassemblies such as motor controller heater elements or magnetic coils. Because of the IEC design philosophy, subcomponent parts failure may sometime require the replacement of entire part, not allowing for replacement of subassemblies, such as coils on a motor starter. The NEMA design approach allows field reparability and parts replacement of parts on equipment such as NEMA design motor starters.

IEC electrical low-voltage power distribution boards can be physically different from their ANSI (panelboards and switchboards) counterparts. The ANSI equipment is respectively governed by NEMA PB 1 and PB 2. Both utilize a system of bolt-on or stab-type connections for the overcurrent devices to a main bus or branch bus. The IEC design presently follows IEC 61499-1. IEC 61439-1 through -6 are under *work in progress* development to supersede the existing IEC 61439 standards. The IEC design is contingent on a number of factors, including the skill of the individuals who may operate the equipment. Those designs are also be affected by many other factors, including *verification of requirements* by testing, calculation/ measurement or by design rules. These replace the previous IEC design verification approaches of TTA and PTA assemblies. Physical designs can vary anywhere from DIN rail mounted overcurrent devices connected to the main bus by insulated conductors to overcurrent device modules stabbed to bus.

References

1. ITU website: http://www.itu.int/net/about/index.aspx.
2. U.S. Standards Strategy Section V. ANSI website: http://publicaa.ansi.org/sites/apdl/documents/standards%20Activities/NSSC/USSS-2005%20-%20FINAL.pdf.
3. Ibid., Section IV.
4. NEMA ICS 2.4-2003, *NEMA and IEC Devices for Motor Services – A Guide for Understanding the Differences*; 2003, Paragraph 1.5.5 and 1.5.2, pages 3, 4. National Electrical Manufacturers Association; Rosslyn, VA.
5. Ibid., Table 2-1, Common Utilization Categories for AC Contactors.
6. Ibid., Section 4.5 Construction; page 18.
7. NEMA Standards Publication, *A Brief Comparison of NEMA 250 – Enclosures for Electrical Equipment (1000 Volts Maximum) and IEC 60529 – Degrees of Protection Provided by Enclosures (IP Code)*; 2002, page 2. National Electrical Manufacturers Association; Rosslyn, VA.
8. Rockwell Automation website: http://www.ab.com/en/epub/catalogs/3377539/5866177/5635113/5640318/; IEC Enclosures Degree of Protection; 2009.
9. NEMA Standards Publication: A *Brief Comparison of NEMA 250 – Enclosures for Electrical Equipment (1000 Volts Maximum) and IEC 60529 – Degrees of Protection Provided by Enclosures (IP Code)*; 2002, page 4. National Electrical Manufacturers Association; Rosslyn, VA.
10. Rockwell Automation website http://www.ab.com/en/epub/catalogs/3377539/5866177/5635113/5640318/; IEC Enclosures Degree of Protection; 2009
11. NEMA Standards Publication: A Brief Comparison of NEMA 250 - Enclosures for Electrical Equipment (1000 Volts Maximum) and IEC 60529 - Degrees of Protection Provided by Enclosures (IP Code); 2002, page 5. National Electrical Manufactures Association; Rosslyn, VA.
12. Rockwell Automation website: http://www.ab.com/en/epub/catalogs/3377539/5866177/5635113/5640318/; IEC Enclosures Degree of Protection; 2009.
13. Duks, Paul, *Electrical Installation Requirements – A Global Perspective*; 1999, pages 1-6. National Electrical Manufacturers Association; Washington, DC.
14. NEMA 250-1991, *Enclosures for Electrical Equipment (1000 Volts Maximum)*; 1991, page 3. National Electrical Manufacturers Association; Washington, DC.
15. United States Standards Strategy; 2005, Part IV; United States Standards Committee, American National Standards Institute; New York, NY.

CHAPTER 3

The Authority Having Jurisdiction (AHJ)

Although it may sound like something from a legal drama, the *Authority Having Jurisdiction (AHJ)* is a very important element in the implementation of codes, standards, and recommended practices. Codes and standards produced in the United States are generally considered voluntary, peer-reviewed consensus documents, with some exceptions. Their implementation is not required, unless mandated by legislative or industry voluntary agreement. The term *authority* is defined as *"A person or group having the right and power to command, decide, rule, or judge; A person with a high degree of knowledge or skill in a particular field …"* [1]. Jurisdiction is defined as *"the right and power to command, decide, rule or judge"* [2].

Once the requirement for the implementation of certain codes, standards, and recommended practices has been mandated, an individual, group, or government agency must be appointed with the authority to enforce the use of the codes, standards, and recommended practice. The AHJ must also have the authority to either waive specific requirements of their use or allow alternatives, should they prove to be an equally safe method, procedure, or means of providing life safety and protection from fire and injury. The individual, group, or agency designated with that authority would be known as the Authority Having Jurisdiction.

The National Fire Protection Association (NFPA) states in NFPA 70®, *National Electrical Code*® (NEC®) in Article 90-4 Enforcement [3]:

> This Code is intended to be suitable for mandatory application by governmental bodies that exercise legal jurisdiction over electrical installations, including signaling and communications systems, and for use by insurance inspectors. The authority having jurisdiction for enforcement of the Code has the responsibility for making interpretations of the rules, for deciding on the approval of equipment and materials, and for granting the special permission contemplated in a number of rules. By special permission the authority having jurisdiction may waive specific requirements in this Code or permit alternate methods, where it is assumed that equivalent objectives can be achieved by establishing and maintaining effective safety.

Based on the NFPA's use of the term Authority Having Jurisdiction (AHJ), it can be a governmental agency or authority that has been legally appointed to enforce a code or codes regarding life safety or other issues. The authority may be a local (municipality or county), state, or federal governmental entity with the legal mandate to enforce governing building codes, safe work practices, life safety issues, workplace safety operational procedures, etc. It could also be an insurance inspector or a Nationally Recognized Testing Laboratory (NRTL) with authority for product certification and testing. As noted above, the AHJ may also have the authority to waive specific requirements or authorize alternative methods to establish equivalent life safety protection. The alternatives must be determined to be reasonable, acceptable, and safe.

NFPA 101®, *Life Safety Code*® (LSC®) defines the Authority Having Jurisdiction as

> An organization, office, or individual responsible for enforcing the requirements of a code or standard, or for approving equipment, materials, an installation, or a procedure. [4]

It also provides a detailed explanation of the term in the Annex A of the LSC. It explains that

> The term Authority Having Jurisdiction, or its acronym AHJ, is used in NFPA documents in a broad manner, since jurisdictions and approval agencies vary, as do their responsibilities. Where public safety is primary, the authority having jurisdiction may be a federal, state, local, or other regional department or individual such as a fire chief; fire marshal; chief of fire prevention bureau, labor department, or health department; building official; electrical inspector; or others having statutory authority. For insurance purposes, an insurance inspection department, rating bureau, or other insurance company representative may be the authority having jurisdiction. In many circumstances, the property owner or his or her designated agent assumes the role of the authority having jurisdiction; at government installations, the commanding officer or department official may be the authority having jurisdiction. [5]

AHJ Adopted Codes and Standards

It was established above that the AHJ or the legislative entity that created it should have the authority to establish or adopt electrical or other codes, standards, and recommended practices. The AHJ will use those documents to enforce and assure life safety, fire protection, or occupational health and safety issues. There are a number of standards documents that the AHJ may choose to enforce, depending upon the task, occupancy or product or service being regulated. Figure 3.1 illustrates the most common general types of electrically related codes

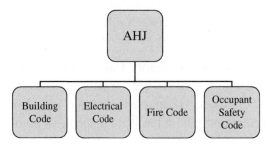

Figure 3.1: Common local, state and municipal AHJ enforced codes

used by local, state or municipal AHJs which involve occupancy construction or renovation. The most common AHJ regulation enforcement involves building codes, which directly affect most individuals. Building codes not only affect people in the homes, apartments, dormitories, condominiums, hotels, etc. in which they may stay, live, and sleep, but also their place of employment, the places where they shop, hospitals and medical buildings, places of entertainment, schools, etc. The construction or renovation of those facilities would be under auspices of municipal, county, or state AHJs.

There are specific personal safety and property protection requirements in each of the general codes and standards in Figure 3.1. They govern electrical design and installation requirements. Each code will generally supplement or support other codes. For instance one code may establish a requirement for a specific life safety system; however, another code will be responsible for the general installation recommendations of that system, supplementing the requirements of the first code.

A review of the states' and municipalities' Fire Marshall and building inspection AHJs in the United States indicates that there are a variety of adopted codes, standards, and recommended practices for various states, cities, and counties. For instance, several large American cities and states have developed their own occupant safety, building, and electrical codes. They may simply use existing national standards with adopted modifications or develop entirely new documents. Some states and municipalities may not have established code enforcement, but may mandate the use of a specific code. Also, the latest edition of a code may not be the edition that has been adopted for enforcement by the AHJ.

Building Codes

During the 1990s decade, there were three generally accepted regional model building codes in use:

Building Officials Code Administrators International (BOCA) – which developed the *BOCA National Building Code (BOCA/NBC)* for the East Coast and Midwest United States;

International Conference of Building Officials (ICBO) – which developed the *Uniform Building Code (UBC)* for the West Coast of the United States; and

Southern Building Code Congress International (SBCCI) – developed the *Standard Building Code (SBC)* for the Southeast United States.

During the latter half of 2003, the three major American building code organizations merged into the *International Code Council (ICC)*, ceasing to develop regional codes. The organization now publishes the *International Building Code*, the *International Residential*

Code, and the *International Fire Code.* As of 2007 those three codes have been adopted as follows [6]:

- the *International Building Code (IBC)* has been adopted at the state or local level in 50 states plus Washington, DC;

- the *International Residential Code (IRC)* has been adopted at the state or local level in 46 states plus Washington, DC; and

- the *International Fire Code* (IFC) has been adopted at the state or local level in 41 states plus Washington, DC.

In 2002, the National Fire Protection Association released NFPA 5000®, *Building Construction and Safety Code*® [7]. It was developed through a consensus process and was accredited by the American National Standards Institute (ANSI). The city of Pasadena, Texas adopted NFPA 5000 in 2003.

> California adopted the NFPA 5000 codes as a baseline for the future California Building Code, but later rescinded the decision and continued to use the IBC. The main driver for this decision was increased costs involved in training architects and engineers to design for a new code, and the disparity that a different code would cause between California and the majority of other states which have adopted IBC. [8]

The Code recognizes NFPA 70, *National Electrical Code* as its electrical section. It also recognizes NFPA 72®, *National Fire Alarm Code*® for fire detection and alarm.

Electrical Code

The most recognized and used electrical code throughout the United States is the National Fire Protection Association's NFPA 70, *National Electrical Code* (NEC). The *National Electrical Manufacturer's Association* (NEMA) reports [9] that 40 states have adopted some edition of the NEC statewide. Ten states have not had statewide adoption of any edition of the NEC; however, municipalities and counties in those states may have a local option adoption. Some states, such as California, and municipalities, such as New York City and Chicago, have established their own electrical code. New York City has adopted a specific edition of the NEC, but includes either an amendment to certain NEC sections and/or chose to not adopt some sections.

In 2006, the City Council of New York City passed the following ordinance [10]:

> §14. Section 27-3024 of the administrative code of the city of New York, as amended by local law number 81 for the year 2003, is amended to read as follows:

> § 27-3024. Adoption of the electrical code technical standards. a. The city of New York hereby adopts the [2002] 2005 edition of the National Fire Protection Association NFPA 70 *National*

Electrical Code as the minimum requirements for the design, installation, alteration or repair of electric wires and wiring apparatus and other appliances used or to be used for the transmission of electricity for electric light, heat, power, signaling, communication, alarm and data transmission in the city subject to the amendments adopted by local law and set forth in section 27-3025 of this subchapter, which shall be known and cited as "the New York city amendments to the [2002] 2005 National Electrical Code". Such [2002] 2005 edition of the National Fire Protection Association NFPA 70 National Electrical Code with such New York City amendments shall together be known and cited as the "electrical code technical standards". The commissioner shall make a copy of the electrical code technical standards available for public inspection at the department of buildings.

An example of one of the amendments to the 2005 NEC adopted by the City [11] is as follows, with the original text of 2005 NEC [12] in italics and the New York City amendment for that section [13] in bold italics at the end of the quote:

2005 NEC Section 210.19(A) (1):

(1) *General—Branch-circuit conductors shall have an ampacity not less than the maximum load to be served. Where a branch circuit supplies continuous loads or any combination of continuous and noncontinuous loads, the minimum branch-circuit conductor size, before the application of any adjustment or correction factors, shall have an allowable ampacity not less than the noncontinuous load plus 125 percent of the continuous load.* ***Conductors of branch circuits shall be sized to allow for a maximum voltage drop of 3% at the last outlet supplying light, heat or power and the maximum voltage drop allowable for feeders and branch circuit combined shall not exceed 5%.***

In describing the functions of the AHJ, local interpretation of codes and standards was noted as one responsibility of that position. Procedures have been established to allow requests for *National Electrical Code* interpretation from the NEC Code Committee. NEC Section 90.6 Formal Interpretations discusses the procedures which may be utilized to assist the AHJ or any member of the National Fire Protection Association. It notes that "formal interpretation procedures have been established and are found in the NFPA Regulations Governing Committee Projects" [14]. The *National Electrical Code Handbook* provides some additional explanations regarding NEC interpretations. It notes:

The authority having jurisdiction is responsible for interpreting Code rules and should attempt to resolve all disagreements at the local level. Two general forms of Formal Interpretations are recognized: (1) those that are interpretations of the literal text and (2) those that are interpretations of the intent of the Committee at the time the particular text was issued. [15]

The *NEC Handbook* notes there are limitations to the Code Committee offering interpretations. It indicates:

Interpretations of the NEC not subject to processing are those that involve (1) determination of compliance of a design, installation, product, or equivalency of protection; (2) a review of plans or specifications or judgment or knowledge that can be acquired only as a result of on-site inspection; (3) text that clearly and decisively provides the requested information; or (4) subjects not previously considered by the Technical Committee or not addressed in the Document ... [16]

Fire Codes

There are several standards providing fire prevention, protection and detection requirements which may be adopted by the AHJ. They include:

Individual state/county/municipal fire codes

NFPA 1, *Uniform Fire Code*™

International Code Council: *International Fire Code*®

NFPA 72®, *National Fire Alarm Code*®

The first three codes in the above list are fire codes. The individual state/county/municipal codes may be entirely written by those entities. The *International Fire Code* or NFPA 1 may be adopted by AHJs, either in their entirety or with amendments and deletions. The last code deals with the installation of fire alarms.

NFPA 1, *Uniform Fire Code*, was jointly written by the Western Fire Chiefs Association (WFCA) and NFPA [17]. It contains provisions and sections from both the (NFPA's) *Fire Prevention Code* and WFCA's *Uniform Fire Code (UFC)*. It contains separate sections for administration and code enforcement. There are also sections on occupancies, processes, equipment, and hazardous materials. To accommodate situations where innovative building solutions may be needed in lieu of specification-based standards, the new Code contains a section on performance-based design. Over 130 NFPA codes and standards are referenced in the document.

> The International Fire Code® is a merger of the provisions in the National Fire Prevention Code, the Standard Fire Prevention Code and the Uniform Fire Code. So while the International Fire Code itself is new, its provisions are not. They are based on fire codes that have been in use in the majority of the United States for decades. [18]

Table 3.1 presented below is a summary of the states that have adopted either the NFPA *Uniform Fire Code*, the International Code Council *International Fire Code*, have their individual statewide fire code, or allow local option for fire code adoption. States may choose to adopt the NFPA-UFC or ICC-IFC either in total or modified/amended with local state changes. The data were obtained from each state's official Fire Marshall website.

TABLE 3.1   United States individual state fire code adoptions

State	Statewide IFC	Statewide NFPA-UFC	State code	Local option
Alabama	X			
Alaska	X			
Arizona			X	
Arkansas	X			
California	X			
Colorado				X
Connecticut	X			
Delaware		X		
Florida		X		
Georgia	X			
Hawaii			1997 Uniform Fire Code	
Idaho	X			
Illinois				X
Indiana	X			
Iowa	X			
Kansas	X			
Kentucky		X		
Louisiana		X		
Maine		X		
Maryland		X		
Massachusetts			X	
Michigan			X	
Minnesota	X			
Mississippi	X			
Missouri				X
Montana		X		
Nebraska		X		
Nevada	X			
New Hampshire		X		
New Jersey			X	
New Mexico		X		
New York	X			
North Carolina	X			
North Dakota	X			
Ohio	X			

(*Continued*)

TABLE 3.1 United States individual state fire code adoptions—cont'd

State	Statewide IFC	Statewide NFPA-UFC	State code	Local option
Oklahoma	X			
Oregon	X			
Pennsylvania	X			
Rhode Island		X		
South Carolina	X			
South Dakota				X
Tennessee	X			
Texas		X		
Utah	X			
Vermont		X		
Virginia	X			
Washington	X			
West Virginia		X		
Wisconsin		X		
Wyoming	X			

The purpose of National Fire Protection Association's NFPA 72, *National Fire Alarm Code* [19]:

> is to define the means of signal initiation, transmission, notification, and annunciation; the levels of performance; and the reliability of the various types of fire alarm systems. This Code defines the features associated with these systems, and also provides the information necessary to modify or upgrade an existing system to meet the requirements of a particular system classification. It is the intent of this code to establish the required levels of performance, extent of redundancy, and quality of installation, but not the methods by which these requirements are to be achieved.

NFPA 72 defines the methods for performance, redundancy, and the quality of installation for fire alarm systems. It does not mandate when or where a fire alarm system should be installed. That requirement is established in the *Life Safety Code*, NFPA 1: *Uniform Fire Code*, and *International Fire Code* or by state/county/municipal fire codes.

Life Safety Code

NFPA 101, *Life Safety Code* is not in itself a fire code. Its purpose

> is to provide minimum requirements, with due regard to function, for the design, operation, and maintenance of buildings and structures for safety to life from fire. Its provisions will also aid life safety in similar emergencies. [20]

Fire detection, notification, and extinguishment requirements are only one small portion of this Code. It also specifies life safety requirements by occupancy type. Some specific life safety areas covered by the Code include fire detection general requirements, when specified; fire suppression, where called for; fire detector locations required in a limited number of occupancies; alarm and notification of occupants; egress lighting and safety; and notification of emergency services. This Code establishes the type of fire detection and suppression systems required by occupancy type, while NFPA 72 provides the standard for installation, operation, and maintenance of the alarm and detection systems Fire suppression equipment installation requirements would be covered under a different code.

NFPA 101, *Life Safety Code*® covers a large number of topics, including structural fireproofing requirements; means of egress; classification of occupancy and content hazards; fire protection features; fire protection equipment and building services; interior furnishings and contents; and occupancy requirements. Of particular interest from an electrical engineering standpoint are the following design areas:

Fire detection, alarm, and communications systems

Emergency lighting and power requirements

Illumination and marking of means of egress

Means of egress smoke control

Egress special locking requirements

Egress door self-closing devices and powered doors

AHJ Process

The process of building a structure begins with structural, mechanical, and electrical plans being developed, by either an architect and/or professional engineer(s). Those plans would provide details of the structure design, its electrical system, plumbing system, HVAC system, and, depending on the size of occupancy type, could include fire detection and suppression systems, elevators, etc. AHJs with specific education, experience, and training would be responsible for plan approval and site inspections of equipment installation in areas of their expertise. Depending on the size of a municipality, a mechanical inspector may have jurisdiction over plumbing, HVAC, and mechanical systems. In large metropolitan areas, each of those discipline areas might have individual AHJs. Once the structure is designed, the detailed design plans would be submitted to the appropriate AHJ for review and eventual inspection. See Figure 3.2 for the general sequence of AHJ inspection and design review.

An initial inspection may involve establishment of temporary power to the construction site. This might involve installation of a temporary power pole with an electric meter pan, circuit breaker panel, and GFCI outlets. A second inspection may be required for hookup of power

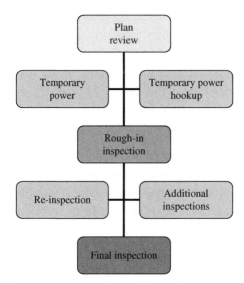

Figure 3.2: General AHJ inspection sequence

from the temporary power pole to the structure or to an onsite construction trailer/office. An electrical rough-in inspection would involve the installation of wiring, raceway, outlet boxes, and panelboards and would be completed before the interior walls are installed or underground trenches with raceway/duct banks are covered. Should a code violation problem be identified, a re-inspection might be scheduled after corrective work is accomplished. The AHJ may require additional inspections, depending upon the project complexity. Final inspection would be completed after all wiring and equipment installation is complete.

Plan review may result in the drawings being returned to the project designer for additional clarification, design changes, or additions. It should be noted that the AHJ design review is not necessarily a detailed design review of calculations and design criteria, but is conducted to verify that the design meets the intent and general requirements of the codes and standards that have been adopted by the AHJ or their legislative authorizer. Determining the adequacy of a branch circuit breaker size to feed a particular load might not necessarily be a review item unless the inspector has access to load data; however, verification that GFCI and arc fault circuit breakers are appropriately specified and correctly installed where required would be a concern of the plan review and site inspections. Once the required changes are implemented, the plans are resubmitted to the AHJ for review and approval. Delays could develop if zoning restrictions or setback and easement requirements conflict with the plans. If the contractor retained to do the installation work is separate from the project design engineer(s), the contractor would be responsible for filing for the building permit and requesting the rough-in and final inspections.

Nationally Recognized Testing Laboratories (NRTL)

A *Nationally Recognized Testing Laboratory (NRTL)* is also considered an Authority Having Jurisdiction (AHJ) for the products or services it certifies. This does not mean that the NRTL would take the place of a municipal building inspector. The authority the NRTL exercises involves the product or service it certifies. When an NRTL certifies that a product meets specified criteria and standards, it affixes its label to that product. It has the authority to reject the product or require changes to meet the criteria it establishes. Random sampling and testing of products and inspection of manufacturing facilities is all part of the NRTLs' monitoring process. The role of NRTLs will be examined in more detail in Chapter 4.

The *National Electrical Code*® notes in Article 110.2 Approval that "The conductors and equipment required or permitted by this Code shall be acceptable only if approved" [21]. The *NEC Handbook* explains:

> All electrical equipment is required to be approved as defined in Article 100 and, as such, to be acceptable to the authority having jurisdiction (also defined in Article 100). Section 110.3 provides guidance for the evaluation of equipment and recognizes listing or labeling as a means of establishing suitability.

> Approval of equipment is the responsibility of the electrical inspection authority, and many such approvals are based on tests and listings of testing laboratories. [22]

Under this scenario, a minimum of three types of AHJs may be involved with an electrical construction design and install project. The first would be the municipal electrical inspector issuing the building permit and inspecting the work. The second would be the NRTL that certifies the electrical equipment and materials being used to construct the project. The third AHJ would be the municipal, state or federal agency charged with employee safety in the workplace.

Owner Authority Having Jurisdiction

An example of an owner Authority Having Jurisdiction can be found in the Reedy Creek Improvement District in Florida. Walt Disney purchased some 27,800 acres of land between Orlando, Florida and Kissimmee, Florida for the construction of Disney World.

> Disney also petitioned with the State of Florida Legislature to give Walt Disney Productions municipal jurisdiction over the land they had acquired. This was to make sure that Walt Disney could have full control over every part of the property, even how the buildings were constructed. Walt was planning new ideas in urban living and did not want the government to interfere. This was the beginning of the Reedy Creek Improvement District (RCID). [23]

To aid in the development of this unique property, Disney created a Department of Building and Safety.

The primary purpose of the Department of Building & Safety is to provide reasonable requirements to safeguard life and property by regulating the design, construction, repair and use of new and existing structures.

Code development began in 1968 and the first EPCOT (Experimental Prototype Community of Tomorrow) Code was adopted in 1970. The District developed codes and standards to both accommodate new and innovative methods and systems, and provide public safety criteria exceeding other available codes and standards. Since that time, RCID has developed and enforced the EPCOT Codes that exceed traditional regulations by setting forth design criteria for such installations as thrill rides and amusement attractions and by requiring complete automatic sprinkler and detector systems in all buildings. Provisions applicable to motion picture and television sound stages, and more extensive than normal requirements for elevators, moving sidewalks and transporting devices have also become a significant part of the EPCOT Codes. [24]

In applications where unique buildings or electrical system applications may not readily fit into a standard specification based code or standard, it may be beneficial for public safety to consider "alternative materials, systems, methods, design calculations or other evidence as an approved alternative" [25] to nationally accepted codes and standards. The EPCOT Codes recognize this alternative use and provide the AHJ with the authority to grant approval of alternatives. However, that approval must be based on documentation which justifies the alternatives.

The AHJ approval of alternative materials and systems has become the hallmark of the Reedy Creek Improvement District (RCID). The District has also established a *Board of Appeals* to consider the unique variances developed through the Disney engineering organization. That Board is staffed by five appointed professionals, with specific training and expertise. The RCID *Department of Building and Safety* consists of:

state licensed and certified inspection personnel and supporting permit processors, [that enforce] the EPCOT Building, Plumbing, Mechanical, Gas, Electrical, Energy Conservation and Accessibility Codes, applicable Florida laws, and other pertinent, local rules and regulations. [26]

Codes and standards developed by the District have a regular review cycle of three years. The Department of Building and Safety has been assigned the task of conducting yearly inspections of all buildings in the District. Those inspections "ensure that all emergency systems are operable and that buildings are maintained in accordance with applicable codes". [27]

The EPCOT Building Code is an extreme example of owner AHJ development. More common examples might involve owner approval of the codes and standards to be used in the development of their property. Of particular interest involving the EPCOT Building Code was its use by the National Fire Protection Association in developing NFPA 5000, *Building Construction and Safety Code*. The first draft of the NFPA Building Code combined the NFPA 101 (Life Safety Code®) and EPCT Building Code. Reedy Creek Improvement Districts is a public corporation in Florida 39 square miles (101.4 square kilometers) in Orange and Osceola Counties. The 30-year-old Experimental Prototype Community of Tomorrow (EPCOT) Building Code is credited for EPCOT's low loss rate. [28]

Federal Authority Having Jurisdiction

There are a number of federal agencies with the power to enforce the use of specific codes and standards or may have legislative authority to regulate, inspect, order recalls, issue fines, etc. They include, but are not limited to, Federal Aviation Administration (FAA), United States Corps of Engineers, US Highway Traffic Safety Administration, US Consumer Product Safety Administration, Minerals Management Service (MMS), United States Coast Guard, Occupational Safety and Health Administration (OSHA), Mine Safety and Health Administration (MSHA), Federal Housing Administration (FHA), etc. Since OSHA is one of the largest federal AHJs, its basic operation will be reviewed in more detail below.

> The Occupational Safety and Health Act of 1970 authorizes the Secretary of Labor through the Occupational Safety and Health Administration (OSHA) to set mandatory occupational safety and health standards applicable to businesses affecting interstate commerce through public rulemaking. [29]

This legislation established OSHA as the Authority Having Jurisdiction:

> To assure safe and healthful working conditions for working men and women; by authorizing enforcement of the standards developed under the Act; by assisting and encouraging the States in their efforts to assure safe and healthful working conditions; by providing for research, information, education, and training in the field of occupational safety and health; and for other purposes. [30]

The Secretary of Labor was ordered by legislative decree to

> promulgate as an occupational safety or health standard any national consensus standard, and any established Federal standard, unless he determines that the promulgation of such a standard would not result in improved safety or health for specifically designated employees. In the event of conflict among any such standards, the Secretary shall promulgate the standard which assures the greatest protection of the safety or health of the affected employees. [31]

The OSH Act of 1970 also established a method for the addition, modification, or rescinding of standards. The ACT provides the Secretary of Labor with the authority to "promulgate, modify, or revoke any occupational safety or health standard" [32].

Should written information be submitted indicating that a rule should be promulgated for occupational health and safety reasons, the Secretary may establish an advisory committee, under Section 7 of the OSH Act, to review the request. Recommendations from that committee are to be submitted to the Secretary. Should those recommendations mandate the need to establish a rule, it *must* be promulgated within 90 days from the date of the appointment of a committee. The Secretary has the authority under the Act to extend or shorten the reporting date, but in accordance with the Act, it cannot be longer than 270 days.

The Secretary is required to "publish a proposed rule promulgating, modifying, or revoking an occupational safety or health standard in the Federal Register and shall afford interested persons a period of thirty days after publication to submit written data or comments" [33]. The Act requires that should the committee submit a recommendation that a rule be promulgated and should their recommendation be approved by the Secretary, then the Secretary is required to publish the proposed rule within 60 days after its submission or the expiration of the period prescribed for the submission by the Secretary.

The Act allows written objection or data supporting or opposing the proposed rule to be filed within 30 days of the publishing of the proposed rule in the Federal Register. The objections must state the grounds for the opposition and request a public hearing on the objections. The Secretary is required to publish a notice in the Federal Register outlining the proposed occupational safety and health standard against which objections have been filed. The Secretary must also specify a time and place for the formal hearings on the objections. This must be done within 30 days after the final date for submitting objections.

The Act sets time constraints for issuing or rejecting "a rule promulgating, modifying, or revoking an occupational safety or health standard or make a determination that a rule should not be issued" [34]. This must occur within 60 days of the expiration of submission of written data or comments on the rule or within 60 days of the completion of any hearing on the rule. The Act allows setting of a grace period, delaying the rule implementation, not to exceed 90 days. The delay necessity will be determined by the Secretary. It is designed "to insure that affected employers and employees will be informed of the existence of the standard and of its terms and that employers affected are given an opportunity to familiarize themselves and their employees with the existence of the requirements of the standard" [35].

The Act allows the Secretary to grant a temporary variance from the standard or any other provision thereof promulgated. An employer must follow specific rules in applying for a temporary variance, including establishment that their inability to comply by the established effective date is based on the lack of the availability "of professional or technical personnel or of materials and equipment needed to come into compliance with the standard or because necessary construction or alteration of facilities cannot be completed by the effective date" [36]. Further, the employer must attest that all necessary and available steps will be taken to safeguard employee health and safety against the hazards governed by the standards. Lastly, the employer must establish that he has established an effective program for the implementation of the standard as soon as it becomes practical.

The employer's variance application must contain the following information:

A description of the standard or the portion thereof for which a variance is being sought.

A detailed statement explaining why compliance cannot be met with supporting statements from knowledgeable, qualified individuals with discrete knowledge of the situation.

Explanation of the steps taken and those that will be taken to ensure employee protection against hazards mitigated by the standard. Specific dates for implementation must be included with the statement.

A statement of when the employer expects to be able to comply with the standard as well as the steps already implemented and those that will be implemented to meet compliance with the standard. Specific dates for implementation must also be included with the statement.

Proof must be provided that the employer has notified their employees of the application for compliance variance. Notification means must include providing copies of the variance application to employee representatives and posting a summary statement describing the application and instructions with a detailed copy of the information that can be made available. The information notice can be placed where employee notices are normally placed and by other appropriate means. Employees must also be notified that they have the right to petition the Secretary for a hearing on the variance. A description of the employee notification steps are required to be included in the variance application.

The Occupational Safety and Health Act allows OSHA, as the Authority Having Jurisdiction, to issue citations against employers:

[when an OSHA] representative believes that an employer has violated a requirement of section 5 of this Act, of any standard, rule or order promulgated pursuant to section 6 of this Act, or of any regulations prescribed pursuant to this Act, he shall with reasonable promptness issue a citation to the employer. [37]

The Act establishes specific requirements and procedures regarding the issuance of citations as follows:

Written format of citation and content directives.

Establishing a maximum reasonable time for abatement implementation.

Issuance of a notice in circumstances where there is no direct or imminent health or safety issue.

Maximum time limit on the issuance of a citation after the violation occurrence.

Before any citation can be issued for safety violations, an extensive inspection, investigation, and recordkeeping process must be followed, as established by the Act. An abbreviated review of those procedures is as follows [38]:

(a) The Secretary of Labor's representative, after presenting appropriate credentials to the authorized business representative shall:

(1) Have authority to enter a workplace, construction site, etc. at reasonable times and without delay, where employees perform work for an employer, and

(2) Reasonably inspect and investigate any place of employment, pertinent conditions, equipment, and materials therein and privately interview any and all appropriate individuals.

(b) Investigative and inspection procedures may require witness testimony and evidence production under oath, through the authority of any United States District Court or any US Court where a witness may live, work, or transact business under the threat of contempt of court for failure to comply.

(c) Establish employer record keeping requirements as:

(1) Employers are required to maintain records prescribed by Department of Labor or Health and Human Services regulations as may be necessary or appropriate for the Act enforcement or for establishing information regarding causes and prevention of accidents and illness with occupational relationships and to keep employees informed by appropriate means, of the employee protection and obligation under this Act.

(2) Require employer recordkeeping and reports on work-related injuries, illness, or deaths, other than those involving minor injuries. Minor injuries are described as those not requiring medical treatment, loss of consciousness, motion or work restrictions, or transfer to another job.

(3) Require employers to maintain records of employee exposure to toxic materials or harmful physical agents; provide notification of employees of exposure; and allow employees or their representatives' access to records.

(d) Protection from undue burden on small businesses providing information under this Act.

(e) Provide OSHA access to both employer and employee representatives during an onsite inspection to aid in the investigation.

(f) Procedures for the disposition of investigations shall:

(1) Allow employees or their representatives to request an inspection by the Department of Labor for health and safety standards violations that have the potential for physical harm or elements of danger. The individual(s) making those requests will not be identified to the employer. The Secretary of Labor will determine if an inspection and investigation are warranted or if there are no reasonable grounds for those requests.

(2) Before or during any workplace inspection, employees or their representatives may present written notice of any workplace violations of safety and/or health standards. Should the Secretary of Labor's representative decide that there are no violations of safety and/or health standards, procedures shall be established for an informal review of that decision, with a written explanation being provided to the employees or their representative of the Secretary's final disposition of the alleged allegations.

(g) Additional duties for the Secretaries mandated by the Act include:
   (1) The Secretaries of Labor and Health and Human Services are authorized to compile, analyze, and publish all reports or information on the investigation. That information can be presented in either a detailed or summary form.
   (2) The Secretaries of Labor and Health and Human Services are authorized to establish any necessary rules and regulations to permit them to implement the enforcement of the OSH Act, including the inspection of an employer's establishment.
(h) Department of Labor employees directly involved in enforcement and investigations of this Act shall not have their work performance evaluation based on the number of citations issued or penalties issued. The Department shall not impose or establish any goals or quotas regarding citations or penalties.

The Occupational Safety and Health Review Commission

The OSH Act establishes a three-member commission entitled the *Occupational Safety and Health Commission (OSHRC)* [39]. The OSHRC

> is an independent Federal agency created to decide contests of citations or penalties resulting from OSHA inspections of American work places. The Review Commission, therefore, functions as an administrative court, with established procedures for conducting hearings, receiving evidence and rendering decisions by its Administrative Law Judges (ALJs). [40]

The duties of the *Administrative Law Judges* as outlined in the Act include:

> [to] hear, and make a determination upon, any proceeding instituted before the Commission and any motion in connection therewith, assigned to such administrative law judge by the Chairman of the Commission, and shall make a report of any such determination which constitutes his final disposition of the proceedings. The report of the administrative law judge shall become the final order of the Commission within thirty days after such report by the administrative law judge, unless within such period any Commission member has directed that such report shall be reviewed by the Commission. [41]

Should any employer be issued a citation or assessed a penalty for violation of any Department of Labor established safety or health standard by the Occupational Safety and Health Review Commission, they

> may obtain a review of such order in any United States Court of Appeals for the Circuit in which the violation is alleged to have occurred or where the employer has its principal office, or in the Court of Appeals for the District of Columbia Circuit, by filing in such court within sixty days following the issuance of such order a written petition praying that the order be modified or set aside. [42]

> The Secretary [of Labor] may also obtain review or enforcement of any final order of the Commission by filing a petition for such relief in the United States court of appeals for the circuit in which the alleged violation occurred or in which the employer has its principal office ... [43]

State Jurisdiction and State Plans

The Act allows any state to enforce health and safety standards not established and enforced under the OSH Act of 1970 and its amendments [44]. It does allow any state to

assume responsibility for development and enforcement therein of occupational safety and health standards relating to any occupational safety or health issue with respect to which a Federal standard has been promulgated under section 6 shall submit a State plan for the development of such standards and their enforcement. [45]

The Act establishes rules, procedures, and actions which must be implemented in order for this to occur. It also establishes procedures and mechanisms [46] by which the Secretary of Labor shall approve or reject a state plan.

As of July, 2008 there are 26 states and jurisdictions operating complete state plans, which cover both private sector and state and local governmental employees. There are four states which cover public employees only. Eight other states were originally approved for the program, but have subsequently withdrawn. Reference Table 3.2 for those states.

There are four stages through which a state must progress, before it can be accredited to assume all OSHA labor health and safety regulatory responsibilities under Section 18 of the OSH Act of 1970. They include:

Developmental Plan

Certification

Operational Status Agreement

Final Approval

Detailed discussion on those stages can be found on the OSHA website: http://www.osha.gov/dcsp/osp/faq.html#oshaprogram; "How does a State establish its own program?"

The *Final Approval* stage is the ultimate accreditation of a state's occupational health and safety plan. Under Section 18(e) of the OSH Act of 1970, in this stage OSHA "relinquishes its authority to cover occupational safety and health matters covered by the State" [48]. It indicates that the state's worker protection regulation is at least as effective as that of OSHA. There are requirements that the state must have 100% compliance with staffing levels and implement a computerized inspection data system before OSHA can grant Final Approval. Table 3.2 illustrates state participation levels in the program, illustrating which states and jurisdictions have received Final Approval [49].

TABLE 3.2   States with OSHA Approved Safety and Health Plans [47]

State	Operational status agreement[1]	Different standards[2]	21(d) on-site consultation agreement[3]	On-site maritime coverage	Date of initial approval	Date certified[4]	Date of 18(e) final approval[5]
Alaska			X		07/31/73	09/09/77	09/28/84
Arizona			X		10/29/74	09/18/81	06/20/85
California	X	X	X	X	04/24/73	08/12/77	
Connecticut[6]			X		10/02/73	08/19/86	
Hawaii		X	X		12/28/73	04/26/78	04/30/84
Indiana			X		02/25/74	09/24/81	09/26/86
Iowa			X		07/20/73	09/14/76	07/02/85
Kentucky					07/23/73	02/08/80	06/13/85
Maryland			X		06/28/73	02/15/80	07/18/85
Michigan	X	X	X		09/24/73	01/16/81	
Minnesota			X	X	05/29/73	09/28/76	07/30/85
Nevada			X		12/04/73	08/13/81	04/18/00
New Jersey[6]			X		01/11/01		
New Mexico	X		X		12/04/75	12/04/84	
New York[6]			X		06/01/84		
North Carolina			X		01/26/73	09/29/76	12/10/96
Oregon	X	X	X	X	12/22/72	09/15/82	05/12/05
Puerto Rico	X				08/15/77	09/07/82	
South Carolina			X		11/30/72	07/28/76	12/15/87
Tennessee			X		06/28/73	05/03/78	07/22/85
Utah			X		01/04/73	11/11/76	07/16/85
Vermont	X		X	X	10/01/73	03/04/77	
Virgin Islands[6]			X		08/31/73	09/22/81	04/17/84[7]
Virginia			X		09/23/76	08/15/84	11/30/88
Washington	X	X		X	01/19/73	01/26/82	
Wyoming			X		04/25/74	12/18/80	06/27/85
Total: 26	7	5	23	5	26	24	17

[1]Concurrent Federal OSHA jurisdiction suspended.
[2]Standards frequently not identical to the Federal.
[3]On-site consultation is available in all states either through 21(d) Agreement or under a State Plan.
[4]Developmental steps satisfactorily completed.
[5]Concurrent Federal jurisdiction relinquished (superseded Operational Status Agreement).
[6]Plan covers state and local government employees only.
[7]Voluntary withdrawal of private sector jurisdiction and retention of public sector jurisdiction on July 1, 2003 (68 FR 4345).

References

1. Roget's II *The New Thesaurus*; 1980, page 65. Houghton Mifflin Company; Boston, MA.
2. Ibid., page 538.
3. NFPA 70, *National Electrical Code*; 2008, Article 90-4. National Fire Protection Association; Quincy, MA.
4. NFPA 101, *Life Safety Code*; 2006, Section 3.2.2. National Fire Protection Association; Quincy, MA.
5. Ibid., Section A3.2.2.
6. ICC website: http://www.iccsafe.org/government/adoption.html.
7. NFPA website: http://www.nfpa.org/assets/files/PDF/C3/FactSheet.pdf.
8. http://en.wikipedia.org/wiki/International_Code_Council.
9. NEMA website: http://www.nema.org/stds/fieldreps/NECadoption/upload/NEC_Adoption_Map.ppt#257,1,Slide 1.
10. NYC website: http://www.nyc.gov/html/dob/downloads/pdf/ll49of2006.pdf.
11. Ibid., § 27-3025 The New York city amendments to the 2005 National Electrical Code.
12. NFPA 70, *National Electrical Code*; 2005, Section 210.19(A) (1). National Fire Protection Association; Quincy, MA.
13. NYC website: http://www.nyc.gov/html/dob/downloads/pdf/ll49of2006.pdf; § 27-3025. The New York city amendments to the 2005 National Electrical Code; Section 219. 10(A) (1).
14. NFPA 70, *National Electrical Code*; 2008, Article 90.6. National Fire Protection Association; Quincy, MA.
15. Earley, Mark W., Sargent, Jeffrey S., Sheehan, Joseph V., and Buss, E. William, *NEC® 2008 Handbook: NFPA 70: National Electrical Code*; 2008, Article 90.6; National Fire Protection Association; Quincy, MA.
16. Ibid.
17. NFPA website: http://www.nfpa.org/itemDetail.asp?categoryID=515&itemID=18190&URL=Codes%20and%20Standards/.
18. http://www.boma.org/Advocacy/Standards/InternationalCodes/InternationalFireCode.htm.
19. NFPA 72, *National Fire Alarm Code*; 1999, Section 1.2.1. National Fire Protection Association; Quincy, MA.
20. NFPA 101, *Life Safety Code*; 2006, Section 1.2. National Fire Protection Association; Quincy, MA.
21. NFPA 70, *National Electrical Code*; 2008, Article 110.2. National Fire Protection Association; Quincy, MA.

22. Earley, Mark W., Sargent, Jeffrey S., Sheehan, Joseph V., and Buss, E. William, *NEC*® *2008 Handbook: NFPA 70: National Electrical Code*; 2008, Article 110.2; National Fire Protection Association; Quincy, MA.

23. EPCOT website: http://www.the-original-epcot.com/2008/05/florida-project.html.

24. Reedy Creek Improvement District website: http://www.rcid.org/Dept_Building_Safety.cfm.

25. Ibid.

26. Ibid.

27. Ibid.

28. "The Reedy Creek Improvement District was established by Disney World to develop building codes which could be used for its unique Florida amusement park"

29. OSHA website: http://www.osha.gov/pls/oshaweb/owadisp.show_document?p_table= FACT_SHEETS&p_id=134.

30. OSHA website: Public Law 91-596; 84 STAT. 1590; 91st Congress, S.2193; December 29, 1970; as amended through January 1, 2004; Section 1.

31. OSH Act of 1970; Occupational Safety and Health Standards; Public Law 91-596 84 STAT. 1590; December 29, 1970; Section 6(a).

32. Ibid.; Section 6(b).

33. Ibid.; Section 6(b)(2).

34. Ibid.; Section 6(b)(4).

35. Ibid.; Section 6(b)(4).

36. Ibid.; Section 6(b)(6)(A).

37. Ibid.; Section 9(a).

38. Ibid.; Section 8.

39. Ibid.; Section 12.

40. OSHRC website: http://www.oshrc.gov/.

41. OSH Act of 1970; Occupational Safety and Health Standards; Public Law 91-596 84 STAT. 1590; December 29, 1970; Section 12(j).

42. OSH Act of 1970; Occupational Safety and Health Standards; Public Law 91-596 84 STAT. 1590; December 29, 1970; Section 11(a).

43. Ibid., Section 11(b).

44. Ibid., Section 18(a).

45. Ibid., Section 18(b).

46. Ibid., Section 18(c) through (h).

47. OSHA website: http://www.osha.gov/dcsp/osp/faq.html#oshaprogram.

48. OSHA website: http://www.osha.gov/dcsp/osp/faq.html#oshaprogram; "How does a State establish its own program?"

CHAPTER 4

Nationally Recognized Testing Laboratories (NRTLs)

Nationally Recognized Testing Laboratories play a significant role in certifying that materials, equipment, and products meet the requirements of the codes and standards by which they were produced. Authorities Having Jurisdiction generally require that material, equipment, or products used must be *Approved/Certified/Listed* by an NRTL. This raises a significant question. How is an organization established as an NRTL?

The United States Department of Labor, Occupational Safety and Health Administration has established a *Final Rule* regarding certification of NRTLs. Directive Number CPL 01-00-003, NRTL Program Policies, Procedures, and Guidelines, was established to certify NRTLs. The scope of that program is covered in Chapter 2 of that document [1] and indicates:

> The NRTL Program recognizes mainly private sector organizations that provide product safety testing and certification services to manufacturers. The testing and certification are done, for purposes of the Program, to U.S. consensus-based product safety test standards. These test standards are not developed or issued by OSHA, but are issued by U.S. standards organizations, such as the American National Standards Institute (ANSI). The range of products covered by the Program is limited to those items for which OSHA safety standards require "certification" by an NRTL. [See Appendix A for a table of these types of products.] The requirements mainly affect electrical products.
>
> (A) Recognition is granted to organizations that meet the requirements established by OSHA for an NRTL. The Program regulations list the requirements, which are summarized as follows:
>
> 1. Capability (including proper testing equipment and facilities, trained staff, written test procedures, and quality assurance programs) to test and evaluate equipment for conformance with *appropriate test standards*.
> 2. Adequate controls for the identification of certified products, conducting follow-up inspections of actual production.
> 3. Complete independence from users (i.e., employers subject to the tested equipment requirements) and from any manufacturers or vendors of the certified products.
> 4. Effective procedures for producing its findings and for handling complaints and disputes.

Listed NRTLs

Through the Program, OSHA has certified 15 NRTLs and as of June, 2008 they include the following [2]:

Canadian Standards Association (CSA)
(also known as CSA International)
416-747-4000
178 Rexdale Boulevard
Etobicoke (Toronto), Ontario M9W 1R3
Canada

Communication Certification Laboratory, Inc. (CCL)
801-972-6146
1940 West Alexander Street
Salt Lake City, Utah 84119

Curtis-Straus LLC (CSL)
978-486-8880
527 Great Road
Littleton, Massachusetts 01460

FM Approvals LLC (FM)
(formerly Factory Mutual Research Corporation)
781-762-4300
1151 Boston-Providence Turnpike
P.O. Box 9102
Norwood, Massachusetts 02062

Intertek Testing Services NA, Inc. (ITSNA)
(formerly ETL, Inchcape)
800-345-3851
3933 US Route 11
Cortland, New York 13045

MET Laboratories, Inc. (MET)
800-638-6057
914 West Patapsco Avenue
Baltimore, Maryland 21230

National Technical Systems, Inc. (NTS)
978-263-2933
1146 Massachusetts Avenue
Boxborough, Massachusetts 01719

NSF International (NSF)
800-673-6275
789 Dixboro Road
Ann Arbor, Michigan 48105

SGS U. S. Testing Company, Inc. (SGSUS)
(formerly US Testing/California Division)
973-575-5252
291 Fairfield Avenue
Fairfield, New Jersey 07004

Southwest Research Institute (SWRI)
210-684-5111
6220 Culebra Road
Post Office Drawer 28510
San Antonio, Texas 78228

TUV America, Inc. (TUVAM)
978-739-7000
5 Cherry Hill Drive
Danvers, Massachusetts 01923

TUV Product Services GmbH (TUVPSG)
49-89-5008-4335
Ridlerstrasse 65, D-80339
Munich, Germany

TUV Rheinland of North America, Inc. (TUV)
203-426-0888
12 Commerce Road
Newtown, Connecticut 06470

Underwriters Laboratories Inc. (UL)
847-272-8800
333 Pfingsten Road
Northbrook, Illinois 60062

Wyle Laboratories, Inc. (WL)
256-837-4411
7800 Highway 20 West
P.O. Box 077777
Huntsville, Alabama 35807

Definitions

Before examining the role of Nationally Recognized Testing Laboratories (NRTL), some terms associated with those types of organizations should be examined. The most common terms associated with NRTL(s) include in alphabetical order:

Accepted (Acceptable)

Approved

Certified

Classified

Identified

Labeled

Listed

Recognized

The US Department of Labor, Occupational Safety and Health Administration's Occupational Safety and Health Standards defines the *Accepted* as follows:

an installation is "accepted" if it has been inspected and found by a nationally recognized testing laboratory to conform to specified plans or to procedures of applicable codes. [3]

NFPA 70®, *National Electrical Code*® (NEC®) [4] defines *Approved* as:

Acceptable to the authority having jurisdiction.

It also notes in Article 110.2 Approval that:

The conductors and equipment required or permitted by this Code shall be acceptable only if approved. [5]

The US Department of Labor Standard 29 CFR 1910.399 defines the term thus:

Approved: Acceptable to the authority enforcing this subpart. The authority enforcing this subpart is the Assistant Secretary of Labor for Occupational Safety and Health. The definition of "acceptable" indicates what is acceptable to the Assistant Secretary of Labor, and therefore approved within the meaning of this subpart. [6]

Many Authorities Having Jurisdiction (AHJ) require that materials or equipment used are to be approved by a Nationally Recognized Testing Laboratory (NRTL). The term *Approved* may or may not appear on a label affixed to equipment or materials. However, a review of the services offered by the major NRTLs in the United States indicates that a variety of NRTL labels may be offered for equipment and materials certified by those organizations. Underwriters' Laboratories, Inc. (UL) is a product certification and testing organization. With reference to the term *Approved*, UL indicates:

"UL approved" is not a valid term used to refer to a UL Listed, UL Recognized or UL Classified products under any circumstance. [7]

Factory Mutual Global (FM) is a global comprehensive commercial and industrial insurer, which offers NRTL product certification. FM considers an *Approved* product or a product *Approval* as

a confirmation and a subsequent listing by FMR [Factory Mutual Research] that a product, roof system or roof assembly has been examined according to FMR's applicable requirements and found suitable for use in all instances subject to any limitations stated in the approval. [8]

The term *Classified* is generally accepted to mean that a product has been evaluated by an NRTL and found to comply with the requirements of some specific standard.

Certified is defined by the US Department of Labor as follows:

Equipment is "certified" if it bears a label, tag, or other record of certification that the equipment:

(1) Has been tested and found by a nationally recognized testing laboratory to meet nationally recognized standards or to be safe for use in a specified manner, or
(2) Is of a kind whose production is periodically inspected by a nationally recognized testing laboratory and is accepted by the laboratory as safe for its intended use. [9]

Identified is defined by the US Department of Labor as follows:

Identified (as applied to equipment). Approved as suitable for the specific purpose, function, use, environment, or application, where described in a particular requirement. [10]

The term *Labeled* is also defined by the US Department of Labor in their Occupational Safety and Health Standards. It defines it as follows:

Equipment is "labeled" if there is attached to it a label, symbol, or other identifying mark of a nationally recognized testing laboratory:

(1) That makes periodic inspections of the production of such equipment, and
(2) Whose labeling indicates compliance with nationally recognized standards or tests to determine safe use in a specified manner. [11]

Listed is defined in Article 100 of the *National Electrical Code* as:

Equipment, materials, or services included in a list published by an organization that is acceptable to the authority having jurisdiction and concerned with evaluation of products or services, that maintains periodic inspection of production of listed equipment or materials or periodic evaluation of services, and whose listing states that the equipment, material, or services either meets appropriate designated standards or has been tested and found suitable for a specified purpose. [12]

The US Department of Labor defines *Listed* as follows:

Equipment is "listed" if it is of a kind mentioned in a list that:

(1) Is published by a nationally recognized laboratory that makes periodic inspection of the production of such equipment, and
(2) States that such equipment meets nationally recognized standards or has been tested and found safe for use in a specified manner. [13]

The term *Recognized* is generally associated with an NRTL's recognition that a manufacturer has demonstrated their ability to produce a component or subassembly that will comply with the NRTL's requirements, and will be factory installed in the manufacturing of an end product under investigation by the NRTL for *Listing* or *Certification*. Similar components or subassemblies produced by other manufacturers would not be classified as *Recognized* by the NRTL for use in the end product, if they have not been investigated and manufactured in compliance with the NRTL's requirements.

NRLT Standards Development

One important role played by Nationally Recognized Testing Laboratories is their development of testing standards. The three NRTL organizations most often referenced for standards in the United States are Underwriters' Laboratories, Inc. (UL); Canadian Standards Association (CSA), and Factory Mutual (FM). All of these organizations develop testing standards for other types of equipment, as well as electrical equipment. From an electrical standpoint, test standards are developed for power generation, transmission, and distribution equipment, as well as tools and appliances.

Anyone wishing to search the standards developed by these organizations can do so at the following websites:

Canadian Standards Association (CSA): http://www.shopcsa.ca/onlinestore/welcome.asp?Language=EN

Factory Mutual Global (FM): http://www.fmglobal.com/page.aspx?id=50030000

Underwriters Laboratories, Inc (UL): http://ulstandardsinfonet.ul.com/

Some NRTLs do not develop their own testing standards, but utilize existing industry accepted consensus standards developed by ANSI, NEMA, ASTM, IEEE, etc.

References

1. OSHA website: http://www.osha.gov/pls/oshaweb/owadisp.show_document?
p_id=2004&p_table=DIRECTIVES, Chapter 2, Section II Scope of Program.
2. OSHA website: http://www.osha.gov/dts/otpca/nrtl/nrtllist.html.
3. US Department of Labor, Occupational Safety and Health Standards, 29 CFR 1910.399 Definitions.
4. NFPA 70, *National Electrical Code*; 2008, Article 100. National Fire Protection Association; Quincy, MA.
5. Ibid., Article 110.2.
6. US Department of Labor, Occupational Safety and Health Standards, 29 CFR 1910.399 Definitions.
7. UL website: http://www.ul.com/faq/terminology.html.
8. http://www.professionalroofing.net/archives/past/sept2000/tech.asp.
9. US Department of Labor, Occupational Safety and Health Standards, 29 CFR 1910.399 Definitions.
10. Ibid.
11. Ibid.
12. NFPA 70, *National Electrical Code*; 2008, Article 100-Definitions. National Fire Protection Association; Quincy, MA.
13. US Department of Labor, Occupational Safety and Health Standards, 29 CFR 1910.339 Definitions.

CHAPTER 5

Common Threads

Electricity is utilized for power, lighting, and control in manufacturing, equipment and facilities. It operates motors, appliances, pumps and compressors, refrigeration equipment, tools, cranes, lighting, HVAC equipment, welding machines, computers, battery chargers, UPS systems, and other equipment. The panelboards, disconnect switches, circuit breakers and fuses, and other power distribution equipment used to feed electrical equipment, could be utilized in any plant, commercial building, or industrial facility including petroleum, chemical, manufacturing, construction, agriculture, utilities, ship building, etc. Electrical codes, standards, and recommended practices governing the use of that power distribution and control equipment is a common thread to all occupancies.

Common Threads

Although the above-noted industries are diverse, there are some common electrical equipment and materials used in each. Low-voltage power distribution and lighting equipment and materials would be one common thread to all occupancies. High-voltage distribution electrical equipment would be another common element to some occupancies and all utilities. Installation of power distribution equipment and wiring suitable for use in areas classified as electrically hazardous because of the presence of combustible or flammable materials would be a common thread to onshore and offshore petroleum production facilities, chemical plants, petroleum and natural gas processing and refining facilities, petroleum and natural gas compression and pipeline facilities, petroleum and chemical storage facilities, manufacturing facilities using petrochemical stock, etc.

Adjustable speed drive equipment may be used to control motors in many settings: petroleum drilling, production, processing/refining, and pipeline facilities; chemical processing plants; manufacturing facilities; HVAC/refrigeration control in buildings and facilities; transportation facilities; food processing facilities; municipal drainage and potable water purification and delivery systems; etc. There are common codes, standards, and recommended practices associated with this type of equipment which are applicable wherever such equipment is used.

Electrical Codes, Standards, Recommended Practices and Regulations; ISBN: 9780815520450

Wiring devices, outlet boxes, switches, receptacles, wiring, raceway, and controls are examples of items having applications in commercial and residential buildings, manufacturing plants, and industrial facilities. The codes, standards, and recommended practices associated with each are common threads in a diverse array of industries and applications.

Emergency generators, UPS systems, battery/battery charger systems, emergency lighting systems, fire detection and alarms, security systems, lightning protection, etc. are all examples that have codes, standards, and recommended practices common threads in different occupancies.

Some common thread codes, standards, and recommended practices are voluntary, consensus-developed by industry groups, professional, standards originating organizations, etc. Others have been promulgated by governmental agencies having the regulatory authority to mandate their implementation. Some were created for public safety, while others have been promulgated for employee safety. Some have been established for fire prevention while others may have been created to prevent electrical shock and personal injury.

NFPA 101, *Life Safety Code* – Common Threads

The National Fire Protection Association's (NFPA) national consensus life safety standard NFPA 101®, *Life Safety Code*® (LSC®) is an excellent example of a code, standard, and recommended practice with common threads through the various occupancy categories covered in that document. The purpose of that code is to address construction, protection, and occupancy features, for a variety of diverse applications, to minimize danger to life from fire and similar emergencies. Table 5.1 illustrates the most common electrical codes and standards referenced in the *Life Safety Code* for a diverse number of occupancies. The codes referenced in the table included:

- NFPA 70®, *National Electrical Code*®
- NFPA 72 ®, *National Fire Alarm Code*®
- NFPA 99, *Health Care Facilities*
- NFPA 110, *Emergency and Standby Power Systems*
- NFPA 111, *Stored Electrical Energy Emergency and Standby Power Systems*
- UL 924, *Emergency Lighting and Power Equipment*

Some explanations are required regarding this table. NFPA 70 is directly referenced in only a few of the specific occupancy chapters in NFPA 101; however, it is specifically referenced in Section 9.6 Fire Detection, Alarm, and Communications Systems. Section 9.6 is referenced in each of the occupancies checked in the table. Section 9.6 also references NFPA 72, *National*

TABLE 5.1   NFPA 101 *Life Safety Code* occupancy common thread codes and standards

New occupancy description	NFPA 70	NFPA 72	NFPA 99	NFPA 110	NFPA 111	UL 924
Special structures and high rise buildings	✔	✔		✔	✔	✔
Assembly occupancies	✔	✔			✔	✔
Educational occupancies	✔	✔			✔	✔
Day care occupancies	✔	✔			✔	✔
Health care occupancies	✔	✔	✔	✔	✔	✔
Ambulatory health care occupancies	✔	✔	✔	✔	✔	✔
Detention and correction occupancies	✔	✔			✔	✔
One- and two-family dwellings	✔	✔				
Lodging and rooming houses	✔	✔				
Hotels and dormitories	✔	✔			✔	✔
Apartment buildings	✔	✔			✔	✔
Residential board and care occupancies	✔	✔			✔	✔
Mercantile occupancies	✔	✔			✔	✔
Business occupancies	✔	✔			✔	✔
Industrial occupancies	✔	✔			✔	✔
Storage occupancies	✔	✔			✔	✔

Fire Alarm Code as do a few of the specific occupancy chapters. Also, only a few of the specific occupancy chapters reference NFPA 110; however, it, along with NFPA 111 and UL 924, are referenced in Section 7.9 Emergency Lighting. When a specific occupancy required compliance with Sections 7.9 and/or 9.6, a check mark was placed in the NFPA 111 column since battery/charger systems are common emergency power sources in almost every occupancy. Check marks for NFPA 110 were only placed in rows in which occupancies specifically mandated emergency generator use. NFPA 99, *Health Care Facilities* has specific requirements for generators in its Essential Electrical System Requirements section and references NFPA 110. Therefore any occupancy mandating compliance with NFPA 99 also had a check placed in the NFPA 111 column.

Several of the NFPA 101 occupancy chapters specifically referenced *smoke control* requirements. Those included Smoke Protected Assembly Seating; Stages; Atriums; Ambulatory Health Care Occupancies; Detention and Correction Occupancies; and Mercantile Occupancies. Smoke management may involve pressurization, such as in Section 7.2.3.10 Activation of Mechanical Ventilation and Pressurized Stair Systems. Smoke management Section 7.2.3.12 Emergency Power Supply System (EPSS) references NFPA 110.

Whenever an occupancy section referenced *High-Rise Buildings* that would mandate that that high-rise occupancy application must comply with either Section 11.8 in its entirety or with some specific referenced portions of that section. Standby power would be required by Section 11.8. That should have mandated a reference to NFPA 110 for those occupancies in Table 5.1; however, it was not included in Table 5.1 because high-rise buildings were in a separate row in the table.

Adoption of NFPA 70, *National Electrical Code*

The National Fire Protection Association's NFPA 70, *National Electrical Code* (NEC) has been adopted as an electrical safety code in all 50 states and all United States Territories. It is also one of the most referenced national consensus standards in other codes, standards, and recommended practices. Tables 5.2 through 5.5 illustrate some of those codes, standards, recommended practices, and state regulatory bodies referencing or adopting NFPA 70. The codes and standards referenced in those tables do not represent all of the standards development organizations which reference NFPA 70 in their documents, nor do they represent all of the documents in which NFPA 70 is referenced. There are many municipal and county Authorities Having Jurisdiction which recognize NFPA 70, either in its entirety or amended, as the applicable electrical safety code for their jurisdiction.

Each specific code, standard, or recommended practice referencing NFPA 70 lists a specific edition of that document. This is readily evident in Table 5.5, with the number of different editions of the *National Electrical Code* that have been adopted throughout the United States. As various codes, standards, and recommended practices are revised, so also may the editions of the Referenced Standards they list.

Tables 5.2 through 5.5 illustrate the extent to which codes, standards, and recommended practices have become interconnected. Not included in the above-mentioned tables, would be the United States governmental regulations, health and safety standards, construction job safety regulations, and energy efficiency standards which also reference the NEC. OSHA 29 CFR 1910 and 1926 are two among several governmental codes which specifically reference that national consensus standard.

Low-Voltage Power Distribution and Service Entrance Equipment

Alternating current (AC) voltage levels are divided into five basic categories including:

- Low Voltage: 0 to 1000 VAC

- Medium Voltage: >1000 to 72,500 VAC

- High Voltage: >72,500 to 242,000 VAC

TABLE 5.2 Some of the NFPA and API codes, standards, and recommended practices referencing NFPA 70®, *National Electrical Code*®

NFPA	Document title
NFPA 30	Flammable and Combustible Liquids Code
NFPA 37	Installation and Use of Stationary Combustion Engines and Gas Turbines
NFPA 70A	National Electrical Code Requirements for One- and Two-Family Dwellings
NFPA 70E®	Electrical Safety in the Workplace®
NFPA 77	Static Electricity
ANSI/NFPA 72	National Fire Alarm Code
NFPA 99	Health Care Facilities
NFPA 101®	Life Safety Code®
NFPA 110	Emergency and Standby Power Systems
NFPA 111	Stored Electrical Energy Emergency and Standby Power Systems
NFPA 230	Standard for the Fire Protection of Storage
NFPA 496	Standard for Purged and Pressurized Enclosures for Electrical Equipment
NFPA 497	Classification of Flammable Liquids, Gases, or Vapors and of Hazardous (Classified) Locations for Electrical Installations in Chemical Process Areas
NFPA 921	Guide for Fire and Explosion Investigations
NFPA 230	Standard for the Fire Protection of Storage
API	
API 14C	Analysis, Design, Installation and Testing of Basic Surface Safety Systems on Offshore Production Platforms
API RP 14F	Recommended Practice for Design and Design and Installation of Electrical Systems for Fixed and Floating Offshore Petroleum Facilities for Unclassified and Class I, Division 1 and Division 2 Locations
API RP 14J	Recommended Practice for Design and Hazards Analysis for Offshore Production Facilities
API RP 500	Recommended Practice for Classification of Locations for Electrical Installations at Petroleum Facilities Classified as Class 1, Division 1 and Division 2
API RP 505	Classification of Locations for Electrical Installations at Petroleum Facilities Classified as Class 1, Zone 0, Zone 1, and Zone 2
API RP 540	Electrical Installations in Petroleum Processing Plants
API RP 2003	Protection Against Ignitions Arising Out of Static, Lightning, and Stray Currents

- Extra-High Voltage: >242,000 <1,000,000 VAC

- Ultra-High Voltage: 1,000,000 VAC and higher

Power distribution voltages of up to 600 Volts are common in many industrial, manufacturing, residential, medical, educational, assembly, and commercial facilities. Those facilities utilize many of the same codes, standards, and recommended practices for

TABLE 5.3 **Some of the NEMA codes, standards and recommended practices referencing NFPA 70®, *National Electrical Code®***

NEMA	Document title
NEMA AB 1	Molded Case Circuit Breakers and Molded Case Switches
NEMA AB 3	Molded Case Circuit Breakers and Their Application
NEMA AB 4	Guidelines for Inspection and Preventive Maintenance of Molded Case Circuit Breakers Used in Commercial and Industrial Applications
NEMA BU 1.1	General Instructions for Proper Handling, Installation, Operation, and Maintenance of Busway Rated 600 Volts or Less
NEMA DC 3	Residential Controls - Electric Wall-Mounted Room Thermostats
NEMA DC 20	Residential Controls - Class 2 Transformers
NEMA EW 1	Electric Arc Welding Power Sources
ANSI/NEMA FB 11	Plugs, Receptacles, and Connectors of the Pin and Sleeve Type for Hazardous Locations
ANSI/NEMA GR 1	Grounding Rod Electrodes and Grounding Rod Electrode Couplings
ICS 1	Industrial Control and Systems - General Requirements
ICS 1.3	Industrial Control and Systems: Preventive Maintenance of Industrial Control and Systems Equipment
ICS 2	Industrial Control and Systems: Controllers, Contactors, and Overload Relays Rated 600 Volts
ICS 2.3	Industrial Control and Systems: Instructions for the Handling, Installation, Operation, and Maintenance of Motor Control Centers Rated Not More than 600 Volts
ICS 4	Industrial Control and Systems: Terminal Blocks
ICS 6	Industrial Control and Systems: Enclosures
ANSI/NEMA ICS 8	Industrial Control and Systems: Crane and Hoist Controllers
ICS 10, Part 1	Industrial Control and Systems Part 1: Electromechanical AC Transfer Switch Equipment
NEMA MG 1	Motors and Generators
NEMA MG 2	Safety Standard for Construction and Guide for Selection, Installation, and Use of Electric Motors and Generators
ANSI/NEMA PB-1.1	Instructions for Safe Installation, Operation and Maintenance for Panelboards
ANSI/NEMA PB 2.1	General Instructions for Proper Handling, Installation, Operation, and Maintenance of Deadfront Distribution Switchboards Rated 600 Volts or Less
NEMA PB 2.2	Application Guide for Ground Fault Protective Devices for Equipment
WD 1	General Color Requirements for Wiring Devices
ANSI/NEMA WD 6	Wiring Devices - Dimensional Requirements
NEMA 250	Enclosures for Electrical Equipment (1000 Volts Maximum)
NEMA 280	Application Guide for Ground Fault Circuit Interrupters
ANSI/NEMA OS 1	Sheet-Steel Outlet Boxes, Device Boxes, Covers, and Box Supports
ANSI/NEMA OS 2	Nonmetallic Outlet Boxes, Device Boxes, Covers, and Box Supports

TABLE 5.4 Some IEEE, ISA, ICC, and UL codes, standards, and recommended practices referencing NFPA 70®, *National Electrical Code*®

IEEE	Document title
ANSI/IEEE 45	Recommended Practice for Electric Installations on Shipboard
IEEE Standard 141	IEEE Recommended Practice for Electric Power Distribution for Industrial Plants
IEEE Standard 142	IEEE Recommended Practice for Grounding of Industrial and Commercial Power Systems
IEEE Standard 241	IEEE Recommended Practice for Electric Power Systems in Commercial Buildings
IEEE Standard 242	IEEE Recommended Practice for Protection and Coordination of Industrial and Commercial Power Systems
IEEE Standard 602	IEEE Recommended Practice for Electric Systems in Health Care Facilities
IEEE Standard 1015	IEEE Recommended Practice for Applying Low-Voltage Circuit Breakers Used in Industrial and Commercial Power Systems
IEEE Standard 1100	IEEE Recommended Practice for Powering and Grounding Electrical Equipment
ISA	
ANSI/ISA RP 12.6	Installation of Intrinsically Safe Systems for Hazardous (Classified) Locations
ASTM	
F 2361	Standard Guide for Ordering Low-Voltage (1000 VAC or Less) Alternating Current Electric Motors for Shipboard Service – Up To and Including Motors of 500 Horsepower
International Code Council (ICC)	
ICC IRC	International Residential Code
ICC ECAP	International Code Council Electrical Code Administrative Provisions
Underwriters Laboratories Inc. (UL)	
UL 498	Standard for Attachment Plugs and Receptacles
UL 507	Standard for Electric Fans
UL 681	Standard for Installation and Classification of Burglar and Holdup Alarm Systems
UL 817	Standard for Cord Sets and Power-Supply Cords
UL 891	Standard for Dead-Front Switchboards
UL 1236	Standard for Battery Chargers for Charging Engine-Starter Batteries

600 V power distribution equipment. Common utility power distribution voltages may be 2400 V to 13,800 V or higher. Those voltages are commonly transformed to 600 V or less at the service entrance to an occupancy. The transformer may or may not be owned by the utility; although in many of the cases, it would be utility owned. The utility feeder from the transformer would be called the service drop for pole-mounted transformers. The

TABLE 5.5   States adoption of NFPA 70®, *National Electrical Code*® (NEC®)

State	NEC Edition	State	NEC Edition
Alabama	Local adoption	Montana	2005
Alaska	2005	Nebraska	2005
Arizona	Local adoption	Nevada	Local adoption
Arkansas	2008	New Hampshire	2008
California	2005	New Jersey	2005
Colorado	2008	New Mexico	2008
Connecticut	2005	New York	2005
Delaware	2005	North Carolina	2008
Florida	2005	North Dakota	2008
Georgia	2005	Ohio	2008
Hawaii	Local adoption	Oklahoma	Local adoption
Idaho	2008	Oregon	2008
Illinois	Local adoption	Pennsylvania	2005
Indiana	2005	Rhode Island	2008
Iowa	2005	South Carolina	2005
Kansas	Local adoption	South Dakota	2008
Kentucky	2005	Tennessee	2002[1]
Louisiana	2005	Texas	—[2]
Maine	2008	Utah	2005
Maryland	Local adoption	Vermont	2005
Massachusetts	2008	Virginia	2005
Michigan	2005	Washington	2005
Minnesota	2008	West Virginia	2005
Mississippi	Local adoption	Wisconsin	2005
Missouri	2005	Wyoming	2008

[1]State unincorporated areas mandated by state to implement 2008 NEC. Incorporated areas have local option.
[2]State legislature has 2008 NEC adoption under consideration.
Source: Data based on NEMA survey dated August 17, 2008 [1]

connection point between the utility service drop and the customer service entrance wiring is called the *service point*.

The customer equipment on the load side of the service point would be called the *service equipment*. Service equipment is defined in the NEC Handbook as:

> The necessary equipment usually consisting of a circuit breaker(s) or switch(es) and fuse(s) and their accessories, connected to the load end of service conductors to a building or other structure, or otherwise designated area, and intended to constitute the main control and cutoff of the supply. [2]

For the purpose of this chapter, two interconnected groups of equipment will be examined. They include service equipment and the equipment that may be directly connected to service equipment. Both equipment groups have common threads of codes, standards, and recommended practices for all occupancies.

A list of some basic equipment that could be considered as service equipment would include:

Kilowatt-hour metering equipment

Surge arresters

Service disconnects

(a) Fused and non-fused switches

(b) Circuit breakers

Ground fault protection devices

Industrial Control Panels

Panelboards, switchgear, and motor control centers

The above-noted equipment may not always be utilized as service equipment. Items may be used as service equipment provided they have been certified or listed by a Nationally Recognized Testing Laboratory for that purpose.

The second group of equipment is that which may be directly connected to or fed by the service entrance equipment. That would include:

Transient voltage surge suppressors

Busways

Transformers

Transfer switches

It should be noted that some of the above equipment may also be utilized as service equipment. Examples of that equipment and the circumstances allowing that will be examined later. Any equipment utilized for service equipment should be certified or listed as such by a Nationally Recognized Testing Laboratory.

Table 5.6 lists the standards for some more common service entrance equipment, as well as some load side equipment, which might be directly connected to service equipment. All of that equipment would be considered *common threads* since the equipment may be applied in a large variety of different occupancies. Note that standards for some common service equipment, including kilowatt-hour meter sockets and surge protectors, are not included in that

TABLE 5.6   Some service entrance and load side connected equipment standards

Developer	Standard No.	Title
IEEE	IEEE C37.13	AC High-Voltage Generator Circuit Breakers Rated on a Symmetrical Current Basis
IEEE	IEEE C37.13.1	IEEE Standard for Definite-Purpose Switching Devices for Use in Metal-Enclosed Low-Voltage Power Circuit Breaker Switchgear
IEEE	IEEE C37.14	Low-Voltage DC Power Circuit Breakers used in Enclosures
IEEE	ANSI/IEEE C37.16	Low-Voltage Power Circuit Breakers and AC Power Circuit Protectors - Preferred Ratings, Related Requirements, and Application Recommendations
IEEE	ANSI/IEEE C37.17	American National Standard for Trip Devices for AC and General Purpose DC Low-Voltage Power Circuit Breakers
IEEE	IEEE C37.20.1	Standard for Metal-Enclosed Low-Voltage Power Circuit Breaker Switchgear
IEEE	IEEE C37.26	Guide for Methods of Power Factor Measurement for Low-Voltage Inductive Test Circuits
IEEE	IEEE C37.27	Application Guide for Low-Voltage AC Non-Integrally Fused Power Circuit Breakers (Using Separately Mounted Current-Limiting Fuses)
IEEE	IEEE C37.29	Low-Voltage AC Power Circuit Protectors
IEEE	ANSI/IEEE C37.50	American National Standard for Switchgear–Low-Voltage AC Power Circuit Breakers Used in Enclosures - Test Procedures
IEEE	ANSI/IEEE C37.51	Conformance Test Procedures For Switchgear – Metal-Enclosed Low-Voltage AC Power Circuit Breaker Switchgear Assemblies
IEEE	ANSI/IEEE C37.52	Test Procedures, Low-Voltage (AC) Power Circuits
IEEE	ANSI C57.13	IEEE Standard Requirements for Instrument Transformers
IEEE	IEEE C57.13.1	IEEE Guide for Field Testing of Relaying Current Transformer
IEEE	IEEE C57.13.2	IEEE Standard Conformance Test Procedure for Instrument Transformers
IEEE	IEEE C57.13.3	IEEE Guide for Grounding of Instrument Transformer Secondary Circuits and Cases
IEEE	IEEE C57.13.6	IEEE Standard for High Accuracy Instrument Transformers
IEEE	IEEE Std. 141	IEEE Recommended Practice for Electric Power Distribution for Industrial Plants
IEEE	ANSI/IEEE Std. 142	Recommended Practice for Grounding Industrial and Commercial Power Systems – IEEE Green Book
IEEE	ANSI/IEEE Std. 242	Recommended Practice for Protection and Coordination. of Industrial and Commercial Power Systems – IEEE Buff Book
IEEE	ANSI/IEEE Std. 446	IEEE Recommended Practice for Emergency and Standby Power Systems for Industrial and Commercial Applications – IEEE Orange Book
IEEE	IEEE 1015	IEEE Recommended Practice for Applying Low-Voltage Circuit Breakers Used in Industrial and Commercial Power Systems (Blue Book)
NEMA	NEMA AB 1	Molded Case Circuit Breakers and Molded Case Switches
NEMA	NEMA AB 3	Molded Case Circuit Breakers and Their Application

TABLE 5.6 Some service entrance and load side connected equipment standards—cont'd

Developer	Standard No.	Title
NEMA	NEMA AB 4	Guidelines for Inspection and Preventive Maintenance of Molded Case Circuit Breakers Used in Commercial and Industrial Applications
NEMA	ANSI C37.50	Low-Voltage AC Power Circuit Breakers Used in Enclosures – Test Procedures
NEMA	ANSI C37.51	For Switchgear – Metal-Enclosed Low-Voltage AC Power Circuit Breaker Switchgear Assemblies – Conformance Test Procedures
NEMA	ANSI C37.52	Test Procedures, Low-Voltage (AC) Power Circuit
NEMA	ANSI C12.11	American National Standard for Instrument Transformers for Revenue Metering 10 kV BIL through 350 kV BIL (0.6 kV NSV through 69 kV NSV)
NEMA	NEMA EI21.1	Instrument Transformers for Revenue Metering (110 kV BIL and less)
NEMA	NEMA EI 21.2	Instrument Transformers for Revenue Metering (125 kV BIL through 350 kV BIL)
NEMA	NEMA KS-1	Enclosed and Miscellaneous Distribution Equipment Switches (600 Volts Maximum)
NEMA	NEMA KS-2	Distribution Equipment Switch Application and Maintenance Guide, A User's Reference
NEMA	PB 1	Panelboards
NEMA	PB 1.1	General Instructions for Proper Installation, Operation, and Maintenance of Panelboards Rated 600 Volts or Less
NEMA	NEMA PB 2	Deadfront Distribution Switchboards
NEMA	PB 2.1	Instructions for Proper Handling, Installation, Operation, and Maintenance of Deadfront Distribution Switchboards Rated 600 Volts or Less
NEMA	PB 2.2	Application Guide for Ground Fault Protective Devices for Equipment
NFPA	NFPA70®	National Electrical Code®
UL	UL 67	Panelboards
UL	UL 98	Enclosed and Dead-Front Switches
UL	ANSI/UL 414	American National Standard for Safety for Meter Sockets
UL	UL 489	Molded-Case Circuit Breakers, Molded-Case Switches and Circuit-Breaker Enclosures
UL	UL 869A	Reference Standard for Service Equipment
UL/CSA/ ANCE	UL 891/CSA-C22.2 No. 244/NMX-J-118/2	Switchboards
UL	UL 977	Fused Power-Circuit Devices
UL	UL 1008	Transfer Switch Equipment
UL	UL 1008M	Transfer Switch Equipment, Meter-Mounted
UL	UL 1053	Ground Fault Sensing and Relaying Equipment
UL	UL 1066	Low-Voltage AC and DC Power Circuit Breakers Used in Enclosures
UL	UL 1429	Pullout Switches
UL	UL 1558	Metal-Enclosed Low-Voltage Power Circuit Breaker Switchgear

TABLE 5.7 Codes, standards, and recommended practices developing organization designations

Designation	Description
AGA	American Gas Association
AGMA	American Gear Manufacturers Association
ANCE	Association of Standardization and Certification (Mexico)
ANSI	American National Standards Institute
APPA	American Public Power Association
ASAE	American Society of Agricultural and Biological Engineers
ASCE	American Society of Civil Engineers
ASME	American Society of Mechanical Engineers
ASTM	ASTM International
ATIS	Alliance of Telecommunications Industry Solutions
AWAP	American Wood Preservers' Association
AWEA	American Wind Energy Association
CEMA	Canadian Electrical Manufacturing Association
CEN	European Committee on Standardization
CFR	Code of Federal Regulations
CGA	Compressed Gas Association
CIP	Critical Infrastructure Protection
CSA	Canadian Standards Association
CSI	Construction Specification Institute
EASA	Electrical Apparatus Service Association, Inc.
EEI	Edison Electric Institute
EEMAC	Electrical and Electronic Manufacturing Association of Canada
EIA	Electronic Industries Association
EGSA	Electrical Generating System Association
ESTA	Entertainment Services & Technology Association
FM	FM Global (formerly Factory Mutual)
GTI	Gas Technologies Institute
IEC	International Electrotechnical Commission
ICEA	Insulated Cable Engineering Association
IEEE	Institute of Electrical and Electronic Engineers
ISA	International Society of Automation (formerly Instrument Society of America)
ISA	International Society of Arboriculture
ISEA	International Safety Equipment Association
ISO	International Organization for Standardization
NACE	NACE International (formerly National Association of Corrosion Engineers)
NECA	National Electrical Contractors Association
NEMA	National Electrical Manufacturers Association

TABLE 5.7 **Codes, standards, and recommended practices developing organization designations—cont'd**

Designation	Description
NERC	North American Electric Reliability Corporation
NETA	International Electrical Testing Association
NFPA	National Fire Protection Association
NMBA	National Materials Advisory Board
NRECA	National Rural Electric Cooperative Association
OSHA	Occupational Safety and Health Administration
RUS	Rural Utility Service (US Department of Agriculture)
SIA	Scaffold Industry Association
SPIB	Southern Pine Inspection Bureau
TIA	Telecommunications Industry Association
WCLIB	West Coast Lumber Inspection Bureau
WECC	Western Electricity Coordinating Council
UL	Underwriters Laboratories, Inc.

table. They are included in Tables 5.8 and 5.9 respectively, and were specifically broken out separately to aid in the discussion of that equipment.

Figure 5.1 represents a typical service entrance schematic and will be used to discuss service entrance equipment. Table 5.7 provides some Standards Development Organization designations to aid in reviewing any of the standards tables presented in this book. Those organizations are accountable for their standards development, approval, revision, reaffirmation, or withdrawal. If a standard has also been adopted as an America National Standard, any changes, revisions, etc. to that document would also require approval by the American National Standards Institute.

Device A in Figure 5.1 represents a transformer that would be used to step down the utility distribution voltage to a more useable and safe level. For residential uses, that transformer might be a single-phase, pole-mounted transformer reducing the distribution voltage to 240/120 Volts or a bank of three single-phase transformers with a wye secondary output of 208/120 Volts, three-phase. In industrial or commercial applications, the transformer might have a secondary output voltage of three-phase 480/277 Volts.

Device B in Figure 5.1 illustrates a kilowatt hour meter, which would be found on residential and small commercial facilities. Larger facilities or industrial facilities might utilize current and potential transformers on the utility service entrance cables, whose outputs would then be connected to a kilowatt hour meter. Kilowatt hour meters can be provided with surge arrestor equipment; however, that is not necessarily standard equipment. Table 5.8 reflects some of the

TABLE 5.8 Kilowatt hour meter sockets and enclosure codes, standards, and recommended practices

Developer	Standard No.	Title
ANSI/NEMA	ANSI C12.1	American National Standard Code for Electricity Metering
ANSI/NEMA	ANSI C12.7	American National Standard Requirements for Watthour Meter Sockets
ANSI/NEMA	ANSI C12.10	American National Standard for Physical Aspects of Watthour Meters – Safety Standard
ANSI/NEMA	ANSI C12.20	For Electricity Meter – 0.2 and 0.5 Accuracy Classes
NEMA	NEMA Standards Publication 250	Enclosures for Electrical Equipment (1000 Volts Maximum)
EEI		*The Handbook for Electricity Metering*, 10th edition, 2002
NFPA	NFPA 70, Article 312	Cabinets, Cutout Boxes, and Meter Socket Enclosures
ANSI/UL	ANSI/UL 50	American National Standard Safety Standard for Electric Cabinets and Boxes
UL/CSA/ ANCE	UL 50/CSA-C22.2 NO. 94.1/NMX-J-235/1-ANCE	Enclosures for Electrical Equipment, Non-Environmental Considerations
UL/CSA/ ANCE	UL 50E/CSA-C22.2 NO. 94.2/NMX-J-235/2-ANCE	Enclosures for Electrical Equipment, Environmental Considerations
ANSI/UL	ANSI/UL 67	American National Standard Safety Standard for Panelboards
ANSI/UL	ANSI/UL 414	American National Standard Safety Standard for Meter Sockets
UL	UL 489	Molded-Case Circuit Breakers, Molded-Case Switches and Circuit-Breaker Enclosures
IEEE	IEEE C37.90.1	IEEE Standard for Surge Withstand Capability (SWC) Tests for Relays and Relay Systems Associated with Electric Power Apparatus
IEEE	IEEE C57.13	IEEE Standard Requirements for Instrument Transformers
IEEE	IEEE C57.13.3	IEEE Guide for Grounding of Instrument Transformer Secondary Circuits and Cases
IEEE	IEEE C5713.1	IEEE Guide for Field Testing of Relaying Current Transformers
IEEE	ANSI/IEEE	Standard Conformance Test Procedure for Instrument Transformers
IEEE	IEEE C57.13.6	IEEE Standard for High Accuracy Instrument Transformers
IEEE	ANSI/IEEE C37.110	IEEE Guide for the Application of Current Transformers Used for Protective Relaying Purposes
IEEE	IEEE C62.41.1	IEEE Guide on the Surge Environment in Low-Voltage (1000 V and less) AC Power Circuits
IEEE	IEEE C62.41.2	IEEE Recommended Practice on Characterization of Surges in Low-Voltage (1000 V and less) AC Power Circuits

TABLE 5.9 Surge-protection devices codes, standards, and recommended practices

Developer	Standard No.	Title
NEMA	NEMA LA 1	Surge Arresters
NEMA	NEMA LS 1	Low-Voltage Surge Protection Devices
IEEE	ANSI/IEEE 1100	IEEE Recommended Practice for Powering and Grounding Electronic Equipment (Emerald Book)
IEEE	IEEE C62.1	IEEE Standard for Gapped Silicon-Carbide Surge Arresters for AC Power Circuits
IEEE	IEEE C62.2	IEEE Guide for the Application of Gapped Silicon-Carbide Surge Arresters for Alternating Current Systems
IEEE	IEEE C62.11	IEEE Standard for Metal-Oxide Surge Arresters for AC Power Circuits (>1 kV)
IEEE	IEEE C62.22	IEEE Guide for the Application of Metal-Oxide Surge Arresters for Alternating-Current Systems
IEEE	IEEE C62.22.1	IEEE Guide for the Connection of Surge Arresters to Protect Insulated, Shielded Electric Power Cable Systems
IEEE	IEEE C62.31	IEEE Standard Test Methods for Low-Voltage Gas-Tube Surge-Protective Device Components
IEEE	IEEE C62.32	IEEE Standard Test Specifications for Low-Voltage Air Gap Surge-Protective Devices (Excluding Valve and Expulsion Types)
IEEE	IEEE C62.33	IEEE Standard Test Specifications for Varistor Surge-Protective Devices
IEEE	IEEE C62.34	IEEE Standard for Performance of Low-Voltage Surge-Protective Devices (Secondary Arresters)
IEEE	IEEE C62.35	IEEE Standard Test Specifications for Avalanche Junction Semiconductor Surge Protective Devices
IEEE	IEEE C62.36	IEEE Standard Test Methods for Surge Protectors Used in Low-Voltage Data, Communications, and Signaling Circuits
IEEE	IEEE C62.37	IEEE Standard Test Specification for Thyristor Diode Surge Protective Devices
IEEE	IEEE C62.37.1	IEEE Guide for the Application of Thyristor Surge Protective Devices
IEEE	IEEE C62.38	IEEE Guide on Electrostatic Discharge (ESD): ESD Withstand Capability Evaluation Methods (for Electronic Equipment Subassemblies)
IEEE	IEEE C62.41	IEEE Recommended Practice on Surge Voltages in Low-Voltage AC Power Circuits
IEEE	IEEE C62.41.1	IEEE Guide on the Surge Environment in Low-Voltage (1000 V and less) AC Power Circuits
IEEE	IEEE C62.41.2	IEEE Recommended Practice on Characterization of Surges in Low-Voltage (1000 V and less) AC Power Circuits
IEEE	IEEE C62.42	IEEE Guide for the Application of Component Surge-Protective Devices for Use in Low-Voltage [Equal to or Less than 1000 V (ac) or 1200 V (dc)] Circuits

(Continued)

TABLE 5.9 Surge-protection devices codes, standards, and recommended practices—cont'd

Developer	Standard No.	Title
IEEE	IEEE C62.43	IEEE Guide for the Application of Surge Protectors Used in Low-Voltage (equal to or less than 1000 V, rms, or 1200 V, DC) Data, Communications, and Signaling Circuits
IEEE	IEEE C62.45	IEEE Guide on Surge Testing for Equipment Connected to Low-Voltage AC Power Circuits
IEEE	IEEE C62.48	IEEE Guide on Interactions Between Power System Disturbances and Surge-Protective Devices
IEEE	IEEE C62.62	IEEE Standard Test Specifications for Surge Protective Devices for Low-Voltage AC Power Circuits
IEEE	IEEE C62.64	IEEE Standard Specifications for Surge Protectors Used in Low-Voltage Data, Communications, and Signaling Circuits
IEEE	IEEE C62.72	IEEE Guide for the Application of Surge-Protective Devices for Low-Voltage (1000 V or Less) AC Power Circuits
NFPA	NFPA 70, Article 280	Surge Arresters, over 1 kV
NFPA	NFPA 70, Article 285	Surge-Protection Devices (SPDs), 1 kV or Less
NFPA	NFPA 780	Standard for the Installation of Lightning Protection Systems
UL	UL 96	Lightning Protection Components
UL	UL 96A	Installation Requirements for Lightning Protection Systems
UL	UL 497	Protectors for Paired-Conductor Communications Circuits
UL	UL 497A	Secondary Protectors for Communications Circuits
UL	UL 497B	Protectors for Data Communications and Fire-Alarm Circuits
UL	UL 497C	Protectors for Coaxial Communications Circuits
UL	UL 1449	Surge Protective Devices
UL/CSA	UL PGA 1950	UL Standard for Safety Practical Application Guidelines for the Third Edition of the Standard for Safety for Information Technology Equipment
CSA/UL	CSA 22.2 No. 950	CSA Standard for Safety Practical Application Guidelines for the Third Edition of the Standard for Safety for Information Technology Equipment

codes, standards, and recommended practices that could be common for low-voltage meter sockets and enclosures in a variety of occupancies. Current transformers standards are included in Table 5.6.

Meter sockets can be installed in a variety of configurations. Single socket device applications are the most common in single-family residential and individual commercial occupancies. Multiple meter sockets are common in multi-family and commercial strip applications. Meter sockets can be installed with integral service disconnect circuit breakers. Panelboards are also available with an integral meter socket, main circuit breaker disconnect, and branch circuit breakers. Meters with integral surge protection are also available.

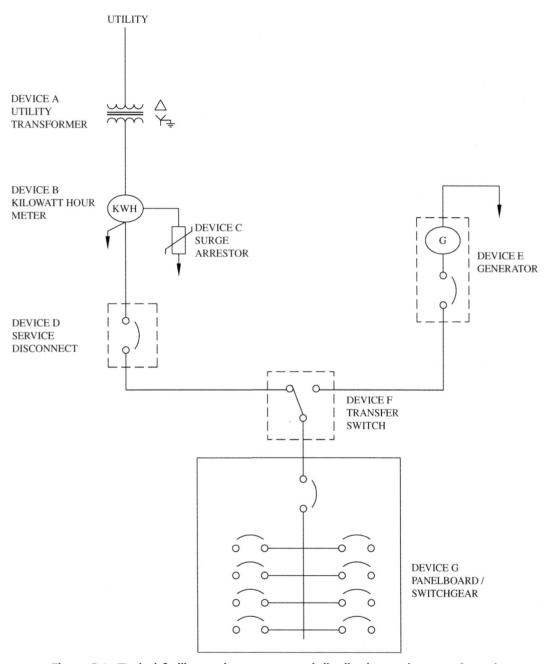

UTILITY

DEVICE A
UTILITY
TRANSFORMER

DEVICE B
KILOWATT HOUR
METER

KWH

DEVICE C
SURGE
ARRESTOR

DEVICE D
SERVICE
DISCONNECT

G

DEVICE E
GENERATOR

DEVICE F
TRANSFER
SWITCH

DEVICE G
PANELBOARD /
SWITCHGEAR

Figure 5.1: Typical facility service entrance and distribution equipment schematic

Surge Protection Devices (SPD)

Electrical distribution lines can sometimes be subjected to transient overvoltages (TOV) and power surge conditions. The creation of those conditions can be related to several causes. Transient overvoltage conditions can be caused by the following [3]:

(a) Line-to-ground fault, particularly on an ungrounded or resistance-grounded system

(b) Loss of neutral ground on a normally grounded system

(c) Sudden loss of load or generator overspeed, or both

(d) Resonance effects and induction from parallel circuits

Direct lightning strikes can inject a significant current surge on a distribution line. Also, lighting strikes within proximity to distribution and transmission power lines can induce a significant overvoltage condition on such lines. Additionally, utility switching surges can also create overvoltage conditions on distribution lines, particularly as a result of the energy stored in transmission lines, long cable circuits, and large capacitor banks.

Surge arresters are used to provide protection for equipment and service drops when there is historical data indicating TOVs and power surges as common on power transmission and distribution lines. Initial technology growth included surge arrester design from the original pellet and expulsion-type arresters to low-voltage gas-tube and valve-type (gap) arresters, which employ both a gap unit and a non-linear material varistor. The varistor material is silicon carbide. Gapless arresters were next developed utilizing zinc-oxide based material for the varistor. The varistor exhibits high resistance characteristics at low voltages and low resistance characteristics at higher voltages. Semiconductor thyristor surge protection devices are also being used on transformer and service entrance protection applications and protection for utility capacitor banks switching surges. This semiconductor device can have two or three leads and can be designed and fabricated to meet different voltage-current protective characteristics.

ANSI/IEEE Standard 141 [4] lists four classes of valve-type surge arresters including:

Station class arrester

Intermediate class arrester

Distribution class arrester

   (a) Heavy duty

   (b) Normal duty

Secondary arrester

Low-voltage surge protection may be placed at the service entrance equipment, or on the utility low-voltage service drop. An example of that can be seen in Figure 5.1. Device C in that diagram illustrates a surge arrester installed at the meter socket wiring. Another location for the surge arrester might be the panelboard main terminal lugs. There are two general types of low-voltage surge protection equipment including surge arresters and transient voltage surge suppressors (TVSS). All are designed to limit surge voltages and divert, limit, discharge, or bypass surge current. They are also required to be capable of repeating those protection functions should future surge or transient voltage events occur, provided they have been properly selected and installed.

Low-voltage surge protection devices must be selected with design ratings suitable for the installation in which they are being installed. Some of the most common codes, standards, and recommended practices associated with surge arresters are presented in Table 5.9. The devices must be selected with design ratings suitable for the energy dissipation they must sustain in the service for which they have been installed. They should be rated for basis of maximum continuous operating voltage (MCOV), temporary overvoltage (TOV) capability, and switching surge capability.

Surge arresters can be installed on both the line side and load sides of a service entrance. They are available for operation on low-voltage, medium-voltage, and high-voltage electrical systems. Transient voltage surge suppressors are installed on circuits 1 kV or less and are listed by Underwriters Laboratories Standard UL 1449 as either Type 2 SPD or Type 3 SPD. A Type 2 SPD must be installed on the load side of a service disconnect overcurrent protection device. UL 1449 requires a Type 3 SPD to be installed on the load side of a branch circuit overcurrent protection device. SPDs can be installed on the line side of service disconnects if they comply with 2008 NEC Articles 230.71(A) and 230.82(8) [5]. Avalanche junction diodes are commonly utilized in TVSS arresters.

Disconnect Switches

Disconnect switches must be rated for service entrance service if they are to be used in that capacity. An example of service disconnect equipment can be seen in Figure 5.1, Device D. Table 5.6 lists standards for some equipment for service entrance applications $\leq$ 1000 VAC. Not included in that table are standards for surge arresters, which can also be utilized in service entrances. Standards for that equipment were included in Table 5.9.

NEC Article 230 Services covers the use of service equipment. There are two basic types of devices that can be used as service disconnects, including switches and circuit breakers. Electrical services at more than 150 V to ground, but not more than 600 V phase-to-phase, rated 1000 amperes or more, are required by the NEC to have ground-fault protection capabilities on the service disconnect equipment. Service disconnect switches with shunt trip devices and ground fault relaying or monitoring devices can be utilized under those circumstances. Fused switches with integral ground-fault protection (GFP) tripping capabilities, used in conjunction with GFP monitoring devices, can also serve as service equipment under that requirement. Equipment utilized for that service must be listed or certified by a Nationally Recognized Testing Laboratory as suitable for that application.

UL 869A Reference Standard for Service Equipment establishes requirements for service equipment. Paragraph 1.1 in the Scope of that document indicates that it

> provides specific service equipment requirements that are intended to be supplemented by requirements for other types of equipment (such as panelboards, switchboards, and the like) that may be used as service equipment. [6]

Disconnect switches may also serve as service disconnects, provided they have been listed or certified for that service. Switches can also be utilized as a switching apparatus, as well as equipment disconnects, provided they are rated for that service. There are four basic types of switches [7], including:

Disconnect switches

Load interrupter

Safety switches

Transfer switches

Disconnect switches, which are used only for circuit and equipment isolation, may not have a load interrupting rating and therefore would not be suitable for use as service equipment. They are not designed to be opened under load, relying on other means to open the circuit while under load or shutting off the load before they can be used to disconnect it. Load break switches are rated for operation under load and when installed with a fuse(s) can qualify as service equipment, should they be listed or certified as such.

Switching devices rated above 600 Volts are normally associated with unit substations or transmission and distribution services or could be utilized to serve specific service applications or overcurrent protection. When opened or tripped under load, an electrical arc will develop across the device opening contacts. That arc must be interrupted to stop the flow of current to the electrical load. There are normally four basic types of arc-interruption mechanisms for switches including:

Air-insulated type

Dielectric fluid immersed type

Vacuum insulated type

Arc-extinguishing gas insulated type

Some switches may also be capable of being equipped with mechanical tripping mechanisms, provided they have been so designed. That would allow them to be used as a shunt device, capable of tripping by remote control, ground fault monitoring, or relaying devices.

Fused safety switches, rated 600 Volts or less, can be used as service equipment provided they are so certified or listed by a Nationally Recognized Testing Laboratory. The last switch type is a transfer switch which can be either an automatic or manual type. They are typically used in conjunction with a service disconnect device.

NEMA Standards Publication KS-1, *Enclosed and Miscellaneous Distribution Equipment Switches (600 Volts Maximum)* [8] lists the following switches as meeting the requirements for service entrance disconnecting applications when marked accordingly:

General duty enclosed switches

Heavy duty enclosed switches

Pullout switches

Fused power circuit devices

Circuit Breakers Operating at 1000 Volts or Less

Circuit breakers are switching or automatic overcurrent protection devices. They are rated by operating voltage and current levels, short-circuit interrupting capacity, construction, switching or load specific use, and phases/poles. Circuit breakers are also classified by their tripping characteristics, i.e. thermal/magnetic, hydraulic-magnetic, and electronic tripping. They can be fixed-mounted or draw out type. Their enclosure construction can be molded case, insulated case, or power type, open construction. They can also be utilized for service disconnects, if appropriately selected and listed. Circuit breakers can also be provided with shunt trip mechanisms to allow remote tripping by external control.

> A molded case circuit breaker is one that is assembled as an integral unit in a supportive and enclosing housing of insulating material. Molded case circuit breakers have factory-calibrated and sealed elements.

> An insulated case circuit breaker is one that is assembled as an integral unit in a supporting and enclosing housing of insulating material and with a stored energy mechanism. Insulated case circuit breakers are certified to the standard for molded case circuit breakers or to the standard for low-voltage power circuit breakers or to both. [9]

The National Electrical Manufacturers Association's NEMA Standard Publication AB 4 [10], *Guidelines for Inspection and Preventive Maintenance of Molded Case Circuit Breakers Used in Commercial and Industrial Applications* lists circuit breakers specific use categories. They include:

Remote controlled

Integrally fused

Current-limiting

Switching duty (SWD)

Instantaneous trip only

Heating, air conditioning, and refrigeration (HACR)

Marine

Naval

Mining

High intensity discharge lighting (HID)

Ground fault circuit interrupter (GFCI)

Circuit breaker with equipment ground fault protection

Classified circuit breakers

Circuit breakers with secondary surge arrester

Circuit breakers with transient voltage surge suppressor

Circuit breakers for use with uninterruptible power supplies

Arc-fault circuit interrupter (AFCI)

Not all of those circuit breakers would be suitable for service entrance equipment. Any circuit breaker that would, would require certification or listing by a Nationally Recognized Testing Laboratory for such use.

Circuit breakers opening under fault or load conditions will develop an arc across their main contacts. Four basic mediums can be utilized as arc-interruption mechanisms including:

Air

Insulating gas (sulfur hexafluoride)

Vacuum

Oil immersion

Low-voltage residential, industrial, and commercial applications utilize air or vacuum as the arc-interrupter mechanisms in circuit breakers. Commercial, industrial, and utility applications, operating at voltages above 1000 V typically may use any of the mediums depending upon the application.

Several methods are commonly used to extinguish an arc developed when a circuit breaker opens under fault or load conditions. They include:

Lengthening the arc path

Deflection of the arc with a differential pressure

Diversion of the arc to secondary contacts

Zero point quenching

Diversion of the arc with a magnetic field (blowout coils)

Intensive cooling of the arc

Use of arc chutes

Use of high speed contacts

Selection of the method utilized can depend on the amount of fault current available and the operating voltage.

Ground Fault Protection Devices

NEC Article 230.95-2008 mandates ground fault protection equipment on service disconnects rated 1000 amperes or more. That requirement is for solidly grounded wye electrical services of greater than 150 V to ground, but not exceeding 600 V phase-to-phase. There are two methods that may be used to monitor for ground fault conditions. They include a ground fault sensor/current transformer surrounding all service conductors and neutral or one monitoring the service entrance bonding jumper.

NEMA PB 1 provides guidance for the use of ground fault protection devices with service entrance disconnects of 1000 amperes or greater. It notes that:

The maximum setting of the ground-fault protection equipment shall be 1200 amperes, and the maximum time delay shall be one second for ground-fault currents equal to or greater than 3000 amperes.

When a ground-fault of a magnitude greater than the ground-fault protection setting occurs, the ground-fault protection equipment shall operate to cause the service disconnecting means to open all ungrounded conductors of the circuit. [11]

NEMA Standards Publication PB 2.2-2004, *Application Guide for Ground Fault Protective Devices for Equipment* provides guidance for the safe and proper use of ground fault protective (GFP) devices. Its scope includes:

> current sensing devices (GFS), relaying equipment (GFR), or combinations of current sensing devices and relaying equipment, or other equivalent protective equipment which will operate to cause a disconnecting means to open all ungrounded conductors at predetermined values of ground fault current and time. GFP devices are intended only to protect equipment against extensive damage from ground faults.

> GFP devices are intended to operate circuit breakers or fusible switches equipped with electrically actuated tripping means. These devices may be supplied as an integral portion of the disconnecting means or as separate devices operating in conjunction with the disconnecting means. GFP devices may or may not require external control power for proper tripping operation. [12]

Electrical Equipment Terms Review

Before proceeding with the examination of additional 1000 V or less service entrance equipment, some electrical equipment terms will be reviewed. Those terms include switchgear, panelboards, and power circuit breakers rated less than 1000 V.

Switchgear involves electrical equipment that is designed to switch and interrupt power to a load. That equipment may stand alone or in combination with other similar equipment. Associated metering, controls, and protective device accessories may also be included with that equipment. Switchgear is normally installed in metallic enclosures and can be rated for either indoor or outdoor service. Those enclosures can be open type, enclosed only by front and rear-mounted metallic covers, or be installed inside specifically designed metal buildings or housings.

The NEC® defines *Metal-Enclosed Power Switchgear* as:

> a switchgear assembly completely enclosed on all sides and top with sheet metal (except for ventilating openings and inspection windows) containing primary power circuit switching, interrupting devices, or both, with buses and connections. The assembly may include control and auxiliary devices. Access to the interior of the enclosure is provided by doors, removable covers, or both. [13]

The NEC® defines a *Switchboard* as:

> a large single panel, frame, or assembly of panels on which are mounted on the face, back, or both, switches, overcurrent and other protective devices, buses, and usually instruments. Switchboards are generally accessible from the rear as well as from the front and are not intended to be installed in cabinets. [14]

A *Panelboard* is defined by the NEC® as:

> a single panel or group of panel units designed for assembly in the form of a single panel, including buses and automatic overcurrent devices, and equipped with or without switches for the control of light, heat, or power circuits; designed to be placed in a cabinet or cutout box placed in or against a wall, partition, or other support; and accessible only from the front. [15]

Switchgear

Power switchgear assemblies are normally associated with larger electrical installations found in industrial or commercial structures or facilities. They can be used as service disconnects to feed and/or control motor control centers; control large horsepower motors; or feed transformers, panelboards, or other power distribution and control equipment.

Metal-enclosed switchgear is available in voltage ratings up to 34.5 kV. Metal-clad switchgear is available from 2.4 kV to 34.5 kV. Table 5.6 lists the most commonly utilized standards for switchgear.

IEEE Standard 141, *IEEE Recommended Practice for Electric Power Distribution for Industrial Plants* lists three types of metal-enclosed power switchgear that may be used in industrial applications. They may also be utilized in large commercial buildings and other large electrical facilities. They include:

Metal-clad switchgear

Metal-enclosed 1000 V and below power circuit breaker switchgear

Metal-enclosed interrupter switchgear

Panelboards

The National Electrical Manufacturers Association's NEMA PB1-2006, *Panelboards* provides requirements for panelboards which are to be used as service disconnect means. It notes in Section 2.8 Suitability for Use as Service Equipment that:

> Panelboards that are intended to be suitable for use as service entrance equipment shall meet the requirements of UL 67 and UL 869A and have provisions for:
>
> a. Connecting to the neutral terminal a grounding electrode conductor the size of which is in accordance with UL 869A.
>
> b. Bonding the enclosure to the grounded conductor (neutral).

c. Disconnecting all ungrounded load conductors from the source of supply by the operation of not more than six service disconnecting means. (For lighting and appliance panelboards, see 2.9.1.)

d. Disconnecting the grounded service conductor when a neutral is provided. [16]

The NEC requirement for panelboard use as service entrance equipment can be found in Article 408.3(C). NEMA Standards Publication PB 1.1, *General Instructions for Proper Installation, Operation, and Maintenance of Panelboards Rated 600 Volts or Less* provides guidance for the installation of panelboards.

Transformers

In residential and small commercial applications, distribution transformers are typically owned and operated by the utility company and the service drop from that device provides electrical energy to a customer's electrical service entranced equipment. In larger commercial and industrial situations, a transformer may be owned and operated by the utility customer and can be connected to service lateral conductors.

Transformers are rated in size in kilovolt-amperes (KVA) or megavolt-amperes (MVA) and are classified in two categories including distribution type (with the range of 3 to 500 kVA) and power type (with all ratings above 500 kVA).[17]. They are also classified by phase (single or three), voltage, insulation (dry, liquid, or combination), enclosure (indoor/outdoor), service (instrument-current/potential, power, lighting, autotransformer, control, power, etc.), winding connections (wye, delta, tertiary, open wye/delta, zigzag, etc.), and other categories.

Liquid-immersed transformers are classified by the type of liquid used, i.e. mineral oil, non-flammable, or low-flammable liquids. Dry type transformers are classified as ventilated, cast coil, totally enclosed non-ventilated, sealed, gas-filled, and vacuum pressure impregnated (VPI). Another classification includes a combination of the mediums including liquid-, vapor-, or gas-filled units [18].

This review will be limited to transformers with 1000 V or less secondary winding output voltage. A more detail discussion of transformers can be found elsewhere in this book. Table 5.10 lists some common standards associated with distribution and power transformers. That table does not represent all standards associated with power and distribution transformers.

Motor Control Center (MCC) – 600 Volts

Industrial and commercial power distribution system applications rely on motor control centers to distribute power and control motors and other electrical loads. Motor control centers

TABLE 5.10 Common distribution and power transformer standards

Developer	Standard No.	Title
NEMA	NEMA 260	Safety Labels for Padmounted Switchgear and Transformers Sited in Public Areas
NEMA	NEMA TR 1	Transformers, Regulators, and Reactors
NEMA	NEMA/ANSI C84.1	Electric Power Systems and Equipment – Voltage Ratings (60 hertz)
IEEE	IEEE C57.12.00	IEEE Standard for Standard General Requirements for Liquid-Immersed Distribution Power and Regulating Transformers
IEEE	IEEE C57.12.01	IEEE Standard General Requirements for Dry-Type Distribution and Power Transformers Including Those with Solid-Cast and/or Resin Encapsulated Windings
IEEE	ANSI/IEEE C57.12.22	American National Standard for Transformers–Pad-Mounted, Compartmental-Type, Self-Cooled Three-Phase Distribution Transformers with High-Voltage Bushings, 2500 kVA and Smaller: High Voltage, 34 500 Grounded Y/19 920 Volts and Below; Low Voltage, 480 Volts and Below
IEEE	IEEE C57.12.25	Requirements for Pad-Mounted Compartmental-Type Self-Cooled Single-Phase Distribution Transformers with Separable Insulated High-Voltage Connectors, High-Voltage, 34 500 Grounded Y/19 920 Volts and Below; Low-Voltage, 240/120; 167 kVA and Smaller
IEEE	IEEE C57.12.29	IEEE Standard for Pad-Mounted Equipment-Enclosure Integrity for Coastal Environments
IEEE	IEEE C57.12.31	IEEE Standard for Pole-Mounted Equipment-Enclosure Integrity
IEEE	ANSI C57.12.70	IEEE Standard Terminal Markings and Connections for Distribution and Power Transformers
IEEE	IEEE C57.12.50	Distribution Transformers 1 to 500 kVA, Single-Phase; and 15 to 500 kVA, Three-Phase with High-Voltage 601-34 500 Volts, Low-Voltage 120-600 Volt, Ventilated Dry-Type
IEEE	ANSI/IEEE C57.12.90	IEEE Standard Test Code for Liquid-Immersed Distribution, Power, and Regulating Transformers
CSA	C22.2 No. 66.1	Low-Voltage Transformers – Part 1: General Requirements (Binational Standard with UL 5085-1)
UL	UL 5085-1	Low-Voltage Transformers – Part 1: General Requirements
UL	ANSI/UL 1561	Dry-Type General Purpose and Power Transformers
NECA	ANSI/NECA 409	Standard for Installing and Maintaining Dry-Type Transformers (ANSI)

can be listed as service entrance equipment. NEMA Standards Publication ICS 18 *Motor Control Centers* defines a MCC as a:

> floor-mounted assembly of one or more enclosed vertical sections typically having a horizontal common power bus and principally containing combination motor-control units.

These units are mounted one above another in the vertical sections. The sections normally incorporate vertical buses connected to the common power bus, thus extending the common power supply to the individual units. Power may be supplied to the individual units by bus bar connections, by stab connection, or by suitable wiring. [19]

The National Electrical Manufacturers Association classifies motor control centers as either Class I or Class II assemblies. Class I assemblies are those which are designed and constructed using a manufacturer's standard design. Class II motor control centers are identical to Class I units, except for manufacturer-furnished interconnecting wiring and interlocks between units, as specified in customer furnished control drawings. Class I-S and Class II-S motor control centers are the same as Class I and II MCCs, except that custom drawings are provided by the customer to the manufacturer, with special device identification and terminal designations specified.

NEMA ICS 18 classifies internal wiring in motor control centers as Types A, B or C. Type A wiring requires the users' field wiring to terminate internally in the MCC, directly to the control or protection device terminals. That Type is only provided in Class I MCCs. Type B wiring can only be used with Size 3 or smaller combination motor starter units. Type B wiring has sub-classification types B–D and B–T. Type B–D allows field wiring to be directly connected to the device terminals that are located adjacent to the vertical wireway in a MCC section. Type B–T wiring provides terminal blocks in or adjacent to the MCC control device for field wiring termination. Type C wiring terminates the field wiring on a master terminal block located at the top or bottom of the MCC vertical section serving that load. Factory wiring interconnects the master terminal block to the MCC control devices. Some of the common standards associated with motor control centers are listed in Table 5.11.

Personal Protective Equipment

An arc flash event can release a substantial amount of thermal energy with an accompanying shockwave. Flash boundary calculations can be used to determine the maximum distance from an arc source in which a second-degree burn can be inflicted.

An arcing fault can occur inside an electrical enclosure. That occurrence is more probable when movement is involved within the enclosure. Opening or closing of a door, a switch, or contactor can sometime initiate that event if a component is defective, under-designed, or is subjected to more fault current than it is designed to withstand. An arc fault can also occur if a conductive tool, being held by a worker, makes inadvertent contact with an exposed, energized source.

Skin and flesh will sustain thermal damage when skin cells temperature are raised sufficiently to damage their structure. The duration of the thermal energy exposure to the cell is directly related to the amount of burn damage sustained. Arc flash protection boundary analysis utilizes

TABLE 5.11 Some common motor control center standards

Developer	Standard No.	Title
NEMA	NEMA ICS 1	Industrial Control and Systems General Requirements
NEMA	NEMA ICS 1.3	Preventative Maintenance of Industrial Control and Systems Equipment
NEMA	NEMA ICS 2.3	Industrial Control and Systems: Controllers, Instructions for the Handling Installation, Operation, and Maintenance of Motor Control Centers
NEMA	NEMA ICS 4	Terminal Blocks
NEMA	NEMA ICS 5	Industrial Control and Systems Control-Circuit and Pilot Devices
NEMA	NEMA ICS 6	Industrial Control and Systems: Enclosures
NENA	NEMA ICS 18	Motor Control Centers
NEMA	NEMA 250	Enclosures for Electrical Equipment (1000 V maximum)
UL	UL 845	Motor Control Centers
IEEE	ANSI/EEE 141	Recommended Practice for Electric Power Distribution for Industrial Plants (Red Book)
NETA	NETA ATS	NETA Acceptance Testing Specifications for Electrical Power Distribution Equipment and Systems
NFPA	NFPA 70E®	Electrical Safety in the Workplace®

a thermal energy exposure of 1.2 calories per square centimeter as the minimum necessary to sustain cell burn damage.

Flame resistant (FR) clothing and personal protective equipment (PPE) can be selected to withstand the incident energy exposure that may be sustained while performing a certain task. Depending upon the task being performed, certain body parts may be closer to an arc event than others. Hands and arms reaching inside an enclosure would be closer to the arc fault event in that area than a worker's legs or upper torso. The probability of that happening would mandate increased protective equipment or clothing to be used on specific body areas when performing certain tasks.

The potential for an arc flash to occur during equipment operation, switching or tripping event has become a greater concern for safety professionals. NFPA 70E®, *Electrical Safety in the Workplace®* establishes criteria for arc flash analysis and personnel protection to assist in the evaluation of the potential for such an event. Calculation methods are presented to establish maximum approach distances based on employee training and the recommended use of personal protective equipment. Arc flash hazard analysis assists in the determination of arc flash protection boundaries. *Hazard/Risk Category Classification Tables and Protective Clothing and PPE Matrixes* [20] help establish recommendations for the use of specific protection levels of equipment and clothing.

The United States Department of Labor, Occupational Safety and Health Administration (OSHA) requires employers, in 29 CFR 1910.132(d)(1) to conduct hazard assessments to determine if their employees may require personal protective equipment. Both NFPA 70E and IEEE 1584, *Guide for Performing Arc-Flash Hazard Calculations* are included in Table 5.12. Although NFPA 70E utilizes the same incident energy level equations as IEEE 1584, the latter goes into a more detailed explanation. Both documents should be used in conjunction with the NEC when conducting an arc-flash analysis.

TABLE 5.12 Some standards for personnel protective equipment and arc-fault analysis

Developer	Standard No.	Title
ASTM	ASTM D 120	Standard Specification for Rubber Insulating Gloves
ASTM	ASTM D 1051	Standard Specification for Rubber Insulating Sleeves
ASTM	ASTM F496	Standard Specification for In-Service Care of Insulating Gloves and Sleeves
ASTM	ASTM F 696	Standard Specification for Leather Protectors for Rubber Insulating Gloves and Mittens
ASTM	ASTM F1117	Standard Specification for Dielectric Overshoe Footwear
ASTM	ASTM F1236	Standard Guide for Visual Inspection of Electrical Protective Rubber Products
ASTM	ASTM F1506	Performance Specification for Flame Resistant Textile Materials for Wearing Apparel for Use by Electrical Workers Exposed to Momentary Electric Arc and Related Thermal Hazards
ASTM	ASTM F1891	Specification for Arc and Flame Resistant Rainwear
ASTM	ASTM F1958/ F1958M	Standard Test Method for Determining the Ignitability of Non-Flame-Resistance Materials for Clothing by Electric Arc Exposure Method Using Mannequins
ASTM	ASTM F1959/1959M	Standard Test Method for Determining the Arc Rating of Materials for Clothing
ASTM	ASTM F2178	Standard Test Method for Determining the Arc Rating of Face Protective Products
ASTM	ASTM F 2412	Standard Test Methods for Foot Protection
ASTM	ASTM F2413	Standard Specification for Performance Requirements for Foot Protection
ASTM	ASTM F2621	Standard Practice for Determining Response Characteristics and Design Integrity of Arc Rated Finished Products in an Electric Arc Exposure
IEEE	IEEE 70E	Standard for Electrical Safety in the Workplace
IEEE	IEEE 1584	Guide for Performing Arc Flash Hazard Calculations
ASSE	ANSI Z87.1	Practice for Occupational and Educational Eye and Face Protection
ISEA	ANSI Z89. 1	Requirements for Protective Headwear for Industrial Workers
OSHA	29 CFR 1910 Subpart I,	Personal Protective Equipment

Busway

Busways can be a part of electrical service entrance equipment; usually in large industrial or commercial applications. Table 5.13 lists common codes, standards, and recommended practices associated with busways. The NEC establishes criteria for the use of busways in Article 368 Busways. It defines a busway as:

> A grounded metal enclosure containing factory-mounted bare or insulated conductors, which are usually copper or aluminum bars, rods, or tubes. [21]

IEEE 141 [22] lists four types of busway, including:

Feeder busway

Plug-in busway

Lighting busway

Trolley busway

Feeder busways generally have an available current range to 6000 A, at 600 VAC. Busways can be used in service entrance, feeders, and branch-circuit applications.

UL 857 is primarily a manufacturing and testing standard and was established for use on busways with ratings of 600 V or less and 6000 A or less. The two NEMA Standards, BU 1.1 and 1.2, develop recommended installation and operation procedures and recommendations

TABLE 5.13 Busway standards

Developer	Standard No.	Title
NECA	NECA 208	American National Standard for Installing and Maintaining Busways
NEMA	NEMA BU1.1	General Instructions for Handling, Installation, Operation, and Maintenance of Busway Rated 600 Volts or Less
NEMA	NEMA BU 1.2	Application Information for Busway Rated 600 Volts or Less
UL	UL 857	Busways
CSI	CSI 16466	Feeder and Plug-in Busway
IEEE	IEEE C37.23	ANSI/IEEE Standard for Metal-Enclosed Bus
IEEE	ANSI//IEEE 141	ANSI/EEE Recommended Practice for Electric Power Distribution for Industrial Plants
NETA	NETA ATS	NETA Acceptance Testing Specifications for Electrical Power Distribution Equipment and Systems, 2007 edition
CSA	CSA C22.2 No. 27	Busways
CSA	CSA C22.2 NO. 201	Metal-Enclosed High-Voltage Busways
NFPA	NFPA 70®, Article 368	Busways

for busways rated 600 V or less and 6000 A or less. They also contain recommended design calculation methods, including voltage drop determination. IEEE Standard C37.23 includes information on ratings, temperature limitations, insulation requirements, dielectric strength requirements test procedures, and information on voltage losses in isolated-phase buses. Busway operating voltage ranges can be between 600 and 38 kV. IEEE C37.23 deals with performance characteristics of metal enclosed busways to 38 kV. CSA Standard CSA C22.2 NO. 201 was prepared for use with busways with an operating voltage range to 46 kV.

References

1. National Fire Protection Association; Field Survey, August 17, 2008; www.nema.org/stds/fieldreps/NECadoption/implement.cfm.
2. Earley, Mark W., Sargent, Jeffrey S., Sheehan, Joseph V., and Buss, E. William, *NEC® 2008 Handbook: NFPA 70: National Electrical Code*; 2008, Article 100; National Fire Protection Association; Quincy, MA.
3. IEEE Std. 141-1993, *IEEE Recommended Practice for Electric Power Distribution for Industrial Plants*; 1993, page 336. Institute of Electrical and Electronic Engineers; New York, NY.
4. ANSI/IEEE Standard 141-1993, IEEE *Recommended Practice for Electric Power Distribution for Industrial Plants*; 2003, page 335. Institute of Electrical and Electronic Engineers; New York.
5. Earley, Mark W., Sargent, Jeffrey S., Sheehan, Joseph V., and Buss, E. William, *NEC 2008® Handbook: NFPA 70: National Electrical Code*; 2008, Article 285.21; National Fire Protection Association; Quincy, MA.
6. UL 889A, *Reference Standard for Service Equipment*; November 10, 2006. Underwriters Laboratories, Inc.; Northbrook, IL.
7. IEEE Std, 141-1986, *Recommended Practice for Electric Power Distribution for Industrial Plants*; 1986, Section 9.2 Switching Apparatus for Power Circuits, page 412. Institute of Electrical and Electronic Engineers; New York, NY.
8. NEMA KS-1-2001 (R2006), *Enclosed and Miscellaneous Distribution Equipment Switches (600 Volts Maximum)*; 2006, pages 68-9. National Electrical Manufacturers Association; Rosslyn, VA.
9. NEMA Standards Publication AB 4-2003, *Guidelines for Inspection and Preventive Maintenance of Molded Case Circuit Breakers Used in Commercial and Industrial Applications*; 2003, Paragraph 3.2, page 3. National Electrical Manufacturers Association; Washington, DC.
10. Ibid., Section 2.3, page 1.
11. NEMA Standards Publication PB 1-2006, *Panelboards*; 2006, Section 2.8.1, page 1. National Electrical Manufacturers Association; Washington, DC.

12. NEMA Standards Publication PB 2.2-2004, *Application Guide for Ground Fault Protective Devices for Equipment*; 2004, page 1. National Electrical Equipment Manufacturers Association; Washington, DC.

13. NFPA 70, *National Electrical Code*; 2008, Article 100. National Fire Protection Association; Quincy, MA.

14. Ibid.

15. Ibid.

16. NEMA Standards Publication PB 1-2006, *Panelboards*; 2006, Section 2.8. National Electrical Manufacturers Association; Washington, DC.

17. IEEE Standard 141-1993, *IEEE Recommended Practice for Electric Power Distribution for Industrial Plants*; 1993, page 503. Institute of Electrical and Electronic Engineers; New York, NY.

18. Ibid., pages 515-16.

19. NEMA Standards Publication ICS 18-2001(R2007), *Motor Control Centers*; 2007, page 3. National Electrical Manufacturers Association; New York.

20. NFPA 70E-2004, *Electrical Safety in the Workplace*; 2004, Article 130. National Fire Protection Association; Quincy, MA.

21. NFPA 70, *National Electrical Code*; 2005, Article 368.2. National Fire Protection Association; Quincy, MA.

22. IEEE 141-1991, *IEEE Recommended Practice for Electrical Power Distribution for Industrial Plants*; 1991, Chapter 12. Institute of Electrical and Electronic Engineers; New York, NY.

CHAPTER 6

CFR 1910 versus CFR 1926

The Unites States Departments of Labor and Energy have played a key role in establishing federal standards affecting a large portion of the American population. Through legislative mandate those Cabinet Level Agencies have produced and enforced significant workplace occupational health and safety standards and residential/commercial/industrial energy efficiency standards. This chapter will examine the most significant of those standards from an electrical engineering perspective.

US Department of Labor

Two of the most far-reaching occupational health and safety standards established by the US Department of Labor are 29 CFR Part 1910, *Occupational Safety and Health Standards* and 29 CFR Part 1926, *Safety and Health Regulations for Construction*. Electrical related sections from both of those standards will be examined, noting similarities. Specific non-electrical related articles will not be examined.

The Occupational Safety and Health Act of 1970 was established as law on December 29, 1970, and has been amended several times. Its purpose was:

> To assure safe and healthful working conditions for working men and women; by authorizing enforcement of the standards developed under the Act; by assisting and encouraging the States in their efforts to assure safe and healthful working conditions; by providing for research, information, education, and training in the field of occupational safety and health; and for other purposes. [1]

The Act authorized

> the Secretary of Labor to set mandatory occupational safety and health standards applicable to businesses affecting interstate commerce, and by creating an Occupational Safety and Health Review Commission for carrying out adjudicatory functions under the Act ... [2]

The Occupational Safety and Health Administration was established to implement that legislation. To assist in its regulatory responsibilities, OSHA produced a set of Occupational Safety and Health Standards in the following general categories:

General Industry

Maritime

Construction

Electrical Codes, Standards, Recommended Practices and Regulations; ISBN: 9780815520450

A list of all of the occupational health and safety standards is provided in Appendix B of this book. This chapter will examine two of those 29 CFR Standards:

Part 1910 Occupational Safety and Health Standards

Part 1926 Safety and Health Regulations for Construction

Part 1910 Occupational Safety and Health Standards (29 CFR 1910) established that:

the Secretary [of Labor] shall, as soon as practicable during the period beginning with the effective date of this Act and ending 2 years after such date, by rule promulgate as an occupational safety or health standard any national consensus standard, and any established Federal standard, unless he determines that the promulgation of such a standard would not result in improved safety or health for specifically designated employees. The legislative purpose of this provision is to establish, as rapidly as possible and without regard to the rule-making provisions of the Administrative Procedure Act, standards with which industries are generally familiar, and on whose adoption interested and affected persons have already had an opportunity to express their views. Such standards are either (1) national consensus standards on whose adoption affected persons have reached substantial agreement, or (2) Federal standards already established by Federal statutes or regulations. [3]

29 CFR 1910 defines a national consensus standard as:

any standard or modification thereof which (1) has been adopted and promulgated by a nationally recognized standards-producing organization under procedures whereby it can be determined by the Secretary of Labor or by the Assistant Secretary of Labor that persons interested and affected by the scope or provisions of the standard have reached substantial agreement on its adoption, (2) was formulated in a manner which afforded an opportunity for diverse views to be considered, and (3) has been designated as such a standard by the Secretary or the Assistant Secretary, after consultation with other appropriate Federal agencies ... [4]

Several sections of the 29 CFR 1910 Occupational Safety and Health Standards are applicable to electrical design and installation procedures. However, the following sections are the most applicable to the purposes of this chapter. Those sections include:

1910.147 – The Control of Hazardous Energy (Lockout/Tagout)

1910.269 – Electric Power Generation, Transmission, and Distribution

1910 Subpart S – Electrical (1910.331 through 1910.335)

It should be noted here that any comparisons made between the above-noted Standard Articles should be reviewed with the fact that each article has a specific area of applicability. The only way in which one Standard Section can apply or be referenced to another Standard Section is if it is specifically referenced as also being jointly applicable.

Section 1910.147 provides detailed safe work practices in the control of all hazardous energy while performing maintenance, repair, or replacement in equipment and machinery. 1910.147(a)(1)(i) defines its scope as pertaining to:

the servicing and maintenance of machines and equipment in which the *unexpected* energization or start up of the machines or equipment, or release of stored energy could cause injury to employees. This standard establishes minimum performance requirements for the control of such hazardous energy. [5]

Section 1910.269 covers "power generation, transmission, and distribution installations that are accessible only to qualified employees". It also covers work on or directly associated with such installations. (See §1910.269(a)(1) for the full scope of §1910.269.) Subpart S covers electric utilization systems and electrical safety-related work practices for all employees working on or near such installations. It also covers electrical safety-related work practices for unqualified employees working near electric power generation, transmission, and distribution installations. (See §§1910.302 and 1910.331 for the full scope and application of Subpart S.) [6]

Part 1926 Safety and Health Regulations for Construction (29 CFR 1926) was promulgated by the Secretary of Labor under section 107 of the Contract Work Hours and Safety Standards Act [7]. The purpose of the Act was:

for construction, alteration, and/or repair, including painting and decorating, that no contractor or subcontractor contracting for any part of the contract work shall require any laborer or mechanic employed in the performance of the contract to work in surroundings or under working conditions which are unsanitary, hazardous, or dangerous to his health or safety, as determined under construction safety and health standards promulgated by the Secretary by regulation. [8]

The Act covers:

(1) Federal contracts requiring or involving the employment of laborers or mechanics (thus including, but not limited to, contracts for construction), and (2) contracts assisted in whole or in part by Federal loans, grants, or guarantees under any statute "providing wage standards for such work". [9]

The national consensus standard for the transmission and distribution of electricity and communications systems is ANSI/IEEE C2, *National Electrical Safety Code* (NESC). The scope of that standard is defined as covering:

supply and communication lines, equipment, and associated work practices employed by a public or private electric supply, communications, railway, or similar utility in the exercise of its function as a utility. They cover similar systems under the control of qualified persons, such as those associated with an industrial complex or utility interactive system. [10]

This standard does not cover buildings, mines, transportation vehicles, aircraft, or marine vessel wiring.

OSHA standard 1910.269 is applicable for "the operation and maintenance of electric power generation, control, transformation, transmission, and distribution lines and equipment" [11]. It should be noted here that the standard clarifies that "Supplementary electric generating equipment that is used to supply a workplace for emergency, standby, or similar purposes only is covered under Subpart S of this Part. (See paragraph (a)(1)(ii)(B) of this section)" [12].

This OSHA standard differs from ANSI/IEEE C2®, NESC® in that it focuses more on performance orientated requirements than ANSI/IEEE C2 the national consensus standard. 1910.269 Appendix E does list ANSI/IEEE C2 as a reference document. However, it does note in that Appendix that

> compliance with the national consensus standards is not a substitute for compliance with the provisions of the OSHA standard. [13]

ANSI/IEEE C2's purpose is stated as

> the practical safeguarding of persons during the installation, operation, or maintenance of electric supply and communication lines and associated equipment. These rules contain the basic provisions that are considered necessary for the safety of employees and the public under the specific conditions. This code is not intended as a design specification or as an instruction manual. [14]

It does not deal with power generation electrical safe work practices.

Facilities and occupancies that are mandated to adhere to the employer/employee health and safety standards of 1910.269 include:

1. Power generation, transmission, and distribution installations, including related equipment for the purpose of communication or metering, which are accessible only to qualified employees;
2. ... installations at an electric power generating station:
   (a) Fuel and ash handling and processing installations;
   (b) Water and steam installations ... providing a source of energy for electric generators;
   (c) Chlorine and hydrogen systems;
3. Test sites where electrical testing involving temporary measurements associated with electric power generation, transmission, and distribution is performed in laboratories, in the field, in substations, and on lines, as opposed to metering, relaying, and routine line work;
4. Work on or directly associated with the installations covered in paragraphs (a)(1)(i)(A) through (a)(1)(i)(C) of this section; and
5. Line-clearance tree-trimming operations ...
   (a) Entire 1910.269 of this Part, except paragraph (r)(1) of this section, applies to line-clearance tree-trimming operations performed by qualified employees ...;

Paragraphs (a)(2), (b), (c), (g), (k), (p), and (r) of this section apply to line-clearance tree-trimming operations performed by line-clearance tree trimmers who are not qualified employees. [15]

Hazardous Energy Control

Hazardous energy control (lockout/tagout) procedures are common to1910.147, 1910.269; 1910 Subpart S Electrical; and 1926.417. Each section will be examined.

Appendix A4 in 29 CFR 1926 provides a flow chart that can be used to determine applicable OSHA standards involving hazardous energy control procedures. Note that standards involve 1910.147, Subpart J General Environmental Controls; 1910.269, Subpart R Special Industries; and 1910.333, Subpart S Electrical. That flow chart is presented in Figure 6.1.

29 CFR 1926.417 also deals with lockout and tagging of circuits involving safeguarding of employees in construction work. The lockout/tagout requirements in this section are the shortest in all of the OSHA Occupational Health and Safety Standards. In fact, the standard

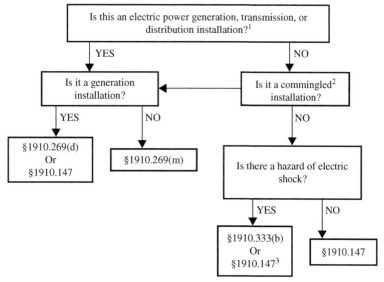

Figure 6.1: Appendix A-4 to 29 CFR 1910.269: "Application of §§1910.147, 1910.269, and 1910.333 to hazardous energy control procedures (lockout/tagout)" [16]
Notes: (1) If the installation conforms to §§1910.303 through 1910.308, the lockout and tagging procedures of 1910.332(b) may be followed for electric shock hazards. (2) Commingled to the extent that the electric power generation, transmission, or distribution installation poses the greater hazard. (3) §1910.333(b)(2)(iii)(D) and (b)(2)(iv)(B) still apply. [16]

only mentions the term *"tagged"*. An OSHA *Standards Interpretation* has been issued explaining 1926.417. It states in part:

> Title 29 CFR 1926 Subpart K addresses electrical safety requirements in construction work. Section 1926.417 ("Lockout and tagging of circuits") states:
>
> (a) Controls. Controls that are to be deactivated during the course of work on energized or deenergized equipment or circuits shall be tagged.
> (b) Equipment and circuits. Equipment or circuits that are deenergized shall be rendered *inoperative* and shall have tags attached at all points where such equipment or circuits can be energized [emphasis added].
>
> In promulgating this section, the Agency used the phrase "rendered inoperative" rather than "locked out". This indicates that methods other than lock-out would be permissible, as long as they rendered the equipment or deenergized circuit inoperative. There are a variety of such methods; two examples are:
>
> (1) Removing a fuse or other circuit element for each phase conductor; or
> (2) Disconnecting the circuit conductors (including disabling plugs for equipment that is plug-connected). [17]

The electrically related health and safety standards covered in 1910.147 and 1910.269 are substantial and have been the topic of many books and articles. The comparison of the electrically related occupational health and safety topics for lockout/tagout in those standards will be limited to the primary areas in Table 6.1.

OSHA 1910.147 defines the purpose of that standard as requiring:

> employers to establish a program and utilize procedures for affixing appropriate lockout devices or tagout devices to energy isolating devices, and to otherwise disable machines or equipment to prevent unexpected energization, start up or release of stored energy in order to prevent injury to employees. [18]

American National Standards Institute/American Society of Safety Engineers [19] Standard ANSI/ASSE Z244.1, *Control of Hazardous Energy – Lockout/Tagout and Alternative Methods*, is the nationally recognized consensus standard for lockout/tagout procedures. However, that consensus standard was not referenced in 1910.6 Incorporation by Reference or 1910.147 The Control of Hazardous Energy (Lockout/Tagout) Standards in September, 2008 on the OSHA Standards website. It was included in 1910 Subpart S Electrical, App. A, Reference Documents.

Upon the request of the American Society of Safety Engineers, OSHA has released a Standards Interpretation of 1910.147(c)(4)(ii) regarding the reference of the National Consensus Standard ANSI/ASSEZ244.1 – *Control of Hazardous Energy – Lockout/Tagout and Alternative Methods in 29 CFR 1910.147*, "The control of hazardous energy (lockout/tagout)".

TABLE 6.1 Lockout/tagout requirements: areas of discussion

Lockout/tagout requirements	29 CFR 1910.269	29 CFR 1910.147
Energy Control Program	1910.269(d)(2)(ii)	1910.147(c)(1)
Energy Control Procedures	1910.269(d)(2)(iii)	1910.147(c)(4)
Protective Material and Hardware	1910.269(d)(3)	1910.147(c)(5)
Periodic Inspection	1910.269(d)(2)(v)	1910.147(c)(6)
Training and Communication	1910.269(d)(2)(vi)	1910.147(c)(7)
Tagout System	1910.269(d)(2)(vii)(A)	1910.147(c)(7)(ii)
Employee Retraining	1910.269(d)(2)(viii)	1910.147(c)(7)(iii)
Energy Isolation/Notification of Employees	1910.269(d)(4) & (5)	1910.147(c)(8) & (9)
Application of Control	1910.269(d)(6)	1910.147(d)
Lockout/Tagout Application	1910.269(d)(6)(iv)	1910.147(d)(4)
Release from Lockout/Tagout	1910.269(d)(7)	1910.147(e)
Additional Requirements	1910.269(d)(8)	1910.147(f)(1)
Group Lockout or Tagout/Outside Personnel (Contractors, etc.)	1910.269(d)(8)(ii)	1910.147(f)(2)

A copy of that interpretation is attached in Appendix D. It concluded that that national consensus standard will be referenced in a revised issue of 29 CFR 1910.

Both 1910.147 Appendix A and ANSI/ASSE Z244.1 have example lockout/tagout procedures. General example procedures such as those are a good reference; however, they will require modifications to be applicable to the specific conditions for the equipment for which the procedure is being prepared. Caution should be exercised when using any generic procedure.

Section 1910.331(a) indicates:

The provisions of 1910.331 through 1910.335 cover electrical safety work practices for both qualified persons (those who have training in avoiding the electrical hazards of working on or near exposed energized parts) and unqualified persons (those with little or no such training) working on, near, or with the following installations:

1. Premises wiring. Installations of electric conductors and equipment within or on buildings or other structures, and on other premises such as yards, carnival, parking, and other lots, and industrial substations;
2. Wiring for connection to supply. Installations of conductors that connect to the supply of electricity; and
3. Other wiring. Installations of other outside conductors on the premises.
4. Optical fiber cable. Installations of optical fiber cable where such installations are made along with electric conductors. [20]

OSHA Regulation (Preamble to Final Rules) Section 6-VI, *Summary and Explanation of the Final Standard ... Hot Topics 1910.147, and 1910.269, 1910.333* states that 29 CFR 1910.331 through 1910.335

> have their own provisions for dealing with lockout/tagout situations, and for controlling employee exposure to hazardous electrical energy by the use of electrical protective equipment. They are based largely on a national consensus standard, NFPA 70E® – Part 11, *Electrical Safety Requirements for Employee Workplaces.* [21]

29 CFR 1910.333 is titled Selection and Use of Work Practices and is part of 1910 Subpart S, Electrical.

OSHA 1910 Subpart S, App A references national consensus standards ANSI/ASSE Z244.1 as well as NFPA 70E®, *Electrical Safety in the Workplace®.* Article 120 in that latter standard establishes lockout/tagout procedures for employers. The NFPA 70E *Handbook* notes in explaining Section 120.2(A):

> Lockout/tagout is only one step in the process of establishing an electrically safe working condition. Installing locks and tags does not ensure that electrical hazards have been removed. Workers must select and use work practices that are identical to working on or near exposed live parts until an electrically safe work condition has been established. [22]

Section 1910.333(b)(2) deals with lockout and tagging requirements and procedures. It states in Note 2 of that section that

> Lockout and tagging procedures that comply with paragraphs (c) through (f) of 1910.147 will also be deemed to comply with paragraph (b)(2) of this section provided that:
>
> (1) The procedures address the electrical safety hazards covered by this Subpart; and
> (2) The procedures also incorporate the requirements of paragraphs (b)(2)(iii)(D) and (b)(2)(iv)(B) of this section. [23]

Paragraphs (c) through (f) of 1910.147 cover all of the lockout/tagout procedures, energy control program requirements, inspection, training, release from lockout/tagout, etc.

The lockout/tagout requirements in Section 1910.147 are not applicable under all circumstances. It notes in Section 1910.147(a)(1)(ii) [24] that:

> This standard does not cover the following:
>
> • Construction, agriculture and maritime employment;
> • Installations under the exclusive control of electric utilities for the purpose of power generation, transmission and distribution, including related equipment for communication or metering; and
> • Exposure to electrical hazards from work on, near, or with conductors or equipment in electric utilization installations, which is covered by Subpart S of this part; and,
> • Oil and gas well drilling and servicing.

Standard 1910.147 is applicable for the control of energy during the maintenance, repair, replacement, or servicing of equipment and machinery. The normal operation of equipment and machinery is not covered under this standard; unless, a guard or other safety devices require bypassing or removal. Also, it is applicable in circumstances where an employee is required to place any part of their body in the area of the equipment or machinery's point of operation or an associated danger zone during the equipment or machinery's normal operating cycle. It is designed to prevent inadvertent starting of that machinery or equipment or the release of stored energy while the employee's work tasks, outside of the normal operation of the equipment, may expose them to the potential for injury or death.

The standard does not apply to:

- Work on cord and plug connected electric equipment for which exposure to the hazards of unexpected energization or start up of the equipment is controlled by the unplugging of the equipment from the energy source and by the plug being under the exclusive control of the employee performing the servicing or maintenance.
- Hot tap operations involving transmission and distribution systems for substances such as gas, steam, water or petroleum products when they are performed on pressurized pipelines, provided that the employer demonstrates that-continuity of service is essential;
- Shutdown of the system is impractical; and
- Documented procedures are followed, and special equipment is used which will provide proven effective protection for employees. [25]

The standard mandates employers to establish

> a program and utilize procedures for affixing appropriate lockout devices or tagout devices to energy isolating devices, and to otherwise disable machines or equipment to prevent unexpected energization, start up or release of stored energy in order to prevent injury to employees. [26]

Stored energy sources may be hydraulic or pneumatic pressure, compressed gas, pressurized liquids, capacitive or inductive electrical devices, coiled springs, potential energy devices, thermal, chemical, etc.

Before examining the energy control program requirements in more detail, a few terms should be examined to assist in understanding the specifics involved with that process.

An *energy isolating device* is defined as:

> A mechanical device that physically prevents the transmission or release of energy, including but not limited to the following: A manually operated electrical circuit breaker; a disconnect switch; a manually operated switch by which the conductors of a circuit can be disconnected from all ungrounded supply conductors, and, in addition, no pole can be operated independently; a line valve; a block; and any similar device used to block or isolate energy. Push buttons, selector switches and other control circuit type devices are not energy isolating devices. [27]

Lockout/tagout procedures are mandated in the energy control program required by 1910.147. Lockout is defined as:

> The placement of a lockout device on an energy isolating device, in accordance with an established procedure, ensuring that the energy isolating device and the equipment being controlled cannot be operated until the lockout device is removed. [28]

Tagout is defined in the standard as:

> The placement of a tagout device on an energy isolating device, in accordance with an established procedure, to indicate that the energy isolating device and the equipment being controlled may not be operated until the tagout device is removed [29].

An *authorized employee* is defined by 29 CFR 1910 as:

> A person who locks out or tags out machines or equipment in order to perform servicing or maintenance on that machine or equipment. An affected employee becomes an authorized employee when that employee's duties include performing servicing or maintenance covered under this section. [30]

An *affected employee* is defined as:

> An employee whose job requires him/her to operate or use a machine or equipment on which servicing or maintenance is being performed under lockout or tagout, or whose job requires him/her to work in an area in which such servicing or maintenance is being performed. [31]

A *qualified person* is defined as

> One who has received training in and has demonstrated skills and knowledge in the construction and operation of electric equipment and installations and the hazards involved. Whether an employee is considered to be a "qualified person" will depend upon various circumstances in the workplace. For example, it is possible and, in fact, likely for an individual to be considered "qualified" with regard to certain equipment in the workplace, but "unqualified" as to other equipment. (See 1910.332(b)(3) for training requirements that specifically apply to qualified persons.)

Key individuals in the discussion of 1910.269 are qualified personnel. 1910.269(x) defines them as:

> knowledgeable in the construction and operation of the electric power generation, transmission, and distribution equipment involved, along with the associated hazards.
>
> Note 1: An employee must have the training required by paragraph (a)(2)(ii) of this section in order to be considered a qualified employee.
>
> Note 2: Except under paragraph (g)(2)(v) of this section, an employee who is undergoing on-the-job training and who, in the course of such training, has demonstrated an ability to perform

duties safely at his or her level of training and who is under the direct supervision of a qualified person is considered to be a qualified person for the performance of those duties. [32]

An employee who is undergoing on-the-job training and who, in the course of such training, has demonstrated an ability to perform duties safely at his or her level of training and who is under the direct supervision of a qualified person is considered to be a qualified person for the performance of those duties.[33]

Anyone desiring to review a complete comparison of the lockout/tagout portions of 1910.147 and 1910.269 should reference Appendix C, Tables C.1 through C.12 in this book. Verbatim comparisons of each of the 12 Standard sections discussed below are presented for review and information.

Energy Control Program

An Energy Control Program is required by 1910.147(c)(1) to consist of two main components:

Lockout/Tagout

(a) 1910.147(c)(2)(i)

(b) 1910.147(c)(2)(ii)

(c) 1910.147(c)(2)(ii)

Full Employee Protection

(a) 1910.147(c)(3)(i)

(b) 1910.147(c)(3)(ii)

In order to insure employee safety in the workplace, employers are required to establish an energy control program. That program is described in 1910.147(c)(1) as:

consisting of energy control procedures, employee training and periodic inspections to ensure that before any employee performs any servicing or maintenance on a machine or equipment where the unexpected energizing, startup or release of stored energy could occur and cause injury, the machine or equipment shall be isolated from the energy source and rendered inoperative. [34]

1910.147(c)(2) Lockout/Tagout

Lockout devices are required to be installed on equipment which is capable of being locked out. This means the equipment has mechanically operable device(s) which can be secured in the de-energized, open, or blocked position with appropriate lockout devices. That securing device should be only capable of being removed with a unique key, other device, or appropriate means by the original installer or their appointed representative or replacement. It should be noted here that circumstances or working requirements may involve one or more individual(s) or group(s) to work equipment. Should that be the case, then multiple lockout devices may be required to be installed. That issue will be addressed in more detail later in this chapter. The energy isolation equipment is also required to have the tagout device(s) installed at the same time as the lockout device(s). If the equipment is not capable of being locked out, the standard requires that a tagout system be developed and utilized, which can be demonstrated to provide a level of safety equivalent to that which may be obtained from a lockout device.

1910.147(c)(3) Full Employee Protection

Where procedures call for the use of only a tagout device on equipment capable of being locked out, the employer must demonstrate that the tagout program will provide a level of safety equivalent to that obtained by using a lockout program. [35]

Additional elements may be required to assure equivalency of procedures with the use of tagout devices only. Those elements

shall include the implementation of additional safety measures such as the removal of an isolating circuit element, blocking of a controlling switch, opening of an extra disconnecting device, or the removal of a valve handle to reduce the likelihood of inadvertent energization. [36]

A far-reaching requirement was established in both 1910.147 and 1910.269 regarding mandatory installation of energy isolating devices on equipment and machinery after January 2, 1990 and November 1, 1994 respectively. Articles 1910.147(c)(3)(iii) and 1910.269(d)(2)(ii)(C) required that

whenever replacement or major repair, renovation or modification of a machine or equipment is performed, and whenever new machines or equipment are installed, energy isolating devices for such machine or equipment shall be designed to accept a lockout device. [37]

A full comparison of the Energy Control Programs between 1910.147 and 1910.269 can be found in Appendix C, Table C.1.

Energy Control Procedures

Before attempting to perform maintenance, repair or replacement on any energized equipment, it is extremely important to isolate all energy sources from those devices, to prevent inadvertent startup or release of stored hazardous energy. To assure that this process is successful, 1910.147 and 1910.269 require the establishment of specific procedures which must be followed in the process of removing all energy sources from the equipment or machinery. Appendix C, Table C.2 in this book contains a complete comparison of those requirements for both 1910.147 and 1910.269.

Those procedures required above must be developed, documented, and utilized by the employer, but need not relate to specific equipment or machinery if *all* of the *Exceptions* to 1910.147(c)(4)(i) are met. If that is not the case, then procedures must be developed. The following items must be incorporated into the written procedures:

Statement of intended use of procedure;

Procedural steps for shutdown, isolation, blocking, and securing of machines or equipment's hazardous energy;

Procedures for the placement, removal, or transfer of lockout/tagout devices and the identification of the responsibility for that task;

Development of requirements for determination and verification of the effectiveness of the lockout/tagout devices and energy control measures by testing of the machines or equipment.

Protective Materials and Hardware

Section 1910.147(c)(5) addresses protective material and hardware. That consists of the lockout devices, locks, tagout devices, tags, chains, wedges, key blocks, self-locking fasteners, etc. which shall be provided by employers to authorized employees. That equipment is required to assist the employee in isolation, securing, or blockage of hazardous energy sources from machines and equipment.

It is essential that lockout and tagout devices be singularly identifiable and utilized solely for those purposes, so that any authorized employee or affected or other employee can readily recognize that equipment is in a state of lockout/tagout and the significance of that state. It is essential that those devices be the only devices used for hazardous energy control, so that their use will not be misunderstood. Should they be capable of other uses that would defeat the purpose of being singularly recognized as lockout/tagout devices.

All lockout/tagout devices are required to meet the specific criteria established in 1910.147(c)(5)(ii), including being:

Capable of withstanding the environment in which they are used, for the time period in which that use is expected;

Tagout devices that are constructed and printed to adhere to the expected environmental conditions and moisture without deterioration or decrease or loss legibility;

Tagout devices that will not deteriorate in any corrosive environment in which they may be used;

Lockout/tagout devices shall be standardized in a facility and shall meet at least one of the following criteria:

Color

Shape

Size

It is also necessary that tagout devices be standardized with respect to print and format to assure easy recognition.

To protect an employee involved with a lockout/tagout event, it is necessary that the lockout device be of substantial construction to prevent removal by anyone without excessive force or other unusual techniques. Those unusual techniques might consist of the use of metal cutting tools or bolt cutters.

In situations where tagout only devices are in use, OSHA mandates that those devices be suitably attached and of substantial construction to prohibit inadvertent or accidental removal. It requires that those devices shall be non-reusable, attachable by hand and self-locking. They also establish a minimum unlocking strength of not less than 50-pounds and shall have design characteristics equivalent to a nylon cable tie, which is one-piece and suitable for all environmental conditions.

One key element in the lockout/tagout procedure is that the person(s) involved with implementing the lockout/tagout procedures be clearly identified on the tagout device. This is necessary to assure that that person(s) is/are the only one(s) who can legally remove those devices, unless that responsibility is transferred to other(s). The last requirement is that the tagout device has adequate, general warnings to cover all possible situations of inadvertent energization of the isolated energy source. That would consist of terms such as *Do Not Start, Do Not Open, Do Not Close, Do Not Energize, Do Not Operate.* A complete comparison of protective material and hardware between 1910.147 and 1910.269 is presented in Appendix C, Table C.7 in this book.

Periodic Inspection

The establishment of an Energy Control Program by employers is mandated in both 1910.147 and 1910.269 Standards. However, implementation of that program is not sufficient in itself, but both standards require *periodic inspection of the energy control procedures* at least annually. This mandate is intended to assure that the procedures and requirements of the Standard are being correctly implemented. That inspection process shall be performed by an authorized employee, who is not using the specific procedure being inspected. An *Authorized Employee* is defined in 1910.147(b) as:

> A person who locks out or tags out machines or equipment in order to perform servicing or maintenance on that machine or equipment. An affected employee becomes an authorized employee when that employee's duties include performing servicing or maintenance covered under this section. [38]

The purpose of the inspection is to determine if there are any deviations or inadequacies in the energy control procedure being evaluated. Use of an impartial authorized employee is required to allow someone with some overall knowledge of energy control procedures to give a third party opinion to its effectiveness. The OSHA Standard goes further in requiring that, where lockout procedures are in use for energy control, the inspector shall interview each authorized employee using that procedure. The Standard also requires the inspector to interview those authorized employees regarding their training, as is mandated in 1910.147(c)(7)(i).

To document the results of those periodic inspections, 1910.147(c)(6)(i)(E) requires the employer to certify that they have been completed. The certification is required to identity the machine or equipment on which the energy control procedure was being used. Also required in the certification is the name of the inspector and employees interviewed and the date of the inspection. A complete verbatim comparison of the Periodic Inspection requirements of 1910.147 and 1910.269 are presented in Appendix C, Table C.3 of this book.

Training and Communication

Section 1910.332 Training, under Subpart S Electrical, provides some insight into the category of employees who are a higher risk in sustaining an electrical accident. This section mandates that these occupational classifications receive training in recognizing the dangers with working on electrical equipment. Table 6.2 illustrates that information.

Standard 1910.147 develops the level of training which employers must provide to their employees. It indicates:

> The employer shall provide training to ensure that the purpose and function of the energy control program are understood by employees and that the knowledge and skills required for the safe application, usage, and removal of the energy controls are acquired by employees. The training shall include the following:
>
> - Each authorized employee shall receive training in the recognition of applicable hazardous energy sources, the type and magnitude of the energy available in the workplace, and the methods and means necessary for energy isolation and control.
> - Each affected employee shall be instructed in the purpose and use of the energy control procedure.

TABLE 6.2 Employees at higher risk of sustaining an electrical accident

Occupation
Blue collar supervisors (1)
Electrical and electronic engineers (1)
Electrical and electronic equipment assemblers (1)
Electrical and electronic technicians (1)
Electricians
Industrial machine operators (1)
Material handling equipment operators (1)
Mechanics and repairers (1)
Painters (1)
Riggers and roustabouts (1)
Stationary engineers (1)
Welders

(1) Workers in these groups do not need to be trained if their work or the work of those they supervise does not bring them or the employees they supervise close enough to exposed parts of electric circuits operating at 50 Volts or more to ground for a hazard to exist.
Source: TABLE S-4. – Typical Occupational Categories of Employees Facing a Higher Than Normal Risk of Electrical Accident [39]

- All other employees whose work operations are or may be in an area where energy control procedures may be utilized, shall be instructed about the procedure, and about the prohibition relating to attempts to restart or reenergize machines or equipment which are locked out or tagged out. [40]

An example of the requirement for proper employee training was provided in a Standards Interpretation of 29 CFR 1910.147 and 1910.147(c)(7)(i) by Richard E. Fairfax, Director; Directorate of Enforcement Programs dated October 24, 2005. A portion of that interpretation and the scenario for which it was written are presented below:

Scenario: Several years ago, we had a rather comprehensive training session on lockout/ tagout. Since that time, a significant number of employees have been reassigned and presently work with different machines. The employees exposed to new machinery have never been trained on how to properly lock out that machinery. We receive a generalized training session once a year during our 1-hour, routine monthly meeting. However, this meeting is not specific to lockout/tagout and includes discussion on Behavior Based Safety, tracking our safety record against targeted safety numbers, and various other topics.

In addition, I have assisted in developing the lockout/tagout procedures for machinery in a new department. Copies of the procedures were distributed to the maintenance employees. However, there has never been any discussion of the procedures. The company has never insured that these employees had a total understanding of these procedures.

Question 1: Is it acceptable to merely distribute copies of lockout/tagout procedures and consider that to be lockout/tagout training? If not, what are the general criteria for lockout/tagout training?

Response: The scenario that you provided appears to address "authorized" employees because the employees in your scenario are locking out equipment and presumably engaging in servicing/maintenance activities. An "authorized" employee is a person who locks out or tags out machines or equipment in order to perform servicing or maintenance work. Paragraph 1910.147(c)(7)(i)(A) of the Lockout/Tagout standard requires that "[e]ach authorized employee shall receive training in the recognition of all potentially hazardous energy sources, the type and magnitude of energy in the workplace, and the methods and means necessary for energy isolation and control." The mere distribution of lockout/tagout procedures will not meet the training requirements of the Lockout/Tagout standard for such employees. Instead, the employer must provide training that will allow each authorized employee to understand the purpose and function of the employer's energy control program and will allow each authorized employee to develop the skills and knowledge necessary to safely apply, use, and remove his/her lockout or tagout device (or its equivalent) and take other necessary steps so as to effectively isolate hazardous energy in every situation in which he/she performs servicing or maintenance activities.

In addition, it appears from your scenario that there have been changes at the worksite that may require additional or supplemental training in order to assure that authorized employees, who may have received adequate training at some point, are able to effectively and safely control hazardous energy in the environment(s) in which they are presently working. While the information contained in your letter does not permit us to determine conclusively whether the changes have occurred at the worksite that would necessitate additional or supplemental training, authorized employees must receive additional or supplemental training when they are exposed to new or additional sources of hazardous energy that are associated with their new work assignments. Likewise, authorized employees must receive additional or supplemental training when using different methods to control the same hazardous energy sources that they have controlled in other contexts. Ultimately, authorized employees must possess the skills and knowledge necessary to understand all relevant provisions of the energy control procedure(s) in order to effectively isolate all sources of hazardous energy to which they (or others) otherwise may be exposed. If prior training is insufficient to allow an authorized employee to follow an energy control procedure and to protect him/herself when servicing or maintaining a machine or piece of equipment, the employer is obligated to provide additional or supplemental training adequate to permit such proficiency. [41]

The question might be asked as to what might constitute adequate employer training for *authorized and affected or other employees.* Training for *authorized employees* must assure that they are able to ascertain the following on any equipment or machinery on which lockout/tagout procedures may be implemented [42]:

All applicable hazardous energy sources present in or around that equipment or machinery;

The type and magnitude of hazardous energy sources which may be present in the work area;

The appropriate methods and procedures required to isolate and control those hazardous energy sources.

Affected and other employees are not trained to institute energy control procedures. However, it is critical that they be trained to recognize when energy control procedures are in use. They should also understand the purpose of those procedures and know implicitly of the importance of not starting up, energizing, or attempt to use equipment or machinery that has been locked out or tagged out. A comparison of training requirements between 1910.147 and 1910.269 is presented in Appendix C, Table C.4 of this book.

Tagout System

In circumstances where lockout procedures are not possible, tagout procedures must be developed for use. Section 1910.147 recognizes the need to include employee training on the proper use of tags in 1910.147(c)(7)(i). It recognizes the limitation of tags in their inability to provide physical restraint from operation of the energy isolation device and they are essentially used as warning devices only. Unqualified or affected employees should understand that a tagout device should never be bypassed, ignored or defeated and should only be removed by the authorized individual responsible for installing the tag.

Employers must emphasize that tags should be clearly legible and understandable to be effective. Their significance must be understood by authorized employees, affected employees, and any employee whose work operation does or may place them in the area of the tag. Tags must also be suitable for the environment in which they are used. Paper tags used in areas susceptible to water spray or rain are examples of improper tag material selection. Emphasis must be placed on the fact that tags can provide a false sense of security, since they can be accidentally removed or inadvertently or accidentally detached during use. A complete comparison of tagout systems requirements between 1910.147 and 1910.269 is presented in Appendix C, Table C.5 of this book.

Employee Retraining

The OSHA Standards Interpretation letter of October 25, 2005 listed above, provided an example of both OSHA's Occupational Health and Safety Standards requirements for training and retraining for hazardous energy control, lockout/tagout. 29 CFR 1910.147(c)(7)(iii) requires employer retraining of employees when any of the following events has taken place:

- Change in job assignment;

- Changes in machines, equipment or processes which present a new hazard; or

- Changes in energy control procedures.

The objective of any employee retraining program should be employee proficiency in new and/ or revised hazardous energy control procedures and methods. Once employee retraining has been accomplished, the employer is responsible for certifying that that training has been completed and is being kept up to date. Records of employee names and training completed must be included in any certification. A complete comparison of retraining requirements is presented in Appendix C, Table C.6 of this book.

Energy Isolation/Notification of Employees

Energy isolation requirements are outlined in 1910.147(c)(8) and (9). Although simple in format, they do form the basis for employee safety with energy isolation on equipment. The first requirement is that only the authorized employees performing the servicing or maintenance work on equipment or machinery have the authority to perform the lockout/ tagout of that equipment.

It is extremely important that the person conducting work in a hazardous energy isolation situation maintain full control over their safety. Surrendering that authority to anyone not directly involved with conducting the work on that equipment or machinery is potentially hazardous. Equally important is the requirement that service personnel or their employer notify all affected employees of the application and removal of lockout/tagout devices. Affected employees would be those employees that work in the area in which the equipment is being locked out. Knowledge that someone is performing hazardous energy removal from equipment in that area is essential to prevent the potential for injury to those working on that equipment or machinery. Notification is required prior to installation of the lockout/tagout devices and after their removal. It is essential that both authorized and affected employees cooperate during the service, repair, maintenance or replacement of equipment and machinery to prevent accidents.

Control Application

Procedures for the application of energy control (the lockout or tagout procedures) cover specific elements and actions and are required to be done in a specific sequence. That sequence involves each of the following tasks along with specific sections in 1910.147 and 1910.269 describing the task:

- Application of control – 1910.147(d) and 1910.269(d)(6)

- Preparation for shutdown – 1910.147(d)(1) and 1910.269(d)(6)(i)

- Machine or equipment shutdown – 1910.147(d)(2) and 1910.269(d)(6)(ii)

- Machine or equipment isolation (de-energization) – 1910.147(d)(3) and 1910.269(d(6)(iii)

- Lockout or tagout device application – 1910.147(d)(4) and 1910.269(d)(6)(iv)

- Stored energy relieved, disconnected, restrained, and otherwise rendered safe – 1910.147(d)(5) and 1910.269(d)(6)(v)

- Possibility of reaccumulation of stored energy – 1910.147(d)(5)(ii) and 1910.269(d)(6)(vi)

- Verification of isolation – 1910.147(d)(6) and 1910.269(d)(6)(vii)

Preparation for shutdown may require substantial planning. Should the shutdown involve process equipment, assembly lines, etc. the procedure would be extremely more complex than only shutting down a single piece of isolated equipment. Shutdown of chemical or petroleum processing trains may involve flaring the product or bypassing equipment, slowly lowering pressures or process flow rates, cooling of process piping or equipment, etc. Shutdown of utility systems may involve switching of energy sources through alternate sources to customers before isolating equipment.

Before shutdown can begin, it is necessary that the authorized employee complete the preparation for that shutdown by identifying the type and magnitude of energy present in the equipment or machinery, the hazards associated with that energy, and have a clear method established to control that energy. Proceeding with a shutdown without that information could present major problems.

Proper energy isolation from and in equipment and machinery is necessary for accident prevention. Relying only on energy isolation through equipment interlocks or stop buttons, in lieu of totally removing and locking out and/or tagging of the available energy source(s), is potentially dangerous. Failure to do this creates the potential for inadvertent startup while an employee is working in a danger zone in or on the equipment.

Paragraph 1910.333(b)(2)(ii)(B), 1910 Subpart S Electrical addresses the problem of using unapproved methods of energy isolation. It notes:

> The circuits and equipment to be worked on shall be disconnected from all electric energy sources. Control circuit devices, such as push buttons, selector switches, and interlocks, may not be used as the sole means for deenergizing circuits or equipment. Interlocks for electric equipment may not be used as a substitute for lockout and tagging procedures. [43]

Although the lockout/tagout process above appears complicated, its most important preparation is the planning and communication between the individuals preparing to do work on equipment and plant personnel in the area of the equipment who might be working or controlling other equipment and processes. Failure to implement that phase of the process could lead to the potential for injuries or deaths involving accidental energization of equipment while under maintenance or repair.

Lockout devices should be affixed to energy isolating devices only by authorized employees. Those devices should be attached in such a manner so that the energy isolation device will be securely held in a "SAFE" or "OFF" position. Tagout devices are also required to be affixed to an energy isolation device in such a manner that it will clearly indicate prohibition of the operation or movement of energy isolation devices from the "SAFE" or "OFF" positions. If tagout devices are utilized on energy isolation devices which can be locked out, they must be attached to the same point at which the lockout device would be installed. If the tagout device cannot be directly affixed to the energy isolation device, it must be installed as close and safely as possible to that device. Its position must be immediately obvious to anyone that might operate the energy isolating device.

Once isolating devices are opened and locked out/tagged out, installation of grounding systems, in the case of electrical distribution or transmission lines or additional safety procedures with other equipment, may be required. Two additional steps are required before actual work can commence on the equipment. Those steps involve removal of any stored energy in the equipment and verification that the equipment has been isolated. Should an isolating switch or device appear to be open, this is not proof in itself that it is actually open. Isolating devices can fail or be bypassed. Checks for the presence of voltage, pressure or other chemical, mechanical or electrical energy sources is essential. The installation of wedges, flanges, blocks, or other mechanical protection means may be necessary to prevent the release of stored potential energy.

The removal of stored energy sources is addressed in 1910.333 indicating:

> Stored electric energy which might endanger personnel shall be released. Capacitors shall be discharged and high capacitance elements shall be short-circuited and grounded, if the stored electric energy might endanger personnel.

> *Note:* If the capacitors or associated equipment are handled in meeting this requirement, they shall be treated as energized. [44]

Stored energy testing is further covered in 1910.333(b)(2)(iv)(B) requiring:

> A qualified person shall use test equipment to test the circuit elements and electrical parts of equipment to which employees will be exposed and shall verify that the circuit elements and equipment parts are deenergized. The test shall also determine if any energized condition exists as a result of inadvertently induced voltage or unrelated voltage backfeed even though specific parts of the circuit have been deenergized and presumed to be safe. If the circuit to be tested is over 600 volts, nominal, the test equipment shall be checked for proper operation immediately after this test. [45]

Release from Lockout/Tagout

Before locked out/tagged out equipment or machinery can be put back into service, a specific sequence of events must be followed. That procedure is outlined in 1910.147(e) and 1910.269(d)(7). The procedure must be conducted by an authorized employee(s). Initially, the work area must be inspected, removing nonessential items and insuring that the machine or equipment components are intact operationally. Also, an inspection of the area must be conducted to assure that all employees have been either removed or relocated to a safe position. The authorized employee(s), or their designated replacement(s), responsible for installing all lockout/tagout devices, will be responsible for removing each (all) lockout/tagout device(s) and notification of all affected employees of that action.

OSHA 1910.147(e)(3) provides additional requirements if the authorized employee(s) originally placing the lockout/tagout devices is not available to remove them. It notes that the

device may be removed under the direction of the employer, provided that specific procedures and training for such removal have been developed, documented and incorporated into the employer's energy control program. The employer shall demonstrate that the specific procedure provides equivalent safety to the removal of the device by the authorized employee who applied it. The specific procedure shall include at least the following elements:

- Verification by the employer that the authorized employee who applied the device is not at the facility:
- Making all reasonable efforts to contact the authorized employee to inform him/her that his/her lockout or tagout device has been removed; and
- Ensuring that the authorized employee has this knowledge before he/she resumes work at that facility. [46]

Additional Requirements

There may be situations where the equipment being serviced or repaired must be temporarily restarted to test or position the machine, equipment, or component, before the lockout/tagout devices are finally removed. That situation is covered in 1910.147(f)(1) and 1910.269(d)(8)(i) and requires a specific sequence of actions. Those actions include:

Removal of nonessential tools and materials from the machine or equipment, ensuring that the machine or equipment components are operationally intact;

Removal or repositioning all employees from the work area;

Removal of each (all) lockout/tagout device(s) from each energy isolating device(s) by the authorized employee(s) responsible for initially applying them. Should that (those) individual(s) not be available, their designated replacement(s) may remove those devices provided the details outlined under the exception to 1910.147(e)(3) are followed.

Energization of the machine or equipment and proceeding with the required testing or positioning; and

De-energization of the equipment or machinery and reapplication of the energy control procedures of 1910.147(d) or 1910.269(d)(6) before continuing the required servicing or maintenance.

Group Lockout or Tagout/Outside Personnel (Contractors, etc.)

Energy control procedures must account for the situations where servicing or maintenance is performed by a department, group, or crew. OSHA regulation 1910.147(f)(3)(i) requires that the employees be afforded the level of protection equal to that which would be provided if the work were performed by a personal lockout/tagout device.

Group lockout/tagout procedures must follow 1910.147(f)(3) or 1910.269(d)(8)(ii). Specific requirements involve the primary lockout/tagout responsibilities vested in an authorized employee or a set number of employees who are working under a group lockout/tagout device (operation lock). The authorized employee shall ascertain the exposure status for all individual group members, under their control and responsibility, with regard to the lockout/tagout of equipment or machinery. If more than one group, craft, department or crew is involved with working on a machine or equipment, overall job-associated lockout/tagout control responsibility shall be assigned to an authorized employee. That employee shall be designated responsibility to coordinate all affected work forces to ensure the continuity of their protection.

Each authorized employee shall have responsibility to affix a personal lockout/tagout device(s) to the group lockout device, group lockout device, or any comparable lockout/tagout mechanism. Those devices shall be installed when work begins on the machinery or equipment and shall be removed when work stops on that equipment or machinery. To accommodate shift or personnel changes, 1910.147(f)(4) requires that procedures shall ensure the continuity of lockout/tagout protection, including the orderly transfer of employee protection between on-coming and off-going employees at shift change. This shall be required to minimize employee hazard exposure from inadvertent and unexpected energization or start-up of the equipment or machine or from stored energy release from that equipment or machinery.

In situations where outside contractors or service personnel are involved with lockout/tagout control application in the repair, maintenance, or servicing of equipment or machinery, the on-site employer and the outside employer shall be informed of their respective lockout/tagout procedures. This is necessary to ensure that all personnel understand and comply with the energy control restrictions and prohibitions in use. The on-site employer has the responsible to ensure that their employees understand and comply with the outside employer's energy control program's restrictions and prohibitions, while they are working on site.

Electric Power Generation, Transmission, and Distribution

Health and Safety Standard 29 CFR 1910.269:

> covers the operation and maintenance of electric power generation, control, transformation, transmission, and distribution lines and equipment. These provisions apply to:

- Power generation, transmission, and distribution installations, including related equipment for the purpose of communication or metering, which are accessible only to qualified employees;
    Note: The types of installations covered by this paragraph include the generation, transmission, and distribution installations of electric utilities, as well as equivalent installations of industrial establishments. Supplementary electric generating equipment that is used to supply a workplace for emergency, standby, or similar purposes only is covered under Subpart S of this Part. (See paragraph (a)(1)(ii)(B) of this section.)
- Other installations at an electric power generating station, as follows:
    1. Fuel and ash handling and processing installations;
    2. Water and steam installations;
    3. Chlorine and hydrogen systems:
- Test sites where electrical testing involving temporary measurements associated with electric power generation, transmission, and distribution is performed in laboratories, in the field, in substations, and on lines, as opposed to metering, relaying, and routine line work;
- Work on or directly associated with the installations covered in paragraphs (a)(1)(i)(A) through (a)(1)(i)(C) of this section; and
- Line-clearance tree-trimming operations. [47]

Paragraph (a)(1)(ii) in 1910.269 indicates that the operation and maintenance of electric power generation, control, transformation, transmission, and distribution lines and equipment does not apply:

> To construction work, as defined in 1910.12; or

> To electrical installations, electrical safety-related work practices, or electrical maintenance considerations covered by Subpart S Electrical in 29 CFR 1910.

The following outline identifies the primary areas covered in 1910.269. This list will be used to discuss the makeup of 1910.269.

- 1910.269(a) General

- 1910.269(b) Medical Services and First Aid

- 1910.269(c) Job Briefing

- 1910.269(d) Hazardous Energy Control (Lockout/Tagout) Procedures

- 1910.269(e) Enclosed Spaces

- 1910.269(f) Excavations

- 1910.269(g) Personal Protection Equipment

- 1910.269(h) Ladders, Platforms, Step Bolts, and Manhole Steps

- 1910.269(i) Hand and Portable Tools

- 1910.269(j) Live-line Tools

- 1910.269(k) Materials Handling and Storage

- 1910.269(l) Working On or Near Exposed Energized Parts

- 1910.269(m) Deenergizing Lines and Equipment for Employee Protection

- 1910.269 (n) Grounding for the Protection of Employees

- 1910.269(o) Testing and Test Facilities

- 1910.269(p) Mechanical Equipment

- 1910.269(q) Overhead Lines

- 1910.269(r) Line-clearance Tree Trimming Operations

- 1910.269(s) Communication Facilities

- 1910.269(t) Underground Electrical Installations

- 1910.269(u) Substations

- 1910.269(v) Power Generation

- 1910.269(w) Special Conditions

- 1910.269(x) Definitions

The 1910.269(a) General section examines the standard's application, training requirements, and existing conditions of a work site. Existing conditions refers to safety considerations at the site on which work is to be performed, including operating nominal voltages; switching transient voltages; potential for induced voltages; protective grounds and equipment grounds which may be present and their condition; condition of poles, facilities, or equipment; environmental conditions which may affect the work; and the presence of other equipment or electrical lines. Its purpose is to identify potential work hazards.

Section 1910.269(b) Medical Services and First Aid establishes the medical training and supplies which must be respectively completed and available before work can be conducted. That includes cardiopulmonary resuscitation (CPR) and first aid kits.

Section 1910.269(c) Job Briefing outlines the extent and number of briefings required before work can begin. It also indicates that an employee who is working alone need not hold a briefing before starting the work; however, their employer has to ensure that the work to be performed has been planned as if a briefing had been conducted.

Section 1910.269(d) Hazardous Energy Control (Lockout/Tagout) Procedures outlines the procedures for the removal or isolation of hazardous energy from equipment or machinery and prevention of energization during the time in which it is under repair, maintenance, or replacement.

The next two sections, 1910.269(e) Enclosed Spaces and 1910.269(f) Excavations, deal with hazardous spaces. Enclosed Spaces establishes general requirements for evaluation of potential hazards, air supply, training, safe work practices, rescue equipment, attendants, ventilation, monitoring, and testing which must be conducted prior to entry into en enclosed space. Excavations references 29 CFR 1910 Subpart P for mandated trenching or other excavation requirements.

An "enclosed space":

> does not apply to vented vaults if a determination is made that the ventilation system is operating to protect employees before they enter the space. This paragraph applies to routine entry into enclosed spaces in lieu of the permit-space entry requirements contained in paragraphs (d) through (k) of 1910.146 ... If, after the precautions given in paragraphs (e) and (t) of this section are taken, the hazards remaining in the enclosed space endanger the life of an entrant or could interfere with escape from the space, then entry into the enclosed space shall meet the permit-space entry requirements of paragraphs (d) through (k) of 1910.146 ... [48]

The next five sections, 1910.269(g) through (k), provide requirements for hand held tools, equipment, and material handling and storage. These are general sections which provide requirements for personal protection equipment (PPE); ladders, platforms, and steps; hand and portable tools; live-line tools; and material handling and storage. Some requirements refer the reader to other sections in 29 CFR 1910, depending upon circumstances and situation at hand. For instance, 1910.269(i)(2)(i) requires cord-and plug-connected equipment supplied by premises wiring to be covered by 1910 Subpart S.

Sections 1910.269(l) through (o) involve safe work practices. Working on or near exposed energized parts in 1910.269(l) covers minimum approach distances; insulation types; working position; making connections; apparel; fuse handling; covered insulators; on-current carrying metal parts; and opening circuits under load. The minimum approach distances in this section were taken from ANSI/IEEE C2, *National Electrical Safety Code*®, Section 441 Energized Conductors. A minimum approach distance is defined in 1910.269 as:

> The closest distance an employee is permitted to approach an energized or a grounded object [49].

NESC® has a slightly different definition for minimum approach distance. It defines it as:

> The closest distance a *qualified* employee is permitted to approach either an energized or a grounded object, *as applicable for the work method being used.* [50] (Italics inserted for emphasis of the definition differences only.)

Those approach distances differ, depending on the level of voltage present.

The minimum approach distances in 1910.269(l)(2) are divided into the following categories in Tables R-6 through R-10:

1. Table R-6. – AC Live-Line Work Minimum Approach Distance

2. Table R-7. – AC Live-Line Work Minimum Approach Distance

With Overvoltage Factor Phase-to-Ground Exposure

3. Table R-8. – AC Live-Line Work Minimum Approach Distance

With Overvoltage Factor Phase-to-Phase Exposure

4. Table R-9. – DC Live-Line Work Minimum Approach Distance

With Overvoltage Factor

5. Table R-9. – DC Live-Line Work Minimum Approach Distance

6. Table R-10. – Altitude Correction Factor

Tables R-6 through R9 have notes indicating that the distances in those tables could only be applied where engineering analysis has been supplied by the employer to determine the maximum anticipated per-unit transient overvoltage It also notes that if the transient overvoltage factor is not known, a factor of 1.8 shall be assumed.

Section 1910.269(l) deals with requirements for the use of insulating gloves and energized bare line sleeves; employee clothing that will not contribute to additional injury should an arc event occur; and assurance that an employee will be trained to not work in a position that might result either in a slip or shock that will bring the employee's body directly in contact with energized, bare conductors or other equipment. The section also deals with the sequence of connection or disconnection to conductors to energized lines, as well as precautions required for opening circuits under load. All of those tasks have the potential to create an arc flash or shock event.

Other safe work practices are covered in 1910.269(m) and deal with the requirements for de-energization of lines and equipment for employee protection. De-energization of a local distribution line might be handled by a utility employee at the scene. However, in transmission lines and substations, a remotely located utility system operator may be charged with the authority and control of that equipment. This section deals with the requirements for coordination of the work between the employee on the scene and the system operator. General protocol for accomplishing that task is covered in this section.

Grounding of transmission and distribution lines and equipment is an important employee safe work practice. Section 1910.269(n) provides specific guidance in three general areas, including when an external ground may or may not be required; general criteria for grounding equipment; and grounding procedures. Grounding procedures provide specific guidance in testing for voltage presence prior to installation; the order of connection and removal of grounds; and the removal of grounds for testing.

Section 1910.269(o) establishes:

> safe work practices for high-voltage and high-power testing performed in laboratories, shops, and substations, and in the field and on electric transmission and distribution lines and equipment. It applies only to testing involving interim measurements utilizing high voltage, high power, or combinations of both, and not to testing involving continuous measurements as in routine metering, relaying, and normal line work. [51]

Those safe work practices include "test area guarding, grounding, and the safe use of measuring and control circuits" [52].

Safe work practices are also established in four general areas for testing and test facilities. They include test cables construction and grounding; equipment with exposed, energized parts; temporary wiring and cable requirements; and provision for test observers with the capability to de-energize the test setup to protect employees. One final safe work practice includes the establishment of safety check procedures prior to and between testing events to assure employee safety.

Section 1910.269(p) deals with mechanical equipment safety. This section establishes requirements for elevating and rotating equipment inspection; vehicle operational procedures and safety equipment; vehicle rollover protection; procedures for operation of equipment with and without outriggers; applied load limits; and safe work practices for operating equipment near energized lines or equipment.

Overhead line safety is addressed in Section 1910.269(q). It requires that the integrity of poles and structures be determined before an employee is allowed to climb onto, install or remove equipment attached thereto. This section also mandates safety precautions when installing or removing poles near energized overhead lines, which also includes protection of employees from falling into any holes created for either installing poles or removing them. Extensive safe work practices for installing or removing overhead lines are also outlined, as are safe work practices involving live-line bare-hand work and bucket truck usage. Work requirements on towers and structures supporting overhead lines are also outlined.

Section 1910.269(r) provides safe work practices for operations and equipment for line-clearance tree trimming operations. This section requires identification of electrical hazards and minimum approach distances from Tables R-6, R-9, and R-10 in 1910.269 before proceeding with the work. It also includes safe work practices when using bush cutters; chemical sprayers, and related equipment; stump cutters; gasoline-engine power saws; backpack power units; and rope. Fall protection, using a climbing rope and safety saddle, is mandated when working in a tree.

Communications facilities safe work practices are outlined in Section 1910.269(s). This section provides electromagnetic radiation safe work practices in microwave facilities. It also requires power line carrier work to be conducted using the requirements of 1910.269 pertaining to work on energized lines. Power line carrier equipment inserts signals, such as those from protective relaying equipment, onto energized power lines for distance relaying and monitoring.

Safe work practices in underground electrical installations, such as manholes, are covered in 1910.269(t). This section involves safe work practices in electrical manholes; use of duct rods; and working on, moving or removing electrical cables.

Section 1910.269(u) pertains to electrical substation safe work practices. It contains the following note pertaining to work done in a substation:

> Guidelines for the dimensions of access and working space about electric equipment in substations are contained in ANSI C2-1987, *National Electrical Safety Code.* Installations meeting the ANSI provisions comply with paragraph (u)(1) of this section. An installation that does not conform to this ANSI standard will, nonetheless, be considered as complying with paragraph (u)(1) of this section if the employer can demonstrate that the installation provides ready and safe access based on the following evidence:
>
> [1] That the installation conforms to the edition of ANSI C2 that was in effect at the time the installation was made,
> [2] That the configuration of the installation enables employees to maintain the minimum approach distances required by paragraph (l)(2) of this section while they are working on exposed, energized parts, and
> [3] That the precautions taken when work is performed on the installation provide protection equivalent to the protection that would be provided by access and working space meeting ANSI C2-1987. [53]

Note that it references the 1987 Edition of the *National Electrical Safety Code*®.

This section has a second similar reference to ANSI/IEEE C2-1987 in 1910.269(u)(5) Guarding of Energized Parts. That reference involves horizontal and vertical clearances around all live parts, without an insulating covering, operating at more than 150 Volts to ground to prevent accidental employee contact. It also deals further with the guarding of energized parts and their removal for operation or maintenance purposes.

Section 1910.269(u) also establishes safe work practices with draw-out-type circuit breakers and grounding requirements for fences. It also provides detailed requirements for guarding of rooms containing electrical supply equipment, as well as substation entry requirements.

Power generation plant safe work practices are covered in 1910.269(v). The first two paragraphs in this section involve basic safety considerations for interlocks and other safety devices and exciter or generator brushes. The section also refers to ANSI/IEEE C2-1987, *National Electrical Safety Code*® regarding access and working space around electric equipment to safely permit operation and maintenance activities.

Guarding of rooms containing electric supply equipment and guarding of energized parts safe work practices are covered in 1910.269(v)(4) and (5) in much the same way as they were in 1910.269(u)(4) and (5) for substations. Safe work practices for cleaning associated with steam and water piping as well as boilers is presented in sections 1910.269(v)(6) and (7). Work practices for power generation equipment and machinery is presented in 1910.269(v)(8) through (12). Equipment covered by those sections include boilers, turbine generators, coal and ash handling equipment and facilities, and hydroplants.

Section 1910.269(w) Special Conditions, provides safe work practices for a variety of equipment, including capacitors, current transformers, and series street lighting. Special personal protection areas are also examined, providing some general requirements, including illumination, drowning protection, employee protection in public work areas, and backfeed.

US Department of Energy (DOE)

The United States Congress passed the *Energy Policy and Conversation Act (EPCA) of 1975*, which was signed into law by United States President Gerald Ford on December 22, 1975. Its purpose was:

(1) to grant specific standby authority to the President, subject to congressional review, to impose rationing, to reduce demand for energy through the implementation of energy conservation plans, and to fulfill obligations of the United States under the international energy program;

(2) to provide for the creation of a Strategic Petroleum Reserve capable of reducing the impact of severe energy supply interruptions;

(3) to increase the supply of fossil fuels in the United States, through price incentives and production requirements;

(4) to conserve energy supplies through energy conservation programs, and, where necessary, the regulation of certain energy uses;

(5) to provide for improved energy efficiency of motor vehicles, major appliances, and certain other consumer products;

(6) to reduce the demand for petroleum products and natural gas through programs designed to provide greater availability and use of this Nation's abundant coal resources;

(7) to provide a means for verification of energy data to assure the reliability of energy data; and

(8) to conserve water by improving the water efficiency of certain plumbing products and appliances. [54]

The Energy Policy and Conservation Act (EPCA) of 1975 was amended by the Energy Policy Act (EPACT) of 2005. That law

was enacted on August 8, 2005. Among the provisions of Subtitle C of Title I of EPACT 2005 are provisions that amend Part B of Title III of the Energy Policy and Conservation Act (EPCA) (42 U.S.C. 6291-6309), which provides for an energy conservation program for consumer products other than automobiles, and Part C of Title III of EPCA (42 U.S.C. 6311-6317), which provides for a program, similar to the one in Part B, for certain commercial and industrial equipment. In addition to provisions directing DOE to undertake rulemakings to promulgate new or amended energy conservation standards for various consumer products and commercial and industrial equipment, Congress itself prescribed new efficiency standards and related definitions for certain consumer products and commercial and industrial equipment. By today's action, DOE is placing in the Code of Federal Regulations (CFR), for the benefit of the public, the energy conservation standards and related definitions that Congress has prescribed for various consumer products and commercial and industrial equipment. In this technical amendment, DOE is not exercising any of the discretionary authority that Congress has provided in EPACT 2005 for the Secretary of Energy to revise, by rule, several of the product or equipment definitions and energy conservation standards.\1\ DOE may exercise this discretionary authority at a later time in rulemakings to establish test procedures or efficiency standards for these products and equipment. [55]

The Secretary of Energy exercised that legislated authority as a final rule for the Distribution Transformers Energy Conservation Standard Rulemaking, 72 FR 58190(October 12, 2007) issuing: 10 CFR Part 431 Energy Conservation Program for Commercial Equipment: Distribution Transformers Energy Conservation Standards; Final Rule. It noted in I. Summary of the Final Rule and Its Benefits. A. *The Standard Levels*:

The standards established in this final rule are minimum efficiency levels. Tables I.1 and I.2 show the standard levels DOE is adopting today. These standards will apply to liquid-immersed and medium-voltage, dry-type distribution transformers manufactured for sale in the United States, or imported to the United States, on or after January 1, 2010. As discussed in section V.C.2 of this notice, any transformers whose kVA rating falls between the kVA ratings shown in Tables I.1 and I.2 shall have its minimum efficiency requirement calculated by a linear interpolation of the minimum efficiency requirements of the kVA ratings immediately above and below that rating. [56]

The efficiency standards set by the Department of Energy's Final Rule Tables I-1 and I-2 are compared here to those listed in NEMA's TP1-2002, *Guide for Determining Energy Efficiency for Distribution Transformers* Tables 4-1 and 4-2 in Tables 6.3 through 6.6 [57–60].

Several points of clarification are needed in reviewing the transformer efficiency tables (Tables 6.3–6.6). First, the terms low voltage and medium voltage should be defined. Low-voltage sources are considered to be distribution transformers that have input voltages of ≤600 V and apply to dry-type transformers. Medium-voltage distribution transformers have input voltages of 601–34,500 V. EPACT 2010 requirements are for medium voltage dry-type and liquid-immersed distribution transformers.

DOE defines a distribution transformer as:

a transformer that—

(1) Has an input voltage of 34.5 kilovolts or less;
(2) Has an output voltage of 600 volts or less; and

TABLE 6.3   Efficiency levels for liquid-filled distribution transformers: single-phase

kVA	NEMA Class 1 TP 1-2002 efficiency	DOE 2010 efficiency requirements	% efficiency improvement
10	98.4	98.62	0.22
15	98.6	98.76	0.16
25	98.7	98.91	0.21
37.5	98.8	99.01	0.21
50	98.9	99.08	0.18
75	99.0	99.17	0.17
100	99.0	99.23	0.23
167	99.1	99.25	0.15
250	99.2	99.32	0.12
333	99.2	99.36	0.16
500	99.3	99.42	0.12
667	99.4	99.46	0.06
833	99.0	99.49	0.49

Note: Efficiency values at 50% of nameplate-rated load.
Source: Data taken from NEMA TP 1-2002, *Guide for Determining Energy Efficiency for Distribution Transformers*, Table 4-1, page 7; and Federal Register, Vol. 72, No. 197, October 12, 2007, Table I.1, page 58191

TABLE 6.4 Efficiency levels for liquid-filled distribution transformers: three-phase

kVA	NEMA Class 1 TP 1-2002 efficiency	DOE 2010 efficiency requirements	% efficiency improvement
15	98.1	98.36	0.26
30	98.4	98.62	0.22
45	98.6	98.76	0.16
75	98.7	98.91	0.21
112.5	98.8	99.01	0.21
150	98.9	99.08	0.18
225	99.0	99.17	0.17
300	99.0	99.23	0.23
500	99.1	99.25	0.15
750	99.2	99.32	.012
1000	99.2	99.36	0.16
1500	99.3	99.42	0.12
2000	99.4	99.46	0.06
2500	99.4	99.49	0.09

Note: Efficiency values at 50% of nameplate-rated load.
Source: Data taken from NEMA TP 1-2002, *Guide for Determining Energy Efficiency for Distribution Transformers*, Table 4-1, page 7; and Federal Register, Vol. 72, No. 197, October 12, 2007, Table I.1, Page 58191

TABLE 6.5 Efficiencies for Dry-type distribution transformers: single-phase

kVA	NEMA TP 1-2002 Low voltage	DOE2010 Req. 20-45 kV	Medium voltage		Medium voltage	
			NEMA TP 1-2002 ≤60 kV BIL	DOE 2010 Req. 46-95 kV	NEMA TP 1-2002 >60 kV BIL	DOE 2010 Req. ≥96 kV
15	97.7	98.10	97.6	97.86	97.6	—
25	98.0	98.33	97.9	98.12	97.9	—
37.5	98.2	98.49	98.1	98.30	98.1	—
50	98.3	98.60	98.2	98.42	98.2	—
75	98.5	98.73	98.4	98.57	98.4	98.53
100	98.6	98.82	98.5	98.67	98.5	98.63
167	98.7	98.96	98.8	98.83	98.7	98.80
250	98.8	99.07	98.9	98.95	98.8	98.91
333	98.9	99.14	99.0	99.03	98.9	98.99
500	—	99.22	99.1	99.12	99.0	99.09
667	—	99.27	99.2	99.18	99.0	99.15
833	—	99.31	99.2	99.23	99.1	99.20

Note: All efficiency values are at 50% of nameplate-rated load.
Source: Data taken from NEMA TP 1-2002, *Guide for Determining Energy Efficiency for Distribution Transformers*, Table 4-2, page 8; and Federal Register, Vol. 72, No. 197, October 12, 2007, Table I.2, pages 58191 and 58192

TABLE 6.6 Efficiencies for dry-type distribution transformers: three-phase

kVA	NEMA TP 1-2002 Low voltage	DOE 2010 Req. 20–45 kV	Medium voltage		Medium voltage	
			NEMA TP 1-2002 ≤60 kV BIL	DOE 2010 Req. 46–95 kV	NEMA TP 1-2002 >60 kV BIL	DOE 2010 Req. ≥96 kV
15	97.0	97.50	96.8	97.18	96.8	—
30	97.5	97.90	97.3	97.63	97.3	—
45	97.7	98.10	97.6	97.86	97.6	—
75	98.0	98.33	97.9	98.12	97.9	—
112.5	98.2	98.49	98.1	98.30	98.1	—
150	98.3	98.60	98.2	98.42	98.2	—
225	98.5	98.73	98.4	98.57	98.4	98.53
300	98.6	98.82	98.6	98.67	98.5	98.63
500	98.7	98.96	98.8	98.83	98.7	98.80
750	98.8	99.07	98.9	98.95	98.8	98.91
1000	98.9	99.14	99.0	99.03	98.9	98.99
1500	—	99.22	99.1	99.12	99.0	99.09
2000	—	99.27	99.2	99.18	99.0	99.15
2500	—	99.31	99.2	99.23	99.1	99.20

Note: All efficiency values are at 50% of nameplate-rated load.

Source: Data taken from NEMA TP 1-2002, *Guide for Determining Energy Efficiency for Distribution Transformers*, Table 4-2, page 8; and Federal Register, Vol. 72, No. 197, October 12, 2007, Table I.2, pages 58191 and 58192

(3) Is rated for operation at a frequency of 60 Hertz; however, the term "distribution transformer" does not include—

(i) A transformer with multiple voltage taps, the highest of which equals at least 20 percent more than the lowest;

(ii) A transformer that is designed to be used in a special purpose application and is unlikely to be used in general purpose applications, such as a drive transformer, rectifier transformer, auto-transformer, Uninterruptible Power System transformer, impedance transformer, regulating transformer, sealed and non-ventilating transformer, machine tool transformer, welding transformer, grounding transformer, or testing transformer; or

(iii) Any transformer not listed in paragraph (3)(ii) of this definition that is excluded by the Secretary by rule because—

(A) The transformer is designed for a special application;

(B) The transformer is unlikely to be used in general purpose applications; and

(C) The application of standards to the transformer would not result in significant energy savings.

Low-voltage dry-type distribution transformer means a distribution transformer that—

(1) Has an input voltage of 600 volts or less;

(2) Is air-cooled; and

(3) Does not use oil as a coolant. [61]

As can be seen above, the efficiency improvements are small; however, the DOE analysis projected an overall energy cost savings and a reduction in cumulative greenhouse gas emissions. Its evaluation of those savings was presented in the October 12, 2005 edition of the Federal Register. It noted in Section I.C *Summary of the Final Rule and Its Benefits – Benefits to Transformer Customers*:

> For liquid-immersed transformers, DOE estimates that approximately 25% of the market incurs a net life-cycle cost from the standard while 75% of the market is either not affected or incurs a net benefit. DOE also investigated how these standards might affect municipal utilities and rural electric cooperatives. While the benefits are positive for municipal utilities, a majority of smaller, pole-mounted transformers for rural electric cooperatives will incur a net life-cycle cost. However, because of a relatively large per-transformer reduction in life-cycle cost for some non-evaluating rural electric cooperatives (i.e., those that do not take into consideration the cost of transformer losses when choosing a transformer) rural electric cooperatives as a whole receive an average life-cycle cost benefit. [62]

A "net life cycle cost" indicates an added cost to the transformer owner; where a "net life cycle cost benefit" indicates costs reduction for the transformer owner.

There are three National Electrical Manufacturers Association Standards Publications regarding liquid-immersed and dry-type distribution transformer energy efficiency. They include:

TP1, *Guide for Determining Energy Efficiencies for Distribution Transformers*

TP 2, *Standard Test for Measuring the Energy Consumption for Distribution Transformers*

TP 3, *Standard for the Labeling of Distribution Transformer Efficiency.*

NEMA issued a press release to explain the purpose of their development of TP-2. It noted:

> The document provides a standardized method for measurement of distribution transformer loss to achieve energy efficiency levels outlined in NEMA publication TP 1, Guide for Determining Energy Efficiency for Distribution Transformers.

> TP 2 was revised to address concerns raised by the Department of Energy with the previous edition. Under the Energy Policy and Conservation Act, DOE was tasked to develop rules to adopt test procedures for measuring the energy efficiency of distribution transformers. These revisions are intended to make TP 2 acceptable to DOE so that it will be adopted as the DOE test procedure. [63]

NEMA TP 2 is a standard that develops the basis for testing procedures of distribution transformer energy efficiency. It was specifically developed for new federal energy efficiency standards and was based on NEMA TP 1 for low-voltage distribution transformers. It is also referenced by state governments and agencies.

The Department of Energy also has a significant number of other energy efficiency standards. They can be found in the Code of Federal Regulations (CFR) (see Table 6.7).

TABLE 6.7 Department of Energy product, equipment and structure standards

CFR	Title
10 CFR 430	Energy Conservation Program for consumer products
10 CFR 431	Energy efficiency program for certain commercial and industrial equipment
10 CFR 434	Energy code for new Federal commercial and multi-family high-rise residential buildings
10 CFR 435	Energy conservation voluntary performance standards for new buildings; mandatory for Federal buildings

References

1. Occupational Safety and Health Act of 1970; Public Law 91-596, 84 STAT. 1590, 91st Congress, S.2193, December 29, 1970, as amended through January 1, 2004; An Act.
2. Occupational Safety and Health Act of 1970; SEC. 2. Congressional Findings and Purpose; Paragraph (b)(3).
3. US Department of Labor, Occupational Safety and Health Standards, 29 CFR 1910; Subpart A General; Section 1910.1(a) Purpose and Scope.
4. Ibid., 1910.2(g).
5. US Department of Labor, Occupational Safety and Health Standards, 29 CFR 1910; Section 147 The Control of Hazardous Energy (Lockout/Tagout); 1910.147(a)(1)(i) Scope.
6. US Department of Labor, Standards Interpretation Letter of 9/26/02, Richard E. Fairfax, Director OSHA Directorate of Enforcement Programs to Marvin B. Moore, ExxonMobil Refining and Supply Co.
7. US Department of Labor, Safety and Health Regulations for Construction, 29CFR1926; Subpart A General; 1926.1 Purpose and Scope.
8. Ibid., Subpart B General Interpretations; 1926.10(a).
9. Ibid., Subpart B General Interpretations; 1926.11(a)(1).
10. ANSI/IEEE C2, *National Electrical Safety Code*; 2007, Section 011. Institute of Electrical and Electronic Engineers; New York, NY.
11. US Department of Labor, Occupational Safety and Health Standards, 29 CFR 1910; Section 269 Electric Power Generation, Transmission, and Distribution; 1910.269(a)(1)(i).
12. Ibid., 1910.269(a)(1)(i)(A).
13. *Referenced Documents*; 29 CFR 1910.269, Appendix E.
14. ANSI/IEEE C-2®, *National Electrical Safety Code*®; 2007, Article 010 Purpose. Institute of Electrical and Electronic Engineers; New York, NY.
15. US Department of Labor, Occupational Safety and Health Standards, 29 CFR 1910; Section 269 Electric Power Generation, Transmission, and Distribution; 1910.269(a)(1)(i)(A) through (E).

16. US Department of Labor, Safety and Health Regulations for Construction, 29CFR1926; Appendix A4.

17. US Department of Labor, Correspondence from Noah Connell, Acting Director, OSHA Directorate of Construction to Bill Principe, 9/28/2006.

18. US Department of Labor, Occupational Safety and Health Standards, 29 CFR 1910; Section 147 The Control of Hazardous Energy (Lockout/Tagout); 1910.147(a)(3)(i).

19. This standard was originally promulgated by the National Safety Council in 1982, which transferred the Standard and Secretariat responsibilities to the American Society of Safety Engineers in 2004.

20. US Department of Labor, Occupational Health and Safety Standards, 29 CFR 1910; Subpart S Electrical; 1910.331(a) Scope.

21. OSHA website: http://www.osha.gov/pls/oshaweb/owadisp.show_document? p_table=PREAMBLES&p_id=1149.

22. Jones, Ray A., Mastrullo, Kenneth G., Jones, Jane G., *NFPA 70E®: Handbook for Electrical Safety in the Workplace*; 2004, page 51. National Fire Protection Association; Quincy, MA.

23. US Department of Labor, Occupational Safety and Health Standards, 29 CFR 1910; Section 333 Selection and Use of Work Practices; 1910.333(b)(2).

24. Ibid., Section 147 The Control of Hazardous Energy (Lockout/Tagout); 1910.147(a)(1)(ii).

25. Ibid., 1910.147(a)(2)(iii).

26. Ibid., 1910.147(a)(3)(i).

27. Ibid., 1910.147(b).

28. Ibid., 1910.147(b).

29. Ibid., 1910.147(b).

30. Ibid., 1910.147(b).

31. Ibid., 1910.147(b).

32. Ibid., Section 269 Electric Power Generation, Transmission, and Distribution; 1910.269(x).

33. Ibid., Subpart S Electrical; Section 1910.399 Definitions.

34. Ibid., Section 147 The Control of Hazardous Energy (Lockout/Tagout); 1910.147(c)(1).

35. Ibid., 1910.147(c)(3)(i).

36. Ibid., 1910.147(c)(3(ii).

37. Ibid., 1910.147(c)(3)(iii); and Section 269 Electric Power Generation, Transmission, and Distribution; 1910.269(d)(2)(ii)(C).

38. Ibid., Section 147 The Control of Hazardous Energy (Lockout/Tagout); 1910.147(b); Definitions.

39. Ibid., Subpart S Electrical; Section 1910.332, Training, Table S-4; 55 FR 32016, Aug. 6, 1990.

40. Ibid., Section 147 The Control of Hazardous Energy (Lockout/Tagout); 1910.147(c)(7)(i).
41. US Department of Labor, OSHA, Standards Interpretation Letter from Richard E. Fairfax, Director, Directorate of Enforcement Programs; October 25, 2005.
42. US Department of Labor, OSHA *Hot Topics – Lockout/Tagout Energy Control Program –* Training and Retraining; OSHA website: http://www.osha.gov/dts/osta/lototraining/hottopics/ht-engcont-2-3.html.
43. US Department of Labor, Occupational Safety and Health Standards, 29 CFR 1910; Subpart S Electrical; Section 1910.333(b)(2)(ii)(B).
44. Ibid., 1910.333(b)(2)(ii)(C).
45. Ibid; 1910.333(b)(2)(iv)(B).
46. Ibid., Section 147 The Control of Hazardous Energy (Lockout/Tagout); 1910.147(e)(3).
47. Ibid., Section 269 Electric Power Generation, Transmission, and Distribution; 1910: 269(a)(1)(i).
48. Ibid., 1910.269(e).
49. Ibid., 1910.269(x).
50. ANSI/IEEE C2-2007®, *National Electrical Safety Code®*; 2007, page 9; Institute of Electrical and Electronic Engineers; New York, NY.
51. US Department of Labor, Occupational Health and Safety Standards, 29 CFR 1910; Section 269 Electric Power Generation, Transmission, and Distribution; 1910:269(o)(1).
52. Ibid., 1910:269(o)(2)(i).
53. Ibid., 1910:269(u)(1).
54. Title 42, Chapter 77 – Energy Conservation Act, Sec. 6201 – Congressional statement of purpose; December 22, 1975.
55. *Federal Register*, October 18, 2005 (Volume 70, Number 200); Rules and Regulations; page 60407-60418; SUPPLEMENTARY INFORMATION: I. BACKGROUND.
56. 10 CFR Part 431 Energy Conservation Program for Commercial Equipment: Distribution Transformers Energy Conservation Standards; Final Rule; I. Summary of the Final Rule and Its Benefits; October 12, 2007.
57. Data taken from NEMA TP 1-2002, *Guide for Determining Energy Efficiency for Distribution Transformers*; Table 4-1, page 7; and Federal Register (Vol. 72, No. 197), October 12, 2007; Table I.1, page 58191.
58. Data taken from NEMA TP 1-2002, *Guide for Determining Energy Efficiency for Distribution Transformers*; Table 4-1, page 7; and Federal Register (Vol. 72, No. 197), October 12, 2007; Table I.1, page 58191.
59. Data taken from NEMA TP 1-2002, *Guide for Determining Energy Efficiency for Distribution Transformers*; Table 4-2, page 8; and Federal Register (Vol. 72, No. 197), October 12, 2007; Table I.2, pages 58191 and 58192.
60. Data taken from NEMA TP 1-2002, *Guide for Determining Energy Efficiency for Distribution Transformers*; Table 4-2, page 8; and Federal Register (Vol. 72, No. 197), October 12, 2007; Table I.2, pages 58191 and 58192.

61. Energy Policy Act of 2005; PUBLIC LAW 109–58–AUG. 8, 2005; 10 CFR Ch. II (1–1–06 Edition); Subpart K–Distribution Transformers; § 431.192 Definitions concerning distribution transformers.

62. Federal Register (Vol. 72, No. 197), October 12, 2007; *I.C. Benefits to Transformer Customers*; page 58192.

63. NEMA News Release, *NEMA Revises Standard for Measuring Distribution Transformer Loss*; 27 October, 2005. National Electrical Manufacturers Association; Rosslyn, VA.

CHAPTER 7

Developing Electrical Safe Work Practices

General

Anyone unaware of the hazards involved with an *electrical arc flash* event should first consider the following. Temperatures sustained in an arc-flash can reach 35,000 °F (19,500 °C). Those extremely high temperatures will easily melt copper conductors. Under those circumstances the melted copper

> expands by a factor of 67,000 times when it turns from a solid to a vapor. The danger associated with this expansion is one of high pressures, sound, and shrapnel. The high pressures can easily exceed hundreds or even thousands of pounds per square foot, knocking workers off ladders, rupturing eardrums, and collapsing lungs. The sounds associated with these pressures can exceed 160 dB. Finally, material and molten metal is expelled away from the arc at speeds exceeding 700 miles pour hour, fast enough for shrapnel to completely penetrate the human body. [1]

The potential for electrical shock is very high when anyone attempts to work on energized electrical equipment or systems without proper training, safety equipment, or locking out and tagging out that equipment. The electrical current levels associated with electrical shock are measured in milliamperes or one-thousandth of an ampere (0.001 Amps). Table 7.1 lists the electrical current levels normally associated with electrical shock and the typical body reaction to those current levels.

An individual's sex, body weight, the electrical current path through body, and skin moisture levels will have a direct affect on the injury sustained from an electrical shock. Body current paths from hand to hand and hand to foot will allow electrical current flow near the area of the heart. Foot-to-foot current paths should not pass directly through the heart region. Electrical current flows through the body's nervous system, muscles, and the blood system. Current flow can damage organs and have long-term effects on the body, which may not be readily visible at the time of the shock event. Electrical shock current levels also can differ between men and women and also between thin and heavier individuals.

Electrical accidents resulting from catastrophic equipment failure can be caused by manufacturing or design problems; improper application of equipment with regards to

TABLE 7.1 Electrical shock current levels

Current range (mA)	Physiological effect	Condition description
1	Threshold of perception	Detect a slight tingling sensation in hands or fingertips
1–9	Let-go threshold	Unpleasant sensation but muscle control not impaired. Women's maximum threshold level is 6 mA
9–25	Muscular contraction	Painful and hard or impossible to release energized object in hand
26–59	Muscular contraction	Breathing difficult
60–100	Ventricular fibrillation	Heart stoppage, respiration inhibition, death possible

Note: Effects at different current levels may differ with body weight and size.

operating conditions, and the available power supply, voltage and current harmonics; problems with supporting safety equipment; improper equipment installation; or failure to maintain the equipment per the manufacturer's recommendations.

Safe Operating Procedures

The development of electrical safe operating procedures requires a detailed evaluation of the facility's electrical system and its relationship to any production process, manufacturing process, chemical process, and transportation facilities. One means of evaluating a facility is through a hazard analysis. The Occupational Safety and Health Administration, United States Department of Labor, Occupational Safety and Health Administration (OSHA), in Title 29, Part 1910, Section 119 (29 CFR 1910:119) provides guidance and requirements for establishing safe operating procedures.

An electrical hazard analysis should identify potential problems which may occur during normal facility operation. Key documentation required in developing this analysis includes, but is not limited to the following:

Electrical One-Line Diagrams

Electrical Hazardous Area Classification Diagrams

Process and Instrumentation Diagrams (P&ID)

Equipment Arrangement Diagrams

Building Plans

Safety Analysis Function Evaluation (SAFE) Charts

Material Safety Data Sheets

Motor Control Wiring Diagrams

Control Panel Wiring Diagrams

Wiring Diagrams for Fire and Safety Panels

Cause and Effect Charts

Equipment Specifications

Equipment Manufacturer's Data Books

Obviously, some of the material listed above may not be always applicable. Hazardous Area Diagrams and P&ID Diagrams would only be applicable for petroleum and chemical processing, production, or transportation facilities.

OSHA, 29 CFR 1910.119, Appendix C.5 defines Operating procedures as:

> tasks to be performed, data to be recorded, operating conditions to be maintained, samples to be collected, and safety and health precautions to be taken. The procedures need to be technically accurate, understandable to employees, and revised periodically to ensure that they reflect current operations.

It further states:

> operating procedures will include specific instructions or details on what steps are to be taken or followed in carrying out the stated procedures. These operating instructions… should include the applicable safety precautions and should contain appropriate information on safety implications …

Safe operating procedures should establish specific tasks that must be performed. They must also establish specific safety procedures for the tasks to be performed. Safe operating procedures should also establish a work permitting process and the necessary procedures to implement it.

The American Petroleum Institute has developed several standards, guides, and recommended practices dealing with facility hazard analysis. Some of those documents include:

API RP 14J: *Recommended Practice for Design and Hazards Analysis for Offshore Production Facilities*

API RP 752: *Management of Hazards Associated with Location of Process Plant Buildings*

API RP 753: *Management of Hazards Associated with Location of Process Plant Portable Buildings*

Work Task Permit Requirements

Permits are a means of coordination and communication of proper safeguards between departments, operating groups, outside contractors or service personnel. These documents establish fixed safety procedures when conducting certain tasks. *Authorization permits* may be written or oral, depending upon the potential hazard of the task. These permits typically include a definition of responsibilities, a fixed time period, emergency procedures, exceptions, the requirements for daily safety meetings, and proper job planning.

Written permits may cover the following tasks involving electrical work:

Fire and Safety Work. This may include welding, grinding, burning, blasting, or cutting of structural members, supports or walls; work on energized conductors or equipment; use of heat or spark producing equipment in classified hazardous areas; or use of vehicles and cranes near power lines.

Confined Space Entry. These permits are normally associated with entry into tanks or other confined spaces. However, this permit would also be required for inspection, repair of electrical equipment in energized switchgear enclosures, vaults, or manholes; entry into assembly line equipment for maintenance or repair.

Switch Opening or Closing. Operation of switches on distribution systems may require a permit to prevent major process equipment shutdown or the removal of power from critical circuits or areas.

Excavation. These permits are essential for digging below grade, particularly in process areas that have been in service for substantial period of times. Sloping and benching systems for trenches greater than 20 feet in depth must be designed by a registered professional engineer. (OSHA, 29 CFR 1926.652[b], App. B and App. F).

Cutting or Breaching of Fire Walls or Dikes. Installation of new cable, conduits, or ducts in existing process areas or buildings will typically require these permits.

Smoking. Establishment of temporary smoking areas for personnel would require issuance of these permits, particularly in or near process areas where combustible or flammable liquids, vapors, gases, dust, powders, solid material, or flyings may be present.

Use of Cranes or Manlifts near Power Lines. Permitting of this work may require additional safety measures such as removing power from and the grounding of lines which may become energized, installation of overhead line insulating sleeves, or the placing of barricades.

Work on Energized Circuits. The NFPA's *Electrical Safety in the Workplace*® (NFPA 70E®-2009), Section 130.(B)(2), describes minimum required elements of a work permit for work on energized electrical circuits. It lists the contents of that permit as:

(a) A description of the circuit and equipment to be worked on and their location

(b) Justification for why the work must be performed in an energized condition [Section 130.1]

(c) A description of the safe work practices to be employed [Section 110.8(B)]

(d) Result of the shock hazardous analysis [Section 110.8(B)(1)(a)]

(e) Determination of shock protection boundaries [Section 130.2(B) and Table 130.2(C)]

(f) Results of the flash hazard analysis [Section 130.3]

(g) The Flash Protection Boundary [Section 130.3(A)]

(h) The necessary personal protective equipment to safely perform the assigned task [Sections 130.3(B), 130.7(C)(9), and Table 130.7(C)(9)]

(i) Means employed to restrict the access of unqualified persons from the work area [Section 110.8(A)(2)]

(j) Evidence of completion of a job briefing, including a discussion of any job-specific hazards [Section 110.7(G)]

(k) Energized work approval (authorizing or responsible management, safety officer, or owner, etc.) signature(s)

Written permits are normally approved only after authorized personnel have checked conditions in the area being permitted, usually immediately prior to initiation of work. Another purpose of the permit is to make supervisory and operating personnel in the vicinity of the permitted area to be aware of the task being done, should that work interfere with the normal operation of that facility. Permits are typically time-limited and list only specific tasks that have been approved. Several signatures may be required on the permit, including the initiator, operating and inspection supervisors, and the person doing the work.

Oral permits might include tasks such as:

Automobile or vehicle entry

Erection or take-down of scaffolds

Minor instrument calibration or repair

Minor machinery adjustment

Product sample collection

Housekeeping

Oral permits should not be utilized where rotating electrical machinery is being inspected. They should be used sparingly and limited to routine tasks that do not involve exposure of energized spark-producing or high-temperature components, static electricity generating tasks, exposure to hazardous gases, vapors, or dust, etc.

Documentation Requirements

Documentation for safe operating procedures can include company-written minimum safety and operating procedures, equipment manufacturers' or vendors' operating and maintenance instructions, written permits, hazard analysis studies, equipment safe operating certification, checklists, reports, or personnel training/certification records. No one enjoys creating or maintaining a paper trail. However, documentation is necessary to assure compliance with safety requirements; establish lines of communication and authority between operating, maintenance, and administrative personnel; verify personnel training for specialized tasks; and ensure loss prevention.

Lockout/Tagout Procedures

Lockout/tagout procedures are critical in assuring safe operating procedures. These procedures assure prevention of accidental energization of de-energized electrical equipment. ANSI Z244.1: *Lockout/Tagout of Energy Sources – Minimum Safety Requirements* provides detailed guidance for establishment of lockout/tagout procedures. OSHA, 29 CFR 1910.147 covers lockout/tagout procedures (see Chapter 6). Appendix A of that document provides a typical minimal lockout procedure.

Failure to implement lockout/tagout procedures can lead to tragic consequences. Failure to observe lockout/tagout procedures during internal inspection of electronically controlled production or assembly line equipment, and instead relying solely on the equipment's computer control scheme, can have serious consequences. The purpose of lockout/tagout procedures is to prevent unintentional energization of equipment while personnel are involved with the operation, repair, or maintenance of electrical equipment. This is accomplished by the removal of mechanical, pneumatic, electrical, hydraulic, or other equipment energy sources and prevents their re-energization while personnel are working inside or outside of that equipment. The process involves placing of a lockout/tagout device(s) on that main energy source disconnect mechanism, preventing its accidental operation. It also includes installing individual locks by all responsible designated personnel involved with the equipment, its repair, maintenance, or installation to prevent accidental energized.

Safety System Bypassing

Often testing of safety systems may require bypassing of some systems to prevent an unplanned process or production line shutdown. An example of this might be the quarterly testing of fire and gas detection systems. Extreme care must be exercised so that the bypassing operation only lasts for a minimum time. Visual or auditory indication, i.e., flashing lights, beacons, annunciator actuation, or signage, should be in place, warning that the safety system has been bypassed. This is necessary to assure that the safety system is maintained in the bypassed mode for only a short period of time.

It is critical that coordination with operational personnel must be maintained during the bypassing of safety systems. Personnel must be made aware of process conditions during bypass, so that abnormalities can be detected prior to the development of an emergency condition. Personnel must be in a position to monitor the safety device's function and manually perform that device's function in an emergency condition. Personnel performing the bypass monitoring work must be specifically trained and certified for that job.

Some electrical safety equipment is designated to prevent accidental shutdown during periodic testing programs. Combustion gas detector systems typically automatically bypass their shutdown circuits during calibration testing with gas samples. Other electrical control systems can have timed bypass circuits that temporarily lockout shutdown devices during operational testing. Consult equipment manufacturer's technical information to determine if this capability is available with a specific piece of equipment.

Operating or Energized Equipment Work Procedures

Electrical work on or near energized conductors or operating equipment requires substantial planning, detailed safety procedures, and equipment, and specific personnel training. These concerns are most often associated with work on high-voltage overhead power distribution systems, covered under ANSI/IEEE C2-2007, *National Electrical Safety Code*® (NESC) [3]. Work on operating electrical equipment that cannot be de-energized for reasons of increased or additional hazards or feasibility would also fall under this category.

Safe operating procedures should establish rules and/or guidelines under which work on energized or operating equipment can be done without de-energizing the equipment. Operating procedures should also establish minimum personnel training requirements, safety equipment, and procedures necessary to complete the task.

The *National Electrical Code*® (NEC®) (NFPA 70®-2008) deals with this safe operating procedure by listing personnel as either qualified or unqualified. NFPA 70, Article 100 Definitions, defines a *Qualified Person* as

> One who has the skills and knowledge related to the construction and operation of the electrical equipment and installations and has received safety training to recognize and avoid the hazards involved.

In a footnote it refers the reader to NFPA 70E to ascertain the electrical safety training requirements. Section 110.6(D)(1) Employee Training – Qualified Person and 110.6(D)(2) Unqualified Persons outline the training requirements for these personnel.

NFPA 70E establishes in Section 110.6(D)(1) the requirement that only a qualified person may work within the limited approach boundary. By means of training and experience these individuals have acquired special knowledge about electricity and its hazards. They have the knowledge to recognize exposed circuits, determine its voltage, and are knowledgeable regarding its construction and operations. Their training gives them the knowledge to determine when and what personal protection equipment (PPE) is required to do the work. They should also be capable of detailed planning of the work to be done and its careful execution, as well as stopping and replanning if circumstances develop which were not previously accounted for.

Safety Inspection and Testing Requirements

Written procedures should be developed for inspection and testing of electrical equipment. Some documents which may be helpful in power distribution equipment testing and inspection for industrial, commercial, and residential facilities are shown in Table 7.2.

Equipment manufacturers testing requirements are ideal sources for testing procedures. Industry standards and recommended practices provide additional assistance in developing test procedures. Frequency of testing, checklists, and permanent test data tracking information are critical components in developing test procedures.

Work Experience and Training Requirements

Operating procedures should specify the minimum requirements for qualifications to operate, maintain, or repair specific electrical equipment. A qualified person is one familiar with the construction and operation of the specific equipment and its associated hazards. Qualification should be through both job experience and formal training. Reliance solely with on-the-job-training can sometimes be unwise. Some job-specific tasks might even require training certification by outside agencies. OSHA provides electrical training requirements, in 29 CFR 1910.332, for personnel in generation, transmission, and distribution installations; communications installations; installation in vehicles (ships, watercraft, aircraft, automobiles, railway rolling stock and automobile vehicles other than mobile homes and recreation vehicles); and railway installations.

Safety Equipment Requirements

Work on some energized electrical equipment may require more rigid safety equipment and procedures than on other equipment, because of their extreme operating condition differences, such as voltage levels. This is particularly true with medium- and high-voltage power distribution equipment when compared to 600 Volt equipment. Operating procedures should provide specific minimum guidance on safety equipment, its certification, and use. Insulating mats, gloves, sleeves, line hoses, blankets, etc. may require periodic recertification. Minimum voltage levels for test equipment and tools should be established and adhered to.

The *National Electrical Code* requires using ground fault interrupter devices on outdoor receptacles. This should be included as a safe operating procedure; including the installation of

TABLE 7.2 Electrical power distribution equipment testing and inspection standards

Developer	Standard No.	Title
ASSE	ANSI Z244.1	American National Standard for Personnel Protection – Lockout/Tagout of Energy Sources – Minimum Safety Requirements
ANSI	ANSI Z535.5	Accident Prevention Tags
ASTM	ASTM F855	Standard Specifications for Temporary Protective Grounds to Be Used on De-energized Electric Power Lines and Equipment
IEEE	IEEE 141	IEEE Recommended Practice for Electric Power Distribution for Industrial Plants (IEEE Red Book)
IEEE	IEEE 241	IEEE Recommended Practice for Electric Power Systems in Commercial Buildings (IEEE Gray Book)
IEEE	IEEE 242	IEEE Recommended Practice for Protection and Coordination of Industrial and Commercial Power Systems (IEEE Buff Book)
IEEE	IEEE Std. 902	Maintenance, Operation, and Safety of Industrial and Commercial Power Systems
NEMA	NEMA AB 4	Guidelines for Inspection and Preventive Maintenance of Molded Case Circuit Breakers Used in Commercial and Industrial Applications
NEMA	ANSI C29.1	Test Methods for Electrical Power Insulators
NEMA	ANSI C37.50	Switchgear Low-Voltage AC Power Circuit Breakers Used in Enclosures – Test Procedures
NEMA	ANSI C37.52	Test Procedures for Low-Voltage AC Power Circuit Protectors Used in Enclosures
NEMA	ANSI C37.54	For Indoor Alternating Current High-Voltage Circuit Breakers Applied as Removable Elements in Metal-Enclosed Switchgear – Conformance Test Procedures
NEMA	ANSI C37.55	American National Standard for Switchgear – Medium-Voltage Metal-Clad Assemblies – Conformance Test Procedures
NEMA	ANSI C37.57	American National Standard for Switchgear – Metal-Enclosed Interrupter Switchgear Assemblies – Conformance Testing
NEMA	ANSI C37.58	American National Standard for Switchgear – Indoor AC Medium-Voltage Switches for Use in Metal-Enclosed Switchgear – Conformance Test Procedures
NEMA	ANSI/NEMA WC 61	Transfer Impedance Testing
NETA	NETA ATS	Acceptance Testing Specifications
NETA	NETA MTS	Maintenance Testing Specifications
NFPA	NFPA 70B	Recommended Practice for Electrical Equipment Maintenance
NFPA	NFPA 70E	Standard for Electrical Safety Requirements for Employee Workplaces
NFPA	NFPA 72	National Fire Alarm Code
UL	UL 1244	Electrical and Electronic Measuring and Testing Equipment

a portable ground fault interrupter device should the outdoor circuit be supplied from only a fuse or circuit breaker. All extension cord sets should be submitted to inspection and testing as is required in NEC Article 590.6(B)(2)(a) and (b). Lockout/tagout equipment should be considered basic essential equipment. NFPA 70E provides specific guidance for safety equipment usage requirements.

Static Electricity Generation Prevention

API RP 2003: *Protection Against Ignition Arising Out of Static, Lightning, and Stray Currents*, provides specific information for use in operating procedures. NFPA 77, *Recommended Practice on Static Electricity* also provides guidance with static electricity. In any facility where the presence of combustible and flammable vapors or dusts may be present, the control of static electricity is crucial. Operating procedures against static electricity buildup is particularly important with tank truck, aircraft, tank car, and marine vessel loading and unloading of petroleum products. Some refined and unrefined petroleum products have properties that tend to maintain static charges, particularly under high flow rates. Operating procedures should provide specific guidance based on the material and the circumstances typically found during its handling.

Fire Watch Requirements

A hot work permit might require the use of fire watch procedures to constantly monitor a permitted hazardous classified area for work which may produce possible heat, arc, or spark-producing events. The monitoring process may include the use of gas detection equipment to monitor for quantities of flammable or combustible gases or vapors. The fire watch might also mandate the use of a constant fire water spray near the area of work should an accidental fire occur. Maintenance of area security and warning signage is also a required of fire watch personnel.

Fire watch procedure development is critical where there is welding, cutting, power grinding, or arc gouging on metal supports and equipment, and when opening energized enclosures in areas designated as hazardous (classified) locations. Areas classified as Class I, Divisions 1 and 2 are most commonly found in petroleum and petrochemical installations and will generally require fire watch procedures for specific types of work tasks. Fire watch procedures should be required as part of the work permitting process.

These procedures should designate the minimum job classification and training of all fire watch personnel involved with the process. Personnel should have been specifically instructed in the use of fire watch equipment and procedures. This will usually require one or more persons equipped with a fire hose and combustible gas monitoring equipment. The procedures should also designate safeguards to assure that other adequate fire fighting equipment is on the job site. The type and quantity of equipment can vary depending on the area and materials being handled. Emergency communications and alarm and shut-in (where applicable) procedures should be part of the fire watch. The procedures should provide guidance for handling small releases of flammable or combustible materials. The fire watch personnel should have no duties other than monitoring the area for the presence of flammable or combustible materials, shutting down the work, and fire fighting.

The fire watch team should monitor the permitted work area prior to initiation of the work and remain on site at least 30 minutes after work completion. Their fire fighting equipment should consist of a charged fire hose, if possible, as well as fire extinguisher equipment. The gas detection monitoring equipment should be portable and suitable for the normal hazardous (classified) area designation of the area in which the work is being performed.

Minimum Lighting Levels

Depending on the type of permitted work being performed, the location and time of day, supplemental lighting may be required. NFPA 70E notes in Article 130.6(C)(1) that

> Employees shall not enter spaces containing live parts unless illumination is provided that enables the employees to perform the work safely.

NFPA 70E does not define adequate illumination. OSHA 29 CFR 1910.333(c)(4) reflects, almost verbatim, the requirements noted above in NFPA 70E. The OSHA requirements also do not define adequate illumination. OSHA does list as a national consensus standard ANSI A11.1: *American National Standard Practice for Industrial Lighting* in the reference chapter of 29 CFR 1910. That document contains recommendations for lighting in the workplace. OSHA 1910 does not specifically reference that document in 1910.333. It should be noted that ANSI now has another standard which appears to have replaced ANS1 A11.1. That document is IESNA BSR RP-7-01: *Recommended Practice for Lighting Industrial Facilities*

API RP540: *Electrical Installation in Petroleum Processing Plants* and API RP 14F: *Design and Installation of Electrical Systems for Offshore Production Platforms* both provide general guidance for minimum recommended lighting levels for safety and visual tasks. There are general area lighting levels and do not reflect recommended levels for permitted areas.

The Illumination Engineering Society of North America (IESNA) provides some guidance in its *Lighting Handbook* for ranges of lighting levels for generic interior activities. Those lighting levels should be analyzed carefully for a specific work task requirement. Some general tasks illumination values are presented in Table 7.3.

Compliance Audits

OSHA 29 CFR 1910(o) and Appendix C, Section 14 of that document provide specific recommendations for compliance audits. An audit is a technique used to gather sufficient facts and information to verify compliance with safety procedures. It should also document areas

TABLE 7.3 Selected Task Lighting Level Ranges

Activity or Task	LUX range
Performance of Visual Tasks of High Contrast or Large Size	107.7 to 538.5
Performance of Visual Tasks of Medium Contrast or Small Size	538.5 to 1077
Performance of Visual Tasks of Low Contrast or Very Small Size	1077 to 2154

where corrective-action is required, as well as areas where process safety management is effective. The corrective-action recommendations should identify required planning, follow-up, and documentation requirements

Safe Work Practices

NFPA 70E and OSHA 29 CFR 1910.331 provide guidelines for implementation of electrical safe work practices. These documents cover the following safety related work practices.

Lockout/Tagout

Lockout/tagout safe work practices assure the safe secured isolation of energy sources that could endanger employees while working in, around, or on machines or equipment during repair, operation, maintenance, adjusting, inspection, or other connected activities. This safe work procedure is defined in detail in ANSI Z244.1: *Standard for Personnel Protection – Lockout/Tagout of Energy Sources – Minimum Safety Requirements.*

OSHA 29 CFR 1910.147 provides additional guidance for lockout/tagout procedures. The Appendix in 1910.147 even has a sample lockout/tagout procedure that can be used or modified. NFPA 70E also has a section on lockout/tagout procedures.

The lockout/tagout procedures basically require that a survey be conducted on the affected equipment or machines to identify and locate all energy sources. All energy sources isolating devices shall also be located and checked against the known energy sources to assure lockout/tagout can be successfully accomplished. Available drawings can be used to aid in that determination. Distinctive lockout/tagout devices should be used only for those procedures. Those devices should not be utilized to normally keep the equipment/process out of service for reasons other than personnel protection.

All personnel directly affected by the equipment/process lockout/tagout should be notified prior to the implementation. A written checklist should be prepared, reflecting the development order of energy isolating device actuation, clearance, release, reactivation, and required personnel approvals. Removal of the lockout/tagout devices is as important as its implementation. Assurance should be made to verify all personnel, tools, equipment, and parts have been removed. All personnel who installed lockout devices should be the individuals to remove those same devices, unless there has been shift change or that responsibility has been officially delegated to someone else. Situations can develop where the equipment must be re-energized for testing before being replaced into service. Such requirements should be coordinated with all affected personnel.

Work on Energized Equipment

Work on energized equipment is sometimes required if removing the equipment from active service presents a greater danger or hazardous condition. Work on energized equipment should only be done by qualified personnel who have received specialized training on this type of equipment or machinery. This work will also require the use of Personal Protective Equipment (PPE), insulating and/or shielding materials, and insulated tools.

Insulated tools and equipment should be utilized to prevent any inadvertent contact between energized parts, tools, and ground. Insulated equipment includes non-conductive, fiberglass, portable ladders; insulated blankets or covers; insulated line sleeves; bucket or lift trucks; etc. Work on energized lines or equipment in confined work spaces requires special precautions to prevent the inadvertent contact of energized parts, elimination of collected water hazards, and the assurance of a reliable and safe air source.

Work on exposed, energized overhead electrical lines is very common, since this is the most common utility distribution and transmission method. NESC provides guidance for working on energized lines and equipment. OSHA 29 CFR 1910.269 provides safe work practice guidance in this area. NFPA 70E also provides detailed safe work practices on energized equipment.

NFPA 70E also requires Flash Hazard Analysis for any work on energized equipment. That analysis is done to protect personnel from an arc flash event. The analysis establishes a Flash Protection Boundary, as well as the PPE required if working within that boundary. The boundary established is the distance from a potential arc fault point that will expose personnel to second degree burns. Reference the NFPA 70E *Handbook* for a more detailed explanation of that process. As of 2008, OSHA 29 CFR 1910 had not yet mandated that a Flash Hazard Analysis would be required before any work is done on energized electrical equipment.

Clearances and Approach Distances

There are many source documents that provide clearances and approach distances for electrical safe work practices. The *National Electrical Code*® deals with minimum clear work space requirements, as is illustrated in Table 7.4.

NESC® establishes minimum approach distances for work on energized parts for both Communications Employees and Utility Employees. Those distances are presented in tables in

TABLE 7.4 NEC clear work space requirements

NEC 2008 Article	Description
Article 110.26	Spaces About Electrical Equipment – 600 Volts, Nominal or Less
Article 110.32	Work Space About Equipment
Article 110.33	Entrance and Access to Work Space
Article 110.33	Work Space and Guarding
Article 110.72	Cabling Work Space – Manholes and Other Electrical Enclosures Intended for Personnel Entry, All Voltages
Article 110.73	Equipment Work Space – Manholes and Other Electrical Enclosures Intended for Personnel Entry, All Voltages
Article 408.18	Clearances – Switchboards

Sections 43 and 44 respectively. OSHA 29 CFR 1910.269 also provides requirements for minimum approach and safe work distances. IEEE Standard 516: *Guide for Maintenance Methods on Energized Power Lines* also provides minimum approach distance data.

Operating voltages determine minimum approach distances and clearances. OSHA 29 CR 1910.269 defines the minimum approach distance as

> the closest distance an employee is permitted to approach an energized or a grounded object.

These distances are designed to allow work to be done safely without the risk of electrical flashover. They also include an additional separation distance factor for inadvertent movement compensation for the worker, relative to an energized part. This adder is called the ergonomic component and is included in the OSHA minimum approach distance. Minimum approach distances also take into account the maximum over-voltages or surges expected on a system from line faults, operation of switches, or breakers, etc.

The OSHA definition of the minimum approach distance includes the term *"grounded object"* as well as *"energized object"*. Personal contact between the energized object and the grounded object will allow electricity to flow through the object making contact, be it animal, human, or a conductive object such as a metal pole. The grounded object referred to can also be an electrically conductive structural support, walkway, enclosure, etc. that may also be come energized during a faulted condition to ground. A line-to-ground fault on a grounded object can impress voltage on that object.

Minimum clearance distances are established for overhead lines by NESC. That clearance reference is actually in both the horizontal and vertical directions. The code establishes minimum horizontal separation distances between adjacent conductors. That distance is a function of voltage potential difference. The vertical separation distances are affected by location and the type of traffic which might pass under the lines.

Alerting Techniques

Alerting techniques include safety signs, tags, alarms, flashing lights, barricades, manual signaling, or attendants. These devices or methods are utilized to warn or protect personnel from exposed hazards such as electrical shock or burns. OSHA 29 CFR 1910.145 provides requirements for accident prevention signs and tags. Barricades are used to limit or impede access of unqualified employees to un-insulated, energized conductors or surfaces.

Energized and De-Energization of Power Circuits

NESC Section 444 De-energizing Equipment or Lines to Protect Employees provides guidance for employee protection when de-energizing and re-energizing equipment or lines. Those procedures include designation of one person to direct the operation of all switches and disconnects. They also require that once de-energized, the switches and disconnects will be rendered inoperable with lockout/tagout devices. Protective grounds should be installed on the disconnected lines or equipment for personnel protection in case of an accidental re-energization. Reference NESC, Section 445 for protective grounds placement requirements. Those same topics are also included in OSHA 29 CFR 1910.269.

A de-energized circuit should not be re-energized until it is determined that the equipment or circuit can be safely re-energized. Re-energization should not occur until all appropriate personnel are notified, temporary ground protection conductors are removed from the lines and equipment, all personnel and equipment have been removed to or beyond minimum approach distances, and lockout/tagout devices have been removed.

Work Near Overhead Power Lines

Work in areas near overhead power lines, not involving those overhead lines, can be extremely hazardous if cranes, booms, or long sections of conductive objects, such as pipe, structural members, etc. are being handled, offloaded, or installed. Electrical safe work practices may require the installation of protective sleeves or guards on overhead power lines. This should be determined in advance of any activity in the area by a site safety survey, so that the overhead line operator can be notified and has time to take protective measures or de-energize the line.

NFPA 70E establishes Limited Approach Boundary distances in Article 130.5(E)(1) for vehicle or mechanical equipment structure which will be elevated near overhead lines. It also provides some exceptions in which the approach distance may be limited to the Restricted Approach Boundary. Minimum distances for aerial lifts, mobile cranes, and derrick trucks from exposed energized overhead lines are presented in 29 CFR 1910.333(c)(3), 1910.67(b)(4)(i), 1910.181(j)(5), and 1910.268(b)(7).

Employees working on the ground, near aerial lifts, mobile cranes, or derrick trucks near energized overhead, exposed lines, should *ALWAYS* consider exposed conductive parts of that equipment or any conductive material being handled by it as if it is energized. Depending on the circumstances, it may be suitable to require personnel working near this equipment to use protective clothing or equipment suitable for the overhead line to ground voltage involved. Consideration should also be given to step voltage potential hazards near this equipment if contact is made with an energized conductor or surface.

Confined Work Spaces

Manholes and transformer vaults are two examples of confined work spaces with electrical equipment. Confined work spaces containing exposed, energized parts require precautions to prevent inadvertent contact with exposed, energized parts. Doors should be secured to prevent accidentally knocking personnel into exposed, energized parts. Supplemental lighting, an exterior fresh air source, and water removal equipment may also be required. Use of personal protection equipment (PPE), including gloves, face shields, flame resistant clothing, etc. is also required. Reference NFPA 70E and OSHA 29 CFR 1910 for specific requirements.

Enclosed spaces in petroleum and chemical facilities may contain hazardous atmospheres from process leaks, particularly if the spaces are located below grade in process areas. The enclosure atmosphere should be tested for the presence of flammable gases and vapors, as well as oxygen content. If flammable gases or vapors are detected or if an oxygen deficiency is found, forced air ventilation will be required, from a clean, unclassified area or air source. A continuous monitoring program should also be established to insure notification should flammable gases or vapors be released and detected. Reference OSHA 29 CFR 1910.269(t) for underground electrical installation work requirements.

Conductive Materials, Equipment, Tools, and Apparel

When working on or near exposed, energized conductors or equipment, special precautions will be required. Conductive articles of jewelry or clothing, including watches, rings, necklaces, cloth with conductive thread, etc., may not be worn. Conductive barrels of screwdrivers, nut drivers, or other tools may require insulation coverings suitable for the applied voltage. Materials such as pipes, ducts, steel tapes, chains, etc. will require special handling near energized conductors or equipment.

Housekeeping Duties

Work in confined spaces or near energized, exposed conductors or equipment will require constant attention to prevent inadvertent contact with stored or handled materials. Unnecessary equipment and materials should be removed from these areas.

General housekeeping and/or janitorial duties should not be performed near energized conductors or equipment unless adequate safeguards have been implemented. Dusting or cleaning should not be performed on energized disconnect switches or circuit breakers. Cleaning materials such as water, conductive cleaning fluids, steel wool, conductive cloth, etc. should not be used near exposed, energized conductors or equipment. General housekeeping duties on or near exposed, energized conductors or equipment should only be scheduled when those conductors or equipment can safely be de-energized, locked out, and tagged out.

Protective Equipment and Tools

Protective equipment will isolate personnel from exposed, energized conductors or equipment. This equipment includes clothing, blankets, head protection, eye and face protection, hand and body protection, line hoses, sleeves, mats, hot sticks, etc. Safe work practices may also require periodic inspection and testing of medium- or high-voltage equipment to assure its integrity. It is also important to assure that the protective equipment is suitable for the operating voltage level present.

Visual inspection should be conducted on all equipment prior to its use. Some medium-voltage tools may even require daily or twice-daily inspections. Extension cords and power service cords should be checked for insulation integrity and suitability for the environment or area in which it is be used. Ground continuity integrity of all extension cords and service cords, including the condition of ground blades on power plugs should also be inspected. Power tools used outdoors should be utilized in conjunction with GFCI protective devices. GFCI devices should be tested each time prior to their use. That test involves depressing the outlet device's "Push-to-Test" button and verifying tripping of the device.

The Standards in Table 7.5 are recommended when selecting, testing or caring for personal protective equipment (PPE) and high-voltage tools and insulation devices.

Installation, Operation, and Maintenance Considerations

Based on years of personal experience in the investigation of electrical accidents, the following information reflects the examination of several work tasks which have resulted in shock, electrocution, or other personal injuries to individuals who failed to follow common safe work practices when dealing with electricity or electrical equipment.

Welding

Welding-related accidents can occur in areas where hydrocarbons may be present if safe work practices are not followed. Welding sparks and hot slag can ignite combustible and flammable gases or vapors in the vicinity of the welding activity hazardous classified areas.

TABLE 7.5   Some high-voltage PPE and equipment testing and service care Standards

Developer	Standard No.	Title
ASTM	ASTM D120	Specifications for Rubber Insulating Gloves
ASTM	ASTM D178	Specifications for Rubber Insulating Matting
ASTM	ASTM D1048	Specifications for Rubber Insulating Blankets
ASTM	ASTM D1049	Specifications for Rubber Insulating Covers
ASTM	ASTM D1050	Specifications for Rubber Insulating Line Hoses
ASTM	ASTM D1051	Specifications for Rubber Insulating Sleeves
ASTM	ASTM F478	Standard Specification for In-Service Care of Insulating Line Hose and Cover
ASTM	ASTM F479	Standard Specification for In-Service Care of Insulating Blankets
ASTM	ASTM F496	Standard Specification for In-service Care of Insulating Gloves and Sleeves
ASTM	ASTM F696	Specification for Leather Protection for Rubber Insulating Gloves and Sleeves
ASTM	ASTM F711	Specification for Fiberglass Reinforced Plastic Rod and Tube Used in Live Line Tools
ASTM	ASTM F887	Specification for Personal Climbing Equipment
ASTM	ASTM F1116	Standard Test Method for Determining Dielectric Strength of Overshoe Footwear
ASTM	ASTM F1117	Specification for Dielectric Overshoe Footwear
ASTM	ASTM F1236	Standard Guide for Visual Inspection of Electrical Protective Rubber Products
ASTM	ASTM STP900	Performance of Protective Clothing
ASTM	ASTM Z87.1	Practice for Occupational and Educational Eye and Face Protection
ASTM	ASTM Z89.1	Protective Headwear for Industrial Workers
ASTM	ANSI/SIA A92.2	American National Standard Vehicle-Mounted Elevating and Rotating Aerial Devices

Electrical shock can be experienced when using AC or DC welding machines if contact is made between insulated energized surface being welded and a grounded conductive surface.

Consider the situation where a 250 ampere DC welding machine, with a 30 VDC full load output was sitting on a steel deck. Equipment pipe flanges were being welded to a steel vessel. The bare steel process vessel was placed on wooden blocks to allow its rotation and movement during the welding operation. The welding machine was set up to weld using its positive lead. Welding has not yet commenced; however, the negative or work lead was clamped to the process vessel, and the diesel-driven welding machine was operating.

A pipefitter simultaneously touches an insulated work piece being welded and a steel deck on which the work is being done, receiving an electrical shock and a second degree burn on his bare hand. During an investigation following the accident with similar circumstances recreated, a no-load voltage potential of 70 VDC is measured between the isolated steel vessel and the steel deck.

ANSI Z49.1, *Safety in Welding, Cutting, and Allied Process* provides guidance in preventing accidents such as these. The investigation revealed that the steel vessel was insulated for the deck and was electrically connected to the welding machine negative or work lead. That setup created a situation where the steel vessel became the negative electrode or plate of a giant capacitor between the insulated steel vessel and the steel deck. The deck became the capacitor's positive electrode or plate. The electrical charge buildup on the steel vessel was proportional to its surface area. When the pipefitter simultaneously made contact with his bare hands and another portion of his body between the steel vessel and the steel deck, the capacitor discharged its electrical energy causing a shock and burn.

The accident could have been prevented by simply grounding the steel process vessel and the welding machine frame to the platform steel deck. This could have been accomplished by placing the vessel directly on the steel deck or by the use of bonding jumpers to connect them.

Electrical welding machine skids are sometimes placed on wooden runners. This is often done to eliminate the electrical arcs sometimes seen between welding machine skid and a grounded steel deck on which it might be located. If the welding machine is a resistance welding type and has a transformer with its secondary grounded to the machine's frame, voltage potential differences can develop if the machine is on a grounded steel deck and its runners are insulated from the deck. Other potential electrical shock hazards possible during the use of an electric arc welding machine include [4]:

Overheating of welding cable insulation, due to the use of welding machines with outputs greater than the ampacity and duty cycle rating of the cables. This can be avoided by selecting arc welding equipment and leads appropriately sized for the job.

Special welding or cutting procedures requiring high open-circuit voltages or frequencies may damage lower-voltage-rated welding cable insulation. Suitably rated welding cable insulation should be used.

A maximum control voltage of 120 V should be used on portable control devices, such as push button stations, normally carried by the operator. If metallic, that control station housing should be electrically grounded with a separate ground conductor in the cable if the control voltage exceeds 50 V.

Inadequate work space can cause an operator to work in a cramped or lying position in physical contact with conductive objects. Conductive objects should be minimized from the near vicinity of the welder.

Water and perspiration on clothing, shoes, and gloves can lead to an electric shock if accidental contact is made with energized objects. Fire resistant (FR) clothing, gloves, and shoes should be used.

Metallic conduits containing energized electrical conductors, when used as a work lead return, may damage the wiring insulation inside the conduit, resulting in an electrical shock or an arc fault/short-circuit event inside the conduit. Never use a conduit as a part of the work lead return path.

Pipelines with threaded, flanged, bolted, or caulked joints, used as part of the work lead return path, can develop hot spots. These hot spots can lead to hidden fires or explosions. Pipelines with these joints should not be used as the work return path.

Electrical current passing through coiled welding cable can lead to overheating and insulation damage. Cables should be uncoiled before use.

Loose or dirty electrical connections can lead to local heating and stray electrical currents. All electrical connections should be tight and clean.

Unused, energized, and exposed metal and carbon electrodes can come in contact with personnel or conductive objects. Electrode holders not in use should be placed out of contact with personnel, conductive objects, or flammable liquid and compressed gas cylinders.

Fuel leaks from the welding machine prime mover can be ignited by welding sparks or slag. Welding machines should be turned off on the detection of a fuel leak.

Unused, operational welding machines can cause electrical shock from inadvertent contact with any of the leads. Unused welding machines should be turned off.

Electrical welding while standing on aluminum or steel ladders could result in inadvertent current flow through the ladder. Nonconductive ladders should be used in these applications.

Operator welding while standing in water could result in electrical shock by inadvertent contact with energized conductors or leads. Suitable protective boots and gloves, in good condition, should be used under these circumstances.

Batteries

Batteries play an integral part in the operation of many offshore facilities. They provide power to operate electrical protective relaying devices, power circuit circuit-breakers, emergency lighting, generator prime mover starters, public address (PA) systems, gas and fire detection equipment, uninterruptible power supply (UPS), etc. Battery maintenance can also cause battery accidents in the production of hydrogen and oxygen gases during over-charging. Sufficient quantities of these gases along with a sufficiently hot heat source or electrical arc can lead to an explosion.

Removal of battery charger or battery leads after charging can be potentially dangerous. Failure to turn off the battery charger before removing the charger leads can produce an arc at the battery terminals. Removal of battery terminals under load will also produce an arc, Dirt accumulation on the top of a battery, covered with water from overfilled cells can leave conductive paths between top post battery terminals. The long-term affect of stray current flow across the top of the battery can lead to ignition of the battery plastic housing. The battery case should be cleaned with bicarbonate of soda, along with wire brushing any corroded terminals.

Long-term terminal corrosion can lead to battery terminal failure, with resulting arcs and possible ignition of the battery plastic housing. Excessive "Float Voltage" from the failure of a battery charger's voltage regulator can lead to excessive generation of hydrogen and oxygen from the battery fluid. Under these conditions, an electrical arc near the battery could cause an explosion.

Battery servicing and replacement present the greatest exposure for personnel injury. Failures to wear protective face shields and fire resistant (FR) clothing could result in greater injury exposure should a simultaneous catastrophic event occur. Arc flash injury is also possible under fault conditions.

Motor Control

Epperly et al. [5] have documented the development of arcing faults in low-voltage, 600 V class motor control centers (MCC). These faults can be the result of excessive fault currents and loose bus-bar or terminal connections with overheating and resulting carbon buildup. Excessive vibration may loosen connections over a long period of time, particularly if bus overheating problems developed from added load or available fault current. Deterioration of bus insulation and airborne contamination of bus insulation, barriers, and supports can lead to phase-to-ground faults. Excessive pitting, misalignment, carbon deposits, and arcing on disconnect switch contacts can result in overheating on the contacts with the potential for arc fault failure to ground.

IEEE 141-1993 (R1999), *IEEE Recommended Practice for Electric Power Distribution for Industrial Plants* recognizes the potential for catastrophic failure in motor control centers. It recommends the following [6]:

> (a) Assure that all circuits are de-energized and locked out in accordance with lock and tag procedures.

(b) Assure that the area around the assembly is kept clean and free of combustibles at all times. This should be part of the day-to-day maintenance.

(c) Inspect buses and connections to be sure that all connections are tight. Look for abnormal conditions that might indicate overheating or weakened insulation. Infrared testing can identify hot-spots caused by loose connections without de-energizing the equipment.

(d) Remove dust and dirt accumulations from bus supports and enclosure surfaces. Use of a vacuum cleaner with a long nozzle is recommended to assist in this cleaning operation. Wipe all bus supports clean with a cloth moistened in a non-toxic cleaning solution. (Refer to manufacturer's instructions for recommended solvent.) Do not use abrasive material for cleaning plated surfaces, since the plating will be removed.

(e) The internal components should be maintained according to the specific instructions supplied for each device.

(f) Secondary wiring connections should be checked to be sure they are tight.

Use of NEMA 7 type explosionproof enclosures outdoors for motor starters, disconnect switches, or control panels can result in a potential for accidents or equipment failure, if the enclosure is not dual-rated NEMA 3R and 7. A NEMA 7 enclosure is designed to breathe, taking in moist air if not specifically designed to prohibit that phenomenon. The use of breathers and drains to prevent moisture accumulation will only work if those devices are periodically cleaned of algae and dirt buildup. Condensation accumulation in these enclosures may result in the development of a conductive path between any phase conductor or terminal and the enclosure. This can also lead to the mechanical failure of contactors and relays.

Another potential accident area with NEMA 7 enclosures with NEMA 1 motor starters in Class I, Division 1 or 2 hazardous areas is when they are opened for inspection and testing while still energized and sufficient accumulation of combustible or flammable gases or vapors develops near the enclosure. This can be avoided with the development of a work permitting system and the implementation of electrical safe work practices.

Medium- and High-Voltage Equipment

Work on energized high-voltage switchgear, switches, or unclassified conductors without the proper equipment and training can lead to accidental electrical shock. Safety-permitting procedures must also be implemented. Failure to properly sign and barricade opened switchgear cabinets can lead to inadvertent contact with distracted or unauthorized personnel. Fire resistant (FR) clothing, safety equipment, minimum safe working clearances, and minimum approach distances must be used while working on or near exposed, energized surfaces, terminals, or conductors. Failure to adhere to OSHA's electrical safety standards 29 CFR 1910.331 through 1910.335, and where applicable 29 CFR 1910.269 could also create the potential for an accident and possible OSHA mandated fines.

Molded Case Circuit Breaker Panels

A potential problem area with molded case circuit-breaker panels involves loose connections or terminals. Loose cable terminations or bus stabs can lead to arcing and carbon deposits at the connection point. It undetected over a long period of time, a glowing contact condition could develop with the development of a high-resistance connection from carbon deposits at the connection. This can cause the connection to become red-hot and the connection could eventually develop into an electrical fault event, with the potential of the development of an arc flash catastrophic event. Low level arcing at the bus/circuit breaker stab may not be detected by the panel's main, overcurrent protection device. In split-bus panels, where there are six or fewer primary circuit breakers, there may not be a main overcurrent protection device protecting the panel's primary bus. A fault of this type on an unprotected bus has the potential to lead to a catastrophic failure event.

An example of a glowing contact can be seen in Figure 7.1. This example involves a receptacle or socket outlet terminal screw and a copper wire terminated to it. The copper has melted and flowed evenly between the two arrows in the photograph. This receptacle fed a window type air conditioner. There was also a fire in the room where this receptacle was located. The remainder of the copper wire attached to the terminal did not melt, indicating localized heating at the screw terminal.

Installation of circuit-breakers with a fault interrupting rating less than the available fault current could lead to a fault inside the circuit breaker with resulting equipment damage and personnel injury. A low-resistance phase-to-phase fault will result in the generation of

Figure 7.1: Glowing contact example – receptacle terminal screw

electromagnetic forces proportional to the square of the peak current (I_p^2). A molded case circuit-breaker with an insufficient fault current interrupting rating can fail catastrophically under fault conditions exceeding its rating.

The use of a non-switch-rated circuit-breaker, for regular switching of fluorescent lighting circuits, could eventually lead to circuit breaker failure, possibly while being switched. Failure to use HACR rated circuit breakers on heating, air conditioning, and refrigeration equipment requiring HACR rated overcurrent devices may result in false tripping problems on electric motor startup,

Permanent removal of enclosure cover bolts from energized NEMA 7 enclosures should be prohibited. Some enclosure designs utilize a large number of nuts and bolts to secure their front covers. Removal and reinstallation of these bolts each time while troubleshooting an equipment problem might create the temptation to secure the enclosure with only a few bolts. Situations like that, coupled with an unexpected release of sufficient combustible or flammable gases or vapors, can lead to the potential for an explosion. Utilization of permitting systems can allow safe opening of energized enclosures for testing under controlled conditions.

Wiring Connections

Pressure-type connectors, known as wire-nuts, are very common splicing devices on small gauge wires, #14 AWG to #8 AWG. These devices can lead to potential failures and ignition if improperly installed. Manufacturers will rate screw type wire connectors in accordance with the number, type and size of the conductors that can be safely connected together. Use of too large or too small a wire connector, or failure to assure that the wire splice surfaces are clean, can create loose or high resistance connections. If allowed to persist, overheating can result, leading to melting of a plastic wire connector. Manufacturer's recommendations should be consulted for rated wire combinations for each connector type.

Stripping of wire insulation and wire composition (copper or aluminum) are important considerations for wire connector selection. Removal of excessive wire insulation for splicing will leave exposed conductor surfaces, creating the potential for short circuits or electrical shock. Cutting wire insulation too short will inhibit the ability of the wire nut to mechanically secure the wires. Cooper/aluminum (CO/ALR) rating is an important consideration when terminating a combination of those conductor types. Use of appropriately listed wiring compounds may also aid in the prevention of overheating of that connection.

Wiring terminals, split-bolt connectors, and other bolted or screwed wiring terminations can develop long-term potential failure areas if improperly torqued. Manufacturers' torque requirements should always be followed. A torque wrench should also be used to assure tightness requirements. If manufacturers' requirements are not available, UL Standard 486B, *Wire Connectors for Use with Aluminum Conductors* and UL 486-A, *Splicing Wire Connectors* will provide some guidance. Also the NEC 2008 *Handbook*, Article 110.14 Electrical Connections – Commentary, pages 45 through 48 provides some bolt torque recommendations.

Cord Sets and Attachment Cords

Abuse of cord sets (extension cords) can be a primary factor in personnel electrical shock and fire initiation potential. Cord sets should not be used to provide power to any fixed electrical load. Electrical cord size ampacity should be compared to equipment or tool electrical current requirements to assure that cord overheating will not occur.

Use of cord sets outdoors with broken or missing ground plugs is prohibited by the NEC. Often times the ground connection may be broken off purposely because of problems with insertion into an ungrounded receptacle. Cord sets should not be used to provide power to any fixed electrical load. This is not uncommon in residences or businesses with insufficient electrical circuits and receptacles.

Cord sets with damaged or cut insulation create the potential for electrical shock. Use of black electrician's tape for cord repair can lead to shock potential because that connection is probably not watertight or mechanically strong enough to protect the cord conductors from further damage. Routing a cord set through a door or window which may be opened or closed constantly can potentially lead to cord stranded conductor damage, leakage current between phase and ground or neutral, and the potential for a catastrophic short-circuit event with subsequent fire potential.

Cord sets used in wet locations create the potential for leakage current to ground from damaged insulation or attachment plugs not rated for damp or wet service. Cord insulation should be inspected prior to use. The use of portable ground fault interrupter (GFCI) devices in conjunction with cord sets will assure personnel shock protection. The GFCI devices should be tested before use each time to assure operability. Cords used outdoors should be marked *SUITABLE FOR USE WITH OUTDOOR APPLIANCES – STORE INDOORS WHILE NOT IN USE.*

Attachment cords for portable power tools or lights should be periodically checked. Damage to cord insulation, loss of ground conductor continuity, and removal of the power cord plug grounding prong all create the potential for electrical shock. Use of power tools, portable lights, or cord sets in wet or damp environments should be limited to only those devices that are listed as being rated for that service. Verification of unfamiliar receptacle (socket outlet) wiring connections should be done with a receptacle (socket outlet) circuit tester prior to use. Reversal of the hot and ground conductors at the receptacle or failure to provide a ground conductor or a broken ground conductor to the receptacle could result in electrical shock. Use of double-insulated tools in situations of excessive body sweat generation could possibly lead to electrical shock from perspiration developing a conductive path between the tool and the user.

The NEC®, in Article 590 Temporary Installations, Section 590.6(B)(2) Assured Equipment Grounding Conductor Program, outlines specific periodic testing programs for cord set ground conductors and plugs, as well as site receptacles not part of permanent building wiring. It also mandates maintenance test records to be made available to the Authority Having Jurisdiction. This program should be utilized on temporary wiring installation during construction, remodeling, maintenance, repair, and demolition activities.

Use of electrical portable power tools in electrical hazardous classified areas for which they are not rated could lead to the potential for an explosion should a simultaneous release of combustible or flammable gases or vapors occur. Receptacles should be inspected to determine the electrical plug configurations requirement necessary for portable equipment. Any electrical appliance inserted into a receptacle should be rated for the environment in which the receptacle is located.

Electrical Receptacles

Damaged, dirty or misused electrical receptacles (socket outlets) can lead to electrical shock or electrical short circuits. Corroded, deformed, and mechanically damaged receptacle stabs can result in potential electrical shock. Loose terminal connections on receptacles can lead to super-heated glowing contacts and catastrophic failure, fire, and/or electrical shock. Receptacles (socket outlets) should be rated for the electrical area classification in which they are used. NEMA 7 circuit-breaking before disconnecting receptacles and plugs assures the safe disconnect of portable appliances in hazardous areas.

The NEC mandates in Article 210.52 that receptacles (socket outlets) are to be installed so that no point along a wall is more than 1.8 meters (6 feet) from a receptacle outlet. Older residences may have an insufficient number of receptacles or socket outlets to handle the number of appliances commonly used in homes in the twenty-first century. Residents sometimes resort to the use of multi-plug outlet devices shown in Figure 7.2 to alleviate that problem. This solution, if not temporary, is a violation of several NEC requirements. A multi-plug outlet strip, like that shown in Figure 7.2 is a temporary wiring device. Anytime one is used as a permanent wiring solution, it is a violation of the National Electrical Code.

Secondly, it is possible that the total current flow of the loads plugged into that device could have exceeded the current rating of the device. Even though there was an integral circuit protective device on the appliance, the cyclic nature of some of the loads may have allowed it to provide power for some time between tripping. One of the appliance loads on the device was a refrigerator, which had a relatively normal running current, but high momentary starting current. Also notice, that the multi-plug outlet device had other 3-way plugs inserted to allow even more appliances to be served by the device. Lastly, the multi-plug outlet device was actually supported in midair by the appliance service cord plugs which were plugged into the device.

Figure 7.2: Example of a multi-outlet device being used as permanent wiring

Light Fixtures

Installation of pendent type light fixtures in high vibration industrial settings without a short length of flexible conduit could create a long-term vibration problem with the fixture and lamp. Floodlights on pivoted stanchions in hazardous classified areas should be de-energized before lowering the stanchion and fixture. Industrial walkway stanchion lights should be adequately supported to prevent excessive vibration. Permanent light stanchions should not be installed on removable handrails unless designed specifically for that purpose with a disconnect plug. Clamping of stanchion supports onto the top handrail without set-back accommodations could lead to loss of balance of personnel on stairs while maneuvering around the stanchion.

Installation of recessed light fixtures into insulated ceilings could lead potentially to a fire, unless the fixture is listed for that service. The NEC normally requires a protective ring structure around a non-rated recessed fixture to prevent insulation accumulation on and around the fixture, which would lead to overheating and possible fixture wiring failure and the constant burning-out of incandescent light bulbs. Installation of light bulbs with wattages above the maximum recommendations of the manufacturer can also lead to fixture overheating.

Rotating Equipment

Rotating equipment accident potential can develop from unintentional energization during maintenance, electrical shock during testing, insulation breakdown, or accidental contact of clothing or appendages on rotating parts during inspection. The absence of electrical bonding jumpers can create the potential for electrical shock. Stored electrical or mechanical energy associated with a motor must be released before any work can commence on the equipment. Devices such as capacitors can cause electrical shock from discharge by accidental contact.

Wiring Considerations

Failure to utilize insulated bushings on cable entry into equipment enclosures can result in damage to wire insulation, leading to the potential for electrical shock and a short-circuit event. The use of flexible conduit can prevent conduit failure from excessive vibration at a terminal box. Suitable bonding jumpers on the flexible conduit or separate ground conductors should provide protection from shock potential. Improper termination of feeder cables at the motor junction box can result in the potential for electrical shock or arcing faults.

Conductors with poly vinyl chloride (PVC) insulation used for direct current (DC) wet service above 40 VDC can deteriorate from electrical endosmosis; that process allows the passage of water through a porous partition, in this case the PVC insulation. This can result in insulation breakdown over the long term, with the potential for short-circuit conditions to develop. Conductors with thermosetting insulation should be used for this service.

Installation of metal-clad (Type MC) cable with a bending radius less that specified by the manufacturer or that specified by the NEC can cause damage to the metal sheathing and conductor insulation. This can result in insulation failure and short-circuit conditions.

Inadequate grounding of cable tray, conduits, and MC cable armor can increase the potential for electrical shock, should damage occur to a conductor's insulation. Pulling wires into conduit can lead to scraping of conductor insulation. Use of galvanized steel water pipe as conduit can damage wires being installed because of metal burrs left from the manufacturing process. Failure to install a pull-type conduit fitting for the equivalent of four 90° bends may require the use of excessive pulling force to install conductors and damage cable insulation.

The cable arc fault ground damage illustrated in Figure 7.3 was caused by fire damage. Similar damage could be caused it the conductor's insulation sustained damage while pulling the conductor and or from failure to use insulated bushings at the cable entry point. Since this was service entrance cable, only protected by the primary fuse cutout on the pole-mounted transformer feeding the service disconnect, it sustained arc damage for a considerable amount of time.

Conduit Seals and Fittings

Failure to install sealing compounds in conduit seals will allow gas vapor or flame front propagation in a conduit with gas leakage through a pressure switch failure. The use of conduit seals with cable fill exceeding 25% can result in seal failure when subjected to internal explosion or high pressure conditions. This would require upsizing of the conduit seal to reduce the conductor fill area to 25%. The use of one manufacturer's sealing compound and fiber in another manufacturer's seal fitting could lead to seal failure under catastrophic conditions. Failure to correctly orientate the seal fitting hub to pour sealing compound can lead

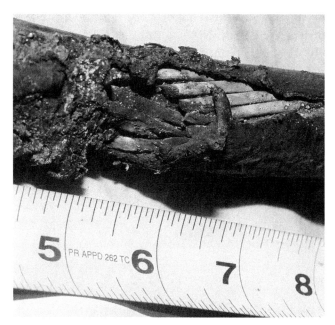

Figure 7.3: Wire insulation failure (from a fire) with resulting short circuit arc damage

to air spaces around cable interstices. Use of vertical run conduit seals in horizontal run applications can also lead to seal failure under catastrophic conditions.

The use of unapproved thread compound on NEMA 7 threaded and flanged covers or Teflon tape on screwed covers can result in enclosure failure under catastrophic conditions. This could prevent the cooling of escaping gases and cause enclosure damage.

Energized Equipment

Work on energized equipment without the proper safety precautions or lockout/tagout procedures can result in personnel injury and equipment. Work on energized equipment will require a safe work permit, the use of trained, qualified personnel, and PPE. Failure to maintain proper working clearances, as recommended in NEC Article 110 can also lead to the potential for electrical shock. Also reference NFPA 70E for other safe work practices.

References

1. Jones, Ray A., Mastrullo, Kenneth G., Jones, Jane G., *NFPA 70E®: Handbook for Electrical Safety in the Workplace*; 2004, Annex K.4, page 285. National Fire Protection Association; Quincy, MA.
2. NFPA 70E, *Electrical Safety in the Workplace®*; 2004. National Fire Protection Association; Quincy, MA.

3. ANSI/IEEE C2, *National Electrical Safety Code*®; 2007. Institute of Electrical and Electronic Engineers; New York, NY.

4. ANSI/ASC Z49.1-1994, *Safety in Welding, Cutting and Allied Processes*; 1994. American Welding Society; Miami, FL

5. Epperly, R.A., Heberlein, G.E., Higgins, J.A., Report on enclosure internal arcing tests; *IEEE Industry Applications Magazine*; Vol. 2, No. 3, May/June 1996, pp 35-41.

6. IEEE 141-1993(R1999), *IEEE Recommended Practice for Electric Power Distribution for Industrial Plants*; 1999; Section 5.9.2.10, pp 302-303. Institute of Electrical and Electronic Engineers, New York, NY.

CHAPTER 8

Motors, Generators, and Controls

The main developer of motor and generator standards in the United States is the National Electrical Manufacturers Association (NEMA). The primary standard for electric motors in the United States is NEMA MG 1, *Motors and Generators*. There are several other NEMA documents which supplement MG 1, including:

- NEMA Standards Publication Condensed MG 1, *Information Guide for General Purpose Industrial AC Small and Medium Squirrel-Cage Induction Motor Standards*

- NEMA Standards Publication No. MG 2, *Safety Standard and Guide for Selection, Installation, and Use of Electric Motors and Generators*

- NEMA Standards Publication No. MG 3, *Sound Level Prediction for Installed Rotating Electrical Machines*

- NEMA Standards Publication No. MG 10, *Energy Management Guide For Selection and Use of Fixed Frequency Medium AC Squirrel-Cage Polyphase Induction Motors*

- NEMA Standards Publication No. MG 11, *Energy Management Guide for Selection and Use of Single-Phase Motors*

NEMA MG 1 also recognizes international motor design concepts for *Index of Cooling (IC) and Index of Protection (IP)*. MG 1, Section 1, Part 5 – *Rotating Electrical Machines – Classification of Degrees of Protection Provided by Enclosures for Rotating Machines* outlines (IP) code designations and tests for the index of protection classifications for motors and generators. MG 1, Section 1, Part 6 – *Rotating Electrical Motors – Methods of Cooling (IC Code)* outlines general definitions and (IC) code designations for the motors and generators.

NEMA MG 2 [1] lists two general categories of motor enclosures: open machines and totally enclosed machines. It defines an *open machine* as:

> one having ventilating openings that permit passage of external cooling air over and around the windings of the machine. The term "open machine," when applied in large apparatus without qualification, designates a machine having no restriction to ventilation other than that necessitated by mechanical construction. [2]

Electrical Codes, Standards, Recommended Practices and Regulations; ISBN: 9780815520450
Copyright © 2010 Elsevier Inc. All rights of reproduction, in any form, reserved.

The standard identifies a *totally enclosed machine* as one that:

> is so enclosed as to prevent the free exchange of air between the inside and outside of the case but not sufficiently enclosed to be termed air-tight and in which dust does not enter in sufficient quantity to interfere with satisfactory operation of the machine. [3]

Both open and totally enclosed machines have subcategories, which include [4]:

1. Open Machine (IP00, IC01)

    A. Drip-Proof Machine – ODP (IP12, IC01)

    B. Splash-Proof Machine – OSP (IP13, IC01)

    C. Semi-Guarded Machine – OSG (IC01)

    D. Guarded Machine (IC01)

    E. Drip-Proof Guarded Machine – ODP$_{(guarded)}$ (IC01)

    F. Open Independently Ventilated Machine – OIV (IC06)

    G. Open-Pipe-Ventilated Machine – OPV

    H. Weather-Protected Machine – WP

       (1) Weather-Protected Type I – (WP-1)

       (2) Weather-Protected Type II – (WP-II)

2. Totally Enclosed Machine

    A. Totally Enclosed Non-Ventilated Machine – TENV (IC410)

    B. Totally Enclosed Fan-Cooled Machine – TEFC

    C. Totally Enclosed Fan Cooled Guarded Machine – TEFC$_{(guarded)}$ (IC411)

    D. Totally Enclosed Pipe-Ventilated Machine – TEPV (IP44)

    E. Totally Enclosed Water-Cooled Machine – TEWC (IP54)

    F. Water-Proof Machine – WP (IP55)

    G. Totally Enclosed Air-to-Water-Cooled Machine – TEAWC (IP54)

    H. Totally Enclosed Air-to-Air Cooled Machine – TEAAC (IP54)

    I. Totally Enclosed Air-Over Machine – TEAO (IP54, IC417)

    J. Explosion-Proof Machine – EP

    K. Dust-Ignition-Proof Machine – DIP

The letter designation above in each motor enclosure description is an abbreviation of the enclosure name. The parenthesis enclosed designations are the international index of protection and motor index of cooling designations for each motor enclosure.

Motors and Generators General Types

The NEMA MG 1, *Motors and Generators* lists the general categories for motors and generators as [5]:

1. Alternating Current Motors

    (a) Induction Motors

        (1) Squirrel-Cage Induction Motors

        (2) Wound-Rotor Induction Motors

    (b) Synchronous Motors

        (1) Direct-Current Excited Synchronous Motors

        (2) Permanent-Magnet Synchronous Motors

        (3) Reluctance Synchronous Motors

2. Alternating Current Generators

    (a) Induction Generators

    (b) Synchronous Generators

3. DC Motors

    (a) Shunt-Wound Motor

        (1) Straight Shunt-Wound Motor

        (2) Stabilized Shunt-Wound Motor

    (b) Series-Wound Motor

    (c) Compound-Wound Motor

    (d) Permanent Magnet Motor

4. DC Generators

    (a) Shunt-Wound Generators

    (b) Compound-Wound Generators

An *induction motor* is an alternating current, rotating asynchronous motor generally consisting of a stationary, primary winding coil or stator and a rotating secondary coil/armature or rotor. The device converts electrical energy to mechanical energy by use of the electromagnetic induction. The rotating electromagnetic field created by the electrical current in the stator coil induces current into the rotor coil by transformer action. That coil is electrically closed on itself. The induced rotor current has a corresponding electromagnetic field, which interacts with the stator electromagnetic field, causing the rotor to turn. The difference between the rotor speed and the speed of the stator electromagnetic field (synchronous speed) is called *slip*. Slip (*s*) is defined as the ratio below:

$$s = (n_s - n_r)/n_s \tag{Eq. 8.1}$$

where n_s is the motor synchronous speed and n_r is the rotor speed.

Induction motor rotors may be of two types, *wound rotor* or *squirrel-cage rotor*. A wound rotor has windings similar to and wound for the same number of poles as the stator. The rotor windings are connected to insulated slip rings mounted on the rotor shaft. Those connections can be connected by carbon brushes and made available externally to the machine. A squirrel-cage rotor has a rotor consisting of conducting bars which are embedded in slots in the rotor iron. Those bars are short-circuited at each end by conducting end rings. Squirrel-cage motors are more commonly used than wound rotor motors.

A *synchronous motor* is a motor whose rotational speed is proportional to the frequency of the alternating current (AC) source powering it. It operates at a constant or synchronous speed when supplied with electrical energy at a fixed frequency. Synchronous speed (n_s) is determined by the following relationship:

$$n_s = 120 \times f/P \tag{Eq. 8.2}$$

where *f* is the alternating current frequency in Hertz and *P* is the number of poles in the motor.

A synchronous motor rotor has the same number of poles as its stator. Direct current (DC) is applied to the rotor with either brush type or brushless exciters. The rotor can also be constructed of permanent magnets. A synchronous motor is started with the use of a squirrel-cage winding inserted in the rotor pole faces. That winding is called an amortisseur or damper winding. The rotor speed comes very near synchronous speed by induction motor action. During that time the motor field winding is unexcited; however, once it nears synchronous speed, it will be necessary to energize the field windings with a DC source, pulling it into synchronous speed. Once at synchronous speed, the motor will produce synchronous torque. Should the connected load requirements exceed synchronous torque, the motor will be pulled out of synchronous speed.

An *induction generator* is an asynchronous device, used primarily in wind turbines and micro-hydroelectric applications. This generator is capable of generating electrical energy with varying rotor speeds above the synchronous speed of the device. External excitation must be applied to the stator coil to induce electrical current into the rotor cage (coil), with an accompanying magnetic field produced in the rotor. The rotor must be driven at a rate faster than the rotating magnetic flux of the stator (synchronous speed) to generate electricity.

A *synchronous generator* is a machine that converts mechanical energy to electrical alternating current energy. A prime mover mechanical power source turns the rotor at a speed equal to the desired frequency of the generator. The generator stationary armature coils (stator) are installed at regular intervals around the generator. A field winding is attached to the rotor and a direct current source, usually from a small DC generator known as an *exciter*, is applied, usually through slip rings. The subsequent mechanically rotating magnetic flux created in the rotor field windings links the stator coils, producing a voltage in accordance with Faraday's Law. Faraday's Law states that an electromagnetic field (emf) is induced in an electric circuit through a rate of change in the magnetic flux linking that circuit. The emf is proportional to the rate of change of the flux linkage.

NEMA MG 1 [6] lists two basic types of *DC motors* including:

1. Series-Wound

2. Shunt-Wound

    A. Straight Shunt-Wound Motor

    B. Stabilized Shunt-Wound Motor

        (1) Compound-Wound

        (2) Permanent Magnet

The armature and field windings in a *series-wound* DC motor are wired in series, so that the current flows through both. The field current is load-dependent providing motor-high speed during no-load conditions and motor low-speed during high load conditions. Applications include hoists, cranes, oil drilling motors, and railway electric locomotives. This configuration has a characteristic high starting torque.

A *shunt-wound* DC motor has its armature and shunt field windings either connected in parallel or its shunt field circuit is connected to a separate excitation voltage source. Each motor pole consists of a large number of fine wire turns and all poles are commonly wired in series. The field current is voltage-dependent as well as dependent on the resistance of the field windings. With constant voltage applied to the armature and varying low field current, the motor characteristically has a relatively constant speed versus applied load. Its applications include reciprocating pumps, conveyors, printing presses, etc. There are two types of shunt-wound motors including *straight shunt-wound* motors and *stabilized shunt-wound* motors.

A *compound-wound* DC motor consists of both series and shunt field windings. It offers characteristics of a relatively high starting torque of a series-wound motor and the speed regulation of a shunt-wound machine. Its applications include elevators, hoists, etc.

A *differential motor* [7] has both series and shunt field windings; however, they oppose each other magnetically. This motor has poor starting torque characteristics and has a limited application.

A *permanent magnet* [8] DC motor consists of six basic components, including the shaft, rotor or armature, stator, commutator, field magnets and brushes. The stator consists of the motor housing on which two or more permanent magnets are mounted in lieu of the field windings. The magnets produce an external magnetic field. The rotor armature windings are mounted on the rotor and are electrically connected to the commutators on the motor shaft. DC voltage is applied to the commutators through carbon brushes. The magnetic field produced by the armature windings interacts with the magnetic fields of the permanent magnets. The rotor aligns itself with the magnet's fields, causing it to rotate. When the rotor begins to turn, the commutators energize the next armature winding creating a new magnetic field which continues the shaft rotation.

A *DC generator* consists of both armature (rotor) and field windings. Armature windings are located on the rotor with generated current removed from them by the use of carbon brushes. The rotor is connected to a constant speed mechanical power source. A voltage is produced on the armature coil by its rotation and has the same waveform as the spatial flux-density distribution in the air gap between the rotor and stator.

The stator voltage produced is an alternating waveform, but not sinusoidal like an AC generator; therefore, requiring rectification. Two methods of rectification may be employed, including external semiconductor rectifiers or mechanical rectification by means of a commutator. The commutator is comprised of two formed copper cylinder segments, insulated from each other, which are mounted on and insulated from the rotor shaft. The commutators connect the armature windings on the rotor and rectify the voltage waveform.

There are several configurations of DC generators including [9]:

Self-Excited Shunt-Wound

Series-Wound

Separately Excited

Compound-Wound

The shunt-wound machine has its field windings electrically connected in parallel with the armature. Its output voltage decreases with load increases. The series-wound generator has its field windings in series with the armature. A separately excited generator has field windings, which are not connected electrically to the armature and which are energized from a separate power source. The compound generator has both series and shunt windings.

NEMA MG 1, *Motors and Generators* provides *Design Letters* for polyphase induction squirrel-cage medium motors. Those Design Letters provide information regarding locked-rotor torque, pull-up torque, and slip values at rated torque and are shown in Table 8.1. Reference MG 1 for specific Design Letter information based on motor horsepower, speed, and frequency. MG 1 also lists Design Letters for single-phase small motors and medium motors. Those letters are also provided in Table 8.1. MG 1 should be referenced for specific information regarding the NEMA Design Letters.

NEMA *Design A* motors have normal starting torque, normal starting current, and low slip. *Design B* motors have normal starting torque, low starting current, and low slip. *Design C* motors have high starting torque and low starting current. *Design D* motors exhibit high starting torque and high slip. *Design N* and *Design O* motors indicate single-phase small motors that are capable of full-voltage starting and locked rotor currents established in NEMA MG 1-2006, Section 12.33. *Design L* and *Design M* motors indicate motors that are capable of full-voltage starting and withstanding locked rotor currents established in NEMA MG 1-2006, Section 12.34.

Single-Phase Induction Motors

There are three basic types of *single-phase induction motors,* including squirrel-cage motors, wound-rotor motors, and universal motors. The most common single-phase induction motors include:

I. Squirrel-Cage Induction Motors

   1. Reactance Split-Phase Motors

   2. Resistance-Start Motors

TABLE 8.1   NEMA motor Design Letters

Polyphase squirrel-cage medium motors	Single-phase motors	
	Small Motors	Medium Motors
Design A	Design N	
Design B	Design O	
Design C		Design L
Design D		Design M

3. Capacitor Motors

    (A) Capacitor-Start Motors

    (B) Permanent-Split Capacitor Motors

    (C) Two-Value Capacitor Motors

4. Shaded-Pole Motors

II. Wound-Rotor Motors

    1. Repulsion Motors

    2. Repulsion-Start Induction Motors

    3. Repulsion-Induction Motors

III. Universal Motors

    1. Series-Wound Motors

    2. Compensated Series-Wound Motors

All of the above noted single-phase induction motors operate on single-phase 50 Hz or 60 Hz AC voltage. However, the *universal motors* are capable of operation on DC voltages at the same RMS voltage levels, with similar speed and horsepower output. The single-phase AC induction motors using starting coils separate from their running induction coils or external capacitors or resistors will switch to the running induction coils when the motors approach operating speed.

Equipment Specification Preparation

When preparing a specification for purchasing a motor or generator, several options may be available including:

Utilization of company standard specifications

Utilization of industry specifications

Utilization of manufacturer-supplied specifications

Utilization of commercially-available specifications

Preparation of a unique specification for a specific application

Care must be exercised when preparing any specification for motors and generators. An equipment manufacturer or supplier must be provided with adequate information regarding the product(s) being purchased. A general specification can be prepared along with an equipment data sheet that would provide application specific information.

Motor and Generator Standards

Good engineering practice mandates as minimum specifying electrical equipment through their application, proposed location, physical constraints, available power sources, motor starting methodology requirements, ambient conditions, and the presence of any caustic or flammable and combustible materials or substances being processed, handled, or transported by that equipment. The codes, standards, and recommended practices applied to any equipment specification should assure that the equipment can operate under those circumstances. Motors, generators, and motor control equipment operating in electrically hazardous (classified) areas should utilize the applicable enclosures and protection schemes presented in Chapter 9.

Motor and generator codes, standards, and recommended practices may be subdivided into several categories including:

- General or Basic
- Industry-specific
- Application-specific
- Supporting documents

General or *basic* codes, standards, and recommended practices consist of those that provide information applicable to equipment in all industries and applications. That would include documents prepared by CSA, NEMA, IEC, NFPA, and IEEE. Examples of those documents are presented in Table 8.2.

Industry-specific codes, standards, and recommended practices are those developed by standards organizations, which might include IEEE, API, IEC, AIChE, etc. and which may be unique for a specific industrial operation. Those standards are designed for specific applications unique to those industries. They may include some of the codes, standards, and recommended practices listed in Table 8.3.

Application-specific codes, standards, and recommended practices are those documents for motor and generator applications that are not covered by the industry-specific category. Examples of application-specific codes, standards, and recommended practices are listed in Table 8.4.

Supporting codes, standards, and recommended practices provide additional information regarding testing, or auxiliary equipment that may be applicable to the other specification categories. Examples of those documents are presented in Table 8.5.

Motor Control and Protection

NFPA 70®, the *National Electrical Code*® (NEC®) provides requirements for motor control in Article 430, Part VII *Motor Controllers*. Control techniques may be as simple as ON–OFF,

TABLE 8.2   Rotating equipment general or basic codes, standards, and recommended practices

Developer	Standard No.	Title
CSA	CSA C22.1	Canadian Electrical Code, part I (21st edition), Safety Standard for Electrical Installations
CSA	CSA C22.2 No. 100	Motors and Generators
IEC	IEC 60034-1	Rotating Electrical Machines – Part 1: Rating and Performance
IEC	IEC 60034-5	Rotating Electrical Machines – Part 5: Degrees of Protection Provided by the Integral Design of Rotating Electrical Machines (IP code) – Classification
IEC	IEC 60034-6	Rotating Electrical Machines – Part 6: Methods of Cooling (IC Code)
IEC	IEC 60034-7	Rotating Electrical Machines – Part 7: Classification of Types of Construction, Mounting Arrangements and Terminal Box Position (IM Code)
IEC	IEC 60034-12	Safety Standard for Construction and Guide for Selection, Installation, and Use of Electric Motors and Generators
IEC	IEC 60050-826	International Electrotechnical Vocabulary – Part 826: Electrical installations
IEC	IEC 60204-1	Safety of Machinery – Electrical Equipment of Machines – Part 1: General Requirements
IEC	IEC 62262	Degrees of Protection Provided by Enclosures for Electrical Equipment against External Mechanical Impacts (IK code)
IEEE	IEEE 1349	IEEE Guide for Application of Electric Motors in Class I, Division 2 Hazardous (Classified) Locations
IEEE	IEEE C37.2	IEEE Standard Electrical Power System Device Function Numbers and Contact Designations
IEEE	IEEE C37.102	IEEE Guide for AC Generator Protection
IEEE	IEEE C50.10	General Requirements for Synchronous Machines

IEEE	IEEE C62.92.1	IEEE Guide for the Application of Neutral Grounding in Electrical Utility Systems – Part 1: Introduction
IEEE	IEEE C62.92.2	IEEE Guide for the Application of Neutral Grounding in Electrical Utility Systems, Part II – Grounding of Synchronous Generator Systems
IEEE	IEEE C62.92.3	IEEE Guide for the Neutral Grounding in Electrical Utility Systems, Part III – Generator Auxiliary Sys
NEMA	NEMA 250	Enclosures for Electrical Equipment (1000 Volts Maximum)
NEMA	ANSI/NEMA C84.1	Electric Power Systems and Equipment – Voltage Ratings (60 Hertz)
NEMA	MEMA MG 1	Motors and Generators
NEMA	NEMA MG 2	Safety Standard for Construction and Guide for Selection, Installation, and Use of Electric Motors and Generators
NEMA	NEMA MG 3	Sound Level Prediction for Installed Rotating Electrical Machines
NEMA	NEMA MG 10	Energy Management Guide for Selection and Use of Fixed Frequency Medium AC Squirrel-Cage Polyphase Induction Motors
NEMA	NEMA MG 11	Energy Management Guide for Selection and Use of Single-Phase Motors
NFPA	NPFA 70®	National Electrical Code
UL	UL 1004	Electric Motors
UL	UL 1004-1	Standard for Rotating Electrical Machines – General Requirements
UL	UL 1004-2	Standard for Impedance Protected Motors
UL	UL 1004-3	Standard for Thermally Protected Motors
UL	UL 1004-4	Electric Generators
UL	UL 2111	UL Standard for Safety Overheating Protection for Motors

TABLE 8.3 Rotating machinery industry-specific codes, standards, and recommended practices

Developer	Standard No.	Title
IEEE	IEEE 11	Standard for Rotating Machinery for Rail and Road Vehicles
IEEE	IEEE 334	IEEE Standard for Qualifying Continuous Duty Class 1E Motors for Nuclear Power Generating Stations
IEEE	ANSI/IEEE 387	IEEE Standard Criteria for Diesel-Generator Units Applied as Standby Power Supplies for Nuclear Power Generating Stations
IEEE	ANSI/IEEE 492	Guide for Operation and Maintenance of Hydro-Generators
IEEE	IEEE 499	RP for Cement Plant Electric Drives and Related Electrical Equipment
IEEE	IEEE 841	Standard for Petroleum and Chemical Industry – Premium Efficiency Severe Duty Totally Enclosed Fan-Cooled (TEFC) Squirrel Cage Induction Motors – Up To and Including 370 kW (500 hp)
IEEE	IEEE 1068	IEEE Recommended Practice for the Repair and Rewinding of Motors for the Petroleum and Chemical Industry
IEEE	IEEE 1095	IEEE Guide for Installation of Vertical Generators and Generator/Motors for Hydroelectric Applications
IEEE	IEEE 1290	IEEE Guide for Motor Operated Valve (MOV) Motor Application, Protection, Control, and Testing in Nuclear Power Generating Stations
IEEE	ANSI/IEEE C50.41	American National Standard for Polyphase Induction Motors for Power Generating Stations
IEEE	ANSI/IEEE C50.49	Polyphase Induction Motors for Power Generating Stations
API	ANSI/API 541	Form-Wound Squirrel-Cage Induction Motors – 250 Horsepower and Larger
API	ANSI/API 546	Brushless Synchronous Machines – 500 kVA and Larger
API	ANSI/API 547	General Purpose Form-Wound Squirrel Cage Induction Motors – 250 Horsepower and Larger
IEC	IEC 60349-1	Electric Traction – Rotating Electrical Machines for Rail and Road Vehicles – Part 1: Machines Other Than Electronic Convertor-Fed Alternating Current Motors
IEC	IEC 60349-2	Electric Traction – Rotating Electrical Machines for Rail and Road Vehicles – Part 2: Electronic Convertor-Fed Alternating Current Motors
IEC	IEC 60050-415	International Electrotechnical Vocabulary – Part 415: Wind Turbine Generator Systems
ASAE	ASAE EP329.2	Single-Phase Rural Distribution Service for Motors and Phase Converters
NEMA	ANSI/NEMA C50.41	American National Standard for Polyphase Induction Motors for Power Generation Stations

TABLE 8.4 Rotating machinery application-specific codes, standards, and recommended practices

Developer	Standard No.	Title
IEC	IEC 60034-2A	Rotating Electrical Machines – Part 2: Methods for Determining Losses and Efficiency of Rotating Electrical Machinery Form Tests (Excluding Machines for Traction Vehicles) – First Supplement: Measurement of Losses by the Calorimetric Method
IEC	IEC 60034-3	Rotating Electrical Machines – Part 3: Specific Requirements for Synchronous Generators Driven by Steam Turbines or Combustion Gas Turbines
IEC	IEC 60034-17	Rotating Electrical Machines – Part 17: Cage Induction Motors When Fed from Converters – Application Guide
IEC	IEC 60034-20-1	Rotating Electrical Machines – Part 20-1: Control Motors - Stepping Motors
IEC	IEC 60034-22	Rotating Electrical Machines – Part 22: AC Generators for Reciprocating Internal Combustion (RIC) Engine Driven Generating Sets
IEC	IEC/TS 60034-25	Rotating Electrical Machines – Part 25: Guidance for the Design and Performance of AC Motors Specifically Designed for Converter Supply
IEC	IEC 60072-1	Dimensions And Output Series For Rotating Electrical Machines – Part 1: Frame Numbers 56 to 400 and Flange Numbers 55 to 1080
IEC	IEC 60072-5	Dimensions and Output Series for Rotating Electrical Machines – Part 2: Frame Numbers 355 to 1000 and Flange Numbers 1180 to 2360
IEEE	ANSI/IEEE 67	Guide for Operation and Maintenance of Turbine Generators
IEEE	IEEE 125	IEEE Recommended Practice for Preparation of Equipment Specifications for Speed-Governing of Hydraulic Turbines Intended to Drive Electric Generators
IEEE	IEEE 252	IEEE Test Procedure for Polyphase Induction Motors Having Liquid in the Magnetic Gap
IEEE	IEEE 286	Recommended Practice for Measurement of Power-Factor Tip-Up of Rotating Machinery Stator Coil Insulation.
IEEE	IEEE 434	Guide for Functional Evaluation of Insulation Systems for Large High-Voltage Machines
IEEE	IEEE 810	IEEE Standard for Hydraulic Turbine and Generator Integrally Forged Shaft Couplings and Shaft Runout Tolerances
IEEE	IEEE 1349	IEEE Guide for Application of Electric Motors in Class I, Division 2 Hazardous (Classified) Locations
IEEE	IEEE C50.12	Standard for Salient-Pole 50 Hz and 60 Hz Synchronous Generators and Generator/Motors for Hydraulic Turbine Applications Rated 5 MVA and Above
IEEE	IEEE C50.13	Standard for Cylindrical Rotor 50 and 60 Hz Synchronous Generators Rated 10 MVA and Above
NECA	NECA 230	Standard for Selecting, Installing, and Maintaining Electric Motors and Motor Controllers (ANSI)
NEMA	NEMA SM 24	Land-Based Steam Turbine Generator Sets 0 to 33 000 kW
UL	ANSI/UL 674	Electric Motors and Generators for Use in Division 1 Hazardous (Classified) Locations
UL	UL 1004-5	Standard for Fire Pump Motors
UL	UL 1836	Electric Motors and Generators for Use in Class I, Divisions 1

TABLE 8.5   Rotating machinery supporting codes, standards and recommended practices

Developer	Standard No.	Title
ASTM	ASTM A288	Standard Specification for Carbon and Alloy Steel Forgings for Magnetic Retaining Rings for Turbine Generators
ASTM	ASTM A289/A289M	Standard Specification for Alloy Steel Forgings for Nonmagnetic Retaining Rings for Generators
ASTM	ASTM A418/A418M	Standard Practice for Ultrasonic Examination of Turbine and Generator Steel Rotor Forgings
ASTM	ASTM A469/A469M	Standard Specification for Vacuum-Treated Steel Forgings for Generator Rotors
ASTM	ASTM A472/A472M	Standard Specification for Heat Stability of Steam Turbine Shafts and Rotor Forgings
ASTM	ASTM A531/A531M	Standard Practice for Ultrasonic Examination of Turbine-Generator Steel Retaining Rings
EASA	ANSI/EASA AR100	Recommended Practice for the Repair of Rotating Electrical Apparatus
IEEE	ANSI/IEEE 1	IEEE Recommended Practice – General Principles for Temperature Limits in the Rating of Electrical Equipment and for the Evaluation of Electrical Insulation
IEEE	IEEE 43	Recommended Practice for Testing Insulation Resistance of Rotating Machinery
IEEE	IEEE 62.2	IEEE Guide for Diagnostic Field Testing of Electric Power Apparatus – Electrical Machinery
IEEE	IEEE 95	IEEE Recommended Practice for Insulation Testing of AC Electric Machinery (2300 V and Above) with High Direct Voltage
IEEE	IEEE 112	IEEE Standard Test Procedure for Polyphase Induction Motors and Generators
IEEE	IEEE 114	IEEE Standard Test Procedures for Single-Phase Induction Motors
IEEE	IEEE 115	Test Procedures for Synchronous Machines
IEEE	IEEE 122	IEEE Recommended Practice for Functional and Performance Characteristics of Control Systems for Steam Turbine-Generator Units
IEEE	IEEE 303	IEEE Std 303 – 2004 IEEE Recommended Practice for Auxiliary Devices for Rotating Electrical Machines in Class I, Division 2 and Zone 2 Locations
IEEE	IEEE 421.1	IEEE Standard Definitions for Excitation Systems for Synchronous Machines
IEEE	IEEE 421.2	IEEE Guide for Identification, Testing, and Evaluation of the Dynamic Performance of Excitation Control Systems
IEEE	IEEE 421.3	IEEE Standard for High-Potential Test Requirements for Excitation Systems for Synchronous Machines
IEEE	IEEE 421.4	IEEE Guide for the Preparation of Excitation System Specifications
IEEE	IEEE 421.5	IEEE Recommended Practice for Excitation System Models for Power System Stability Studies
IEEE	IEEE 522	IEEE Guide for Testing Turn Insulation of Form-Wound Stator Coils for Alternating-Current Electric Machines
IEEE	IEEE 620	Guide for the Presentation of Thermal Limit Curves for Squirrel Cage Induction Machines

IEEE	IEEE 1043	IEEE Recommended Practice for Voltage-Endurance Testing of Form-Wound Bars and Coils
IEEE	IEEE 1110	Guide for Synchronous Generator Modeling Practices and Applications in Power System Stability Analyses
IEEE	IEEE 1255	Guide for Evaluation of Torque Pulsations During Starting of Synchronous Motors
IEEE	IEEE 1415	Guide for Induction Machinery Maintenance Testing and Failure Analysis
IEEE	IEEE 1434	Guide to Measurement of Partial Discharges in Rotating Machinery
IEEE	IEEE C37.96	Guide for AC Motor Protection
IEEE	IEEE C37.102	IEEE Guide for AC Generator Protection
IEEE	IEEE C62.21	Guide for the Application of Surge Voltage Protective Equipment on AC Rotating Machinery 1000 Volts and Greater
IEC	IEC 60034-8	Rotating Electrical Machines – Part 8: Terminal Markings and Direction of Rotation
IEC	IEC 60034-9	Rotating Electrical Machines – Part 9: Noise Limits
IEC	IEC 60034-11	Rotating Electrical Machines – Part 11: Thermal Protection
IEC	IEC 60034-12	Rotating Electrical Machines – Part 12: Starting Performance of Single-speed Three-phase Cage Induction Motors
IEC	IEC 60034-14	Rotating Electrical Machines – Part 14: Mechanical Vibration of Certain Machines with Shaft Heights 56 mm and Higher – Measurement, Evaluation and Limits of Vibration Severity
IEC	IEC60034-15	Rotating Electrical Machines – Part 15: Impulse Voltage Withstand Levels of Rotating AC Machines with Form-Wound Stator Coils
IEC	IEC 60034-16-1	Rotating Electrical Machines – Part 16: Excitation Systems for Synchronous Machines – Chapter 1: Definitions
IEC	IEC 60034-16-2	Rotating Electrical Machines – Part 16: Excitation Systems for Synchronous Machines – Chapter 2: Models for Power System Studies
IEC	IEC 60034-16-3	Rotating Electrical Machines – Part 16: Excitation Systems for Synchronous Machines – Section 3: Dynamic Performance
IEC	IEC 60034-18-1	Rotating Electrical Machines – Part 18: Functional Evaluation of Insulation Systems – Section 1: General Guidelines
IEC	IEC 60034-18-21-am1	Amendment 1 – Rotating Electrical Machines – Part 18: Functional Evaluation of Insulation Systems – Section 21: Test Procedures for Wire-Wound Windings – Thermal Evaluation and Classification
IEC	IEC 60034-18-21-am1	Amendment 2 – Rotating Electrical Machines – Part 18: Functional Evaluation of Insulation Systems – Section 21: Test Procedures for Wire-Wound Windings – Thermal Evaluation and Classification
IEC	IEC 60034-18-22	Rotating Electrical Machines – Part 18-22: Functional Evaluation of Insulation Systems – Test Procedures for Wire-Wound Windings – Classification of Changes and Insulation Component Substitutions

Continued

TABLE 8.5 Rotating machinery supporting codes, standards and recommended practices—cont'd

Developer	Standard No.	Title
IEC	IEC 60034-18-31-am1	Amendment 1 – Rotating Electrical Machines – Part 18: Functional Evaluation of Insulation Systems – Section 31: Test Procedures for Form-Wound Windings – Thermal Evaluation and Classification of Insulation Systems Used in Machines Up To and Including 50 MVA and 15 kV
IEC	IEC 60034-18-32	Rotating Electrical Machines – Part 18: Functional Evaluation of Insulation Systems – Section 32: Test Procedures for Form-Wound Windings – Electrical Evaluation of Insulation Systems Used in Machines Up To and Including 50 MVA and 15 kV
IEC	IEC 60034-18-33	Rotating Electrical Machines – Part 18: Functional Evaluation of Insulation Systems – Section 33: Test Procedures for Form-Wound Windings – Multifactor Functional Evaluation – Endurance Under Combined Thermal and Electrical Stresses of Insulation Systems Used in Machines Up To and Including 50 MVA and 15 kV
IEC	IEC 60034-18-34	Rotating Electrical Machines – Part 18-34: Functional Evaluation of Insulation Systems – Test Procedures for Form-Wound Windings – Evaluation of Thermomechanical Endurance of Insulation Systems
IEC	IEC 60034-18-41	Rotating Electrical Machines – Part 18-41: Qualification and Type Tests For Type I Electrical Insulation Systems Used In Rotating Electrical Machines Fed From Voltage Converters
IEC	IEC 60034-18-42	Rotating Electrical Machines – Part 18-42: Qualification and Acceptance Tests for Partial Discharge Resistant Electrical Insulation Systems (Type II) Used in Rotating Electrical Machines Fed from Voltage Converters
IEC	IEC 60034-19	Rotating Electrical Machines – Part 19: Specific Test Methods for DC Machines on Conventional and Rectifier-Fed Supplies
IEC	IEC 60034-23	Rotating Electrical Machines – Part 23: Specification for the Refurbishing of Rotating Electrical Machines
IEC	IEC 60034-26	Rotating Electrical Machines – Part 26: Effects of Unbalanced Voltages on the Performance of Three-Phase Cage Induction Motors
IEC	IEC 60034-27	Rotating Electrical Machines – Part 27: Off-Line Partial Discharge Measurements on the Stator Winding Insulation of Rotating Electrical Machines
IEC	IEC 60034-28	Rotating Electrical Machines – Part 28: Test Methods for Determining Quantities of Equivalent Circuit Diagrams for Three-Phase Low-Voltage Cage Induction Motors
IEC	IEC 60034-29	Rotating Electrical Machines – Part 29: Equivalent Loading and Superposition Techniques – Indirect Testing to Determine Temperature Rise
IEC	IEC 61922	High-Frequency Induction Heating Installations – Test Methods for the Determination of Power Output of the Generator
NEMA	NEMA CB 1	Brushes for Electrical Machines
UL	UL 3200	Performance Testing of Engine and Turbine Generators

Hand–OFF–Auto, sequence control, or simple speed control. More complex control schemes may require the use of Programmable Logic Controllers (PLC), Variable Frequency Drives (VFD), etc.

Control systems may be mechanical, electrical, pneumatic, hydraulic, electropneumatic, electrohyraulic or electronic. Control can be defined as a means of governing the performance of a rotating electrical apparatus. A control device is one that executes a control function. A control system is one:

in which deliberate guidance or manipulation is used to achieve a prescribed value of a variable. [10]

An electric motor controller is:

A device or group of devices that serve to govern in some predetermined manner the electric power delivered to the motor. [11]

Mechanical means of motor control may simply include a mechanical switch or a circuit breaker as the ON–OFF controlling device. An electromechanical means may be through the use of electrical contactor. Electrical means might include auto-transformers, part-windings, and other means in conjunction with electromagnetic contactors. Electronic motor control may utilize silicon controller rectifiers (SCR), thyristors, insulated-gate bipolar transistors (IGBT), or other devices.

Control circuits may consist of individual control elements such as relays, contacts, switches, or other control devices, in addition to the main motor control device. A control circuit allows the use of low voltage/current devices to control large horsepower or high-voltage motors.

Control circuits may operate at the rated voltage of the motor they are controlling or at lower control voltages. In the United States, 120 V, 60 Hz is a common control voltage, as is 24 DC. Control power may be obtained from the same source on which the motor is operating or from independent source or uninterruptable power source. Use of a fused control power transformer (CPT) is common practice when obtaining control voltage from the motor feeder circuit.

Table 8.6 presents a list of some of the codes, standards, and recommended practices for motor control and protection utilized in the United States. Motor control and protection codes, standards, and recommended practices utilized in Canada are listed in Table 8.7. Motor protection can be as simple as a molded case branch circuit breaker in a panelboard for fractional horsepower motors or the use of molded case circuit breakers, power circuit breakers, fuses, or vacuum circuit breakers and protective relaying, depending on the application. Overcurrent/shift-circuit protection may include motor thermal devices implanted in the motor windings, motor overloads in a motor starter device, thermal/magnetic circuit breakers, protective relaying etc.

TABLE 8.6 Motor control and protection codes, standards, and recommended practices

Developer	Standard No.	Title
NEMA	NEMA AB 1	Molded Case Circuit Breakers and Molded Case Switches
NEMA	NEMA AB 3	Molded Case Circuit Breakers and Their Application
NEMA	NEMA AB 4	Guidelines for Inspection and Preventive Maintenance of Molded Case Circuit Breakers Used in Commercial and Industrial Applications
NEMA	NEMA BU 1.1	General Instructions for Proper Handling, Installation, Operation, and Maintenance of Busway Rated 600 Volts or Less
NEMA	NEMA ICS 1	Industrial Control and Systems – General Requirements
NEMA	NEMA ICS 1.1	Safety Guidelines for the Application, Installation, and Maintenance of Solid State Control
NEMA	NEMA ICS 1.3	Industrial Control and Systems: Preventive Maintenance of Industrial Control and Systems Equipment
NEMA	NEMA ICS 2	Industrial Control and Systems: Controllers, Contactors, and Overload Relays Rated 600 Volts
NEMA	NEMA ICS 2, Part 8	Industrial Control and Systems: Controllers, Contactors, and Overload Relays Rated 600 Volts Part 8: Disconnect Devices for Use in Industrial Control Equipment
NEMA	NEMA ICS 2, Part 9	Industrial Automation Control Products and Systems: Starters, Contactors, and Overload Relays Rated Not More Than 2000 Volts AC or 750 Volts DC – Part 9 – AC Vacuum-Break Magnetic Controllers Rated 1500 Volts AC
NEMA	NEMA ICS 2.3	Industrial Control and Systems: Instructions for the Handling, Installation, Operation, and Maintenance of Motor Control Centers Rated Not More than 600 Volts
NEMA	NEMA ICS 2.4	NEMA IEC Devices for Motor Service – A Guide for Understanding the Differences
NEMA	NEMA ICS 3	Industrial Control and Systems: Medium-Voltage Controllers Rated 2001 to 7200 Volts AC
NEMA	NEMA ICS 4	Industrial Control and Systems: Terminal Blocks
NEMA	NEMA ICS 6	Industrial Control and Systems: Enclosures
NEMA	ANSI/NEMA ICS 8	Industrial Control and Systems: Crane and Hoist Controllers
NEMA	NEMA ICS 10, Part 1	Industrial Control and Systems – Part 1: Electromechanical AC Transfer Switch Equipment
NEMA	NEMA ICS 10, Part 2	Industrial Control and Systems: AC Transfer Equipment – Part 2: Static AC Transfer Equipment
NEMA	NEMA ICS 12.1	Profiles of Networked Industrial Devices – Part 1: General Rules
NEMA	NEMA ICS 14	Application Guide for Electric Fire Pump Controllers
NEMA	NEMA ICS 16	Industrial Control and Systems – Motion/Position Control Motors, Controls, and Feedback Devices
NEMA	NEMA ICS 18	Motor Control Centers

NEMA	NEMA ICS 19	Industrial Control and Systems: Diagrams, Device Designations, and Symbols
NEMA	NEMA ICS 61800-1	Adjustable Speed Electric Power Drive Systems – Part 1: General Requirements – Rating Specifications for Low-Voltage Adjustable Speed DC Power Drive Systems
NEMA	NEMA ICS 61800-2	Adjustable Speed Electric Power Drive Systems – Part 2: General Requirements – Rating Specifications for Low-Voltage Adjustable Frequency AC Power Drive Systems
NEMA	NEMA ICS 61800-4	Adjustable Speed Electric Power Drive Systems – Part 1: General Requirements – Rating Specifications for AC Power Drive Systems Above 1000 VAC and not Exceeding 35 kV
NEMA	ANSI/NEMA PB-1.1	Instructions for Safe Installation, Operation and Maintenance for Panelboards
NEMA	ANSI/NEMA PB 2.1	General Instructions for Proper Handling, Installation, Operation, and Maintenance of Deadfront Distribution Switchboards Rated 600 Volts or Less
NEMA	NEMA PB 2.2	Application Guide for Ground Fault Protective Devices for Equipment
NEMA	NEMA 250	Enclosures for Electrical Equipment (1000 Volts Maximum)
NEMA	NEMA 280	Application Guide for Ground Fault Circuit Interrupters
NEMA	ANSI/NEMA C93.1	Requirements for Power-Line Carrier Coupling Capacitors and Coupling Capacitor Voltage Transformers (CCVT)
NEMA	ANSI/NEMA C84.1	American National Standard for Electrical Power Systems and Equipment – Voltage Ratings (60 Hertz)
NEMA/IEEE	ANSI C37.50	American National Standard for Switchgear – Low-Voltage AC Power Circuit Breakers Used in Enclosures – Test Procedures
NEMA	ANSI C37.51	For Switchgear – Metal-Enclosed Low-Voltage AC Power Circuit Breaker Switchgear Assemblies – Conformance Test Procedures
NEMA/IEEE	ANSI C37.52	Test Procedures, Low-Voltage (AC) Power Circuit
NEMA	ANSI C37.54	For Indoor Alternating Current High-Voltage Circuit Breakers Applied as Removable Elements in Metal-Enclosed Switchgear – Conformance Test Procedures
NEMA	ANSI C37.55	American National Standard for Switchgear – Medium-Voltage Metal-Clad Assemblies – Conformance Test Procedures
NEMA	ANSI C37.57	American National Standard for Switchgear – Metal-Enclosed Interrupter Switchgear Assemblies – Conformance Testing
NEMA	ANSI C37.58	American National Standard for Switchgear – Indoor AC Medium-Voltage Switches for Use in Metal-Enclosed Switchgear – Conformance Test Procedures
NEMA	MEMA DC 20	Residential Controls – Class 2 Transformers

Continued

TABLE 8.6 **Motor control and protection codes, standards, and recommended practices—cont'd**

Developer	Standard No.	Title
NEMA	NEMA FU 1	Low-Voltage Cartridge Fuses
NEMA	NEMA KS 1	Enclosed and Miscellaneous Distribution Equipment Switches (600 Volts Maximum)
NEMA	PB 1	Panelboards
NEMA	PB 1.1	General Instructions for Proper Installation, Operation, and Maintenance of Panelboards Rated 600 Volts or Less
NEMA	NEMA PB 2	Deadfront Distribution Switchboards
NEMA	PB 2.1	Instructions for Proper Handling, Installation, Operation, and Maintenance of Deadfront Distribution Switchboards Rated 600 Volts or Less
UL	UL 67	Panelboards
UL	UL 98	Enclosed and Dead-Front Switches
UL	UL 198M	Mine-Duty Fuses
UL	UL 248-1	Low-Voltage Fuses – Part 1: General Requirements
UL	UL 248-2	Low-Voltage Fuses – Part 2: Class C Fuses
UL	UL 248-3	Low-Voltage Fuses – Part 3: Class CA and CB Fuses
UL	UL 248-4	Low-Voltage Fuses – Part 4: Class CC Fuses
UL	UL 248-5	Low-Voltage Fuses – Part 5: Class G Fuses
UL	UL 248-6	Low-Voltage Fuses – Part 6: Class H Non-Renewable Fuses
UL	UL 248-7	Low-Voltage Fuses – Part 7: Class H Renewable Fuses
UL	UL 248-8	Low-Voltage Fuses – Part 8: Class J Fuses
UL	UL 248-9	Low-Voltage Fuses – Part 9: Class K Fuses
UL	UL 248-10	Low-Voltage Fuses – Part 10: Class L Fuses
UL	UL 248-11	Low-Voltage Fuses – Part 11: Plug Fuses
UL	UL 248-12	Low-Voltage Fuses – Part 12: Class R Fuses
UL	UL 248-13	Low-Voltage Fuses – Part 13: Semiconductor Fuses
UL	UL 248-14	Low-Voltage Fuses – Part 14: Supplemental Fuses
UL	UL 248-15	Low-Voltage Fuses – Part 15: Class T Fuses
UL	UL 248-16	Low-Voltage Fuses – Part 16: Test Limiters

UL	UL 347B	Medium-Voltage Motor Controllers, Up To 15 kV
UL	ANSI/UL 414	American National Standard for Safety for Meter Sockets
UL	UL 489	Circuit Breakers, Molded-Case Switches and Circuit-Molded-Case Breaker Enclosures
UL	UL 508A	Standard for Industrial Control Panels
UL	UL 508E	IEC TYPE "2" Coordination Short Circuit Tests of Electromechanical Motor Controllers in Accordance with IEC Publication 947-4-1
UL	UL 845	Motor Control Centers
UL	UL 869A	Reference Standard for Service Equipment
UL/CSA/ ANCE	UL 891/CSA-C22.2 No. 244/NMX-J-118/2	Standard for Dead-Front Switchboards
UL	UL 977	Fused Power-Circuit Devices
UL	UL 1008	Transfer Switch Equipment
UL	UL 1008M	Transfer Switch Equipment, Meter-Mounted
UL	UL 1053	Ground Fault Sensing and Relaying Equipment
UL	UL 1066	Low-Voltage AC and DC Power Circuit Breakers Used in Enclosures
UL	UL 1429	Pullout Switches
UL	UL 1558	Metal-Enclosed Low-Voltage Power Circuit Breaker Switchgear
UL	UL 2111	Standard for Overheating Protection for Motors
UL	UL 4248-1	Fuseholders – Part 1: General Requirements
UL	UL 4248-4	Standard for Safety for Fuseholders – Part 4: Class CC
UL	UL 4248-5	Standard for Safety for Fuseholders – Part 5: Class G
UL	UL 4248-6	Standard for Safety for Fuseholders – Part 6: Class H
UL	UL 4248-8	Standard for Safety for Fuseholders – Part 8: Class J
UL	UL 4248-9	Standard for Safety for Fuseholders – Part 9: Class K
UL	UL 4248-11	Standard for Safety for Fuseholders – Part 11: Type C (Edison Base) and Type S Plug Fuse
UL	UL 4248-12	Standard for Safety for Fuseholders – Part 12: Class R
UL	UL 4248-13	Standard for Safety for Fuseholders – Part 15: Class T
UL	UL 60947-1	Standard for Safety for Low-Voltage Switchgear and Controlgear – Part 1: General Rules

Continued

TABLE 8.6   Motor control and protection codes, standards, and recommended practices—cont'd

Developer	Standard No.	Title
UL	UL 60947-1	Low-Voltage Switchgear and Controlgear – Part 1: General rules
UL	UL 60947-4-1A	Low-Voltage Switchgear and Controlgear – Part 4-1: Contactors and Motor-Starters – Electromechanical Contactors and Motor-Starters
UL	UL 60947-5-2	Standard for Low-Voltage Switchgear and Controlgear – Part 5-2: Control Circuit Devices and Switching Elements – Proximity Switches
UL	UL 60947-7-1	Standard for Low-Voltage Switchgear And Controlgear – Part 7-1: Ancillary equipment – Terminal Blocks for Copper Conductors
UL	UL 60947-7-2	Standard for Low-Voltage Switchgear and Controlgear – Part 7-2: Ancillary Equipment – Protective Conductor Terminal Blocks for Copper Conductors
UL	UL 60947-7-3	Standard for Low-Voltage Switchgear and Controlgear – Part 7-3: Ancillary Equipment – Safety Requirements for Fuse Terminal Blocks
NFPA	NFPA 70®	National Electrical Code®
NFPA	NFPA 79	Electrical Standard for Industrial Machinery
NFPA	NFPA 110	Emergency and Standby Power Systems
NFPA	NFPA 111	Stored Electrical Energy Emergency and Standby Power Systems
IEEE	IEEE 649	IEEE Standard for Qualifying Class 1E Motor Control Centers for Nuclear Power Generating Stations
IEEE	IEEE 1290	IEEE Guide for Motor Operated Valve (MOV) Motor Application, Protection, Control, and Testing in Nuclear Power Generating Stations
IEEE	IEEE C37.013	IEEE Standard for AC High-Voltage Generator Circuit Breakers Rated on a Symmetrical Current Basis
IEEE	IEEE C37.13	IEEE Standard for Low-Voltage AC Power Circuit Breakers Used in Enclosures
IEEE	IEEE C37.103	IEEE Guide for Differential and Polarizing Relay Circuit Testing
IEEE	IEEE C37.110	IEEE Guide for the Application of Current Transformers Used for Protective Relaying Purposes
IEEE	ANSI/IEEE C27.20.1	Standard for Metal-Enclosed Low-Voltage Power Circuit Breaker Switchgear
IEEE	IEEE C37.27	IEEE Standard Application Guide for Low-Voltage AC Nonintegrally Fused Power Circuit Breakers (Using Separately Mounted Current-Limiting Fuses)

IEEE	IEEE C37.46	American National Standard Specifications for Power Fuses and Fuse Disconnecting
IEEE	IEEE C37.47	American National Standard Specifications for Distribution Fuse Disconnecting Switches, Fuse Supports, and Current-Limiting Fuses
IEEE	IEEE C37.90	IEEE Standard for Relays and Relay Systems Associated with Electric Power Apparatus
IEEE	IEEE C37.90.1	IEEE Standard for Surge Withstand Capability (SWC) Tests for Relays and Relay Systems Associated with Electric Power Apparatus
IEEE	IEEE C37.90.2	IEEE Standard for Withstand Capability of Relay Systems to Radiated Electromagnetic Interference from Transceivers
IEEE	IEEE C37.90.3	IEEE Standard Electrostatic Discharge Tests for Protective Relays
IEEE	IEEE C37.92	IEEE Standard for Analog Inputs to Protective Relays from Electronic Voltage and Current Transducers
IEEE	IEEE C37.96	IEEE Guide for AC Motor Protection
IEEE	IEEE Std C37.103	IEEE Std C37.103 – 2004 IEEE Guide for Differential and Polarizing Relay Circuit Testing
IEEE	IEEE C37.105	IEEE Standard for Qualifying Class 1E Protective Relays and Auxiliaries for Nuclear Power Generating Stations
IEEE	IEEE C37.110	IEEE Guide for the Application of Current Transformers Used for Protective Relaying Purposes
IEEE	IEEE C37.112	IEEE Standard Inverse-Time Characteristic Equations for Overcurrent Relays
IEEE	IEEE C37.117	IEEE Guide for the Application of Protective Relays Used for Abnormal Frequency Load Shedding and Restoration
IEEE	IEEE C37.235	IEEE Guide for the Application of Rogowski Coils Used for Protective Relaying Purposes
IEEE	IEEE C57.13	IEEE Standard Requirements for Instrument Transformers
IEEE	IEEE C57.13.1	IEEE Guide for Field Testing of Relaying Current Transformers
IEEE	IEEE C57.13.2	IEEE Standard C57.13.2 – 2005 IEEE Standard Conformance Test Procedure for Instrument Transformers
IEEE	IEEE C57.13.3	IEEE Standard C57.13.3 – 2005 IEEE Guide for Grounding of Instrument Transformer Secondary Circuits and Cases
IEEE	IEEE C57.13.6	IEEE Standard for High-Accuracy Instrument Transformers

TABLE 8.7 Canadian motor control and protection codes, standards, and recommended practices

Developer	Standard No.	Title
CSA	CAN/CSA C22.1	Canadian Electrical Code, Part I, Safety Standard for Electrical Installations
CSA	CAN/CSA-C22.2 NO. 0	General Requirements – Canadian Electrical Code, Part II
CSA	CAN/CSA-C22.2 NO. 4 [12]	Enclosed and Dead-Front Switches
CSA	CAN/CSA-C22.2 NO. 5 [13]	Molded-Case Circuit Breakers, Molded-Case Switches and Circuit-Breaker Enclosures
CSA	CAN/CSA-C22.2 NO. 14	Industrial Control Equipment
CSA	C22.2 NO. 29	Panelboards and Enclosed Panelboards
CSA	C22.2 NO. 31	Switchgear Assemblies
CSA	C22.2 NO. 39	Fuseholder Assemblies
CSA	C22.2 NO. 66.3 [14]	Low-Voltage Transformers – Part 3: Class 2 and Class 3 Transformers
CSA	C22.2 NO. 77	Motors with Inherent Overheating Protection
CSA	C22.2 NO. 106	HRC – Miscellaneous Fuses
CSA	C22.2 NO. 107.3-05	Uninterruptible Power Systems
CSA	C22.2 NO. 156	Solid-State Speed Controls
CSA	C22.2 NO. 178	Automatic Transfer Switches
CSA	C22.2 NO. 178.1	Requirements for Transfer Switches
CSA	C22.2 NO. 178.2	Requirements for Manually Operated Generator Transfer Panels
CSA	C22.2 NO. 244 [15]	Switchboards
CSA	CAN/CSA-C22.2 NO. 248.1 [16]	Low-Voltage Fuses – Part 1: General Requirements
CSA	CAN/CSA-C22.2 NO. 248.2	Low-Voltage Fuses – Part 2: Class C Fuses
CSA	CAN/CSA-C22.2 NO. 248.3	Low-Voltage Fuses – Part 3: Class CA and CB Fuses
CSA	CAN/CSA-C22.2 NO. 248.4	Low-Voltage Fuses – Part 4: Class CC Fuses
CSA	CAN/CSA-C22.2 NO. 248.5	Low-Voltage Fuses – Part 5: Class G Fuses
CSA	CAN/CSA-C22.2 NO. 248.6	Low-Voltage Fuses – Part 6: Class H Non-Renewable Fuses
CSA	CAN/CSA-C22.2 NO. 248.7	Low-Voltage Fuses – Part 7: Class H Renewable Fuses
CSA	CAN/CSA-C22.2 NO. 248.8	Low-Voltage Fuses – Part 8: Class J Fuses
CSA	CAN/CSA-C22.2 NO. 248.9	Low-Voltage Fuses – Part 9: Class K Fuses
CSA	CAN/CSA-C22.2 NO. 248.10	Low-Voltage Fuses – Part 10: Class L Fuses
CSA	CAN/CSA-C22.2 NO. 248.11	Low-Voltage Fuses – Part 11: Plug Fuses
CSA	CAN/CSA-C22.2 NO. 248.12	Low-Voltage Fuses – Part 12: Class R Fuses
CSA	CAN/CSA-C22.2 NO. 248.13	Low-Voltage Fuses – Part 13: Semiconductor Fuses
CSA	CAN/CSA-C22.2 NO. 248.14	Low-Voltage Fuses – Part 14: Supplemental Fuses
CSA	CAN/CSA-C22.2 NO. 248.15	Low-Voltage Fuses – Part 15: Class T Fuses
CSA	CAN/CSA-C22.2 NO. 248.16	Low-Voltage Fuses – Part 16: Test Limiters
CSA	C22.2 NO. 254 [17]	Motor Control Centers

TABLE 8.7 Canadian motor control and protection codes, standards, and recommended practices—cont'd

Developer	Standard No.	Title
CSA	CAN/CSA-C22.2 NO. 4248.1 [18]	Fuseholders – Part 1: General Requirements
CSA	CAN/CSA-C22.2 NO. 4248.4	Fuseholders – Part 4: Class CC
CSA	CAN/CSA-C22.2 NO. 4248.5	Fuseholders – Part 5: Class G
CSA	CAN/CSA-C22.2 NO. 4248.6	Fuseholders – Part 6: Class H
CSA	CAN/CSA-C22.2 NO. 4248.8	Fuseholders – Part 8: Class J
CSA	CAN/CSA-C22.2 NO. 4248.9	Fuseholders – Part 9: Class K
CSA	CAN/CSA-C22.2 NO. 4248.12	Fuseholders – Part 12: Class R
CSA	CAN/CSA-C22.2 NO. 4248.15	Fuseholders – Part 15: Class T
CSA	CAN/CSA-C60044-1	Instrument Transformers – Part 1: Current Transformers
CSA	CAN/CSA-C60044-2	Instrument Transformers – Part 2: Inductive Voltage Transformers
CSA	CAN/CSA-C60044-3	Instrument Transformers – Part 3: Combined Transformers
CSA	CAN/CSA-C60044-5	Instrument Transformers – Part 5: Capacitor Voltage Transformers
CSA	CAN/CSA-C60044-6	Instrument Transformers – Part 6: Requirements for Protective Current Transformers for Transient Performance
CSA	CAN/CSA-C60044-7	Instrument Transformers – Part 7: Electronic Voltage Transformers
CSA	CAN/CSA-C60044-8	Instrument Transformers – Part 8: Electronic Current Transformers
CSA	CAN/CSA-C22.2 NO. 60947-1 [19]	Low-Voltage Switchgear and Controlgear – Part 1: General Rules
CSA	CAN/CSA-C22.2 NO. 60947-4-1 ***	Low-Voltage Switchgear and Controlgear – Part 4-1: Contactors and Motor-Starters – Electromechanical Contactors and Motor-Starters

***Tri-National standard, with UL 60947-1A and NMX-J-290-ANCE

Motor, motor conductors, and motor controller protection is primarily governed in the United States by Article 430, Motors, Motor Circuits, and Controllers in NFPA 70, *National Electrical Code*® [20]. Motor circuits are reduced into several component parts in Figure 430.1 of that document including:

Motor Feeder

Motor Feeder-Short-Circuit and Ground-Fault Protection

Motor Disconnecting Means

Motor Branch-Circuit Short-Circuit and Ground-Fault Protection

Motor Conductor

Motor Controller

Motor Control Circuits

Motor Overload Protection

Motor

Motor Thermal Protection

Secondary Controller and Secondary Conductors

Secondary Resistor

The sections applicable to those component parts are also listed in NEC® Figure 430.1. Conductor ampacity requirements will vary with the motor service and type. Motors can be rated for continuous or non-continuous duty. Requirements for selection of conductors for Duty-Cycle Service for some motor applications are provided in NEC® Table 430.22(E). Motors with wound-rotor secondaries will have their control conductors sized in accordance with NEC® Article 430.23 and Table 430.23(C). Feeder demand factors may apply under certain circumstances noted in NEC® Article 430.26.

Operating voltage levels are defined for the United States in ANSI/NEMA C84.1 *Electric Power Systems and Equipment – Voltage Ratings (60 Hertz)*. A similar document used in Canada is CAN3-C235-83, *Preferred Voltage Levels for AC Systems, 0 to 50 000 V*. Voltage operating levels established by the International Electrotechnical Commission can be found in IEC 60038 *Standard Voltages*.

The most common low-voltage motor starting technique is through the use of a magnetic contactor. NEMA Standard ICS 2, *Industrial Control and Systems: Controllers, Contactors, and Overload Relays Rated 600 Volts* sets requirements for control devices. That standard is also used in conjunction with NEMA ICS 1, *Industrial Control and Systems: General Requirements*. NEMA ICS 2 lists three classifications for manual or magnetic controllers, based on their interrupting medium and rating. Those classifications include Class A, B, and V. Controllers are only rated for interruption during operating overload and not while under circuit fault conditions. NEMA ICS 1 defines an *Operating Overload* as:

> The overcurrent to which electric apparatus is subjected in the course of the normal operating conditions that it may encounter. [21]

> Class A controllers are alternating current air-break, vacuum-break or oil-immersed manual or magnetic controllers for service on 600 volts or less. They are capable of interrupting operating overloads but not short circuits or faults beyond operating overloads.

Class B controllers are direct current air-break manual or magnetic controllers for service on 600 volts or less. They are capable of interrupting operating overloads but not short circuits or faults beyond operating overloads.

Class V controllers are alternating current vacuum-break magnetic controllers for service on 1500 volts or less, and capable of interrupting operating overloads but not short circuits or faults beyond operating overloads. [22]

NEMA ICS 2, Part 2 lists several AC non-combination magnetic motor controller classification starting methods. Those methods are specifically for squirrel-cage and wound-rotor induction motors rated 600 V or less, 50–60 Hertz. Non-combination indicates that the magnetic controller is not used in conjunction with an integral circuit breaker. The motor controller classifications include:

1. Full-Voltage Controller

2. Full-Voltage Part-Winding Controller

3. Full-Voltage Two-Speed Motor Controller

4. Reduced-Voltage Controller

    (a) Reactor or Resistor

    (b) Wye-Delta (Wye Start/Delta Run)

        (1) Open-Circuit Transition

        (2) Closed-Circuit Transition

    (c) Reduced-Voltage Part-Winding

    (d) Reduced-Voltage Autotransformer

Full-voltage part-winding controllers utilize only part of the motor windings for starting. The remaining windings are connected once the motor comes to speed, with the use of both a starting and running contactor and a time delay control circuit. Starting winding arrangements for $\frac{1}{2}$-wye or $\frac{1}{2}$-delta or $\frac{2}{3}$-wye or $\frac{2}{3}$-delta motor configurations are normally used.

NEMA MG 1, Part 14.38 – *Characteristics of Part-Winding-Start Polyphase Induction Motors* provides some additional information. It notes that a motor may only develop approximately 50% of its normal locked-rotor torque and approximately 60% of its locked-rotor current in that starting configuration. The determination of the motor driven load requirements becomes critical when using this motor starting configuration to assure that the motor will start. This control scheme may also result in higher starting current in the energized portion of the windings, causing increased temperature rise in that winding insulation.

Full-voltage, two-speed motor controllers are available in two motor-winding configurations:

Single-winding two-speed motors

Two-winding two-speed motors

Because of the potential for short-circuit-induced currents in any unused portion of a delta configured motor winding, the controller must be designed to open one corner of each unused delta winding not operating in the circuit.

NEMA ICS 2 establishes the range of operating voltages for contactor/controller coils. AC controller coils should be capable of operating without sustaining damage between 85% and 110% of their rated voltage. It also requires that should a control-circuit transformer be used to supply the controller coil and have its primary winding connected to the supply side of the controller, then that controller should be capable of sustaining 110% of rated voltage without permanent damage. It should also be capable of continuous operation at 90% of the rated voltage of the supply circuit.

NEMA ICS 1 establishes insulation classes for the controller coils in Section 8.3.2.1. The magnetic controller coil insulation classes include:

1. Class A Insulation – Materials or combinations of materials such as cotton, silk, and paper where suitably impregnated or coated or where immersed in a dielectric liquid such as oil …

2. Class B Insulation – Materials or combinations of materials such as mica, glass fiber, etc., with suitable bonding substances ….

3. Class C Insulation – Insulation that consists entirely of mica, porcelain, glass, quartz, and similar inorganic materials …

4. Class O Insulation – Materials or combinations of materials such as cotton, silk, and paper without impregnation …

5. Class H Insulation – Materials or combinations of materials such as silicone elastomer, mica, glass fiber, etc., with suitable bonding substances such as appropriate silicone resins …[23]

Overload Relays

Overload relays provide overcurrent protection on motor circuits. NEMA ICS 2, Part 4 – *Overload Relays* provides guidance in their utilization. NEC® Article 430.37 defines the number of overload relays required for motor overload protection. The NEC also requires in Article 430.40 that the overloads be protected by circuit breakers, fuses, or motor short-circuit protectors if the overloads are not capable of opening under short circuit or ground fault conditions. Overload protection device sizing requirements are specified in NEC® Articles 430.32 and 430.33. Overload devices can be mounted separately or utilized in conjunction with motor controllers. Overload relay classifications include *instantaneous overcurrent relays* and *inverse-time overload relays*.

Inverse-time overload relays are described by time-current characteristics which are designated by a Class Number. The Class Number represents the maximum operating or tripping time that the device will operate within, carrying a current equal to 600% of its current rating. Class 10, 20, and 30 overloads will operate or trip within 10, 20, and 30 seconds or less respectively.

Inverse-time overload relays have an operating memory characteristic which may be either volatile or nonvolatile. That characteristic has the capability to consider the cumulative heating effect of the motor current because of motor operation or overload conditions. The categories associated with overload relay memory are characterized below from NEMA ICS 2, Part 4 – *Overload Relays* [24]:

Category	*Capability*
A	Nonvolatile Operating Memory
B	Volatile Operating Memory
C	No Operating Memory

An instantaneous trip overload relay is one that will trip without delay when subjected to sufficient current levels. They are utilized in applications where a motor must be removed instantly when subjected to overcurrent conditions. An instantaneous overload relay must not operate during normal motor starting conditions. That may require that it be temporarily disabled during normal motor start-up.

Overload relays include the following types outlined in NEMA ICS 2, Part 4:

1. Devices Responding to Current Heating Effects

    (A) *Thermal Relays* – line current produces heat within the device

    (B) *Solid State* – monitors current and determines heating effect

2. Current Magnitude

    (A) *Magnetic*

    (B) *Induction-Disc*

    (C) *Solid-State Relays*

Overload relays also have a limit of self-protection which is defined for device operation at 40 °C. NEMA ICS 2 defines it as:

> The maximum current value that an overload relay can respond to without sustaining damage that will impair its function. [25]

Inverse-time overload relays also have time-current characteristic curves for operation at 40 °C ambient conditions. Those curves express:

> the maximum operating times in seconds under the designated conditions associated with the current rating, at current values corresponding to multiples of the current rating. [26]

Overload relays have contacts which are controlled by the relay operation. Those contacts can be utilized to perform a control function, such as opening a motor controller or contactor. ICS 2-2000 [27] designates that the overload relay contacts shall meet the requirements of NEMA ICS 5, Table 1-4-1 for AC control circuits; Table 1-4-2 for DC control circuits; and Tables 1-4-4 and 1-4-5 for semiconductor control circuit switching elements. NEMA ICS 5-2000, *Industrial Control and Systems Control-Circuit and Pilot Devices* establishes the criteria for control contacts in Part 1, Sections 4.1 AC Mechanical Switching Elements; 4.2.1 DC Contact Rating Designations; and 4.3 AC/DC Solid State Switching Elements Ratings.

DC Manual and Magnetic Controllers

NEMA ICS 2, Part 5 – *DC General-Purpose Constant-Voltage Controllers* establishes criteria for the utilization of Class B manual and magnetic controllers for 600 Volts or less DC motors. NEMA requires magnetic controller DC coils in NEMA ICS 2-2000 (R2005), Part 1, Paragraph 8.2.1 to withstand continuously, without sustaining permanent damage, 110% of the controller rated voltage. Magnetic controllers are also required to close (operate) at 80% of their rated voltage.

Shunt or compound DC motors can be controlled by application of a constant voltage source to the motor armature windings and variation of the voltage level applied to the field excitation windings. This technique allows speed control over a range of about 4 or 5 to 1 [28] by placing a rheostat in series with the shunt field. NEMA ICS 2, Part 5 discusses that speed control technique in more detail.

AC Combination Motor Controllers

A *combination motor controller* (600V or less) is:

> A magnetic controller with additional externally operable disconnecting means contained in a common enclosure. The disconnecting means may be a circuit breaker or a disconnect switch. [29]

Part 6 of NEMA ICS 2-2000 (R2005) is applicable to the following combination motor controllers [30]:

a. Full voltage, nonreversing

b. Full voltage, reversing

c. Full voltage, two speed, two winding

d. Full voltage, two speed, one winding

e. Part winding, nonreversing

f. Wye delta, single step, nonreversing

g. Reduced voltage, single step, nonreversing

The National Electrical Code® specifies motor branch-circuit and ground-fault protection in Article 430, Part IV. NEMA ICS 2, Part 8 – *Disconnect Devices for Use in Industrial Control Equipment* provides additional requirements for motor control disconnects. The following standards are applicable for motor controllers:

1. NEMA ICS 1, *Industrial Control and Systems: General Requirements*

2. NEMA ICS 2, *Industrial Control and Systems: Controllers, Contactors, and Overload Relays, Rated Not More Than 2000 Volts AC or 750 Volts DC*

3. NEMA ICS 3, *Industrial Control and Systems: Medium-Voltage Controllers Rated 2001 to 7200 Volts AC*

4. NEMA ICS 6, *Industrial Control and Systems: Enclosures*

5. NEMA KS 1, *Enclosed and Miscellaneous Distribution Equipment Switches (600 Volts Maximum)*

6. NEMA FU 1, *Low-Voltage Cartridge Fuses*

7. UL 489, *Molded-Case Circuit Breakers, Molded Case Switches, and Circuit-Breaker Enclosures*

Any short-circuit protection device, fuse or circuit breaker, used in a combination motor controller must be capable of interrupting the available short-circuit currents at its connection point. The short-circuit protection device *short-circuit interrupting rating* is defined as:

A rating based on the highest rms AC current that the short-circuit protective device (SCPD) is required to interrupt under conditions specified. For a SCPD, such as a circuit breaker, the short-circuit interrupting rating is expressed as maximum available fault current in rms symmetrical amperes and maximum nominal application voltage. [31]

The rating or setting of a short-circuit/overcurrent protection device is defined in NEC Article 430.52 Rating or Setting for Individual Motor Circuit. NEC Article 430.55 Combined Overcurrent Protection permits short-circuit, ground-fault, and overload protection to be combined into a single protective device.

Combination motor controller short-circuit protection devices (SCPD) and overload protection devices are respectively designed to open under short-circuit and overload conditions only. The SCPD must be capable of being manually opened to disconnect the motor it supplies. Fuse disconnects must be provided with access to remove and replace their fuses. Circuit breakers must be provided with a means to reset that device should it trip. Overload protection devices may be either automatically or manually reset. With a combination motor controller mounted in an enclosure, a mechanical means must be provided to allow opening the circuit breaker and fused switch, resetting the circuit breaker, and manually resetting a non-automatic resetting overload relay from outside an enclosure. Padlocking provisions must also be provided to allow locking of the SCPD in the "OPEN" position.

There are two basic types of overcurrent trips for circuit breakers used in combination motor controllers. They include inverse-time and instantaneous tripping elements. An instantaneous trip circuit breaker is a magnetic-only device, which must be utilized in conjunction with a circuit overload device to provide motor overcurrent and short-circuit protection. The instantaneous trip device may have an adjustable trip provision. NEMA AB 3, *Molded Case Circuit Breakers and Their Application* provides information on the use of circuit breakers. Electromechanical circuit breakers utilize thermal devices to trip under overcurrent conditions. Electronic circuit breakers utilize current level sensing devices.

Adjustable Speed Drives

The NEMA Standards Publication NEMA ICS 7, *Industrial Control and Systems, Adjustable-Speed Drives* defines a motor drive system as:

> An interconnected combination of equipment which provides a means of adjusting the speed of a mechanical load coupled to a motor. A drive system typically consists of a drive and auxiliary electrical apparatus. [32]

Table 8.8 presents a list of *Adjustable Speed Drive (ASD)* applicable standards used in the United States. The term adjustable speed drive is often used interchangeability with *Variable Frequency Drive (VFD)* and *Variable Speed Drive (VSD)*. The IEC adjustable speed drive standards are listed in Table 8.9.

The National Electrical Code® defines an Adjustable Speed Drive as:

> A combination of the power converter, motor and motor-mounted auxiliary devices such as encoders, tachometers, thermal switches and detectors, air blowers, heaters, and vibration sensors. [33]

TABLE 8.8 Some adjustable speed drives United States standards

NEMA	ANSI/NEMA ICS 61800-1	Adjustable Speed Electrical Power Drive Systems, Part 1: General Requirements – Rating Specifications for Low-Voltage Adjustable Speed DC Power Drive Systems
NEMA	NEMA ICS 61800-2	Adjustable Speed Electrical Power Drive Systems, Part 2: General Requirements – Rating Specifications for Low-Voltage Adjustable Frequency AC Power Drive Systems
NEMA	NEMA ICS 61800-4	Adjustable Speed Electrical Power Drive Systems: Part 4: General Requirements – Rating Specifications for AC Power Drive Systems Above 1000 V AC and not Exceeding 35 kV
NEMA	NEMA ICS 3.1	Safety Standards for Construction and Guide for Selection, Installation and Operation of Adjustable Speed Drive Systems
NEMA	NEMA ICS 7	Industrial Control and Systems: Adjustable-Speed Drives
NEMA	NEMA ICS 7.1	Safety Standards for Construction and Guide for Selection, Installation, and Operation of Adjustable-Speed Drive Systems
IEEE	IEEE 519	IEEE Recommended Practices and Requirements for Harmonic Control in Electric Power Systems
IEEE	IEEE Std 958™	IEEE Guide for the Application of AC Adjustable-Speed Drives on 2400 to 13 800 V Auxiliary Systems in Electric Power Generating Stations -Description
IEEE	IEEE Std 1566	IEEE Standard for Performance of Adjustable Speed AC Drives Rated 375 kW (500 HP) and Larger
IEEE	IEEE Std C37.96	IEEE Guide for AC Motor Protection -Description
UL	UL 508C	Power Conversion Equipment

NEC® Article 430, Part X, Adjustable Speed Drives provides requirements for conductor selection, overload protection, motor overtemperature protection, and disconnecting means on system 600 Volts and less. Parts I through IX are also applicable, unless modified or supplemented by Part 8. Part XI in Article 430 similarly deals with adjustable speed drives over 600 Volts nominal.

There are five basic methods of speed control for an induction motor. Those include [34]:

  a. Changing the number of poles

  b. Varying the line frequency

  c. Varying the line voltage

  d. Varying the rotor resistance

  e. Inserting rotor voltages of the appropriate frequency

TABLE 8.9 **Some IEC adjustable speed drive standards**

Developer	Standard No.	Title
IEC	IEC 61800-1	Adjustable Speed Electrical Power Drive Systems – Part 1: General Requirements – Rating specifications for low-voltage adjustables speed DC power drive systems
IEC	IEC 61800-2	Adjustable Speed Electrical Power Drive Systems – Part 2: General Requirements – Rating specifications for low-voltage adjustable frequency AC power drive systems
IEC	IEC 61800-3	Adjustable Speed Electrical Power Drive Systems – Part 3: EMC Requirements and Specific Test Methods
IEC	IEC 61800-4	Adjustable Speed Electrical Power Drive Systems – Part 4: General Requirements – Rating specifications for AC power drive systems above 1000 VAC and not exceeding 35 kV
IEC	IEC 61800-5-1	Adjustable Speed Electrical Power Drive Systems – Part 5-1: Safety Requirements – Electrical, thermal and energy
IEC	IEC 61800-5-2	Adjustable Speed Electrical Power Drive Systems – Part 5-2: Safety Requirements – Functional
IEC	IEC 61800-7-1	Adjustable Speed Electrical Power Drive Systems – Part 7-1: Generic Interface and Use of Profiles for Power Drive Systems – Interface definition
IEC	IEC 61800-7-201	Adjustable Speed Electrical Power Drive Systems – Part 7-201: Generic Interface and Use of Profiles for Power Drive Systems – Profile type 1 specification
IEC	IEC 61800-7-202	Adjustable Speed Electrical Power Drive Systems – Part 7-202: Generic Interface and Use of Profiles for Power Drive Systems – Profile type 2 specification
IEC	IEC 61800-7-203	Adjustable Speed Electrical Power Drive Systems – Part 7-203: Generic Interface and Use of Profiles for Power Drive Systems – Profile type 3 specification
IEC	IEC 61800-7-204	Adjustable Speed Electrical Power Drive Systems – Part 7-204: Generic Interface and Use of Profiles for Power Drive Systems – Profile type 4 specification
IEC	IEC 61800-7-301	Adjustable Speed Electrical Power Drive Systems – Part 7-301: Generic Interface and Use of Profiles for Power Drive Systems – Mapping of profile type 1 to network technologies
IEC	IEC 61800-7-302	Adjustable Speed Electrical Power Drive Systems – Part 7-302: Generic Interface and Use of Profiles for Power Drive Systems – Mapping of profile type 2 to network technologies
IEC	IEC 61800-7-303	Adjustable Speed Electrical Power Drive Systems – Part 7-303: Generic Interface and Use of Profiles for Power Drive Systems – Mapping of profile type 3 to network technologies
IEC	IEC 61800-1-304	Adjustable Speed Electrical Power Drive Systems – Part 7-304: Generic Interface and Use of Profiles for Power Drive Systems – Mapping of profile type 4 to network technologies

Motor speed can be determined by the following relationship:

$$s = 120 f/P \hspace{4cm} \text{(Eq. 8.3)}$$

where s is the speed in RPM, f is the line frequency (Hertz) and P is the number of motor poles.

From that mathematical relationship, it is evident that motor speed can be controlled by changing the line frequency. VFD are basically AC to DC to AC converters that control the motor feeder voltage frequency electronically. A simplified description of an alternating current (AC) VFD device would include one which has three basic segments, including an input rectifier section, a dc bus section and an inverter output section.

Rectifiers commonly convert a source sinusoidal waveform into a rectified DC half-wave or full-wave waveform. The *rectifier* section in a VFD converts the input sinusoidal voltage waveform into a series of half-wave or full-wave pulses, depending upon the rectifier configuration. The rectified waveforms are composed of a fundamental frequency component plus harmonic components. The amount of harmonic components contained in the waveform can be directly affected by the rectifier section electronic components utilized, i.e. diodes, thyristors, silicon controlled rectifiers (SCR) or IGBT transistors. The harmonic component content is also affected by the number of rectifiers used per phase. With a three-phase rectifier section, three rectifiers would be the minimum number used. However, a minimum of two rectifiers per phase are normally used. That concept is normally described as a 6-pulse VFD. VFDs are available in multiples of six, i.e. 6-pulse, 12-pulse, 18-pulse, etc. The higher the pulse level, the less *total harmonic distortion* (THD) will be present in the VFD output waveform.

A Fourier Analysis of a VFD-rectified waveform would reveal that its harmonic components are based on multiples of the basic input frequency or *fundamental frequency*. For a 60 Hz input waveform, the fundamental frequency would be 60 Hz and its harmonic components would be multiples of 60 Hz, with an infinite number of frequency components theoretically possible. However, the rectifier section voltage typically contains the fundamental frequency plus odd harmonic components including the 5th, 7th, 11th, and 13th harmonics.

Harmonics that are multiples of the 2nd (2, 4, 6, etc.) and 3rd (3, 6, 9, etc.) order harmonics tend to cancel out their respective 2nd and 3rd harmonic components. The 5th, 7th, 11th, 13th, etc. harmonic components are the most damaging to electrical circuit components; however, the magnitude of those harmonics drops substantially with the 13th and higher order components. Devices such as transformers, power factor correction capacitors and sensitive electronic components can be adversely affected by voltage and current harmonic components. Neutral cable overheating and derating of cable ampacity can also be the result of power system harmonic content.

The VFD *DC bus* section contains inductive and capacitance elements which filter the rectified DC power waveform, smoothing out the rippled DC waveform from the rectifier.

The *inverter* section converts the DC bus voltage into an alternating current (AC) voltage waveform. There are several methods that can be utilized to develop a sinusoidal waveform thought the inverter. One common device is an *insulated gate bipolar transistor* (IGBT). *Pulse width modulation* (PWM) switching of an IGBT can be utilized to create a sinusoidal waveform for motor control by adjusting the inverter output voltage waveform frequency and amplitude.

Starting currents can be substantially reduced using VFDs by producing a low frequency output voltage waveform for starting. This allows a VFD to be used as a motor starter, limiting the motor starting current to the motor's full load current. Across-the-line full voltage motor starting can create starting currents 800% or more of the motor full load current. Reduced voltage starters and solid state starters produce motor starting currents of approximately 400–500% and 200–350% respectively of the motor full load current.

Three-phase, four-wire circuits supplying nonlinear loads, such as rectifier and capacitor power supplies, can generate excessive current on the neutral conductor.

> In three-phase circuits, the triplen harmonic neutral currents (e.g., odd-order multiples of three such as the 3rd, 9th, and 15th) add together on the neutral instead of cancelling, so unexpectedly high neutral currents may exist where line-to-neutral connected nonlinear loads are in use on a four-wire, wye-connected supply system. [35] … Excessive current in the neutral path occurs as the triplen currents (e.g., odd-ordered multiples of 3 times the fundamental power frequency) are additive on the neutral path since they are both in-phase and spaced apart by 120 electrical degrees. Therefore, under worst-case conditions, the true-rms neutral current can approach 1.73 times (e.g., times) the phase current .,. but its signature will also be predominantly (but not exclusively) at 3 times the fundamental frequency, or 180 Hz. [36]

The National Electrical Code® Handbook recognizes harmonic distortion problems on neutral and ground conductors after its discussion of Article 220.61 Feeder or Service Neutral Load [37]:

> If the system also supplies nonlinear loads…, the neutral is considered a current-carrying conductor if the load of the electric-discharge lighting, data-processing, or similar equipment on the feeder neutral consists of more than half the total load, in accordance with 310.15(B)(4)(c). Electric-discharge lighting and data-processing equipment may have harmonic currents in the neutral that may exceed the load current in the ungrounded conductors. It would be appropriate to require a full-size or larger feeder neutral conductor, depending on the total harmonic distortion contributed by the equipment to be supplied (see 220.61, FPN No.2)…In some instances, the neutral current may exceed the current in the phase conductors. See the commentary following 310.15(B)(4)(c) regarding neutral conductor ampacity

The non-linear loads described in NEC Article 220.61 could be static power inverters which convert AC to DC, DC to DC, DC to AC, or AC to AC. Other harmonic producing equipment might include [38] power rectifiers, variable speed drives, switch-mode power supplies, and uninterruptible power supplies.

The current and voltage harmonics generated by the use of non-linear devices can also be introduced into the utility system service point providing power. Harmonic distortion damage can be mitigated on the customer electrical distribution system feeding non-linear electrically controlled equipment by the use of harmonic filters, line reactors, etc. IEEE 519, *IEEE Recommended Practices and Requirements for Harmonic Control in Electrical Power Systems* was developed to assist in the prevention of the introduction of customer generated harmonic voltage components into a utility system at the customer service point. That standard limits [39] the maximum voltage total harmonic distortion (V_{THD}) for general electrical systems to 5%, with no more than 3% for any individual harmonic component. Hospitals, airports, and other specialty applications are limited to a V_{THD} of 3%. On dedicated systems in which there are no non-linear loads, IEEE 519 allows a V_{THD} of 10%.

Voltage total harmonic distortion (V_{THD}) is defined by IEEE 519 as the ratio of the root-sum-square value for the harmonic content of the line voltage to the root-mean-square (RMS) value of the fundamental voltage V_1. The mathematical relationship for V_{THD}:

$$V_{THD} = \frac{100\% \times \sqrt{V_2^2 + V_3^2 + V_4^2 + V_5^2 + ...}}{V_1} \qquad \text{(Eq. 8.4)}$$

where V_1 is the fundamental frequency voltage component and V_2, V_3, V_4, V_5 ... are the magnitudes of the harmonic components.

Harmonic Mitigation

There are several methods used for the mitigation of the harmonics produced by non-linear loads such as variable frequency drives. Some methods are required because of the type of rectifier used in the VFD, the type of DC to AC inverter used, failure to utilize adequate harmonic filtering, etc. Some of the harmonic mitigation methods include:

Use of 18 pulse VFD drives

AC drives with active front ends – transistor rectifiers with a microprocessor controlled gate circuit

AC drives with active shunt filters

Use of AC input reactors

Use of phase shift transformers with the VFD rectifiers

Use of K-Factor transformers or derating transformers

Use of DC link reactors

Use of tuned LC (inductive/capacitive) or trap filters

Insertion of delta-wye transformers in the feeder to minimize harmonic currents

The use of K-Factor transformers or derating transformers for use with nonlinear loads is not in itself a harmonic mitigation method. They are, however, a damage mitigation method for transformers subjected to harmonic loading. ANSI/IEEE C57.12.00, *IEEE Standard for Standard General Requirements for Liquid-Immersed Distribution, Power, and Regulating Transformers* indicates:

> that power transformers should not be expected to carry load currents with harmonic factor in excess of 5% of rating. [40]

IEEE C57.12.01, *IEEE Standard General Requirements for Dry-Type Distribution and Power Transformers Including Those with Solid-Cast and/or Resin Encapsulated Windings* similarly provides recommendations for dry type transformers.

IEEE Standard C57.110, *IEEE Recommended Practice for Establishing Liquid-Filled and Dry-Type Power and Distribution Transformer Capability When Supplying Nonsinusoidal Load Currents* was developed to deal with the problem of harmonic generating nonlinear loads. The UL Standards used in association with K-Factor transformers are UL 1561, *Dry-Type General Purpose and Power Transformers* and UL 1562, *Standard for Transformers, Distribution, Dry-Type – Over 600 Volts*. Another standard associated with harmonic generating loads for transformers is IEEE Standard C57.18.10, *IEEE Standard Practices and Requirements for Semiconductor Power Rectifier Transformers*. Transformer use with harmonic generating loads is also addressed in the IEEE Emerald Book, IEEE 1100, *IEEE Recommended Practice for Powering and Grounding Electronic Equipment*.

K-Factor is a mathematical representation to characterize a transformer's ability to withstand overheating from harmonic loading without loss of normal life expectancy. UL 1561 and 1562 list two equations for determining K-Factors [41, 42]. The first is as follows:

$$K = \sum_{h-1}^{\infty} (I_h)2h2 \qquad \qquad \text{(Eq. 8.5)}$$

where K is the unit-less weighing factor (K-Factor), I_h is the per unit harmonic current component related to the fundamental frequency, and h is the harmonic order number. A K-Factor value of 1.0 would indicate a liner load with no harmonics. As the value of

K increases, so does the effect of harmonic heating. The per unit current I_h is expressed such that the total RMS current is one ampere, i.e.:

$$\sum_{h-1}^{max}(I_h)^2 = 1.0 \qquad \text{(Eq. 8.6)}$$

The mathematical expression I_h could be determined for harmonic components from the first harmonic to some very high harmonic value. It would be difficult to perform this calculation without accepting some level of harmonic as the normally accepted maximum effective harmonic level above which harmonic order components are very small in magnitude. The square of the harmonic component number can become a large number. There have been suggestions limiting the calculation to the 25th or 50th harmonic component. Many of the available harmonic analyzer equipment produce harmonic current reading in the per unit format, making insertion of any collected current data into the mathematical relationship in Eq. 8.6 above very simple.

The second equation used by UL to establish the K-Factor for a transformer is shown in Equation 8.7.

$$K = \frac{\sum_{h=1}^{max} f_h^2 \times h^2}{\sum_{H=1}^{max} f_h^2} \qquad \text{(Eq. 8.7)}$$

where f_h is the frequency in Hertz and h is the harmonic order component. Underwriters Laboratories recognizes K-Factors of 4, 9, 13, 20, 30, 40, and 50 as standard transformer ratings.

Phase shift transformers are sometimes used in variable frequency drive rectifier sections to minimize harmonic effects. IEEE C57.153, *IEEE Guide for the Application, Specification, and Testing of Phase-Shifting Transformers* and IEC 62032 Ed. 1, *Guide for the Application, Specification, and Testing of Phase-Shifting Transformers* (IEEE Standard C57.135) provide guidance with the utilization of phase shift transformers.

Long cables, connecting a power source to a motor/VFD, contain a series of natural self-inductive components and shunt distributed natural capacitive components. Natural resonant conditions can develop from those inductive and capacitive components, should the proper excitement conditions develop. Resonant conditions development can occur as a result of the presence of 5th, 7th, 11th, 13th, etc. harmonic components feeding the motor load. The only natural damping factor for the created resonant voltage and current waveforms is the resistance of the line conductors feeding the motor. Resonant conditions can produce insulation damaging voltage levels and overheat motor connections.

The introduction of a series connected inductive choke is one method to attenuate or eliminate the resonant waveform. The choke will act as a low-pass filter, attenuating the higher harmonic components and passing the lower fundamental frequency components to the motor. Three percent and 5% line reactors are commonly used to accomplish that task.

VFD generated harmonic waveform components can also cause motor winding and bearing heating problems. Because of that potential, the use of motors with a 1.15 service factor, energy efficient motors, or VFD rated motors is recommended.

Alternating current VFD-driven excited motors, operating with large inertia loads, can in some circumstances act as an induction generator, causing the voltage on the DC bus to rise above normal operating levels. To protect the VFD rectifier section from damage a *braking resistor*, can be inserted in parallel with the line to ground capacitor filter in that section. The braking resistor can be activated by an electronic component called a *brake chopper*. That component will conduct, tying the braking resistor to the circuit neutral when DC Bus voltage levels reach a predetermined point. That action will divert motor generated current from the DC bus, preventing damage to the VFD.

PWM VFDs can also cause current flow into motor rotor bearings by capacitive coupling. It can induce a rotor shaft voltage of up to 30 Volts [43]. The high switching frequency of an IGBT inverter can result in inducing current pulses in the motor bearings, if the rotor is not properly grounded. Large motors can develop circulating current between the rotor, shaft bearings, and the stator frame because of motor stator winding capacitive leakage current. The leakage current will eventually overcome the impedance of the bearing lubrication film in a process called *bearing fluting*. That process will result in a rhythmic pattern of pitting and gouges on the bearing race. Current flow through the bearings can eventually result in bearing overheating and failure.

There are several methods to minimize VFD induced motor bearing failures. They include:

Proper selection of motor feeder cable and minimizing its length

Insertion of a filter at the motor terminal end of the cable

Use of motor insulated bearings

Use of non-conductive mechanical couplings in the motor

Addition of a motor shaft grounding device

Ensure proper grounding of a motor and VFD

Selection of VFD-rated motors manufactured with insulation meeting the requirements of NEMA Standard MG1 Part 31; Paragraph 40.4.2

In situations where single-phase to ground connected harmonic generating nonlinear loads are fed over a three-phase, four-wire feeder circuit, IEEE 1100 [44] recommends the use of a delta-wye three-phase transformer on that feeder. The triplen harmonics from the nonlinear loads will be trapped in the transformer primary (delta) windings, reducing the introduction of those harmonics to other parts of the electrical distribution system. The delta-wye transformer selected for that task must be listed or certified for that service.

References

1. NEMA MG 2 Revision 1 2007, *Safety Standard and Guide for Selection, Installation, and Use of Electric Motors and Generators*; Section 4 Environmental Protection and Methods of Cooling; 2007, page 4. National Electrical Manufacturers Association; Washington, DC.
2. Ibid., Paragraph 4.1, page 4.
3. Ibid., Paragraph 4.2, page 7.
4. Ibid, page i.
5. NFPA MG 1-2006, *Motors and Generators*; 2006, page i. National Fire Protection Association; Quincy, MA.
6. NEMA MG 1-1993, *Motors and Generators* Section 1.21; 1993. National Electrical Manufacturers Association; Washington, DC.
7. Pender, Harold and Del Mar, William, *Electrical Engineers' Handbook: Electric Power*; 4th Edition, 1949, Section 8-19. John Wiley & Sons, Inc.; New York.
8. Polka, David; *Motors & Drives: A Practical Technology Guide*; c2003, Chapter 3. Research Triangle Park, NC: ISA.
9. Pender, Harold and Del Mar, William, *Electrical Engineers' Handbook: Electric Power*; 4th Edition, 1949, Section 8-2. John Wiley & Sons, Inc.; New York.
10. NEMA ICS 1-2000, *Industrial Control and Systems General Requirements*; 2000, page 5. National Electrical Manufacturers Association; Rosslyn, VA.
11. Ibid., page 6.
12. Tri-National Standard, with ANCE NMX-J-162-2004 and UL 98.
13. Tri-National Standard, with UL 489 and NMX-J-266-ANCE.
14. Bi-National Standard, with UL 5085-3.
15. Tri-National Standard, with UL 891 and ANCE NMX-J-118/2.
16. Tri-National Standard, with UL 248-1 and NMX-J-009/248/1-ANCE.
17. Tri-National Standard, with UL 845-x and NMX-J-353/x-ANCE.
18. Tri-National standard, with UL 4248-x and NMX-J-009/4248/x-ANCE.
19. Tri-National Standard, with UL 60947-x and NMX-J-XXX/x-ANCE.
20. NFPA 70, *National Electrical Code® (NEC)*; 2008, Article 430. National Fire Protection Association; Quincy, MA.

21. ICS 1-200 (R2005, R2008), *Industrial Control and Systems: General Requirements*; 2008, page 10. National Electrical Manufacturers Association, Rosslyn, VA.

22. ICS 2-2000 (R2005), *Industrial Control and Systems: Controllers, Contactors and Overload Relays Rated 600 Volts*; *General Requirements*; 2008, page 1-2. National Electrical Manufacturers Association, Rosslyn, VA.

23. ICS 2-2000 (R2005, R2008), *Industrial Control and Systems: Controllers, Contactors, and Overload Relays Rated 600 Volts*; 2008, pages 37-38. National Electrical Manufacturers Association; Rosslyn, VA.

24. NEMA ICS 2-2000 (R2005), *Industrial Control and Systems Controllers, Contactors and Overload Relays Rated 600 Volts*; 2005, Part 4, Paragraph 3.1.2, page 4-3. National Electrical Manufacturers Association; Rosslyn, VA.

25. Ibid., Section 2 Definitions page 4-2.

26. Ibid., Paragraph 4.3, page 4-4.

27. Ibid., Paragraphs 4.6 and 4.7, page 4-5.

28. Fitzgerald, A.E. and Kingsley, Charles, Jr., *Electric Machinery: The Dynamics and Statics of Electromechanical Energy Conversion*; 2nd Edition, 1961, page 173. McGraw-Hill, New York.

29. NEMA ICS 2-2000 (R2005), *Industrial Control and Systems Controllers, Contactors and Overload Relays Rated 600 Volts*; 2005, Part 6, Paragraph 2.0, page 6-1. National Electrical Manufacturers Association; Rosslyn, VA.

30. Ibid., Paragraph 3.0, page 6-1.

31. NEMA ICS 2, Part 8 *Disconnect Devices for Use in Industrial Control Equipment*; Section 2.0, page 8-2. National Electrical Manufacturers Association; Rosslyn, VA.

32. NEMA ICS 7-2006, *Industrial Control and Systems, Adjustable-Speed Drives*; 2006, Paragraph 2, page 1-1. National Electrical Manufacturers Association; Rosslyn, VA.

33. NFPA 70, *National Electrical Code*; 2005, Paragraph 430.2. National Fire Protection Association; Quincy, MA.

34. Fitzgerald, A.E. and Kingsley, Charles, Jr., *Electric Machinery: The Dynamics and Statics of Electromechanical Energy Conversion*; 2nd Edition, 1961, page 484. McGraw-Hill, New York.

35. IEEE 1100-1999, *IEEE Recommended Practice for Powering and Grounding Electronic Equipment*; 1999, Section 4.5.3.1, page 99. Institute of Electrical and Electronic Engineers; Piscataway, NJ.

36. Ibid., Section 4.5.4.2, page 105.

37. Earley, M.W., Sargent, J.S., Sheehan, J.V., Caloggero, J.M., NFPA 70, *National Electrical Code®*; 2005. National Fire Protection Association; Quincy, MA.

38. Hoevenaars, Tony P., LeDoux, Kirt, and Colosino, Matt; *Interrupting IEEE Std 519 and Meeting Its Harmonic Limits in VFD Applications*; May 6, 2003, page 2. IEEE Paper No. PCIC-2003-15.

39. Evans, Ivan, Methods of Mitigation; *Middle East Electricity Magazine*, December, 2002, pages 25-26. IIR Middle East; Dubai.

40. Jayasinghe, NR, Lucus, JR, and Perera, KBIM; *Power System Harmonic Effects on Distribution Transformers and New Design Considerations for K Factor Transformers*; IEE Sri Lanka Annual Sessions – September 2003; Section 4.0.

41. UL 1561, *Standard for Dry-Type General Purpose and Power Transformers.* Underwriters Laboratories, Inc.; Northbrook, IL.

42. UL 1562, *Standard for Transformers, Distribution, Dry-Type – Over 600 Volts.* Underwriters Laboratories, Inc.; Northbrook, IL.

43. Bulington, E.J., Abney, S., and Skibinski, G.L., Cable alternatives for PWM AC drive applications; Petroleum and Chemical Industry Conference, 1999; Industry Applications Society 46th Annual; pages 247-259.

44. IEEE 1100-1999, *IEEE Recommended Practice for Powering and Grounding Electronic Equipment; Institute of Electrical and Electronic Engineers*; 1999, Section 8.3.3.1, page 282. Piscataway, NJ.

CHAPTER 9

Electrical Hazardous (Classified) Area Design and Safe Work Practices

The *National Electrical Code*® (NEC®), NFPA 70® establishes the requirements for electrical installations in areas that are "classified as hazardous locations due to materials handled, processed, or stored" [1]. It does not classify any areas involved with the manufacture, handling, storage, transportation, or use of explosive materials. Explosive materials would consist of blasting powder, dynamite, ammunition, etc. Standards involving those materials can be found in NFPA 495, *Explosive Materials Code.*

Hazardous (classified) locations are defined by two separate methods: *Division Classification* and *Zone Classification*. These classification methods are defined in NEC Articles 500 and 505. The Division Classification system uses three descriptors to define the presence of combustible or flammable materials. Those descriptors include:

Class: I, II, or III

Division: 1 or 2

Material Group: A, B, C, or D (Class I); E, F, or G (Class II)

Class designation describes the type of materials that may be present, including gas, liquid-produced vapors, dusts, fibers, or flying. The Division designation defines the degree of probability that those materials will be present to form flammable or combustible fuel–air mixtures. Group designation is based on the physical characteristics of the flammable or combustible properties of the gases, liquid-produced vapors or dusts.

NEC Article 500.5 establishes the criteria for defining Division classification for hazardous locations. Areas are classified by the class of the vapors, liquids, gases, combustible dusts, flyings or fibers which may be present, as well as the likelihood that a sufficient ignitable quantity or concentration of flammable or combustible materials will be present. The following are the definitions of those classes:

- *Class I* – areas in which flammable gases or vapors are either present, or may be present, in sufficient quantities to produce an explosive or ignitable mixture.

- *Class II* – areas which are considered hazardous because of the presence of combustible dust.

- *Class III* – areas which are considered hazardous because of the presence of easily ignitable fibers or flyings. However, those fibers or flyings are considered not likely to be suspended in air in sufficient quantities to produce ignitable fuel-air mixtures.

The Zone designations are based on NEC, Article 505 – Class I, Zone 0, 1, and 2 Locations. Like Division designations, Zones are used to identify the probability of the presence of ignitable mixtures of flammable or combustible materials in the environment. Zone designations include Zones 0, 1, or 2.

The following is a description of the conditions for each Division/Zone Classification location:

A *Class I, Division 1* hazardous location is one:

1. In which ignitable concentrations of flammable gases or vapors can exist under normal operating conditions; or
2. In which ignitable concentrations of such gases or vapors may exist frequently because of repair or maintenance operations or because of leakage; or
3. In which breakdown or faulty operation of equipment or processes might release ignitable concentrations of flammable gases or vapor and might also cause simultaneous failure of electrical equipment to become a source of ignition. [2]

A *Class I Division 2* location is one:

1. In which volatile flammable liquids or flammable gases are handled, processed, or used, but in which the liquids, vapors, or gases will normally be confined within closed containers or closed systems from which they can escape only in case of accidental rupture or breakdown of such containers or systems or in case of abnormal operation of equipment; or
2. In which ignitable concentrations of gases or vapors are normally prevented by positive mechanical ventilation and which might become hazardous through failure or abnormal operation of the ventilating equipment; or
3. That is adjacent to a Class I, Divisions 1 location, and to which ignitable concentrations of gases or vapors might occasionally be communicated unless such communication is prevented by adequate positive-pressure ventilation from a source of clean air and effective safe-guards against ventilation failure are provided. [3]

A *Class II, Division 1* location is one:

1. In which combustible dust is in the air under normal operating conditions in quantities sufficient to produce explosive or ignitable mixtures; or
2. Where mechanical failure or abnormal operation of machinery or equipment might cause such explosive or ignitable mixtures to be produced, and might also provide a source of ignition through simultaneous failure of electric equipment, operation of protection devices, or from other causes; or
3. In which Group E combustible dusts may be present in quantities sufficient to be hazardous. [4]

A *Class II, Division 2* location is one:

1. In which combustible dust due to abnormal operations may be present in the air in quantities sufficient to produce explosive or ignitable mixtures; or
2. Where combustible dust accumulations are present but are normally insufficient to interfere with the normal operation of electrical equipment or other apparatus, but could as a result of infrequent malfunctioning of handling or processing equipment become suspended in the air; or
3. In which combustible dust accumulations on, in, or in the vicinity of the electrical equipment could be sufficient to interfere with the safe dissipation of heat from electrical equipment, or could be ignitable by abnormal operation or failure of electrical equipment. [5]

A *Class III, Division 1* location is one in which:

easily ignitable fibers or materials producing combustible flyings are handled, manufactured, or used. [6]

A *Class III, Division 2* location is one in which:

easily ignitable fibers are stored or handled other than in the process of manufacturing. [7]

Class I, Zone 0, 1, and 2 locations are those where flammable gases or liquid-produced vapors are present or may be present in sufficient quantities to mix with air and produce explosive or ignitable concentrations. They are described as follows:

A *Class I, Zone 0* location is one:

1. In which ignitable concentrations of flammable gases or vapors are present continuously; or
2. Ignitable concentrations of flammable gases or vapors are present for long period of time. [8]

Class I, Zone 1 location is one:

1. In which ignitable concentrations of flammable gases or vapors are likely to exist under normal operating conditions; or
2. In which ignitable concentrations of flammable gases or vapors may exist frequently because of repair or maintenance operations or because of leakage; or
3. In which equipment is operated or processes are carried on, of such a nature that equipment breakdown or faulty operations could result in the release of ignitable concentrations of flammable gases or vapors and also cause simultaneous failure of electrical equipment in a mode to cause the electrical equipment to become a source of ignition; or
4. That is adjacent to a Class I, Zone 0 location from which ignitable concentrations of vapors could be communicated, unless communication is prevented by adequate positive pressure ventilation from a source of clean air and effective safeguards against ventilation failure are provided. [9]

A *Class I, Zone 2* location is one:

1. In which ignitable concentrations of flammable gases or vapors are not likely to occur in normal operation and, if they do occur, will exist only for a short period; or

2. In which volatile flammable liquids, flammable gases, or flammable vapors are handled, processed, or used but in which the liquids, gases, or vapors normally are confined within closed containers of closed systems from which they can escape, only as a result of accidental rupture or breakdown of the containers or system, or as a result of the abnormal operation of the equipment with which the liquids or gases are handled, processed, or used; or

3. In which ignitable concentrations of flammable gases or vapors normally are prevented by positive mechanical ventilation but which may become hazardous as a result of failure or abnormal operation of the ventilation equipment; or

4. That is adjacent to a Class I, Zone 1 location, from which ignitable concentrations of flammable gases or vapors could be communicated, unless such communication is prevented by adequate positive-pressure ventilation from a source of clean air and effective safeguards against ventilation failure are provided. [10]

Zone classification is also utilized for combustible dusts, fibers, and flyings and is described in NEC Article 506. There are three zone designations in use which are listed in Article 506.3 as:

- Zone 20 Hazardous (Classified) Location

- Zone 21 Hazardous (Classified) Location

- Zone 22 Hazardous (Classified) Location

A *Zone 20* location is one in which:

(a) Ignitable concentrations of combustible dust or ignitable fibers or flyings are present continuously.

(b) Ignitable concentrations of combustible dust or ignitable fibers or flyings are present for long periods of time. [11]

A *Zone 21* location is one :

(a) [In which] ignitable concentrations of combustible dust or ignitable fibers or flyings are likely to exist occasionally under normal operating conditions; or

(b) In which ignitable concentrations of combustible dust or ignitable fibers or flyings may exist frequently because of repair or maintenance operations or because of leakage; or

(c) In which equipment is operated or processes are carried on, of such a nature that equipment breakdown or faulty operations could result in the release of ignitable concentrations of combustible dust, or ignitable fibers or flyings and also cause simultaneous failure of electrical equipment in a mode to cause the electrical equipment to become a source of ignition; or

(d) That is adjacent to a Zone 20 location from which ignitable concentrations of dust or ignitable fibers or flyings could be communicated, unless communication is prevented by adequate positive pressure ventilation from a source of clean air and effective safeguards against ventilation failure are provided. [12]

A *Zone 22* location is one:

(a) [In which] ignitable concentrations of combustible dust or ignorable fibers or flyings are not likely to occur in normal operation and if they do occur, will only persist for a short period; or

(b) In which combustible dust, or fibers, or flyings are handled, processed, or used but in which the dust, fibers, or flyings are normally confined within closed containers of closed systems from which they can escape only as a result of abnormal operation of the equipment with which the dust, or fibers, or flyings are handled, processed, or used; or

(c) That is adjacent to a Zone 21 location, from which ignitable concentrations of dust or fibers or flyings could be communicated, unless such communication is prevented by adequate positive pressure ventilation from a source of clean air and effective safeguards against ventilation failure are provided. [13]

The third Division classification designator is *Material Group*. Group designations are determined experimentally. Controlled samples of those materials are mixed with air and ignited in an enclosure to evaluate explosive pressures and maximum safe clearances between a test enclosure's clamped joint. Multiple tests are conducted varying the test conditions to facilitate comparison of data. Table 9.1 presents the Material Group designations provided in NEC Articles 500.6 and 505.6 and NFPA 499, *Recommended Practice for the Classification*

TABLE 9.1 Class I and II Material Group designations

	Group	Materials
500.6 [14]		
	A	Acetylene
	B	Flammable gas, flammable liquid-produced vapor, or combustible liquid-produced vapor mixed with air that may burn or explode, having either a maximum experimental safe gap (MESG) value less than or equal to 0.45 mm or a minimum igniting current ratio (MIC ratio) less than or equal to 0.4 (see Exceptions 1 and 2)
	C	Flammable gas, flammable liquid-produced vapor, or combustible liquid-produced vapor mixed with air that may burn or explode, having either a maximum experimental safe gap (MESG) value greater than 0.45 mm and less than or equal to 0.75 mm, or a minimum igniting current ratio (MIC ratio) greater than 0.40 and less than or equal to 0.80
	D	Flammable gas, flammable liquid-produced vapor, or combustible liquid-produced vapor mixed with air that may burn or explode, having either a maximum experimental safe gap (MESG) value greater than 0.75 mm or a minimum igniting current ratio (MIN ratio) greater than 0.80

(Continued)

TABLE 9.1   Class I and II Material Group designations—cont'd

	Group	Materials
505.6 [15]		
	IIC	Atmospheres containing acetylene, hydrogen, or flammable gas, flammable liquid-produced vapor, or combustible liquid-produced vapor mixed with air that may burn or explode, having either a maximum experimental safe gap (MESG) value less than or equal to 0.50 mm or minimum igniting current ratio (MIC ratio) less than or equal to 0.45
	IIB	Atmospheres containing acetaldehyde, ethylene, or flammable gas, flammable liquid-produced vapor, or combustible liquid-produced vapor mixed with air that may burn or explode, having either maximum experimental safe gap (MESG) values greater than 0.50 mm and less than or equal to 0.90 mm or minimum igniting current ratio (MIC ratio) greater than 0.45 and less than or equal to 0.80
	IIA	Atmospheres containing acetone, ammonia, ethyl alcohol, gasoline, methane, propane, or flammable gas, flammable liquid-produced vapor, or combustible liquid-produced vapor mixed with air that may burn or explode, having either a maximum experimental safe gap (MESG) value greater than 0.90 mm or minimum igniting current ratio (MIC ratio) greater than 0.80
499 [16]		
	E	Atmospheres containing combustible metal dusts, including aluminum, magnesium, and their commercial alloys, or other combustible dusts whose particle size, abrasiveness, and conductivity present similar hazards in the use of electrical equipment
	F	Atmospheres containing combustible carbonaceous dusts that have more than 8 percent total entrapped volatiles (See ASTM D3175, Standard Test Method for Volatile Matter in the Analysis Sample of Coal and Coke, for coal and coke dusts) or that have been sensitized by other materials so that they present an explosion hazard, coal, carbon black, charcoal, and coke dusts are examples of carbonaceous dusts
	G	Atmospheres containing other combustible dusts including flour, grains, or a hybrid mixture that may burn, flame, or explode
		Note: Group IIA is equivalent to Class I, Group D

Exception No. 1: Group D equipment shall be permitted to be used for atmospheres containing butadiene, provided all conduit runs into explosionproof equipment are provided with explosionproof seals installed within 450 mm (18 in.) of the enclosure.

Exception No. 2: Group C equipment shall be permitted to be used for atmospheres containing allyl glycidyl ether, n-butyl glycidyl ether, ethylene oxide, propylene oxide, and acrolein, provided all conduit runs into explosionproof equipment are provided with explosionproof seals installed within 450 mm (18 in.) of the enclosure.

of Combustible Dusts and of Hazardous (Classified) Locations for Electrical Installations in Chemical Process Areas.

The following Tables in NFPA documents can be referenced for information on specific Materials Group designations and physical properties:

NFPA 70 *National Electrical Code® Handbook*, Article 500.6: Commentary Table 5.1 *Selected Chemicals*

NFPA 70 *National Electrical Code® Handbook*, Article 500.7: Commentary Table 5.2 *Selected Combustible Materials*

NFPA 499, Table 4.5.2: *Selected Combustible Materials*

Two defining terms are used to describe the physicals properties that designate material groups. They include Maximum Experimental Safety Gap (MESG) and Minimum Igniting Current Ratio (MIC). *MESG* is defined in NFPA 497 [17] as:

> The maximum clearance between two parallel metal surfaces that has been found, under specified test conditions, to prevent an explosion in a test chamber from being propagated to a secondary chamber containing the same gas or vapor at the same concentration.

MIC is defined [18] as:

> The ratio of the minimum current required from an inductive spark discharge to ignite the most easily ignitable mixture of a gas or vapor, divided by the minimum current required from an inductive spark discharge to ignite methane under the same test conditions.

Table 9.2 lists the experimentally derived values for Class I, Division and Zone materials groups. The value for Group A MIC ratio was not listed in NEC 500.6(A)(1). The MESG and MIC ratio values for acetylene were obtained from the NEC *Handbook*, 2005 Edition, Commentary Table 5.1. The experimental values for MESG were developed in the 1960s when Underwriters Laboratories developed the Westerberg Explosion Test Vessel (WETV). That test device allows the ignition of a specific gas or vapor and measurement of the resulting maximum explosive pressure and flame transmission.

The NEC *Handbook* 2005, Commentary Table 5.1, presents a list of selected chemicals with their characteristics, including the Material Group designations. That information is also available in NFPA 497 *Recommended Practice for the Classification of Flammable Liquids, Gases, or Vapors and of Hazardous (Classified) Locations for Electrical Installations in Chemical Process Areas* and other sources.

TABLE 9.2   Division and Zone Class I Material Group MESG and MIC ratio values [19]

Group	Division Classification NEC Article 500.6		Group	Zone Classification NEC Article 505.6	
	MESG	MIC ratio		MESG	MIC ratio
A	0.25 mm	0.28[1]	IIA	MESG >0.90 mm	MIC >0.80
B[2]	MESG ≤0.45 mm	MIC ≤ 0.4	IIB	0.50<MESG≤0.90 mm	0.45<MIC≤0.75
C	0.45<MESG≤0.75 mm	0.40<MIC≤0.80	IIC	MESG ≤0.50 mm	MIC≤ 0.45
D	MESG >0.75 mm	MIC >0.80			

Note 1: Group A consists of acetylene.
Note 2: See exceptions in NEC Article 500.6(A)(2) for use of Class C and D equipment with certain Group B chemicals.

Summarizing, Division area classification designations for NEC Article 500 contain the following information:

Class I, Division (1 or 2), Group (A, B, C, or D)

Class II, Division (1 or 2), Group (E, F, or G)

Class III, Division (1 or 2)

There are no Group designations for Class III locations. NEC Article 505 Zone Area Classification Designations contains the following information:

Class I, Zone (0, 1, or 2), Group (IIC, IIB, or IIA)

NEC Article 506, for combustible dusts or ignitable fibers and flyings, zone area classification designations are grouped as the following:

Zone 20, Zone 21, Zone 22

Area Classification Boundaries

Neither NFPA 70 Article 500.5 or Article 505.5 defines the extent to which area classification boundaries extend from equipment. Table 9.3 contains a list of standards that can be referenced to assist in determining the area classification extent in petroleum and chemical facilities.

TABLE 9.3   Area classification standards

Facility type	Standard No.	Title
Storage, Handling, and Use of Flammable and Combustible Liquids	NFPA 30	Flammable and Combustible Liquids Code
Fuel Dispensing Facilities and Garages	NFPA 30A	Code for Motor Fuel Dispensing Facilities and Repair Garages
Spraying Facilities	NFPA 33	Standard for Spray Application Using Flammable or Combustible Materials
Dipping and Coating Facilities	NFPA 34	Standard for Dipping and Coating Processes Using Flammable or Combustible Liquids
Dipping, Coating, and Spraying Facilities	NFPA 91	Standard for Exhaust Systems for Air Conveying of Vapors, Gases, Mists, and Noncombustible Particulate Solids
Chemical Process Areas	NFPA 497	Recommended Practice for the Classification of Flammable Liquids, Gases, or Vapors and of Hazardous (Classified) Locations for Electrical Installations in Chemical Process Areas

TABLE 9.3 Area classification standards—cont'd

Facility type	Standard No.	Title
Chemical Process Areas	NFPA 499	Recommended Practice for the Classification of Combustible Dusts and of Hazardous Locations for Electrical Installations in Chemical Process Areas
Oil and Gas Wells	API RP 54	Recommended Practice for Occupational Safety for Oil and Gas Well Drilling and Servicing Operations
Petroleum Facilities	ANSI/API RP 500	Recommended Practice for Classification of Locations for Electrical Installations at Petroleum Facilities Classified as Class I Division 1 and Division 2
Petroleum Facilities	ANSI/API RP 505	Recommended Practice for Classification of Locations for Electrical Installations at Petroleum Facilities Classified as Class I, Zone 0, Zone 1, and Zone 2
Petroleum Facilities	IEC 60079-10	Explosive Atmospheres – Part 10-1: Classification of Areas – Explosive Gas Atmospheres
Offshore Petroleum Facilities	IEC 61892-7	Mobile and Fixed Offshore units – Electrical Installations – Part 7: Hazardous Areas
Gas Utility Areas	AGA XF0277	Classification of Gas Utility Areas for Electrical Installations
Agricultural and Chemical Facilities	ISA 12.10	Area Classification in Hazardous (Classified) Dust Locations
Petroleum Facilities	ISA RP 12.24.01 (IEC 79-10 Mod)	Recommended Practice for Classification of Locations for Electrical Installations Classified as Class I, Zone 0, Zone 1, or Zone 2: ISA

Area classification boundaries establish critical information regarding the class of equipment that may be installed in that location. Any electrical equipment operating within those boundaries is required to be suitable for that service. Classification boundaries establish the need for safe work practices. Any heat- or spark-producing activity, such as welding, which may occur within that area, will require specific safety precautions. Those precautions might necessitate acquiring work permits from appropriate operating personnel before any work is initiated; fire watches; standby water spraying equipment; prohibition of motor vehicles to operate in the area; etc.

For any fire or explosion to occur within a classified (hazardous) area, four components are required to simultaneously occur. They include the presence of:

A fuel source

An oxidizing agent

A heat source

Establishment of an uninhibited chemical chain reaction

If any of the above components are removed or are not present, combustion cannot occur. Fuel and the oxidization agent must be present in a specific range of concentrations, i.e. percentage of fuel present per volume of air. Fuel-air mixtures in ratios above or below that range will not ignite. The limits of those ranges are defined as the *lower explosive limit* (LEL) and the *upper explosive limit* (UEL). The heat source component can be in the form of an electrical arc or spark, an open flame, a chemical reaction, or heated surfaces.

Any heat source must have an adequate energy level and sufficiently high temperature for combustion to occur. That includes temperatures above the auto-ignition temperature of the fuel source and sufficient energy to initiate ignition. The fuel autoignition temperature is the minimum temperature at which material will ignite in air without a spark or flame. For additional information on combustion science reference NFPA 921, *Guide for Fire and Explosion Investigations.*

Equipment Temperature

Identifying maximum operating surface temperature is important when selecting equipment for use in hazardous (classified) areas. NEC Article 500.8(B)(4) requires that the Temperature Class (T Codes) be included on heat-producing electrical equipment identification labels, along with the ambient temperature at which it was rated. That information is not required on non-heat-producing equipment labels, such as junction boxes. NEC Table 500.8(B), *Classification of Maximum Surface Temperatures* provides the maximum temperatures that can be expected from heat producing equipment surfaces with T Codes T1 through T6 for Division hazardous (classified) areas.

Temperature codes for Zone classified systems are covered in the NEC under Article 505.9(D)(1). To better understand the zone classification for temperature codes, we should look first at Material Group classifications. Zone classification materials are broken into several groups as shown in Table 9.4 [20].

A comparison between Class and Zone temperature codes can be easily seen in Table 9.5.

The American Petroleum Institute Recommended Practice RP 2216, *Ignition Risk of Hydrocarbon Liquids and Vapors by Hot Surfaces in Open Air* addresses the probability of hot surface ignition of liquids or vapors. It notes in Paragraph 1.1 that:

> Hydrocarbon liquids, when heated sufficiently, can ignite without the application of a flame or spark. The ignition of hydrocarbons by hot surfaces may occur when oil is released under pressure and sprays on a hot surface or is spilled and lies on a hot surface for a period of time … [23]

TABLE 9.4   Zone Material Group classifications

Zone Material Group	Group description	Class equivalent
Group I	Atmospheres containing firedamp in mines (firedamp – methane mixture and other gases)	
Group II		
IIC	Acetylene, hydrogen, or flammable/combustible gas or vapor w/MESG $\leq$ 0.50 mm or MIC $\leq$0.45	Combination of Class I, Group A and B.
IIB	Acetaldehyde, ethylene, or flammable/combustible gas or vapor w/ 0.50 mm <MESG $\leq$0.90 mm or 0.45 < MIC $\leq$0.80	Class I, Group C
IIA	Acetone, ammonia, ethyl alcohol, gasoline, methane, propane, flammable or combustible gas or vapor w/ MESG > 0.90 mm or MIC >0.80	Class I, Group D

The American Society of Testing and Materials Standard ASTM E659, *Standard Test Method for Autoignition Temperature of Liquid Chemicals* provides testing requirements for the determination of material autoignition temperatures. NFPA 497, *Recommended Practice for the Classification of Flammable Liquids, Gases, or Vapors and of Hazardous (Classified) Locations for Electrical Installations in Chemical Process Areas* provides

TABLE 9.5   Comparison of Class and Zone temperature codes

Class temperature codes [21]		Zone temperature codes [22]	
T Code	Maximum temperature	T Code	Maximum temperature
T1	450° C	T1	$\leq$450°
T2	300° C	T2	$\leq$300°
T2A	280° C		
T2B	260° C		
T2C	230° C		
T2D	215° C		
T3	200° C	T3	$\leq$200°
T3A	180° C		
T3B	165° C		
T3C	160° C		
T4	135° C	T4	$\leq$135°
T4A	120° C		
T5	100° C	T5	$\leq$100°
T6	85° C	T6	$\leq$85°

materials' physical characteristic data, including the ignition temperatures of some gases and vapors.

API RP 2216 indicates:

> ignition of flammable hydrocarbon vapors by a hot surface at the minimum ignition temperatures (for the specific hydrocarbon) is not likely. Experimental studies, testing and practical experience have shown that hot surfaces must typically be hundreds of degrees above published minimum ignition temperatures to ignite freely moving hydrocarbon vapor in the open air … Whether or not flames will develop when a hydrocarbon contacts a hot surface depends not only on the surface temperature, but also on the extent (size) of the hot surface, its geometry and the ambient conditions [24]

Codes sometimes limit the maximum equipment operating surface temperature in hazardous (classified) areas to not exceed 80% of the ignition temperature in degrees Celsius for the gas or vapor that may be exposed to the surface of that equipment while it is continuously energized at the maximum rated ambient temperature. The equipment T Code printed on the identification label provides information as to the equipment's maximum safe operating temperature. The equipment rating information for Class, Division, and Group also provide the necessary data to allow selection of equipment for the environment in which it operates.

Hazardous Area Equipment

NEC Article 500.7 provides a list of Division classification electrical equipment protection techniques which can be employed for equipment operating in hazardous (classified) areas. The equipment protection techniques include those in Table 9.6.

The use of combustible gas detection systems has some restrictions when used as an equipment protection technique in hazardous (classified) areas. The NEC limits its use to locations with restricted public access. It also mandates certain maintenance and supervision requirements. The gas detection equipment must be listed for use in detection of the specific gas or vapor it will encounter. The NEC also requires documentation on the gas detection equipment, its listing, the location where it will be installed, and the shutdown procedures in place. An equipment calibration schedule is also mandated. The NEC also places some restrictions on the equipment minimum hazardous (classified) ratings when inadequate ventilation is the cause for the use of the gas detector protection technique.

Other circumstances under which gas detection protection techniques are acceptable for equipment protection include buildings located within a Class I, Division 2 area or with one of their openings into that area [27]. If the building interior does not contain a flammable gas or vapor source, then unclassified electrical equipment may be installed when the gas detection technique is employed. The latest edition of the National Electrical Code and/or API RP 500/505 should be referenced for future changes or additions. Class I Division 2

TABLE 9.6 Electrical and electronic equipment division protection techniques [25]

Protection technique	Suitable hazardous (classified) area locations
Explosionproof Apparatus	Class I, Division 1 or 2
Dust-Ignitionproof	Class II, Division 1 or 2
Dusttight	Class II, Division 2 or Class III, Division 1 or 2
Purged and Pressurized [26]	Any hazardous designation identified
Intrinsic Safety	Class I, Divisions 1 or 2; Class II, Divisions 1 or 2; Class III, Divisions 1 or 2 (see NEC Article 504, Intrinsically Safe Systems)
Nonincendive Circuit	Equipment in Class I, Division 2; Class II, Division 2; Class III, Divisions 1 or 2
Nonincendive Equipment	Equipment in Class I, Division 2; Class II, Division 2; Class III, Divisions 1 or 2
Nonincendive Component	Equipment in Class I, Division 2; Class II, Division 2; Class III, Divisions 1 or 2
Oil Immersion	Current Limiting Contacts in Class I, Division 2
Hermetically Sealed	Class I, Division 2; Class II, Division 2; Class III, Divisions 1 or 2
Combustible Gas Detection System	See NEC Article 500.7(K)
Other Protection Techniques	Reference equipment listing by Nationally Recognized Testing Laboratory.

classified equipment can be used in conjunction with a combustible gas detection system, when the equipment is mounted in the interior of a control panel that contains instruments that utilize or measure flammable gases, liquids, or vapors.

Zone classification equipment protection techniques are slightly different from the Division classification techniques noted above. Equipment protection techniques acceptable in Zone hazardous (classified) locations are presented in Table 9.7.

Zone classified equipment techniques, like Division classified equipment techniques, allow the use of a combustible gas detection system as an acceptable means to protect equipment. The restrictions for its use in Zone classified areas are presented in NEC Article 505.8(I). Some of the standards applicable for combustible gas detection systems are listed in Table 9.8.

Table 9.9 presents a list of the North American protection techniques for equipment operating in Class I hazardous (classified) locations. Also listed are many of the corresponding applicable codes, standards, and recommended practices with which the equipment must comply. To assist in evaluating that table, the information presented below in *Definitions – Flammable and Combustible Gases and Vapors Equipment Protection Techniques* should be consulted.

TABLE 9.7 Zone classification equipment protection techniques

Protection technique	Suitable hazardous (classified) area locations
Flameproof "d"	Class I, Zone 1 or 2
Purged and Pressurized	Class I, Zone 1 or 2
Intrinsic Safety	Class I, Zones 0, 1, or 2 (as listed)
Type of Protection "n"	Class I, Zone 2 ("nA", "nC", "nR")
Oil Immersion "o"	Class I, Zone 1 or 2
Increased Safety "e"	Class I, Zone 1 or 2
Encapsulation "m"	Class I, Zone 1 or 2
Powder Filling "q"	Class I, Zone 1 or 2
Combustible Gas Detection System	See NEC Article 505.8(I)

Definitions: Flammable and Combustible Gases and Vapors Equipment Protection Techniques

The following definitions describe some of the most common terms used in flammable and combustible gases and vapors protection techniques listed in Tables 9.9, Table 9.13, and Table 9.14. Some definitions are only applicable to those equipment protection techniques utilized in

TABLE 9.8 Standards governing use of gas detection equipment

Developer	Standard No.	Title
ISA	ANSI/ISA RP 12.13.01	Performance Requirements, Combustible Gas Detectors
ISA	ISA RP 12.13.02	Installation, Operation, and Maintenance of Combustible Gas Detection Instruments
API	ANSI/API RP 500	Recommended Practice for Classification of Locations for Electrical Installations at Petroleum Facilities Classified as Class I Division 1 and Division 2
API	ANSI/API RP 505	Recommended Practice for Classification of Locations for Electrical Installations at Petroleum Facilities Classified as Class I, Zone 0, Zone 1, and Zone 2
IEC	IEC 60079-29-1	Explosive atmospheres – Part 29-1: Gas Detectors – Performance Requirements of Detectors for Flammable Gases
IEC	IEC 60079-29-2	Explosive Atmospheres – Part 29-2: Gas Detectors – Selection, Installation, Use and Maintenance of Detectors for Flammable Gases and Oxygen
CSA	CAN/CSA C22.2 No. 152	Performance of Combustible Gas Detection Instruments
FM Global	FM6310/6320	Approval Standard for Combustible Gas Detectors Class Number 6310, 6320

TABLE 9.9   USA and Canada Class I combustible gas and vapor equipment protection techniques

Protection technique/ Marking	Suitable hazardous (classified) area locations	Standard No.	Title
Explosionproof			
(XP)	Class I, Division 1 or 2	NFPA 70	National Electrical Code Article 501 Class I Locations
		NFPA 30	Flammable and Combustible Liquids Code
		NFPA 497	Classification of Flammable Liquids, Gases, or Vapors and of Hazardous (Classified) Locations for Electrical Installations in Chemical Process Areas
	Class I, Division 2	ANSI/ISA RP12.12.03	Recommendations for Portable Electronic Products Suitable for Use in Class I and II, Division 2, Class I, Zone 2, and Class III, Division 1 and 2 Hazardous (Classified) Locations
	Class I, Division 2	ANSI/ISA-TR12.06.01	Electrical Equipment in a Class I, Division 2/Zone 2 Hazardous Location
	Class I, Division 1 or 2	FM3600	Electrical Equipment for Use in Hazardous (Classified) Locations, General Requirements
	Class I, Division 1	FM3615	Explosionproof Electrical Equipment, General Requirements
		CAN/CSA C22.2 No 30	Explosion-Proof Enclosures for Use in Class I Hazardous Locations Industrial Products
	Class I, Division 1	UL 1203	Explosion-Proof and Dust-Ignition-Proof Electrical Equipment for Use in Hazardous (Classified) Locations
Purged and Pressurized			
(Type X), (Type Y), or (Type Z); AEx px, AEx py, or AEx pz	Class I, Division 1 and 2; Class I, Zones 0, 1, and 2	NFPA 496	Standard for Purged and Pressurized Enclosures for Electrical Equipment
(Type X), (Type Y), or (Type Z); AEx px, AEx py, or AEx pz	Class I, Division 1 and 2; Class I, Zones 0, 1, and 2	ISA RP 12.4	Pressurized Enclosures
AEx px, AEx py, or AEx pz	Class I, Zone 1 and 2	ANSI/ISA-12.04.01 (IEC 60079-2 Mod)	Electrical Apparatus for Explosive Gas Atmospheres – Part 2 Pressurized Enclosures "p"

(*Continued*)

TABLE 9.9   USA and Canada Class I combustible gas and vapor equipment protection techniques—cont'd

Protection technique/ Marking	Suitable hazardous (classified) area locations	Standard No.	Title
AEx px, AEx py, or AEx pz	Class I, Zone 1 and 2	CAN/CSA E60079-2	Electrical Apparatus for Explosive Gas Atmospheres – Part 2: Pressurized Enclosures "p"
(Type X),(Type Y), or (Type Z); AEx px, AEx py, or AEx pz	Class I, Division 1 and 2	FM3600	Electrical Equipment for Use in Hazardous (Classified) Locations, General Requirements
(Type X), (Type Y), or (Type Z); AEx px, AEx py, or AEx pz	Class I, Division 1 and 2	FM3620	Purged and Pressurized Electrical Equipment for Hazardous (Classified) Locations
Intrinsic Safety			
		NFPA 70®	National Electrical Code Article 504, Intrinsically Safe Systems
AEx ia	Class I, Zone 0	ANSI/ISA-60079-11 (12.02.01)	Electrical Apparatus for Use in Class I, Zones 0, 1 and 2 Hazardous (Classified) Locations – Intrinsic Safety "i"
AEx ib	Class I, Zone 1	ANSI/ISA-60079-11 (12.02.01)	Electrical Apparatus for Use in Class I, Zones 0, 1, and 2 Hazardous (Classified) Locations – Intrinsic Safety "i"
AEx ic	Class I, Zone 2	ANSI/ISA-60079-11 (12.02.01)	Electrical Apparatus for Use in Class I, Zones 0, 1, and 2 Hazardous (Classified) Locations – Intrinsic Safety "i"
AEx ia	Class I, Zone 0	ANSI/ISA 60079-27(12.02.05)	Electrical Apparatus For Use in Class I, Zones 0, 1, and 2 Hazardous (Classified) Locations – Fieldbus Intrinsically Safe Concept (FISCO) and Fieldbus Non-Incendive Concept (FNICO)
AEx ib	Class I, Zone 1	ANSI/ISA 60079-27(12.02.05)	Electrical Apparatus For Use in Class I, Zones 0, 1, and 2 Hazardous (Classified) Locations – Fieldbus Intrinsically Safe Concept (FISCO) and Fieldbus Non-Incendive Concept (FNICO)
AEx ic	Class I, Zone 2	ANSI/ISA 60079-27 (12.02.05)	Electrical Apparatus for Use in Class I, Zones 0, 1, and 2 Hazardous (Classified) Locations – Fieldbus Intrinsically Safe Concept (FISCO) and Fieldbus Non-Incendive Concept (FNICO)

TABLE 9.9 USA and Canada Class I combustible gas and vapor equipment protection techniques—cont'd

Protection technique/ Marking	Suitable hazardous (classified) area locations	Standard No.	Title
—	—	ISA-TR12.2	Intrinsically Safe System Assessment Using the Entity Concept
AEx ia	Class I, Zone 0	ANSI/ISA RP 12.06.01	Wiring Methods for Hazardous (Classified) Locations Instrumentation – Part 1: Intrinsic Safety
AEx ib	Class I, Zone 1	ANSI/ISA RP 12.06.01	Wiring Methods for Hazardous (Classified) Locations Instrumentation – Part 1: Intrinsic Safety
AEx ic	Class I, Zone 2	ANSI/ISA RP 12.06.01	Wiring Methods for Hazardous (Classified) Locations Instrumentation – Part 1: Intrinsic Safety
(IS)	Class I, Divisions 1 and 2	FM3610	Intrinsically Safe Apparatus and Associated Apparatus for Use in Class I, II and III, Division 1 Hazardous (Classified) Locations
AEx ia	Class I, Zone 0	FM3610	Intrinsically Safe Apparatus and Associated Apparatus for Use in Class I, II, and III, Division 1 Hazardous (Classified) Locations
AEx ib	Class I, Zone 1	FM3610	Intrinsically Safe Apparatus and Associated Apparatus for Use in Class I, II, and III, Division 1 Hazardous (Classified) Locations
Ex ia	Class I, Zone 0	CAN/CSA E60079-11	Electrical Apparatus for Explosive Gas Atmospheres Part 11: Intrinsic Safety "i"
Ex ib	Class I, Zone 1	CAN/CSA E60079-11	Electrical Apparatus for Explosive Gas Atmospheres Part 11: Intrinsic Safety "i"
(IS)	Class I, Division 1 and 2	CAN/CSA 22.2 No. 157	Intrinsically Safe and Non-Incendive Equipment for Use in Hazardous Locations
(IS)	Class I, Divisions 1 and 2	UL 913	Intrinsically Safe Apparatus and Associated Apparatus for Use in Class I, II, and III, Division I, Hazardous (Classified) Locations

(Continued)

TABLE 9.9   USA and Canada Class I combustible gas and vapor equipment protection techniques—cont'd

Protection technique/ Marking	Suitable hazardous (classified) area locations	Standard No.	Title
Ex ia or Ex ib	Class I, Division 0 or 1	UL 913	Intrinsically Safe Apparatus and Associated Apparatus for Use in Class I, II, and III, Division I, Hazardous (Classified) Locations
Oil Immersion			
AEx o	Class I, Zone 1	ANSI/ISA-60079-6 (12.26.01)	Electrical Apparatus for Use in Class I, Zone 1 Hazardous (Classified) Locations: Type of Protection – Oil-Immersion "o"
AEx o	Class I, Zone 1	ANSI/UL 60079-6	Electrical Apparatus for Explosive Gas Atmospheres – Part 6: Oil-Immersion "o"
Ex o	Class I, Zone 1	CAN/CSA E60079-6	Electrical Apparatus for Explosive Gas Atmospheres – Part 6: Oil-Immersion "o"
Encapsulation			
AEx m	Class I, Zone 1	ANSI/ISA 60079-18 (12.23.01)	Electrical Apparatus for Use in Class I, Zone 1 Hazardous (Classified) Locations Type of Protection - Encapsulation "m"
AEx ma	Class I, Zone 0	ANSI/UL 60079-18	Electrical Apparatus for Explosive Gas Atmospheres – Part 18: Encapsulation "m"
AEx m	Class I, Zone 1	ANSI/UL 60079-18	Electrical apparatus for explosive gas atmospheres – Part 18: Encapsulation "m"
Ex m	Class I, Zone 1	CAN/CSA E60079-18	Electrical Apparatus for Explosive Gas Atmospheres – Part 18: Encapsulation "m"
Flameproof			
AEx d	Class I, Zone 1	ANSI/ISA-60079-1 (12.22.01)	Electrical Apparatus for Use in Class I, Zone 1 Hazardous (Classified) Locations: Type of Protection – Flameproof "d"
AEx d	Class I, Zone 1	ANSI/UL 60079-1	Electrical Apparatus for Explosive Gas Atmospheres – Part 1: Flameproof Enclosures "d"
Ex d	Class I, Zone 1	CAN/CSA E60079-1	Electrical Apparatus for Explosive Gas Atmospheres – Part 1: Flameproof Enclosures "d"

TABLE 9.9 USA and Canada Class I combustible gas and vapor equipment protection techniques—cont'd

Protection technique/ Marking	Suitable hazardous (classified) area locations	Standard No.	Title
Increased Safety			
AEx e	Class I, Zone 1	ANSI/ISA-60079-7 (12.16.01)	Electrical Apparatus for Use in Class I, Zone 1 Hazardous (Classified) Locations: Type of Protection – Increased Safety "e"
AEx e	Class I, Zone 1	ANSI/UL60079-7	Explosive Atmospheres – Part 7: Equipment protection by increased safety "e"
Ex e	Class I, Zone 1	CAN/CSA E60079-7	Electrical Apparatus for Explosive Gas Atmospheres – Part 7: Increased Safety "e"
Powder Filling			
AEx q	Class I, Zone 1	ANSI/ISA-60079-5 (12.25.01)	Electrical Apparatus for Use in Class I, Zone 1 Hazardous (Classified) Locations: Type of Protection - Powder Filling "q"
AEx q	Class I, Zone 1	ANSI/UL 60079-5	Electrical Apparatus for Explosive Gas Atmospheres – Part 5: Powder Filling "q"
Ex q	Class I, Zone 1	CAN/CSA E60079-5	Electrical Apparatus for Explosive Gas Atmospheres – Part 5: Powder Filling "q"
Non-sparking			
AEx nA	Class I, Zone 2	ANSI/ISA-60079-15 (12.12.02)	Electrical Apparatus for Use in Class I, Zone 2 Hazardous (Classified) Locations – Type of Protection "n"
AEx nA	Class I, Zone 2	ANSI/UL 60079-15	Electrical Apparatus for explosive gas atmospheres – Part 15: Type of protection "n"
EX nA	Class I, Zone 2	CAN/CSA E60079-15	Electrical Apparatus for Explosive Gas Atmospheres – Part 15: Type of Protection "n"
Hermetically Sealed			
AEX nC	Class I, Zone 2	ANSI/UL 60079-15	Electrical Apparatus for Explosive Gas Atmospheres – Part 15: Electrical Apparatus with Type of Protection "n"
Ex nC	Class I, Zone 2	CAN/CSA E60079-15	Electrical Apparatus for Explosive Gas Atmospheres – Part 15: Type of Protection "n"

(Continued)

TABLE 9.9   USA and Canada Class I combustible gas and vapor equipment protection techniques—cont'd

Protection technique/ Marking	Suitable hazardous (classified) area locations	Standard No.	Title
(HS)	Class I, Division 2	UL 1604	Standard for Electrical Equipment for Use in Class I and II, Division 2, And Class III Hazardous (Classified) Locations
	Class I, Division 2	CAN/CSA 22.2 No. 213	
Restrictive Breathing			
AEx nR	Class I, Zone 2	ANSI/UL 60079-15	Electrical Apparatus for Explosive Gas Atmospheres – Part 15: Electrical Apparatus with Type of Protection "n"
Ex nR	Class I, Zone 2	CAN/CSA E60079-15	Electrical Apparatus for Explosive Gas Atmospheres – Part 15: Type of Protection "n"
Nonincendive Equipment, Components, and Circuits			
(NI)	Class I, Division 2	ANSI/ISA 12.12.01	Nonincendive Electrical Equipment for Use in Class 1 and II, Division 2 and Class III, Divisions 1 and 2 Hazardous (Classified) Locations
(NI)	Class I, Division 2	UL 1604	Electrical Equipment for Use in Class I and II, Division 2, and Class III Hazardous (Classified) Locations
(NI)		CAN/CSA-C22.2 NO. 157	Intrinsically Safe and Non-Incendive Equipment for Use in Hazardous Locations
(NI)	Class I, Division 2	CAN/CSA C22.2 No 213	Non-Incendive Electrical Equipment for Use in Class I, Division 2 Hazardous Locations Industrial Products
(NI)	Class I, Division 2	FM3611	Nonincendive Electrical Equipment for Use in Class I and II, Division 2, and Class III, Divisions 1 and 2, Hazardous (Classified) Locations
Combustible Gas Detection System			
—	Class I, Division 1 and 2	ANSI/API RP 500	Recommended Practice for Classification of Locations for Electrical Installations at Petroleum Facilities Classified as Class I, Division 1 or Divisions 2

TABLE 9.9 USA and Canada Class I combustible gas and vapor equipment protection techniques—cont'd

Protection technique/ Marking	Suitable hazardous (classified) area locations	Standard No.	Title
—	Class I, Zone 0, 1, or 2	ANSI/ISA 12.13.01(IEC 61779-1 through 5 Mod)	Performance Requirements for Combustible Gas Detectors
—	Class I, Zone 0, 1, or 2	ANSI/ISA-RP12.13.02 (IEC 61779-6 Mod)	Recommended Practice for the Installation, Operation, and Maintenance of Combustible Gas Detection Instruments
—	Class I, Zone 0, 1, or 2	ANSI/API RP 505	Recommended Practice for Classification of Locations for Electrical Installations at Petroleum Facilities Classified as Class I, Zone 0, Zone 1, and Zone 2

United States and/or Canadian applications. Others are applicable for those areas and in International Electrotechnical Commission (IEC) and European Union (EU) equipment applications.

Explosionproof

Explosionproof refers to Class I, Groups A, B, C, or D equipment enclosures, conduit fittings, junction boxes, etc. which are capable of withstanding internal pressures resulting from an explosion caused by the accumulation of specified gas/vapor–air mixtures inside the enclosure and the subsequent ignition of the gas–air mixture from an energy source in that enclosure. The protection technique design requires that the enclosure sustain that explosion event intact and prevent the resulting flames, hot gases, and burning debris from escaping from the enclosure and igniting external flammable or combustible gases or vapors. This is accomplished by cooling the escaping gases as they pass over the threaded or flanged surfaces on the enclosure, entry way, doors, or hubs. It is also required to prohibit the equipment external surfaces temperatures to rise to levels capable of igniting explosive gas-air mixtures in the surrounding atmosphere. The National Electrical Manufacturers Association Standard NEMA 250, *Enclosures for Electrical Equipment (1000 Volts Maximum)* NEMA Type 7 designation is an example of an explosionproof enclosure.

Initially, explosionproof enclosures were constructed of cast metal with a designed wall thickness to sustain an internal explosion without external damage or increases in equipment surface temperatures to sufficient levels to ignite the surrounding vapor/gas–air mixture in which they are installed. However, newer designs have incorporated non-metallic materials and factory-sealed components and equipment. These designs are capable of maintaining their integrity during an internal explosion by decreasing their internal volumetric size, therefore limiting the amount of entrapped gas that can be ignited. That technique will cause a reduction in the explosive forces produced, allowing for thinner enclosures and the use of non-metallic materials. Any hot gases expelled from the enclosure are cooled by increasing the path through which they must travel and by using high temperature resistant materials on the path, such as sintered bronze plates.

Purged and Pressurized

Purging is the process of applying sufficient non-combustible gas to an enclosure, at a specified flow rate and positive pressure, sufficient to reduce entrapped flammable or combustible gases or vapors to acceptable levels below the lower explosive limits. *Pressurization* is the process in which an enclosure is supplied with a clean, non-hazardous air source at a maintained positive pressure level to prevent the entry of flammable or combustible gases or vapors, combustible dust or ignitable fibers. The most notable standard governing this process is NFPA 496 *Standard for Purged and Pressurized Enclosures for Electrical Equipment*. The enclosure may consist of a junction box, equipment enclosure, a room, or an entire building. Requirements may also include a minimum rate of air changes per hour and a minimum outward air/gas velocity should the enclosure contain any doors which would be opened during normal operation.

Intrinsically Safe Circuit

An *intrinsically safe* circuit is defined by the National Electrical Code (NEC 2005) as:

> A circuit in which any spark or thermal effect is incapable of causing ignition of a mixture of flammable or combustible material in air under prescribed test conditions. [46]

This indicates that the circuit design is such that any spark or arc fault event or overload condition, with resulting thermal energy production, cannot produce sufficient energy, measured in milliJoules (mJ), to cause the temperature of a specified gas or vapor to exceed its ignition temperature, while operating under normal or abnormal conditions. Note that the definition indicates that the circuit has been designed for specific test conditions. Intrinsically safe circuits are also utilized in atmospheres containing flammable or combustible gases or vapors and combustible concentrations of dusts, fibers, or flyings.

The ignition temperature of selected gases and vapors can be found in NFPA 497, Recommended Practice for the Classification of Flammable Liquids, Gases, or Vapors and of Hazardous (Classified) Locations for Electrical Installations in Chemical Process Areas. Not all low-energy devices are considered as intrinsically safe devices; although, intrinsically safe devices operate at low energy levels. NFPA 921, Guide for Fire and Explosion Investigations [47] Table 21.13.3.1, Typical Explosive Characteristics, provides some general information as to the minimum ignition energy in milliJoules to cause ignition in several classes of flammable and combustible materials. They include the general material categories of:

Lighter-than-air gases: 0.17–0.25 mJ

Heavier-than-air gases: 017–0.25 mJ

Liquid vapors: 0.25 mJ

Dusts: 10–40 mJ

NEC® Article 504, *Intrinsically Safe Systems* should be referenced for additional information on intrinsically safe apparatus, circuits, and systems. Additional information can also be obtained from the Intrinsic Safety standards listed in Tables 9.9 and 9.10 for gases, vapors, dusts, fibers, and flyings.

Nonincendive Circuits

Nonincendive circuits "nC" are defined in NEC Article 500.2 [48] as:

> A circuit, other than field wiring, in which any arc or thermal effect produced under intended operating conditions of the equipment is not capable, under specified test conditions, of igniting the flammable gas–air, vapor–air, or dust–air mixture.

The protection techniques employed in a nonincendive circuit, to prevent ignition of flammable or combustible gases or vapors and combustible dust, fibers, or flyings, are only permitted for use in the hazardous (classified) areas noted in Tables 9.9 and 9.10 respectively. Nonincendive components and field wiring contained in a nonincendive circuit are also incapable of igniting a specified flammable gas–air or vapor–air or dust–air mixture. It should be noted that a nonincendive equipment enclosure is not designed to prevent the entry of flammable or combustible materials.

Nonincendive equipment and circuits are considered low-energy, much like intrinsically safe equipment and circuits. Unlike intrinsically safe equipment and circuits, the nonincendive devices are not tested under abnormal conditions like the intrinsically safe devices. Nonincendive "field wiring" is normally not part of the approval process by a nationally recognized testing laboratory. Field wiring for those devices requires caution in its design, specifically regarding wiring mutual inductance and capacitance stored energy potentials. Reference ISA 12.12.01, Section 7 Nonincendive Circuits and Nonincendive Field Wiring for additional information on field wiring requirements.

Encapsulation

Encapsulation "m" indicates an equipment protection technique that encases electrical parts in a compound that prevents ignition of an explosive atmosphere from either thermal or sparking

TABLE 9.10 USA and Canada combustible dust, fibers, and flyings equipment protection techniques [32-35]

Protection technique/Marking	Suitable hazardous (classified) area locations	Standard No.	Title
General Requirements			
—	Class II, Division 1 and 2	FM3600	Electric Equipment for Use in Hazardous (Classified) Locations: General Requirements
—	Class II, Division 1 and 2	CAN/CSA C22.2 No. 0	General Requirements – Canadian Electrical Code, Part II
—	Class III, Division 1 and 2	FM3600	Electric Equipment for Use in Hazardous (Classified) Locations General Requirements
—	Class III, Division 1 and 2	FM3600	Electric Equipment for use in Hazardous (Classified) Locations: General Requirements
Ex	Zone 20, 21, and 22	ANSI/ISA 61241-0 (12.10.02)	Electrical Apparatus for Use in Zone 20, Zone 21, and Zone 22 Hazardous (Classified) Locations – General Requirements
Dust-Ignitionproof			
(DIP)	Class II, Division 1 and 2	NFPA 70®	National Electrical Code Article 502 Class II Locations
(DIP)	Class II, Division 1	FM3610	Intrinsically Safe Apparatus and Associated Apparatus for Use in Class I, II, and III, Division 1, Hazardous (Classified) Locations
(DIP)	Class II, Division 1and 2	C22.2 NO. 25	Enclosures for Use in Class II Groups E, F and G Hazardous Locations
—	Class II, Division 1 and 2	CAN/CSA-E61241-1-1	Electrical Apparatus for Use in the Presence of Combustible Dust – Part 1-1: Electrical Apparatus Protected by Enclosures and Surface Temperature Limitation – Specification for Apparatus
—	Class II, Division 1	UL 1203	Explosion-Proof and Dust-Ignition-Proof Electrical Equipment for Use in Hazardous (Classified) Locations
Dust-Protected (Tight)			
(NI)	Class II, Division 2	FM3610	Intrinsically Safe Apparatus and Associated Apparatus for Use in Class I, II, and III, Division 1, Hazardous (Classified) Locations
—	Class II, Division 1 and 2	CAN/CSA C22.2 No. 157	Intrinsically Safe and Non-Incendive Equipment for Use in Hazardous Locations

TABLE 9.10 USA and Canada combustible dust, fibers, and flyings equipment protection techniques [32-35]—cont'd

Protection technique/Marking	Suitable hazardous (classified) area locations	Standard No.	Title
—	Class II, Division 1 and 2	CAN/CSA-E61241-1-1	Electrical Apparatus for Use in the Presence of Combustible Dust – Part 1-1: Electrical Apparatus Protected by Enclosures and Surface Temperature Limitation – Specification for Apparatus
—	Class II , Division 2	ANSI/ISA 12.12.01	Nonincendive Electrical Equipment for Use in Class I and II, Division 2 and Class III, Divisions 1 and 2 Hazardous (Classified) Locations
—	Class II, Divisions 2	UL1604	Electrical Equipment for Use in Class I and II, Division 2, and Class III Hazardous (Classified) Locations
—	Zone 20, 21, and 22	ANSI/ISA-61241-0 (12.10.02)	Electrical Apparatus for Use in Zone 20, Zone 21, and Zone 22 Hazardous (Classified) Locations-General Requirements
AEx tD	Zone 20, 21, and 22	ANSI/ISA 61241-1(12.10.03)	Electrical Apparatus for Use in Zone 21 and Zone 22 Hazardous (Classified) Locations – Protection by Enclosures "tD"
—	Zone 20, 21, and 22	ANSI/ISA-61241-2 (12.10.05)	Electrical Apparatus for Use in Zone 20, Zone 21, and Zone 22 Hazardous (Classified) Locations – Classification of Zone 20, Zone 21, and Zone 22 Hazardous (Classified) Locations
Protection by Enclosure			
AEx tD	Zone 21 and Zone 22	ANSI/ISA 61241-1(12.10.03)	Electrical Apparatus for Use in Zone 21 and Zone 22 Hazardous (Classified) Locations – Protection by Enclosures "tD"
DIP A21	Zone 21	CAN/CSA-E61241-1-1	Electrical Apparatus for Use in the Presence of Combustible Dust – Part 1-1: Electrical Apparatus Protected by Enclosures and Surface Temperature Limitation – Specification for Apparatus
DIP A22	Zone 22	CAN/CSA-E61241-1-1	Electrical Apparatus for Use in the Presence of Combustible Dust – Part 1-1: Electrical Apparatus Protected by Enclosures and Surface Temperature Limitation – Specification for Apparatus

(Continued)

TABLE 9.10 USA and Canada combustible dust, fibers, and flyings equipment protection techniques [32-35]—cont'd

Protection technique/Marking	Suitable hazardous (classified) area locations	Standard No.	Title
Fiber and Flying Protected			
DIP	Class III, Division 1 and 2	FM3611	Nonincendive Electrical Equipment for Use in Class I and II, Division 2, and Class III, Divisions 1 and 2, Hazardous (Classified) Locations
—	Class III, Division 1 and 2	CAN/CSA 22.1	Canadian Electrical Code, Part I
Encapsulation			
AEx maD	Zone 20	ANSI/ISA 61241-18 (12.10.07)	Electrical Apparatus for Use in Zone 20, Zone 21, and Zone 22 Hazardous (Classified) Locations – Protection by Encapsulation "mD"
AEx mbD	Zone 21	ANSI/ISA 61241-18 (12.10.07)	Electrical Apparatus for Use in Zone 20, Zone 21, and Zone 22 Hazardous (Classified) Locations – Protection by Encapsulation "mD"
Pressurization			
Type X	Class II, Division 1	FM3620	Purged and Pressurized Electrical Equipment for Hazardous (Classified) Locations
Type X, Type Y, and Type Z	Class II, Division 1 and 2	NFPA 496	Standard for Purged and Pressurized Enclosures for Electrical Equipment
Type Y	Class II, Division 1	FM3620	Purged and Pressurized Electrical Equipment for Hazardous (Classified) Locations
Type Z	Class II, Division 2	FM3620	Purged and Pressurized Electrical Equipment for Hazardous (Classified) Locations
AEx pD	Zone 21	ANSI/ISA 61241-2 (12.10.06)	Electrical Apparatus for Use in Zone 21 and Zone 22 Hazardous (Classified) Locations – Protection by Pressurization "pD"
—	Class I, II, III	ISA RP 12.4	Pressurized Enclosures
Intrinsic Safety			
(IS)	Class II, Division 1	FM3610	Intrinsically Safe Apparatus and Associated Apparatus for Use in Class I, II, and III, Division 1, Hazardous (Classified) Locations

TABLE 9.10 USA and Canada combustible dust, fibers, and flyings equipment protection techniques [32-35]—cont'd

Protection technique/Marking	Suitable hazardous (classified) area locations	Standard No.	Title
(IS)	Class II, Division 1	CAN/CSA C22.2 No. 157	Intrinsically Safe and Non-Incendive Equipment for Use in Hazardous Locations
(IS)	Class III, Division 1	FM3610	Intrinsically Safe Apparatus and Associated Apparatus for Use in Class I, II, and III, Division 1, Hazardous (Classified) Locations
(IS)	Class III, Division 1	CAN/CSA C22.2 No. 157	Intrinsically Safe and Non-Incendive Equipment for Use in Hazardous Locations
AEx iaD	Zone 20	ANSI/ISA 61241-10 (12.10.05)	Electrical Apparatus for Use in Zone 20, Zone 21, and Zone 22 Hazardous (Classified) Locations – Classification of Zone 20, Zone 21, and Zone 22 Hazardous (Classified) Locations
(IS)	Class II, Division 1	UL 913	Standard for Intrinsically Safe Apparatus and Associated Apparatus for Use in Class I, II, III, Division 1, Hazardous (Classified) Locations
Nonincendive			
(NI)	Class II, Division 2; Class III, Division 1 and 2	ANSI/ISA 12.12.01	Nonincendive Electrical Equipment for Use in Class I and II, Division 2 and Class III, Divisions 1 and 2 Hazardous (Classified) Locations
(NI)	Class II, Division 2; Class III, Division 1 and 2	FM3611	Nonincendive Electrical Equipment for Use in Class I and II, Division 2, and Class III, Divisions 1 and 2, Hazardous (Classified) Locations
(NI)	Class II, Division 2	UL 1604	Standard for Electrical Equipment for Use in Class I and II, Division 2, and Class III Hazardous (Classified) Locations
Hermetically sealed			
(HS)	Class II, Division 2; Class III, Division 1 and 2	ANSI/ISA 12.12.01	Nonincendive Electrical Equipment for Use in Class I and II, Division 2 and Class III, Divisions 1 and 2 Hazardous (Classified) Locations
(HS)	Class II, Division 2	UL 1604	Standard for Electrical Equipment for Use in Class I and II, Division 2, and Class III Hazardous (Classified) Locations

energy sources. The encapsulation prohibits an explosive mixture from mitigating into the equipment or component in sufficient quantities to allow ignitable mixtures. Standards applicable to encapsulation of electrical components and equipment are listed in Tables 9.9 through 9.12. Encapsulation insulating materials:

> may be any thermosetting, thermoplastic, epoxy, resin (cold curing) or elastomeric material with or without fillers and/or additives, in their solid state ... [49]

TABLE 9.11 EU combustible dust and fibers equipment protection techniques [36–38]

Protection technique/ Marking	Suitable hazardous (classified) area locations	Standard No.	Title
General Requirements			
Ex	Category 1D, 2D, or 3D	EN 61241-0	Electrical Apparatus for Use in the Presence of Combustible Dust – Part 0: General Requirements
Ex	Category 1D, 2D, or 3D	EN 60079-0	Electrical Apparatus for Explosive Gas Atmospheres. General Requirements
Protection by Enclosure			
Ex tD	Category 2D	EN 61241-1	Electrical Apparatus for Use in the Presence of Combustible Dust – Part 1: Protection by Enclosures "tD"
Ex ta	Category 1D	EN 60079-31	Explosive Atmospheres – Part 31: Equipment Dust Ignition Protection by Enclosure "t"
Ex tb	Category 2D	EN 60079-31	Explosive Atmospheres – Part 31: Equipment Dust Ignition Protection by Enclosure "t"
Ex tc	Category 3D	EN 60079-31	Explosive Atmospheres – Part 31: Equipment Dust Ignition Protection by Enclosure "t"
Encapsulation			
Ex maD	Category 1D	EN 61241-18	Electrical Apparatus for Use in the Presence of Combustible Dust – Part 18: Protection by Encapsulation "m"
Ex mbD	Category 2D	EN 61241-18	Electrical Apparatus for Use in the Presence of Combustible Dust – Part 18: Protection by Encapsulation "md"
Pressurization			
Ex pD	Category 2D	EN 61241-4	Electrical Apparatus for Use in the Presence of Combustible Dust. Type of Protection "pD"
Intrinsic Safety			
Ex iaD	Category 1D	EN 61241-11	Electrical Apparatus for Use in the Presence of Combustible Dust. Protection by Intrinsic Safety "iD"

TABLE 9.12 IEC combustible dust and fibers equipment protection techniques [39–41]

Protection technique/ Marking	Suitable hazardous (classified) area locations	Standard No.	Title
General Requirements			
Ex	Zone 20, 21, or 22	IEC 61241-0	Electrical Apparatus for Use in the Presence of Combustible Dust – Part 0: General Requirements
Ex	ELP Ga, Gb, or Gc	IEC 60079-0	Explosive Atmospheres – Part 0: Equipment – General Requirements
Protection by Enclosure			
Ex tD	EPL Db	IEC 61241-1	Electrical Apparatus for Use in the Presence of Combustible Dust. Part 1: Protection by Enclosures "tD"
Ex ta	EPL Da	IEC 60079-31	Explosive Atmospheres – Part 31: Equipment Dust Ignition Protection by Enclosure "t"
Ex tb	EPL Db	IEC 60079-31	Explosive Atmospheres – Part 31: Equipment Dust Ignition Protection by Enclosure "t"
Ex tc	EPL Dc	IEC 60079-31	Explosive Atmospheres – Part 31: Equipment Dust Ignition Protection by Enclosure "t"
Encapsulation			
Ex maD	EPL Da	IEC 61241-18	Electrical Apparatus for Use in the Presence of Combustible Dust – Part 18: Protection by Encapsulation "mD"
Ex mbD	EPL Db	IEC 61241-18	Electrical Apparatus for Use in the Presence of Combustible Dust – Part 18: Protection by Encapsulation "mD"
Pressurization			
Ex pD	EPL Db	IEC 61241-4	Electrical Apparatus for Use in the Presence of Combustible Dust. Type of Protection "pD"
Intrinsic Safety			
Ex iaD	EPL Da	IEC 61241-11	Electrical Apparatus for Use in the Presence of Combustible Dust. Protection by Intrinsic Safety "iD"
Ex ibD	EPL Db	IEC 61241-11	Electrical Apparatus for Use in the Presence of Combustible Dust. Protection by Intrinsic Safety "iD"

Flameproof

Flameproof "d" equipment protection techniques involve enclosures so designed to withstand the internal ignition of flammable or combustible air mixtures without sustaining damage or allowing the propagation of the flames, burning gases, or debris through enclosure seams,

TABLE 9.13 EU flammable and combustible gases and vapor equipment protection techniques [42, 43]

Protection Technique/ Marking	Suitable Hazardous (Classified) Area Locations	Standard No.	Title
General Requirements			
Ex	Category 1G, 2G, or 3G	EN 60079-0	Electrical Apparatus for Explosive Gas Atmospheres. General Requirements
Ex	Category 1G, 2G, or 3G	EN 60079-14	Electrical Apparatus for Explosive Gas Atmospheres. Electrical Installations in Hazardous Areas (Other Than Mines)
Increased Safety			
Ex e or Ex eb	Category 2G	EN 60079-7	Electrical Apparatus for Explosive Gas Atmospheres. Increased Safety "e"
Non-Sparking			
Ex nA or Ex nAc	Category 3G	EN 60079-15	Electrical Apparatus for Explosive Gas Atmospheres. Type of Protection "n"
Nonincendive			
Ex nC or Ex nCc	Category 3G	EN 60079-15	Electrical Apparatus for Explosive Gas Atmospheres. Type of Protection "n"
Flameproof			
Ex d or Ex db	Category 2G	EN 60079-1	Explosive Atmospheres. Equipment Protection by Flameproof Enclosures "d"
Powder-Filled			
Ex q or Ex qb	Category 2G	EN 60079-5	Explosive Atmospheres. Equipment Protection by Powder Filling "q"
Enclosed Break			
Ex nC or Ex nCc	Category 3G	EN 60079-15	Electrical Apparatus for Explosive Gas Atmospheres. Type of Protection "n"
Intrinsic Safety			
Ex ia	Category 1G	EN 60079-11	Explosive Atmospheres. Equipment Protection by Intrinsic Safety "i"
Ex ib	Category 2G	EN 60079-11	Explosive Atmospheres. Equipment Protection by Intrinsic Safety "i"
Ex ic	Category 3G	EN 60079-11	Explosive Atmospheres. Equipment Protection by Intrinsic Safety "i"
Limited Energy			
Ex nL or Ex nLc	Categoryv3G	EN 60079-15	Electrical Apparatus for Explosive Gas Atmospheres. Type of Protection "n"
Pressurization			
Ex px or Ex pxb	Category 2G	EN 60079-2	Explosive Atmospheres. Equipment Protection by Pressurized Enclosure "p"

TABLE 9.13 EU flammable and combustible gases and vapor equipment protection techniques [42, 43]—cont'd

Protection Technique/ Marking	Suitable Hazardous (Classified) Area Locations	Standard No.	Title
Ex py or Ex pyb	Category 2G	EN 60079-2	Explosive Atmospheres. Equipment Protection by Pressurized Enclosure "p"
Ex pz or Ex pzc	Category 3G	EN 60079-2	Explosive Atmospheres. Equipment Protection by Pressurized Enclosure "p"
Restricted Breathing			
Ex nR or Ex nRc	Category 3G	EN 60079-15	Electrical Apparatus for Explosive Gas Atmospheres. Type of Protection "n"
Encapsulation			
Ex ma	Category 1G	EN 60079-18	Electrical Apparatus for Explosive Gas Atmospheres. Construction, Test and Marking of Type of Protection Encapsulation "m" Electrical Apparatus
Ex mb	Category 2G	EN 60079-18	Electrical Apparatus for Explosive Gas Atmospheres. Construction, Test and Marking of Type of Protection Encapsulation "m" Electrical Apparatus
Oil Immersion			
Ex o or Ex ob	Category 2G	EN 60079-6	Explosive Atmospheres. Equipment Protected by Oil Immersion "o"

joints, or structural openings and igniting exterior flammable or combustible gases or vapors. Flameproof enclosures are designed for operation in atmospheres containing specific flammable and combustible materials and are provided with markings identifying those materials. Cooper-Crouse Hinds IEC Digest [50] indicates that individual flameproof enclosures are subjected to routine testing at 1.5 times the enclosure design pressure before leaving the factory.

Increased Safety

Increased Safety "e" is an equipment protection technique which prohibits the development of arcs or sparks under normal operating conditions. It also prohibits their production under specific abnormal conditions. Increased safety features are added to compensate for specified abnormal conditions to assure that increased surface temperatures, arcs, or sparks do not develop under those conditions. Increased safety measures are generally applied to terminal boxes, light fixtures, transformers, instruments, and motors.

TABLE 9.14 **IEC flammable and combustible gases and vapor equipment protection techniques [44, 45]**

Protection technique/Marking	Suitable hazardous (classified) area locations	Standard No.	Title
General Requirements			
Ex	EPL Ga, Gb or Gc	IEC 60079-0	Explosive Atmospheres – Part 0: Equipment – General Requirements
Ex	EPL Ga, Gb or Gc	IEC 60079-14	Explosive Atmospheres – Part 14: Electrical Installations Design, Selection and Erection
Increased Safety			
Ex e or Ex eb	EPL Gb	IEC 60079-7	Explosive Atmospheres – Part 7: Equipment Protection by Increased Safety "e"'
Nonincendive			
Ex nC	EPL Gc	IEC 60079-15	Electrical Apparatus for Explosive Gas Atmospheres – Part 15: Construction, Test and Marking of Type of Protection "n" Electrical Apparatus
Non-Sparking			
Ex nA or Ex nAc	EPL Gc	IEC 60079-15	Electrical Apparatus for Explosive Gas Atmospheres – Part 15: Construction, Test and Marking of Type of Protection "n" Electrical Apparatus
Flameproof			
Ex d or Ex db	EPL Gb	IEC 60079-1	Explosive Atmospheres – Part 1: Equipment Protection by Flameproof Enclosures "d"
Ex d or Ex db	EPL Gb	IEC 60079-1-1	Electrical Apparatus for Explosive Gas Atmospheres – Part 1-1: Flameproof Enclosures "d" – Method of Test for Ascertainment of Maximum Experimental Safe Gap
Powder-Filled			
Ex q or Ex be	EPL Gb	IEC 60079-5	Explosive Atmospheres – Part 5: Equipment Protection by Powder Filling "q"
Enclosed Break			
Ex nC or Ex neck	EPL Gc	IENC60079-15	Electrical Apparatus for Explosive Gas Atmospheres – Part 15: Construction, Test and Marking of Type of Protection "n" Electrical Apparatus
Intrinsic Safety			
Ex air	EPL Ga	IEC 60079-11	Explosive Atmospheres – Part 11: Equipment Protection By Intrinsic Safety 'I'
Ex big	EPL Gb	IEC 60079-11	Explosive Atmospheres – Part 11: Equipment Protection by Intrinsic Safety "i"

TABLE 9.14 IEC flammable and combustible gases and vapor equipment protection techniques [44, 45]—cont'd

Protection technique/Marking	Suitable hazardous (classified) area locations	Standard No.	Title
Ex ic	EPL Gc	IEC 60079-11	Explosive Atmospheres – Part 11: Equipment Protection by Intrinsic Safety "i"
Limited Energy			
Ex nL or Ex nLc	EPL Gc	IEC 60079-15	Electrical Apparatus for Explosive Gas Atmospheres – Part 15: Construction, Test and Marking of Type of Protection "n" Electrical Apparatus
Pressurization			
Ex px or Ex pxb	EPL Gb	IEC 60079-2	Explosive Atmospheres – Part 2: Equipment Protection by Pressurized Enclosures "p"
Ex py or Ex pyb	EPL Gb	IEC 60079-2	Explosive Atmospheres – Part 2: Equipment Protection by Pressurized Enclosures "p"
Ex pz or Ex pzc	EPL Gc	IEC 60079-2	Explosive Atmospheres – Part 2: Equipment Protection by Pressurized Enclosures "p"
Restricted Breathing			
Ex nR or Ex nRc	EPL Gc	IEC 60079-15	Electrical Apparatus for Explosive Gas Atmospheres – Part 15: Construction, Test and Marking of Type of Protection "n" Electrical Apparatus
Encapsulation			
Ex ma	EPL Ga	IEC 60079-18	Electrical Apparatus for Explosive Gas Atmospheres – Part 18: Construction, Test and Marking of Type of Protection Encapsulation '"m" Electrical Apparatus
Ex mb	EPL Gb	IEC 60079-18	Electrical Apparatus for Explosive Gas Atmospheres – Part 18: Construction, Test and Marking of Type of Protection Encapsulation "m" Electrical Apparatus
Oil Immersion			
Ex o or Ex ob	EPL Gb	IEC 60079-6	Explosive Atmospheres – Part 6: Equipment Protection by Oil Immersion "o"

Equipment listed as compliant with Increased Safety requirements by the use of quality insulation. The equipment enclosures are specifically designed for air line leakage and creepage distances. Stove-enameled sheet steel, stainless sheet steel and polyester materials can be used to construct those enclosures. Cable entry details also require conformity, particularly in integrating cable shield bonding. Also, the electrical connections inside the enclosure are adequately secured to prevent loosening. Each component temperature class is observed, identifying the hottest spot temperatures.

Powder Filling

Powder Filling "q" is an equipment protection technique that has fixed electrical components that are totally encased or surrounded by filling material preventing ignition by excessive surface temperatures of any enclosed or external explosive gas/vapor–air mixtures. The fill material is normally quartz sand or solid glass beads. This application is typically limited to small electrical components without moving parts, such as small transformers, capacitors, and other similar components. There are limitations regarding stored energy in all capacitors in powder filled electrical apparatus. There are also restrictions on current, voltage, and VA rating in electrical components and apparatus.

Type of Protection "n" Techniques

Type of Protection "n" equipment protection techniques prohibit electrical equipment from igniting explosive vapor/gas–air mixtures. This technique assumes that any explosive initiating event occurring in a component or apparatus inside the enclosure is considered unlikely. There are several "n" protection techniques including:

Encapsulated "nC"

Enclosed-break "nC"

Energy-limited "nL" ("nC" in USA)

Hermetically sealed "nC"

Nonincendive "nC"

Non-sparking "nA"

Restricted breathing "nR"

Sealed "nC"

Oil Immersion

Oil Immersion "o" is an equipment protection technique that immerses electrical equipment in a protective liquid, preventing ignition of an explosive atmosphere either inside or outside of the equipment enclosure. Oil immersion equipment normally includes heavy duty switchgear, motor starters, and transformers. Means are required for the inclusion of sight glasses and/or liquid level gauges to assure proper equipment immersion. Also required is a means for removing and installing fluid and the addition of inert gas or creation of a vacuum above the liquid. Safety items which may be required include pressure relief valves and pressure switches. Some electrical/coolant immersion oils include:

Naphthenic mineral oil

Silicon-based or fluorinated hydrocarbons

Synthetic pentaerythritol tetra fatty acid esters

Natural ester fluid

Table 9.15 contains a list of some of the codes or standards associated with dielectric/coolant fluids. Biodegradable natural ester fluids are manufactured from seed oil, along with the addition of additives.

Definitions – Combustible Dust, Fibers, and Flyings Protection Techniques

A list of equipment protection techniques utilized for combustible dust, fibers, and flyings are presented in Tables 9.10 through 9.12. The following definitions describe some of the terms utilized in those tables. Some definitions are only applicable to some equipment protection techniques utilized in United States and/or Canadian applications. Others are applicable to apparatus in those areas and in International Electrotechnical Commission (IEC) and European Union (EU) equipment applications.

Dust-ignitionproof

Dust-ignitionproof (DIP) is an equipment protection technique that utilizes an enclosure to exclude dust from electrical equipment. The enclosure prohibits arcs, sparks, or thermal energy created inside an enclosure from causing ignition of specified dust on or in the vicinity of that enclosure. NEMA 9 is the common designation for dust-ignition proof enclosures in the United States.

Dusttight

A *dusttight* protection technique prevents dust from entering the enclosure under specified test conditions. Testing for dusttight enclosures is listed in ANSI/ISA 12.12.01, *Nonincendive Electrical Equipment for Use in Class I and II, Division 2 and Class III, Divisions 1 and 2 Hazardous (Classified) Locations* and UL 1604. UL 50, *Enclosures for Electrical Equipment, Non-Environmental Considerations* covers general-purpose dusttight enclosures. NEMA 250, *Enclosures for Electrical Equipment (1000 Volts Maximum)* lists

TABLE 9.15 Dielectric/coolant oil codes, standards, and recommended practices

Developer	Standard No.	Title
FM Global	FM 3990	Approval Standard for Less or Nonflammable Liquid-Insulated Transformers
IEEE	ANSI/IEEE 979	IEEE Guide to Substation Fire Protection
IEEE	ANSI/IEEE C57.12.00	General Requirements for Liquid-Immersed Distribution, Power, and Regulating Transformers
IEEE	ANSI/IEEE C57.12.90	Test Code for Liquid-Immersed Distribution, Power, and Regulating Transformers
IEEE	IEEE C57.100	IEEE Standard Test Procedure for Thermal Evaluation of Liquid-Immersed Distribution and Power Transformers
IEEE	IEEE C57.106	IEEE Guide for Acceptance and Maintenance of Insulating Oil in Equipment
IEEE	IEEEC57.140	IEEE Guide for the Evaluation and Reconditioning of Liquid-Immersed Power Transformers
IEEE	IEEE C57.147	Guide for Acceptance and Maintenance of Natural Ester Fluids in Transformers
ASTM	ASTM D3300	Standard Test Method for Dielectric Breakdown Voltage of Insulating Oils of Petroleum Origin Under Impulse Conditions
ASTM	ASTM D6871	Standard Specification for Natural (Vegetable Oil) Ester Fluids Used in Electrical Apparatus
UL	ANSI/UL 60079-0	Electrical Apparatus for Explosive Gas Atmospheres – Part 0: General Requirements
UL	ANSI/UL60079-6	Electrical Apparatus for Explosive Gas Atmospheres – Part 6: Oil-Immersion "o"
IEC	IEC 60079-0	Explosive Atmospheres – Part 0: Equipment – General Requirements
IEC	IEC 60079-6	Explosive Atmospheres – Part 6: Equipment Protection by Oil Immersion "o"
IEC	IEC 60156	Insulating Liquids – Determination of the Breakdown Voltage at Power Frequency – Test Method
IEC	IEC 60247	Insulating Liquids – Measurement of Relative Permittivity, Dielectric Dissipation Factor (tan δ) and DC Resistivity
IEC	IEC 60296	Fluids for Electrotechnical Applications – Unused Mineral Insulating Oils for Transformers and Switchgear
IEC	IEC 60567	Oil-Filled Electrical Equipment – Sampling of Gases and of Oil for Analysis of Free and Dissolved Gases – Guidance
IEC	IEC 60588-2	Askarels for Transformers and Capacitors – Part 2: Test Methods
IEC	IEC 60599	Mineral Oil-Impregnated Electrical Equipment in Service – Guide to the Interpretation of dissolved and Free Gases Analysis
IEC	IEC 60836	Specifications for Unused Silicone Insulating Liquids for Electrotechnical Purposes

dusttight enclosures as NEMA Types 3, 4, 4X, 12, 12K, and 13. NEMA 12, 13 and 13K are indoor enclosures.

Hermetically Sealed

Hermetically sealed protection techniques prevent the entrance of atmospheric combustible materials by the use of fusion techniques on electrical components in an enclosure. Soldering, brazing, welding, and glass fused to metal are techniques that will prevent the entry of combustible materials.

Hazardous (Classified) Area Equipment Standards

Selected standards applicable for electrical equipment utilized in hazardous (classified) locations are presented in Table 9.16. This table represents some of the most commonly referenced standards. That table does not contain the standards previously listed in Table 9.15 regarding hazardous (classified) location protection techniques. Those standards would be included in specifying any equipment for use in hazardous (classified) locations.

North American Equipment Markings

The National Electrical Code mandates the suitability of equipment for the hazardous (classified) location in which it is installed. That suitability is determined by [51]:

1. Equipment listing or labeling
2. Evidence of equipment evaluation from a qualified testing laboratory or inspection agency concerned with product evaluation
3. Evidence acceptable to the authority having jurisdiction such as a manufacturer's self-evaluation or an owner's engineering judgment.

The NEC provides a *marking* requirement in Article 505.9(C). It mandates that equipment labels shall identify the hazardous (classified) location for which it has been approved. That *marking* is to be in either a Zone or Division format. The *Division* format must provide the following information:

Class: I, II, or III

Division: 1 or 2

Group: A, B, C, or D

Temperature Code: T1, T2, T2A, T2B, T2C, T2D,T3, T3A, T3B, T3C, T4, T4A, T5, or T6

TABLE 9.16 Selected hazardous (classified) location equipment standards

Developer	Standard No.	Title
UL	UL 515	Electric Resistance Heat Tracing for Commercial and Industrial Applications
UL	UL 583	Electric-Battery-Powered Industrial Trucks
UL	UL 674	Electric Motors and Generators for Use in Division 1 Hazardous (Classified) Locations
UL	UL 698A	Industrial Control Panels Relating to Hazardous (Classified) Locations
UL	UL 779	Electrically Conductive Floorings
UL	UL 783	Electric Flashlights and Lanterns for Use in Hazardous (Classified) Locations
UL	UL 823	Electric Heaters for Use in Hazardous (Classified) Locations
UL	UL 844	Electric Lighting Fixtures for Use in Hazardous (Classified) Locations
UL	UL 913	Intrinsically Safe Apparatus and Associated Apparatus for Use in Class I, II, and III, Division 1, Hazardous (Classified) Locations
UL	UL 1010	Standard for Receptacle-Plug Combinations for Use in Hazardous (Classified) Locations
UL	UL 1067	Electrically Conductive Equipment and Materials for Use in Flammable Anesthetizing Locations
UL	UL 1203	Explosion-Proof and Dust-Ignition-Proof Electrical Equipment for Use in Hazardous (Classified) Locations
UL	UL 1207	Sewage Pumps for Use in Hazardous (Classified) Locations
UL	UL 1604	Electrical Equipment for Use in Class I and II, Division 2, and Class III Hazardous (Classified) Locations
UL	UL 2225	Metal-Clad Cables and Cable-Sealing Fittings for Use in Hazardous (Classified) Locations
API	API RP 14F	Design and Installation of Electrical Systems for Fixed and Floating Offshore Petroleum Facilities for Unclassified and Class I, Division 1, and Division 2 Locations
API	API RP 14FZ	Design and Installation of Electrical Systems for Fixed and Floating Offshore Petroleum Facilities for Unclassified and Class I, Zone 0, Zone 1, and Zone 2 Locations
API	AP RP 11S3	Electric Submersible Pump Installations
API	API RP 11S5	Recommended Practice for the Application of Electric Submersible Cable Systems
API	API RP 11S6	Recommended Practice for Testing of Electric Submersible Pump Cable Systems
API	ANSI/API RP 500	Recommended Practice for Classification of Locations for Electrical Installations at Petroleum Facilities Classified as Class I, Division 1 and Division 2
API	ANSI/API RP 505	Recommended Practice for Classification of Locations for Electrical Installations at Petroleum Facilities Classified as Class I, Zone 0, Zone 1 and Zone 2

TABLE 9.16 Selected hazardous (classified) location equipment standards—cont'd

Developer	Standard No.	Title
API	API RP 540	Electrical Installations in Petroleum Processing Plants
API	API Std 541	Form-Wound Squirrel-Cage Induction Motors 500 Horsepower and Larger
API	API Std 546	API Brushless Synchronous Machines – 500 kVA and Larger
API	API Std 547	General-purpose Form-wound Squirrel Cage Induction Motors – 250 Horsepower and Larger
API	API Std 2003	Protection Against Ignitions Arising Out of Static, Lightning, and Stray Currents
IEEE	IEEE 45	Recommended Practice for Electrical Installation on Shipboard
IEEE	IEEE 515	Standard for the Testing, Design, Installation, and Maintenance of Electrical Resistance Heat Tracing for Industrial Applications
IEEE	IEEE 576	RP for Installation, Termination, and Testing of Insulated Power Cable as Used in Industrial and Commercial Applications
IEEE	IEEE 739	RP for Energy Management in Commercial and Industrial Facilities Bronze Book
IEEE	IEEE 841	Standard for the Petroleum and Chemical Industry-Severe Duty Totally Enclosed Fan-Cooled (TEFC) Squirrel-Cage Induction Motors – Up To and Including 500 hp
IEEE	IEEE 844	Recommended Practice for Electrical Impedance, Induction, and Skin Effect Heating of Pipelines and Vessels
IEEE	IEEE 1068	RP for the Repair and Rewinding of Motors for the Petroleum and Chemical Industry
IEEE	IEEE 1202	Standard for Flame Testing of Cables for Use in Cable Tray in Industrial and Commercial Occupancies
IEEE	IEEE 1242	Guide for Specifying and Selecting Power, Control and Purpose Cable for Petroleum and Chemical Plants
IEEE	IEEE 1349	IEEE Guide for Application of Electric Motors in Class I, Division 2 Hazardous (Classified) Locations
ISA	ISA 12.13.01	Performance Requirements for Combustible Gas Detectors
ISA	ISA 12.13.04	Performance Requirements for Open Path Combustible Gas Detectors
ISA	ISA 12.27.01	Requirements for Process Sealing Between Electrical Systems and Flammable or Combustible Process Fluids
ISA	ISA 92.00.01	Performance Requirements for Toxic Gas Detectors
ISA	ISA 92.00.04	Performance Requirements for Open Path Toxic Gas Detectors
NEMA	ANSI C80.1	Electrical Rigid Steel Conduit (ersc)
NEMA	ANSI C80.5	Electrical Rigid Aluminum Conduit (erac)
NEMA	ANSI/NEMA FB 1	Fittings, Cast Metal Boxes, and Conduit Bodies for Conduit, Electrical Metallic Tubing and Cable

(Continued)

TABLE 9.16 Selected hazardous (classified) location equipment standards—cont'd

Developer	Standard No.	Title
NEMA	NEMA FB 2.20	Selection and Installation Guidelines for Fittings for Use with Flexible Electrical Conduit and Cable
NEMA	ANSI/NEMA FB 11	Plugs, Receptacles, and Connectors of the Pin and Sleeve Type for Hazardous Locations
NEMA	NEMA MG 1	Motors and Generators
NEMA	NEMA RV 1	Application and Installation Guidelines for Armored Cable and Metal-Clad Cable
NEMA	NEMA RV 3	Application and Installation Guidelines for Flexible and Liquidtight Flexible Metal Conduits
NEMA	NEMA WC 71/ ICEA S-96659	Standard for Nonshielded Cables Rated 2001–5000 Volts for Use in the Distribution of Electric Energy
NEMA	ANSI/NEMA WC 74/ICEA S-93-639	5–46 kV Shielded Power Cable for Use in the Transmission and Distribution of Electric Energy
CSA	CSA PLUS 2203 HAZLOC-01	Hazardous Locations Guide for the Design, Testing, Construction, and Installation of Equipment in Explosive Atmospheres

Zone Equipment Markings

Zone equipment markings require slightly different information. Examples of those requirements are presented in Figures 9.1 through 9.5. Zone equipment is required to have the following marking:

Class

Zone

Symbol: "Ex" (Canada) or "AEx" (USA)

Protection Technique(s) [52]

Gas Classification Group(s) [53]

Temperature Classification [54]

There is an exception for Zone marking for an associated apparatus that is not suitable for hazardous (classified) location service. An example of that might be an intrinsically safe component, installed as part of an intrinsically safe system, but located outside of the hazardous (classified) area. The NEC allows this component to be marked without Class and Zone designation information, if it is not listed for use in hazardous (classified) locations. However, it requires inclusion of the United States (A) designation, explosion protected

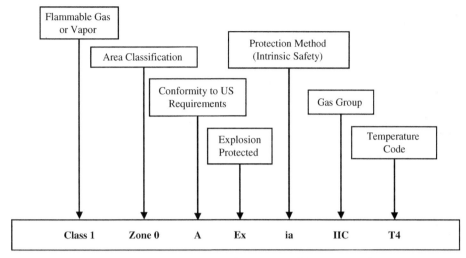

Figure 9.1: Hazardous (classified) locations equipment marking – United States: Option A (Zone system)

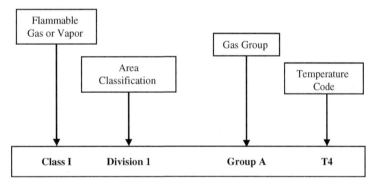

Figure 9.2: Hazardous (classified) locations equipment marking: Option B (Division system)

designation (Ex) and protection method, and gas group information. It also requires bracketing of the first three items to signify their non-suitability for location in a hazardous (classified) area, i.e. *[AEx ia] IIC*. The bracketed items indicate that the equipment is not suitable for installation in hazardous (classified) locations (see Figure 9.4).

There is also an exception for a *simple apparatus*, which does not require a marked operating temperature or temperature class (T Code). A simple apparatus is defined as:

An electrical component or combination of components of simple construction with well-defined electrical parameters that does not generate more than 1.5 volts, 100 milliamperes, and 25 milliwatts, or a passive component that does not dissipate more than 1.3 watts and is compatible with the intrinsic safety of the circuit in which it is used. [55]

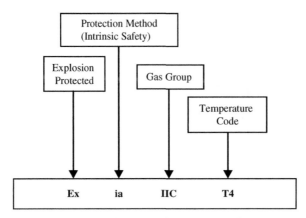

Figure 9.3: Hazardous (classified) locations equipment marking – Canada: Option A (Zone system)

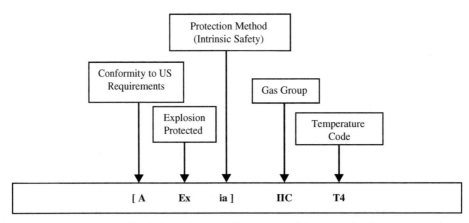

Figure 9.4: Hazardous (classified) locations equipment marking – United States: Option A (Zone system) w/I. S. Apparatus Not Rated for Hazardous Service

The NEC® lists certain protection techniques that can be referenced to specific Group designations as noted in Table 9.17 and NEC® 505.9(C)(2).

Figure 9.1 illustrates the *marking designation* for equipment used in the United States with the Zone designation. In that example, the equipment is rated for use in the United States (AEx) and in a Class I, Zone 0 hazardous (classified) location. It is also rated for Group IIC gas service and has a T Code of T4 (≤135°C). Since no further temperature information was provided on the label, it was assumed that the T4 rating was referenced for operating in a temperature range of −20°C and +40°C. If a special ambient temperature range was required the symbols "Ta" or "Tamb" would have been used to designate the difference from standard ambient conditions. For example, −30°C ≤ Ta ≤ +50°C.

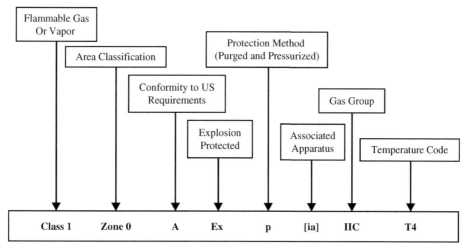

Figure 9.5: Hazardous (Classified) Locations Equipment Marking – United States: Option A (Zone system) w/Associated Apparatus

Where equipment is rated to operate at ambient temperatures higher than +40°C, the NEC requires that the equipment shall be marked with the maximum ambient temperature and the temperature class or operating temperature at the higher ambient temperature. There are exceptions for non-heat-producing equipment, such as conduit fittings and also for heat producing equipment with a maximum operating temperature of 100°C. Under that exception [56], the Temperature Class or operating temperature is not required to be included on the equipment *marking*. There is also another exception for Class I, Division 1 or 2 equipment. NEC Articles 505.20(B) and 505.20(C) permit such equipment to be T Code marked in accordance with the Class I, and II designations in NEC 500.8(B).

TABLE 9.17 Protection techniques for specific gas group designations

Protection technique designation	Group designation
"e", "m", "p", or "q"	II
"d", "ia", "ib", "[ia]", or [ib]"	IIA, IIB, IIC, or specific gas or vapor
"n"	II
"n"	IIA, IIB, IIC, or specific gas or vapor (See Note 1)
Other electrical equipment protection techniques	II (See Note 2)

Note 1: This designation is required if an apparatus also contains enclosed-break devices, nonincendive components, or energy limited equipment or circuits [56].
Note 2: Group IIA, IIB, IIC, or specific gas or vapor markings may be mandated by the protection technique used in equipment.

Figure 9.2 illustrates an example for the Division hazardous (classified) location equipment marking. NEC Article 505.9(C)(1) permits that equipment to also have Zone marking information. That additional information, if included, would involve the following designations:

Class I, Zone 1 or Class I, Zone 2, whichever is applicable.

Gas classification group designation in accordance with NEC Table 505.9(C).

Temperature (T Code) designation in accordance with NEC Table 505.9(D)(1).

Figure 9.3 illustrates the Canadian hazardous (classified) location equipment marking. Note the difference between it and the example of the United States marking requirements is the "Ex" designation for Canada and the deletion of the Class and Zone information. Figure 9.5 illustrates an example where equipment utilizes two different protection techniques. The example illustrates "p" and "ia" equipment protection techniques.

References

1. Earley, Mark W., Sargent, Jeffrey S., Sheehan, Joseph V., and Caloggero, John M., Eds, *NEC® 2005 Handbook: NFPA 70: National Electrical Code*; 2005, Article 500.1, page 647. National Fire Protection Association; Quincy, MA.
2. NFPA 70, *National Electrical Code®*; 2005, Article 500.5(B)(1). National Fire Protection Association, Quincy, MA.
3. Ibid., Article 500.5(B)(2).
4. Ibid., Article 500.5(C)(1).
5. Ibid., Article 500.5(C)(2).
6. Ibid., Article 500.5(D)(1).
7. Ibid., Article 500.5(D)(2).
8. Ibid., Article 505.5(B)(1).
9. Ibid., Article 505.5((B)(2).
10. Ibid., Article 505.5(B)(3).
11. Ibid., Article 506.5(B)(1).
12. Ibid., Article 506.5(B)(2).
13. Ibid., Article 506.5(B)(3).
14. NFPA 70, *National Electrical Code®*; 2005, Article 500.6. National Fire Protection Association; Quincy, MA.
15. Ibid., Article 505.6.
16. NFPA 499, *Recommended Practice for the Classification of Combustible Dusts and of Hazardous (Classified) Locations for Electrical Installations in Chemical Process Areas*; 2008, Paragraph 3.3.4. National Fire Protection Association; Quincy, MA.

17. NFPA 497, *Recommended Practice for the Classification of Flammable Liquids, Gases, or Vapors and of Hazardous (Classified) Locations for Electrical Installations in Chemical Process Areas*; 2008, Paragraph 3.3.9. National Fire Protection Association; Quincy, MA.

18. NFPA 497, *Recommended Practice for the Classification of Flammable Liquids, Gases, or Vapors and of Hazardous (Classified) Locations for Electrical Installations in Chemical Process Areas*; 2008, Paragraph 3.3.10. National Fire Protection Association; Quincy, MA.

19. NFPA 70, *National Electrical Code*®; 2005, Article 500.6(A)(1) and 505.6(A), (B), and (C). National Fire Protection Association, Quincy, MA.

20. NFPA 70, *National Electrical Code*®; 2008, Article 505.6. National Fire Protection Association; Quincy, MA.

21. Ibid., Article 500.8(C)(4), Table 500.8(C).

22. Ibid., Article 505.9(D)(1), Table 505.9(D)(!)

23. API RP 2216-2003, *Ignition Risk of Hydrocarbon Liquids and Vapors by Hot Surfaces in Open Air*. American Petroleum Institute; Washington, DC.

24. Ibid., Paragraph 3.1.

25. NFPA 70, *National Electrical Code*; 2005, Article 500.7. National Fire Protection Association; Quincy, MA.

26. See NFPA 496, *Standard for Purged and Pressurized Enclosures for Electrical Equipment for Class I and II Hazardous (Classified) Locations*. 2008; National Fire Protection Association; Quincy, MA.

27. NFPA 70, *National Electrical Code*®; 2005, Article 500.7(K)(2). National Fire Protection Association; Quincy, MA.

28. NFPA 70, *National Electrical Code*®; 2005, Article 505.8. National Fire Protection Association; Quincy, MA.

29. NFPA 70, *National Electrical Code*®; 2005, Articles 500.7 and 505.8. National Fire Protection Association; Quincy, MA.

30. FM Approvals Guide to Hazardous Locations – Explosive Gas Atmospheres; FM Global Group;, Johnston, RI;Website:http://www.fmglobal.com/assets/pdf/FM_EXGas_HazardPoster.pdf.

31. Underwriters Laboratories; Northbrook, IL; 2008. UL website: http://www.ul.com/hazloc/tech_docs/CI_protection_methods.pdf

32. NFPA 70, *National Electrical Code*®; 2005, Articles 506.8. National Fire Protection Association; Quincy, MA.

33. FM Approvals Guide to Hazardous Locations – Explosive Dust Atmospheres; FM Global Group; Johnston, RI;Website:http://www.fmglobal.com/assets/pdf/FM_EXDust_HazardPoster.pdf.

34. Underwriters Laboratories; Northbrook, IL; 2008. UL website: http://www.ul.com/hazloc/tech_docs/CII_protection_methods.pdf.

35. Underwriters Laboratories; Northbrook, IL; 2008; UL website: http://www.ul.com/hazloc/tech_docs/CIII_protection_methods.pdf.

36. Underwriters Laboratories; Northbrook, IL; 2008;UL website: http://www.ul.com/hazloc/tech_docs/CII_protection_methods.pdf.

37. FM Approvals Guide to Hazardous Locations – Explosive Dust Atmospheres; FM Global Group;, Johnston, RI;Website:http://www.fmglobal.com/assets/pdf/FM_EXDust_HazardPoster.pdf.

38. Underwriters Laboratories; Northbrook, IL; 2008; UL website: http://www.ul.com/hazloc/tech_docs/CIII_protection_methods.pdf.

39. Underwriters Laboratories; Northbrook, IL; 2008; UL website: http://www.ul.com/hazloc/tech_docs/CII_protection_methods.pdf.

40. FM Approvals Guide to Hazardous Locations – Explosive Dust Atmospheres; FM Global Group;, Johnston, RI;Website:http://www.fmglobal.com/assets/pdf/FM_EXDust_HazardPoster.pdf.

41. Underwriters Laboratories; Northbrook, IL; 2008; UL website: http://www.ul.com/hazloc/tech_docs/CIII_protection_methods.pdf.

42. Underwriters Laboratories; Northbrook, IL; 2008; UL website: http://www.ul.com/hazloc/tech_docs/CI_protection_methods.pdf.

43. FM Approvals Guide to Hazardous Locations – Explosive Gas Atmospheres; FM Global Group; Johnston, RI;Website:http://www.fmglobal.com/assets/pdf/FM_EXGas_HazardPoster.pdf.

44. Underwriters Laboratories; Northbrook, IL; 2008; UL website: http://www.ul.com/hazloc/tech_docs/CI_protection_methods.pdf.

45. FM Approvals Guide to Hazardous Locations – Explosive Gas Atmospheres; FM Global Group; Johnston, RI;Website:http://www.fmglobal.com/assets/pdf/FM_EXGas_HazardPoster.pdf.

46. NFPA 70, *National Electrical Code*®; 2005, Article 504.2. National Fire Protection Association; Quincy, MA.

47. NFPA 921-2004, *Guide for Fire and Explosion Investigations*; Table 21.13.3.1. National Fire Protection Association; Quincy, MA.

48. NPFA 70, *National Electrical Code*®; 2005, Article 500.2. National Fire Protection Association; Quincy, MA.

49. Cooper Crouse Hinds 2004 Ex Digest; Chapter 4, Paragraph 4.8.1, page 22. Cooper Crouse Hinds; Syracuse, NY.

50. Cooper Crouse Hinds 2004 Ex Digest; Chapter 4, page 16. Cooper Crouse Hinds, Syracuse, NY.

51. NFPA 70, *National Electrical Code*®; 2008, Article 505.9(A). National Fire Protection Association, Quincy, MA.

52. Protection techniques are listed in NFPA 70-2005, *National Electrical Code*®, Table 505.9(C)(2)(4).

53. Reference NFPA 70-2005, *National Electrical Code*®, Table 505.9(C) for Gas classification group designations.

54. Reference NFPA 70-2005, *National Electrical Code*®, Table 505.9(D)(1) for temperature class (T-Code).

55. NFPA 70, *National Electrical Code*; 2005, Article 504.2. National Fire Protection Association; Quincy, MA.

56. NFPA 70, *National Electrical Code*; 2005, Article 505.9(C)(2). National Fire Protection Association; Quincy, MA.

CHAPTER 10

Wire, Cable, and Raceway

General

Electrical conductor materials can be classified in three groups based on their resistivity (ohm-centimeter) properties. They include:

Conductors – 1.59×10^{-6} to 400×10^{-6} ohm-centimeters

Semiconductors – 10^{-3} to 10^{9} ohm-centimeters

Dielectrics (Insulators) – 10^{10} to $<10^{20}$ ohm-centimeters

Electrical wiring is generally composed of two of those components, i.e. a conductor (#1) plus insulator (#3) in the above list.

Definitions

Before examining wire and cable in more detail, two words sometimes used interchangeably, cable and wire, should be examined. A *wire* can be defined as a slender rod or filament of drawn metal. This describes a solid wire verses a stranded wire which is composed of several strands of solid wires, intertwined. Wire with an insulation covering would be called an *insulated wire*. A *cable* can be defined as a single-stranded-conductor or a combination of individually insulated (multiple) conductors, with an outer jacket of insulation.

Wire and cable sizes are measured by the conductor area. One unit of measurement often used in the United States is a *mil*. One *mil* is defined as one-thousandth of an inch. A *circular mil* is defined as a unit of area equal to $\pi/4 = 0.7854$ of a square mil. Therefore, the cross-sectional area of a circle is:

$$A = (d)^2 \tag{Eq. 10.1}$$

where A is the area and d is the conductor diameter in mils. There are approximately 1974 circular mils in a square millimeter (mm^2). There are 1,000,000 circular mils in a conductor circular inch.

Electrical Codes, Standards, Recommended Practices and Regulations; ISBN: 9780815520450

Wiring in the United States is sized by the American Wire Gauge (AWG) sizes and circular mils (kcmil). Some common American wire sizes are shown in Table 10.1. Each wire number corresponds to a specific circular mils cross-sectional area. The wire area increases as the AWG wire number becomes smaller. The National Electrical Code uses the American Wire Gauge (AWG) designation for all copper, aluminum or copper-clad aluminum conductors up to size 4/0 AWG. Conductors 250 through 2000 are designated in thousand circular mils (kcmil). The AWG or kcmil wire numbers relate to the wire or cable conductor and not the insulation thickness. Stranded and solid conductors with the same AWG or kcmil number have the same conductor cross-sectional area. It should be noted that the term *kcmil* was originally referred to by the Roman numerical designation MCM. However, in 1990 that designation was changed to the kcmil designation by NFPA 70®, *National Electrical Code*® (NEC®), IEEE, and UL.

Wire number identification in the metric system utilizes the cross-sectional conductor size in mm^2 as the cable or wire designator. IEEE/ASTM SI 10, *Standard for Use of the International System of Units (SI): The Modern Metric System* provides conversion from circular mils to square millimeters (mm^2). Conversion can be made by using

$$mm^2 = 5.067075 \times 10^{-10} \times \text{(Number of circular mils)} \qquad \text{(Eq. 10.2)}$$

Conversely, conversion from circular mils to mm^2 can be obtained by

$$\text{Circular mil area} = 1973.53 \times \text{(Number of } mm^2) \qquad \text{(Eq. 10.3)}$$

The NEC requires that all conductors and cables shall be marked with either AWG or circular mil area in Article 310.11 Marking.

TABLE 10.1 Some common American wire sizes

AWG	Circular mils	AWG	Circular mils	Circular mils kcmil (MCM)[1]	Circular mils kcmil (MCM)*
24	404	6	26,240	250	900
22	642.4	4	41,740	300	1,000
20	1,022	2	52,620	350	1,250
18	1,620	2	66,360	400	1,500
16	2,580	1	83,690	500	1,750
14	4,110	1/0	105,600	600	2,000
12	6,530	2/0	133,100	700	—
10	10,380	3/0	167,800	750	—
8	16,510	4/0	211,600	800	—

[1]See text.

TABLE 10.2 Common cable ampacity (current capacity) standards

Standard No.	Title
IEC 60096-0-1	Guide to the Design of Detailed Specifications – Section One: Coaxial Cables – Current Carrying Capacity (Ampacity)
IEC 60287	Electric Cables – Calculation of the Current Rating: Current Rating Equations (100% load factor) and Calculation of Losses
IEC 61196-1-121	Coaxial Communication Cables – Part 1-121: Electrical Test Methods – Test Method for Ampacity
IEC TR 62095	Electric Cables Calculations for Current Ratings Finite Element Method
ICEA P-43-457	Conductor Resistance and Ampacities at 400 and 800Hz
ICEA P-46-426/IEEE S-135-1	IEEE/ICEA Power Cable Ampacity Tables
ICEA P-48-426	Paper Cable Ampacities at AEIC Temperatures
IEEE Std 835	IEEE Standard Power Cable Ampacity Tables, Part 1 through Part 5
ANSI/IEEE 848	IEEE Standard Procedure for the Determination of the Ampacity Derating of Fire-Protected Cables
NEMA WC50 ICEA P-53-426	Ampacities, 15–69kV 1/c Power Cable Including Effect of Shield Losses (Solid Dielectric)
NEMA WC51 ICEA P-54-440	Ampacities of Cables in Open-Top Trays
NFPA 70®, Article 310	Conductors for General Wiring
ANSI/SCTE 32	Ampacity of Coaxial Telecommunications Cables

Conductor Material

There are several basic materials utilized for electrical wiring conductors. The most commonly used include:

1. Copper
   a. Soft-annealed
   b. Medium hard-drawn
   c. Hard-drawn
2. Copper clad steel
3. Aluminum
   a. Soft-annealed
   b. Hard-drawn
4. Aluminum clad steel

Aluminum has approximately 60% of the conductivity of copper.

There are many factors involved with the selection of wire and cable. Conductor selection begins with determining the application and the voltage rating at which it will be operating. Also required would be the selection of the number, size, and type of conductors needed as well as its method of installation, i.e. cable or conductors in raceway. Determination must also be made if any of the conductors are required to be twisted pairs or shielded. Also, consideration must be given to the conductor insulation and outer jacket materials. Conductor operating temperatures and ambient conditions must also be identified.

Shielding is a grounded, conducting medium which surrounds an insulated conductor or cable. This practice confines its electric field to the inside of the cable. There are several purposes for shielding insulated power cables including [1]:

1. To obtain symmetrical radial stress distribution within the insulation and to control tangential and longitudinal stresses or discharges on the surfaces of the insulation.
2. To protect cables connected to overhead lines or otherwise subject to induced potentials,
3. To provide increased safety to human life.

When selecting cable, it must be determined if a metal sheath or armor is required, as well as any outer jacket and covering. The cable use, such as direct buried, exposure to direct sunlight, ambient temperature, conductor operating temperature limitations, corrosive conditions, exposure to damp or wet locations will also require determination. All of the above determined information is necessary for the selection of conductor(s) insulation, cable outer jacket material, and the need for shielding, armor or sheathing.

Insulation Material

NEC® Table 310.13(A) *Conductor Application and Insulations Rated 600 Volts* list several insulation types used on insulated wires and cables. They include the following:

- Ethylene tetrafluoro-ethylene
- Fluorinated ethylene propylene
- Magnesium oxide (mineral insulation)
- Paper
- Perfluoro-alkoxy
- Extended polytetra-fluoro-ethylene
- Silicon

- Thermoplastic

  - Flame-retardant, moisture-resistant

  - Flame-retardant, moisture and heat resistant

  - Flame-retardant and heat-resistant

- Thermoset

  - Flame-retardant

  - Flame-retardant, heat-resistant

Thermoplastic and thermoset plastics encompass a large number of insulation classes in use today with power and control conductors. They include insulation type letters MTW, RHH, RHW, RHW-2, SIS, THHN, THHW, THW, THW-2, THWN, THWN-2, TW, XHH, XHHW, and XHHW-2. The -2 suffix designation indicates the cable is permitted to be utilized at a continuous 90 °C (194 °F) operating temperature in wet or dry conditions. Reference NEC® Table 310.13(A) for an explanation of the conductor insulation letters.

Thermoplastic and thermosetting plastic are two general groups of polymers.

> Thermoplastic polymers are capable of being repeatedly softened by increased temperatures and hardened by decreasing temperatures, the change upon heating being physical rather than chemical. Thermosetting polymers are capable of being changed into a substantially infusible or insoluble product when cured by the application of heat or chemical means. [2]

One important fact regarding thermoplastic insulation should be noted. Because of its property of softening under heating, it may deform if simultaneously subjected to pressure at points of support. Also, it has the tendency to stiffen at temperatures less than –10 °C (+14 °F). Problems can also develop if thermoplastic insulation is utilized in DC circuits under wet conditions. Under those circumstances electroendosmosis may develop between the conductor and its insulation [3].

Ampacity

Ampacity is defined by the 2008 NEC in Article 100 as:

> The current, in amperes, that a conductor can carry continuously under the conditions of use without exceeding its temperature rating. [4]

A conductor's ampacity can be affected by many factors including:

- Ambient temperature

- Number of current carrying conductors in close proximity

- Harmonic content of a load

- Temperature compatibility with equipment

- Load current I^2R losses in conductors

- Heat sources external from adjacent conductors

- Hysteresis losses from steel conduit

- Cable location: buried; in raceway; in cable tray; or overhead aerial

There are several sources for cable and wire ampacities (see Table 10.2). Some include cable manufacturer's data; NEC Article 310; and IEEE Std 835, *IEEE Standard Power Cable Ampacity Tables* – Part 1 thru 5. The Code also allows the use of the Neher–McGrath formula under engineering supervision. That relationship can be found in NEC Article 310.15(C) Engineering Supervision. That formula is a heat transfer equation, taking into account all possible heat sources, thermal resistances, heat transfer mechanisms, and free air. Its most common use is to calculate cable ampacities in underground electrical-duct. It can also be applicable to all other cable installation ampacity calculations.

NFPA 70 recognizes several situations in which cable and wire ampacities must be de-rated. That is outlined in NEC Article 310.15(B)(2) Adjustment Factors. Table 310.15(B)(2)(a) provides adjustment factors for the number of current carrying conductors in a raceway or cable. Correction factors for ambient temperatures other than 30°C (86°F) can be found on the ampacity table in Article 310.

Power and Control Cables

The most common conductor materials used in electrical power and control cables are copper, aluminum, or alloys of them. Copper wire can be soft-annealed, medium hard-drawn, or hard-drawn. Aluminum has approximately 60% of the conductivity of copper. Because of its excellent conductivity, copper is the most commonly used insulated conductor in industrial, commercial, and residential applications for power and control service. Copper's high conductivity and lower line resistivity reduces heating losses and has higher ampacity than equivalent sizes of aluminum conductors.

Aluminum has a low-tensile strength when compared to copper. It also has a higher coefficient of thermal expansion than copper. Although this physical characteristics comparison would seem to indicate that copper would be the better material for overhead electrical transmission and main distribution conductors that is actually not the case. Aluminum is more commonly used, with the addition of a center steel core such as in aluminum-conductor-steel reinforced (ACSR) conductors. The steel core offers conductor structural support and coupled with lower material costs than copper, makes this conductor the material of choice for utility overhead, un-insulated conductor use.

Common codes, standards, and recommended practices for copper and aluminum conductors are presented in Table 10.3. Standards for the three basic forms of copper are ASTM B1, B2, and B3. They respectively represent: hard drawn copper; medium hard drawn

TABLE 10.3 Copper and aluminum electrical conductor standards

Developer	Standard No.	Title
Copper Conductors		
ASTM	ASTM B1	Standard Specification for Hard-Drawn Copper Wire
ASTM	ASTM B2	Standard Specification for Medium-Hard-Drawn Copper Wire
ASTM	ASTM B3	Standard Specification for Soft or Annealed Copper Wire
ASTM	ASTM B8	Standard Specification for Concentric-Lay Stranded Copper Conductors, Hard, Medium-Hard, or Soft
ASTM	ASTM B33	Specification for Tinned Soft or Annealed Copper Wire for Electrical Purposes
ASTM	ASTM B49	Specification for Copper Rod Drawing Stock for Electrical Purposes
ASTM	ASTM B105	Standard Specification for Hard-Drawn Copper Alloy Wires for Electrical Conductors
ASTM	ASTM B170	Standard Specification for Oxygen-Free Electrolytic Copper
ASTM	ASTM B172	Specification for Rope-Lay Stranded Copper Conductors Having Bunch-Stranded Members, for Electrical Conductors
ASTM	ASTM B173	Standard Specification for Rope-Lay-Stranded Copper Conductors Having Concentric-Stranded Members, for Electrical Conductors
ASTM	ASTM B174	Specification for Bunch-Stranded Copper Conductors for Electrical Conductors
ASTM	ASTM B189	Standard Specification for Lead-Coated and Lead-Alloy-Coated Soft Copper Wire for Electrical Purposes
ASTM	ASTM B193	Test Method for Resistivity of Electrical Conductor Materials
ASTM	ASTM B226	Standard Specification for Cored, Annular, Concentric-Lay-Stranded Copper Conductors
ASTM	ASTM B227	Standard Specification for Hard-Drawn Copper-Clad Steel Wire
ASTM	ASTM B228	Standard Specification for Concentric-Lay-Stranded Copper-Clad Steel Conductors
ASTM	ASTM B229	Standard Specification for Concentric-Lay-Stranded Copper and Copper-Clad Steel Composite Conductors
ASTM	ASTM B230	Standard Specification for Aluminum 1350
ASTM	ASTM B230M	Standard Specification for Aluminum 1350
ASTM	ASTM B246	Standard Specification for Tinned Hard-Drawn and Medium-Hard-Drawn Copper Wire for Electrical Purposes

(Continued)

TABLE 10.3 Copper and aluminum electrical conductor standards—cont'd

Developer	Standard No.	Title
ASTM	ASTM B258	Specification for Standard Nominal Diameters and Cross-Sectional Areas of AWG Sizes of Solid Round Wires Used as Electrical Conductors
ASTM	ASTM B298	Specification for Silver-Coated Soft or Annealed Copper Wire
ASTM	ASTM B324	Standard Specification for Nickel-Coated Soft or Annealed Copper Wire
ASTM	ASTM B355	Standard Specification for Nickel-Coated Soft or Annealed Copper Wire
ASTM	ASTM B496	Standard Specification for Compact Round Concentric-Lay-Stranded Copper Conductors
ASTM	ASTM B738	Standard Specification for Fine-Wire Bunch-Stranded and Rope-Lay Bunch-Stranded Copper Conductors for Use as Electrical Conductors
ICEA	ICEA P-51-432	Copper Conductors, Bare and Weather Resistant
ICEA	ICEA P-52-469	Copper-Covered Steel and Copper Composite Conductors
Aluminum Conductors		
ASTM	ASTM B230	Standard Specification for Aluminum 1350-H19 Wire for Electrical Purposes
ASTM	ASTM B230M	Standard Specification for Aluminum 1350-H19 Wire for Electrical Purposes (Metric)
ASTM	ASTM B231	Standard Specification for Concentric-Lay-Stranded Aluminum 1350 Conductors
ASTM	ASTM B231M	Standard Specification for Concentric-Lay-Stranded Aluminum 1350 Conductors (Metric)
ASTM	ASTM B232	Standard Specification for Concentric-Lay-Stranded Aluminum Conductors, Coated-Steel Reinforced (ACSR)
ASTM	ASTM B232	Standard Specification for Concentric-Lay-Stranded Aluminum Conductors, Coated-Steel Reinforced (ACSR) [Metric]
ASTM	ASTM B233	Standard Specification for Aluminum 1350 Drawing Stock for Electrical Purposes
ASTM	ASTM B236	Standard Specification for Aluminum Bars for Electrical Purposes (Bus Bars)
ASTM	ASTM B236M	Standard Specification for Aluminum Bars for Electrical Purposes (Bus Bars) Metric
ASTM	ASTM 371	Standard Specification for Aluminum-Alloy Extruded Bar, Rod, Tube, Pipe, Structural Profiles, and Profiles for Electrical Purposes (Bus Conductor)
ASTM	ASTM 371M	Standard Specification for Aluminum-Alloy Extruded Bar, Rod, Tube, Pipe, Structural Profiles, and Profiles for Electrical Purposes (Bus Conductor) [Metric]
ASTM	ASTM B324	Standard Specification for Aluminum Rectangular and Square Wire for Electrical Purposes
ASTM	ASTM B398	Standard Specification for Aluminum-Alloy 6201-T81 Wire for Electrical Purposes
ASTM	ASTM B398M	Standard Specification for Aluminum-Alloy 6201-T81 Wire for Electrical Purposes (Metric)

TABLE 10.3 Copper and aluminum electrical conductor standards—cont'd

Developer	Standard No.	Title
ASTM	ASTM B399	Standard Specification for Concentric-Lay-Stranded Aluminum-Alloy 6201-T81 Conductors
ASTM	ASTM B399M	Standard Specification for Concentric-Lay-Stranded Aluminum-Alloy 6201-T81 Conductors (Metric)
ASTM	ASTM B401	Standard Specification for Compact Round Concentric-Lay-Stranded Aluminum Conductors, Steel-Reinforced (ACSR/COMP)
ASTM	ASTM B415	Standard Specification for Hard-Drawn Aluminum-Clad Steel Wire
ASTM	ASTM B416	Standard Specification for Concentric-Lay-Stranded Aluminum-Clad Steel Conductors
ASTM	ASTM B502	Standard Specification for Aluminum-Clad Steel Core Wire for Aluminum Conductors, Aluminum-Clad Steel Reinforced
ASTM	ASTM B524	Standard Specification for Concentric-Lay-Stranded Aluminum Conductors, Aluminum-Alloy Reinforced (ACAR, 1350/6201)
ASTM	ASTM B524M	Standard Specification for Concentric-Lay-Stranded Aluminum Conductors, Aluminum-Alloy Reinforced (ACAR, 1350/6201) (Metric)
ASTM	ASTM B549	Standard Specification for Concentric-Lay-Stranded Aluminum Conductors, Aluminum-Clad Steel Reinforced (ACSR/AW)
ASTM	ASTM B566	Standard Specification for Copper-Clad Aluminum Wire
ASTM	ASTM B609	Standard Specification for Aluminum 1350 Round Wire, Annealed and Intermediate Tempers, for Electrical Purposes
ASTM	ASTM B609M	Standard Specification for Aluminum 1350 Round Wire, Annealed and Intermediate Tempers, for Electrical Purposes (Metric)
ASTM	ASTM B701	Standard Specification for Concentric-Lay-Stranded Self-Damping Aluminum Conductors, Steel Reinforced (ACSR/SD)
ASTM	ASTM B701M	Standard Specification for Concentric-Lay-Stranded Self-Damping Aluminum Conductors, Steel Reinforced (ACSR/SD) [Metric]
ASTM	ASTM B711	Standard Specification for Concentric-Lay-Stranded Aluminum-Alloy Conductors, Steel Reinforced (AACSR) (6201)
ASTM	ASTM B736	Standard Specification for Aluminum, Aluminum Alloy and Aluminum-Clad Steel Cable Shielding Stock
ASTM	ASTM B778	Standard Specification for Shaped Wire Compact Concentric-Lay-Stranded Aluminum Conductors (AAC/TW)
ASTM	ASTM B778M	Standard Specification for Shaped Wire Compact Concentric-Lay-Stranded Aluminum Conductors (AAC/TW) [Metric]
ASTM	ASTM B779	Standard Specification for Shaped Wire Compact Concentric-Lay-Stranded Aluminum Conductors, Steel-Reinforced (ACSR/TW)
ASTM	ASTM B800	Standard Specification for 8000 Series Aluminum Alloy Wire for Electrical Purposes-Annealed and Intermediate Tempers
ASTM	ASTM B801	Standard Specification for Concentric-Lay-Stranded Conductors of 8000 Series Aluminum Alloy for Subsequent Covering or Insulation

(Continued)

TABLE 10.3 **Copper and aluminum electrical conductor standards—cont'd**

Developer	Standard No.	Title
ASTM	ASTM B812	Standard Test Method for Resistance to Environmental Degradation of Electrical Pressure Connections Involving Aluminum and Intended for Residential Applications
ASTM	ASTM B836	Standard Specification for Compact Round Stranded Aluminum Conductors Using Single Input Wire Construction
ASTM	ASTM B856	Standard Specification for Concentric-Lay-Stranded Aluminum Conductors, Coated Steel Supported (ACSS)
ASTM	ASTM B857	Standard Specification for Shaped Wire Compact Concentric-Lay-Stranded Aluminum Conductors, Coated-Steel Supported (ACSS/TW)
ASTM	ASTM B872	Standard Specification for Precipitation-Hardening Nickel Alloys Plate, Sheet, and Strip
ASTM	ASTM B888	Standard Specification for Copper Alloy Strip for Use in Manufacture of Electrical Connectors or Spring Contacts
ASTM	ASTM B901	Standard Specification for Compressed Round Stranded Aluminum Conductors Using Single Input Wire Construction
ASTM	ASTM B911	Standard Specification for ACSR Twisted Pair Conductor (ACSR/TP)
ASTM	ASTM B911M	Standard Specification for ACSR Twisted Pair Conductor (ACSR/TP)[Metric]
ASTM	ASTM B941	Standard Specification for Heat Resistant Aluminum-Zirconium Alloy Wire for Electrical Purposes
ASTM	ASTM D2519	Standard Test Method for Bond Strength of Electrical Insulating Varnishes by the Helical Coil Test
ASTM	ASTM F1593	Standard Test Method for Trace Metallic Impurities in Electronic Grade Aluminum by High Mass-Resolution Glow-Discharge Mass Spectrometer
ASTM	ASTM F1845	Standard Test Method for Trace Metallic Impurities in Electronic Grade Aluminum-Copper, Aluminum-Silicon, and Aluminum-Copper-Silicon Alloys by High-Mass-Resolution Glow Discharge Mass Spectrometer
ICEA	ICEA P-50-431	Aluminum Conductors, Bare and Weather Resistant
NEMA	ANSI/NEMA C119.4	Standard for Electric Connectors for Use Between Aluminum-to-Aluminum or Aluminum-to-Copper Conductors (DRAFT STANDARD)
NEMA	ANSI/NEMA C119.6	Standard For Electric Connectors-Non-Sealed, Multiport Connector Systems Rated 600 Volts or Less for Aluminum and Copper Conductors
SCTE	ANSI/SCTE 100	Specification for 75 Ohm Smooth Aluminum Subscriber Access Cable
UL	ANSI/UL 486E	Standard for Equipment Wiring Terminals for Use with Aluminum and/or Copper Conductors

copper wire; and soft or annealed copper. ASTM provides several standards specifically regarding copper for electrical use.

ASTM also provides standards for electrical use aluminum conductors. Aluminum is heavily utilized by utilities in overhead, un-insulated, medium- and high-voltage applications. Because of aluminum's softness and other properties, the overhead aluminum conductors are commonly used with steel cores. The aluminum conductor, steel-reinforced (ACSR) cable standards are covered by Standards ASTM B232/232M, ASTM B401, ASTM B549, ASTM B701/701M and ASTM B779.

Other reinforced forms of aluminum aerial conductors include: Aluminum Conductor, Aluminum-Alloy Reinforced (ACAR); Aluminum-Alloy Conductor, Steel Reinforced (AACSR); Aluminum Conductors, Coated Steel Supported (ACSS); and Aluminum Conductors, Aluminum-Clad-Steel Reinforced (ASCR/AW). The reinforced conductors allow longer spans between supports and provide added support for sustained wind forces.

A power cable consists of one or more line or phase conductors, with or without a grounded (e.g. neutral) conductor, or a grounding conductor feeding a single electrical load; a group of electrical loads; a building, structure, industrial facility or commercial facility; an electrical substation, or other loads. The cable can consist of two or more insulated conductors rated for the voltage and temperature in which they are operating and an insulated or bare ground conductor, all enclosed by an outer insulating jacket. The cable should be suitable for use in the environment in which it is located. It may or may not have an outer mechanical protection jacket or armor. In cases of medium- and high-voltage cables, they may be supplied with a conductive electrical shield under the outer jacket.

A control cable is one consisting of two or more insulated electrical conductors with an outer insulated jacket, used in conjunction with switches; relays; indicator lights; level, flow and pressure switches; limit switches; electrical solenoids; electronic control devices; etc, for the control of motors; alarms; valves; lights; and other electrical equipment and loads. The conductor insulation should be rated for the voltage and service for which they are used.

Many of the standards associated with power and control cables are listed in Table 10.4. Power and control cables can consist of a variety of cable and wire types. Power cables can be rated low-voltage, medium-voltage, and high-voltage. Control cables are typically low-voltage, but can even be fiber-optic with signal converters and controls at both ends, for long distance control applications. Individual power conductors can be run in conduit or cable tray to supply electrical loads. Individual control conductors can also be run in cable tray or raceway or can also be run in the same raceway with the power conductors, provided they have equivalent insulation voltage ratings, and are suitable for similar environmental and temperature conditions.

The most distinguishing factor between insulated power and control cables may be conductor size. Insulated power cables are commonly included in one of five insulation groups, including [5]:

Mineral insulation (MI)

Paper laminated tapes

Thermosetting compounds

TABLE 10.4 Power and control cable standards

Developer	Standard No.	Title
CSA	CSA22.5 No. 35	Extra-Low-Voltage Control Circuit Cables, Low-Energy Control Cable, and Extra-Low-Voltage Control Cable
CSA	CSA 22.5 No. 38	Thermoset-Insulated Wires and Cables
CSA	CSA 22.5 No. 48	Nonmetallic Sheathed Cable
CSA	CSA 22.5 No. 49	Flexible Cords and Cables
CSA	CSA 22.5 No. 51	Armoured Cables
CSA	CSA 22.5 No. 52	Underground Service-Entrance Cables
CSA	CSA22.5 No. 75	Thermoplastic-Insulated Wires and Cables
CSA	CSA 22.5 No. 96	Portable Power Cables
CSA	CSA 22.5 No. 123	Metal Sheathed Cables
CSA	CSA 22.5 No. 124	Mineral Insulated Cable
CSA	CSA 22.5 No. 129	Neutral Supported Cable
CSA	CAN/CSA 22.5 No. 130	Requirements for Electrical Resistance Heating Cables and Heating Device Sets
CSA	CSA 22.5 No. 174	Cables and Cable Glands for Use in Hazardous Locations
CSA	CSA 22.5 No. 230	Tray Cables
CSA	CSA 22.5 No. 239	Control and Instrumentation Cables
CSA	CSA C49.2	Compact Aluminum Conductors Steel Reinforced (ACSR)
CSA	CSA C49.3	Aluminum Alloy 1350 Round Wire, All Tempers, for Electrical Purposes
CSA	CSA C49.5	Aluminum Alloy 1350 Round Wire, All Tempers, for Electrical Purposes
CSA	CSA C68.1	Specifications for Impregnated Paper-Insulated Metallic-Sheathed Cable, Solid Type
CSA	CSA C68.3	Shielded and Concentric Neutral Power Cables Rated 5-46kV
CSA	CSA C68.5	Primary Shielded and Concentric Neutral Cable for Distribution Utilities
CSA	CAN/CSA C60104	Aluminium-Magnesium-Silicon Alloy Wire for Overhead Line Conductors (Adopted CEI/IEC 60104:1987, second edition, 1987-12, with Canadian deviations)
CSA	CAN/CSA C60888	Zinc-Coated Steel Wires for Stranded Conductors (Adopted CEI/IEC 888:1987, first edition, 1987-12, with Canadian deviations)
CSA	CAN/CSA C60889	Hard-Drawn Aluminum Wire for Overhead Line Conductors (Adopted CEI/IEC 889:1987, first edition, 1987-11, with Canadian deviations)
CSA	CAN/CSA C61089	Round Wire Concentric Lay Overhead Electrical Stranded Conductors (Adopted CEI/IEC 1089:1991, first edition, 1991-05, with Canadian deviations)
CSA	CAN/CSA C61232	Aluminium-Clad Steel Wires for Electrical Purposes (Adopted CEI/IEC 1232:1993, first edition, 1993-06)
CSA/CEI/IEC	CAN/CSA-CEI/IEC 1000-2-1	Electromagnetic Compatibility (EMC) - Part 2: Environment - Section 1: Description of the Environment – Electromagnetic Environment for Low-Frequency Conducted Disturbances and Signalling in Public Power Supply Systems (Adopted CEI/IEC 1000-2-1:1990)

TABLE 10.4 Power and control cable standards—cont'd

Developer	Standard No.	Title
CSA	CAN/CSA-C61000-2-2	Electromagnetic Compatibility (EMC) – Part 2-2: Environment – Compatibility Levels for Low-Frequency Conducted Disturbances and Signalling in Public Low-Voltage Power Supply Systems (Adopted CEI/IEC 61000-2-2:2002, second edition, 2002-03, with Canadian deviations)
CSA	CAN/CSA-CEI/IEC 61000-2-4	Electromagnetic Compatibility (EMC) – Part 2-4: Environment – Compatibility Levels in Industrial Plants for Low-Frequency Conducted Disturbances
CSA	CAN/CSA-CEI/IEC 61000-2-6	Electromagnetic Compatibility (EMC) – Part 2: Environment. Section 6: Assessment of the Emission Levels in the Power Supply of Industrial Plants as Regards Low-Frequency Conducted Disturbances
CSA	CAN/CSA-C61000-2-12	Electromagnetic Compatibility (EMC) – Part 2-12: Environment – Compatibility Levels for Low-Frequency Conducted Disturbances and Signalling in Public Medium-Voltage Power Supply Systems (Adopted CEI/IEC 61000-2-12:2003, first edition, 2003-04, with Canadian deviations)
ICEA	ANSI/ICEA S-97-682	Utility Shielded Power Cables Rated 5 Through 46 kV
ICEA	ANSI/ICEA S-108-720	Extruded Insulation Power Cables Rated Above 46 kV Through 345 kV
IEEE	IEEE 383™	Standard for Qualifying Class 1E Electric Cables and Field Splices for Nuclear Power Generating Stations
IEEE	IEEE 404™	IEEE Standard for Extruded and Laminated Dielectric Shielded Cable Joints Rated 2 500 V to 500 000 V
IEEE	IEEE 515	IEEE Recommended Practice for the Testing, Design, Installation, and Maintenance of Electrical Resistance Heat Tracing for Industrial Applications
IEEE	IEEE 524	IEEE Guide to the Installation of Overhead Transmission Line Conductors
IEEE	IEEE Std 525	IEEE Guide for the Design and Installation of Cable Systems in Substations – Description
IEEE	IEEE 576	Recommended Practice for Installation, Termination, and Testing of Insulated Power Cable as Used in Industrial and Commercial Application
IEEE	ANSI/IEEE 592	IEEE Standard for Exposed Semiconducting Shields on High-Voltage Cable Joints and Separable Insulated Connectors
IEEE	IEEE 635™	Guide for Selection and Design of Aluminum Sheaths for Power Cable
IEEE	ANSI/IEEE 690	IEEE Standard for the Design and Installation of Cable Systems for Class 1E Circuits in Nuclear Power Generating Stations
IEEE	IEEE Std 835	IEEE Standard Power Cable Ampacity Tables – Part 1 thru 5
IEEE	ANSI/IEEE 1018	IEEE Recommended Practice for Specifying Electric Submersible Pump Cable – Ethylene-Propylene Rubber Insulation
IEEE	ANSI/IEEE 1019	IEEE Recommended Practice for Specifying Electric Submersible Pump Cable – Polypropylene Insulation

(Continued)

TABLE 10.4 Power and control cable standards—cont'd

Developer	Standard No.	Title
IEEE	ANSI/IEEE 1120	IEEE Guide for the Planning, Design, Installation, and Repair of Submarine Power Cable Systems
IEEE	IEEE 1142	IEEE Guide for the Design, Testing, and Application of Moisture-Impervious, Solid Dielectric, 5-35 kV Power Cable Using Metal-Plastic Laminates
IEEE	ANSI/IEEE 1185	IEEE Guide for Installation Methods for Generating Station Cables
IEEE	ANSI/IEEE 1235	IEEE Guide for the Properties of Identifiable Jackets for Underground Power Cables and Ducts
IEEE	ANSI/IEEE 1242	IEEE Guide for Specifying and Selecting Power, Control, and Special-Purpose Cable for Petroleum and Chemical Plants
IEEE	ANSI/IEEE 1300	IEEE Guide for Cable Connections for Gas-Insulated Substations
IEEE	IEEE P1410™	Guide for Improving the Lightning Performance of Electric Power Overhead Distribution Lines
IEEE	IEEE 1580™	IEEE Recommended Practice for Marine Cable for Use on Shipboard and Fixed or Floating Platforms
IEEE	ANSI/IEEE 1617	IEEE Guide for Detection, Mitigation, and Control of Concentric Neutral Corrosion in Medium-Voltage Underground Cables
NEMA	NEMA WC 26	Binational Wire and Cable Packaging Standard
NEMA	ANSI/NEMA WC 57 ICEA S-73-532	Standard for Control, Thermocouple Extension, and Instrumentation Cables
NEMA	NEMA WC 58 ANSI/ICEA S-75-381	Portable and Power Feeder Cables for Use in Mines and Similar Applications
NEMA	NEMA WC 65	A Reasoned Approach to Solving Solderability Problems with Tin-Coated and Nickel-Coated Stranded Conductors In High Performance Wire and Cable Applications
NEMA	ANSI/NEMA WC 70 ANSI/ICEA S-95-658	Non-shielded Power Cable 2000 V or Less for the Distribution of Electrical Energy
NEMA	ANSI/NEMA WC 71 ANSI/ICEA S-96-659	Standard for Nonshielded Cables Rated 2001–5000 Volts for Use in the Distribution of Electric Energy
NEMA	NEMA WC 72	Continuity of Coating Testing for Electrical Conductors
NEMA	NEMA WC 73	Wire Selection Guidelines for Wires Rated at 200–450 Degrees C
NEMA	NEMA WC 74	5–46 kV Shielded Power Cable for Use in the Transmission and Distribution of Electric Energy
NFPA	NFPA 70, Article 310	Conductors for General Wiring
NFPA	NFPA 70, Article 320	Armored Cable: Type AC
NFPA	NFPA 70, Article 322	Flat Cable Assemblies: Type FC
NFPA	NFPA 70, Article 324	Flat Conductor Cable: Type FCC
NFPA	NFPA 70, Article 326	Integrated Gas Space Cable: Type IGS
NFPA	NFPA 70, Article 328	Medium-Voltage Cagle: Type MV
NFPA	NFPA 70, Article 330	Metal-Clad Cable: Type MC
NFPA	NFPA 70, Article 332	Mineral-Insulated, Metal-Sheathed Cable: Type MI

TABLE 10.4 Power and control cable standards—cont'd

Developer	Standard No.	Title
NFPA	NFPA 70, Article 334	Nonmetallic-Sheathed Cable Types: NM, NMC, and NMS
NFPA	NFPA 70, Article 336	Power and Control Tray Cable: Type TC
NFPA	NFPA 70, Article 338	Service-Entrance Cable: Type SE and USE
NFPA	NFPA 70, Article 340	Underground Feeder and Branch-Circuit Cable: Type UT
NFPA	ANSI/NFPA 262	Standard Method of Test for Flame Travel and Smoke of Wires and Cables for Use in Air-Handling Spaces
UL	UL 4	Armored Cable
UL	UL 44	Thermoset-Insulated Wires and Cables
UL	UL 62	Flexible Cord and Fixture Wire
UL	UL 83	Thermoplastic-Insulated Wires and Cables
UL	UL 486A-486B	Wire Connectors
UL	UL 486C	Splicing Wire Connectors
UL	UL 486D	Sealed Wire Connector Systems
UL	UL 486E	Standard for Equipment Wiring Terminals for Use with Aluminum and/or Copper Conductors
UL	UL 493	Standard for Thermoplastic-Insulated Underground Feeder and Branch-Circuit Cables
UL	UL 515	Standard for Electrical Resistance Heat Tracing for Commercial and Industrial Applications
UL	UL 719	Nonmetallic-Sheathed Cables
UL	UL 854	Service-Entrance Cables
UL	UL 1072	Medium-Voltage Power Cables
UL	UL 1277	Electrical Power and Control Tray Cables with Optional Optical-Fiber Members
UL	UL 1462	Standard for Mobile Home Pipe Heating Cable
UL	UL 1569	Metal-Clad Cables
UL	UL 1673	Standard for Electric Space Heating Cables
UL	UL 2049	Standard for Residential Pipe Heating Cable
UL	UL 2225	Metal-Clad Cables and Cable-Sealing Fittings for Use in Hazardous (Classified) Locations

Thermoplastic compounds

Varnished cloth laminated tapes

Insulation voltage ratings must be selected for the cable application. Insulation suitability for damp, moist, or wet applications must be verified. Conductors exposed to damp or wet ambient conditions will require insulation types which has different characteristics than insulation suitable for dry locations. A conductor's insulation and/or outer jacket's suitability for exposure to sunlight and ozone; gasoline or chemicals; or high or low temperatures may require determination depending upon the environment in which it will be located.

Section 12.2.3 *Insulation*, in IEEE *Recommended Practice for Electric Power Distribution for Industrial Plants*, IEEE 141-1993 (R1999) provides some of the most common low-voltage insulation parameters and lists the cable insulation types which are best suited for those conditions. They include:

1. Relative Heat Resistance – This characteristic is important for conductors operating under overload conditions for any length of time, areas with high ambient temperatures and any other heat producing application. Thermosetting plastics are better with this application than thermoplastic insulation conductors. Insulation will tend to harden and crack under prolonged thermal exposure.

2. Heat Aging – Long term heat aging can cause a conductor's insulation or a cable's jacket to elongate, reducing the insulation thickness.

3. Ozone and Corona Discharge Resistance – Can lead to insulation failure. The following insulation types offer resistance to these phenomenon:

   a. Silicon insulation

   b. Rubber insulation

   c. Polyethylene insulations

   d. Cross-linked polyethylene (XLPE) insulations

   e. Polyvinyl chloride (PVC) insulation

   f. Ethylene propylene rubber (EPR)

   Corona discharge is normally not associated with low-voltage cables. However EPR insulation is better suited to that condition.

4. Moisture Resistance – Cable and conductors can easily be exposed to moisture if run outdoor in cable tray or in conduit and other raceway or indoor in areas exposed to moisture. Cable insulations best suited for that environment include:

   a. Cross-linked polyethylene (XLPE)

   b. Polyethylene

   c. Ethylene propylene rubber (EPR)

5. Treeing – Is an insulation degradation phenomenon associated with the presence of moisture. It is reportedly more prevalent with XLPE and polyethylene than EPR insulation [6].

Cable insulation associated with medium- and high-voltage applications include [7]:

- Paper – (solid type)

- Varnished cambric

- Polyethylene (natural)

- SBR rubber

- Butyl rubber

- Oil-based rubber

- Cross-linked polyethylene (XLPE)

- Ethylene propylene rubber (EPR)

- Chlorosulfonated polyethylene (CPE)

- Polyvinyl chloride (PVC)

- Silicon rubber

- Ethylene tetrafluoroethylene

IEEE 141-1993(R1999), Section 12.2.4 *Cable Design* presents the major factors for consideration when choosing cable insulation. They include:

Electrical Characteristics – including conductor size, insulation thickness, operating voltage, etc.

Thermal Conditions – cable suitability with current overload conditions, ambient and conductor operating temperatures, etc.

Mechanical – considerations include mechanical impact, abrasion, moisture resistance, flexibility for ease of installation

Chemical – insulation exposure to oil, flames, ozone, sunlight, acids, alkalis

Flame Resistance – area of operation may require this type of insulation for personnel safety and system operation under fire conditions

Low Smoke – cable that is flame retardant and which has low smoke producing characteristics under fire conditions is desirable from an occupant and fire fighting standpoint

Toxicity – cable which is resistant to the release of toxic gases while under fire conditions is desirable from an occupant and fire fighting standpoint

The National Electrical Code® provides *"Uses Permitted and Uses Not Permitted"* in the description and application of power and control cables noted in the Articles listed in Table 10.5. Those NEC Articles provide the information needed to aid in the selection of the proper cables for specific applications. Also included in the noted NEC Articles is information on conductor minimum bending radius restrictions necessary for installation.

Many of the International Electrotechnical Commission (IEC) standards on power and control cables are presented in Table 10.6. One of the obvious differences between the IEC standard cables and those meeting ANSI requirements is the difference in conductor sizes. Both IEC and North American standards classify their cable by the cable cross-sectional area. IEC categorizes their cable by mm^2 area. North American standards utilize American Wire Gauge (AWG) for common sizes 24 through 4/0. Cable sizes 250 MCM and larger are listed by kcmil (thousand circular mils).

Wire and cable insulation thickness by voltage rating is also different between IEC standards and North American Standards because of differences in operating voltages and other factors. IEC cable standards are based solely on IEC codes and standards. IEC standards including as IEC 60228 *Conductors of Insulated Cables* are included in Table 10.10 below. North American Standards are based on AEIC, ICEA, NEMA, IEEE, CSA, and UL standards. There are some differences in testing requirements between IEC and North American standards. Inspection requirements can also differ, depending upon the cable application and service.

The Association of Edison Illuminating Companies (AEIC) was established in 1885 by members and managers of Edison Companies to exchange views and experiences and pool their resources to improve the operation of their companies. The organization established a committee structure, composed of members from each Edison Company. Membership was

TABLE 10.5 2008 NEC power and control cables uses permitted/not-permitted

Cable type	NEC Article
Armored Cable (AC)	Article 320
Flat Cable Assemblies (FC)	Article 322
Flat Conductor Cable (FCC)	Article 324
Integrated Gas Spacer Cable (IGS)	Article 326
Medium-Voltage Cable (MV)	Article 328
Metal-Clad Cable (MC)	Article 330
Mineral-Insulated, Metal-Sheathed Cable (MI)	Article 332
Nonmetallic-Sheathed Cable (NM), (NMC), and (NMS)	Article 334
Power and Control Tray Cable (TC)	Article 336
Service-Entrance Cable (SE) and (USE)	Article 338
Underground Feeder and Branch-Circuit Cable (UF)	Article 340

TABLE 10.6 IEC cable standards

Developer	Standard No.	Title
IEC	IEC 60050-461	International Electrotechnical Vocabulary – Part 461: Electric Cables
IEC	IEC 60092-350	Electrical Installations in Ships – Part 350: General Construction and Test Methods of Power, Control and Instrumentation Cables for Shipboard and Offshore Applications
IEC	IEC 60092-351	Electrical Installations in Ships – Part 351: Insulating Materials for Shipboard and Offshore Units, Power, Control, Instrumentation, Telecommunication and Data Cables
IEC	IEC 60092-352	Electrical Installations in Ships – Part 352: Choice and Installation of Electrical Cables
IEC	IEC 60092-353	Electrical Installations in Ships – Part 353: Single and Multi-Core Non-Radial Field Power Cables with Extruded Solid Insulation for Rated Voltages 1 kV and 3 kV
IEC	IEC 60092-354	Electrical Installations in Ships – Part 354: Single- and Three-Core Power Cables with Extruded Solid Insulation for Rated Voltages 6 kV (Um = 7.2 kV) Up To 30 kV (Um = 36 kV)
IEC	IEC 60092-376	Electrical Installations in Ships – Part 376: Cables for Control and Instrumentation Circuits 150/250 V (300 V)
IEC	IEC 60227-1	Polyvinyl Chloride Insulated Cables of Rated Voltages Up To and Including 450/750 V – Part 1: General Requirements
IEC	IEC 60227-3	Polyvinyl Chloride Insulated Cables of Rated Voltages Up To and Including 450/750 V – Part 3: Non-Sheathed Cables for Fixed Wiring
IEC	IEC 60227-6	Polyvinyl Chloride Insulated Cables of Rated Voltages Up To and Including 450/750 V – Part 6: Lift Cables and Cables for Flexible Connections
IEC	IEC 60227-7	Polyvinyl Chloride Insulated Cables of Rated Voltages Up To and Including 450/750 V – Part 7: Flexible Cables Screened and Unscreened with Two or More Conductors
IEC	IEC 60228	Conductors of Insulated Cables
IEC	IEC 60230	Impulse Tests on Cables and Their Accessories
IEC	IEC 60287-1-3	Electric Cables – Calculation of the Current Rating – Part 1-3: Current Rating Equations (100% Load Factor) and Calculation of Losses – Current Sharing Between Parallel Single-Core Cables and Calculation of Circulating Current Losses
IEC	IEC 60287-2-2	Electric Cables – Calculation of the Current Rating – Part 2: Thermal Resistance – Section 2: A Method for Calculating Reduction Factors for Groups of Cables in Free Air, Protected from Solar Radiation
IEC	IEC 60287-3-2	Electric Cables – Calculation of the Current Rating – Part 3: Sections on Operating Conditions – Section 2: Economic Optimization of Power Cable Size
IEC	IEC 60331-11	Tests for Electric Cables Under Fire Conditions – Circuit Integrity – Part 11: Apparatus – Fire Alone at a Flame Temperature of at Least 750°C
IEC	IEC 60331-12	Tests for Electric Cables Under Fire Conditions – Circuit Integrity – Part 12: Apparatus – Fire with Shock at a Temperature of at Least 830°C

(Continued)

TABLE 10.6 IEC cable standards—cont'd

Developer	Standard No.	Title
IEC	IEC 60331-21	Tests for Electric Cables Under Fire Conditions – Circuit Integrity – Part 21: Procedures and Requirements – Cables of Rated Voltage Up To and Including 0,6/1,0 kV
IEC	IEC 60331-23	Tests for Electric Cables Under Fire Conditions – Circuit Integrity – Part 23: Procedures and Requirements – Electric Data Cables
IEC	IEC 60331-31	Tests for Electric Cables Under Fire Conditions – Circuit Integrity – Part 31: Procedures and Requirements for Fire with Shock – Cables of Rated Voltage Up To and Including 0,6/1 kV
IEC	IEC 60332-2-1	Tests on Electric and Optical Fibre Cables Under Fire Conditions – Part 2-1: Test for Vertical Flame Propagation for a Single Small Insulated Wire or Cable – Apparatus
IEC	IEC 60332-2-2	Tests on Electric and Optical Fibre Cables Under Fire Conditions – Part 2-2: Test for Vertical Flame Propagation for A Single Small Insulated Wire or Cable – Procedure for Diffusion Flame
IEC	IEC 60332-3-21	Tests on Electric Cables Under Fire Conditions – Part 3-21: Test for Vertical Flame Spread of Vertically-Mounted Bunched Wires or Cables – Category A F/R
IEC	IEC 60364-1	Low-Voltage Electrical Installations – Part 1: Fundamental Principles, Assessment of General Characteristics, Definitions
IEC	IEC 60364-4-43	Low-Voltage Electrical Installations – Part 4-43: Protection for Safety – Protection Against Overcurrent
IEC	IEC 60446	Basic and Safety Principles for Man–Machine Interface, Marking and Identification – Identification of Conductors By Colours or Alphanumerics
IEC	IEC 60465	Specification for Unused Insulating Mineral Oils for Cables with Oil Ducts
IEC	IEC 60502-SER	Power Cables with Extruded Insulation and Their Accessories for Rated Voltages from 1 kV (Um = 1.2 kV) Up To 30 kV (Um = 36 kV) – ALL PARTS
IEC	IEC 60502-1	Power Cables with Extruded Insulation and Their Accessories for Rated Voltages from 1 kV (Um = 1,2 kV) Up To 30 kV (Um = 36 kV) – Part 1: Cables for Rated Voltages of 1 kV (Um = 1,2 kV) and 3 kV (Um = 3,6 kV)
IEC	IEC 60502-2	Power Cables with Extruded Insulation and Their Accessories for Rated Voltages from 1 kV (Um = 1.2 kV) Up To 30 kV (Um = 36 kV) – Part 2: Cables for Rated Voltages from 6 kV (Um = 7.2 kV) Up To 30 kV (Um = 36 kV)
IEC	IEC 60502-4	Power Cables with Extruded insulation and their accessories for rated voltages from 1 kV (Um = 1.2 kV) Up To 30 kV (Um = 36 kV) – Part 4: Test Requirements on Accessories for Cables with Rated Voltages from 6 kV (Um = 7.2 kV) Up To 30 kV (Um = 36 kV)
IEC	IEC 60544-5	Electrical Insulating Materials – Determination of the Effects of Ionizing Radiation – Part 5: Procedures for Assessment of Ageing in Service
IEC	IEC 60684-3-340	Flexible Insulating Sleeving – Part 3: Specifications for Individual Types of Sleeving – Sheets 340 to 342: Expandable Braided Polyethylene Terephthalate Textile Sleeving
IEC	IEC 60702-1	Mineral Insulated Cables and Their Terminations with A Rated Voltage Not Exceeding 750 V – Part 1: Cables

TABLE 10.6 IEC cable standards—cont'd

Developer	Standard No.	Title
IEC	IEC 60702-2	Mineral Insulated Cables and Their Terminations with A Rated Voltage Not Exceeding 750 V – Part 2: Terminations
IEC	IEC 60754-1	Test on Gases Evolved During Combustion of Materials from Cables – Part 1: Determination of the Amount of Halogen Acid Gas
IEC	IEC 60754-2	Test on Gases Evolved During Combustion of Electric Cables – Part 2: Determination of Degree of Acidity of Gases Evolved During the Combustion of Materials Taken from Electric Cables by Measuring pH and Conductivity
IEC	IEC 60811-4-1	Insulating and Sheathing Materials of Electric and Optical Cables – Common Test Methods – Part 4-1: Methods Specific To Polyethylene and Polypropylene Compounds – Resistance to Environmental Stress Cracking – Measurement of the Melt Flow Index – Carbon Black and/or Mineral Filler Content Measurement in Polyethylene By Direct Combustion – Measurement of Carbon Black Content by Thermogravimetric Analysis (TGA) – Assessment of Carbon Black Dispersion in Polyethylene Using a Microscope
IEC	IEC 60811-4-2	Insulating and Sheathing Materials of Electric and Optical Cables – Common Test Methods – Part 4-2: Methods Specific to Polyethylene and Polypropylene Compounds – Tensile Strength and Elongation at Break After Conditioning at Elevated Temperature – Wrapping Test After Conditioning at Elevated Temperature – Wrapping Test After Thermal Ageing in Air – Measurement of Mass Increase – Long-Term Stability Test – Test Method for Copper-Catalyzed Oxidative Degradation
IEC	IEC 60840	Power Cables with Extruded Insulation and Their Accessories for Rated Voltages Above 30 kV (Um = 36 kV) Up To 150 kV (Um = 170 kV) – Test Methods and Requirements
IEC	IEC 60885-1	Electrical Test Methods for Electric Cables. Part 1: Electrical Tests for Cables, Cords and Wires for Voltages Up To and Including 450/750 V
IEC	IEC 60885-2	Electrical Test Methods for Electric Cables. Part 2: Partial Discharge Tests
IEC	IEC 60885-3	Electrical Test Methods for Electric Cables. Part 3: Test Methods for Partial Discharge Measurements on Lengths of Extruded Power Cables
IEC	IEC 61034-1	Measurement of Smoke Density of Cables Burning Under Defined Conditions – Part 1: Test Apparatus
IEC	IEC 61034-2	Measurement of Smoke Density of Cables Burning Under Defined Conditions – Part 2: Test Procedure and Requirements
IEC	IEC 61084-2-2	Cable Trunking and Ducting Systems for Electrical Installations – Part 2-2: Particular Requirements – Cable Trunking Systems and Cable Ducting Systems Intended for Underfloor and Flushfloor Installations
IEC	IEC 61138	Cables for Portable Earthing and Short-Circuiting Equipment
IEC	IEC 61219	Live Working – Earthing or Earthing and Short-Circuiting Equipment Using Lances as a Short-Circuiting Device – Lance Earthing
IEC	IEC 61238-1	Compression and Mechanical Connectors for Power Cables for Rated Voltages Up To 30 kV (Um = 36 kV) – Part 1: Test methods and requirements

(Continued)

TABLE 10.6 IEC cable standards—cont'd

Developer	Standard No.	Title
IEC	IEC 61316	Industrial Cable Reels
IEC	IEC 61442 .	Test Methods for Accessories for Power Cables with Rated Voltages from 6 kV (Um = 7.2 kV) Up To 30 kV (Um = 36 kV)
IEC	IEC 61892-4	Mobile and Fixed Offshore Units – Electrical Installations – Part 4: Cables
IEC	IEC 61892-6	Mobile and Fixed Offshore Units – Electrical Installations – Part 6: Installation
IEC	IEC 61918	Industrial Communication Networks – Installation of Communication Networks in Industrial Premises
IEC	IEC 61935-1	Testing of Balanced Communication Cabling in Accordance with ISO/IEC 11801 – Part 1: Installed Cabling
IEC	IEC 61935-2	Testing of Balanced Communication Cabling in Accordance with ISO/IEC 11801 – Part 2: Patch Cords and Work Area Cords
IEC	IEC 61935-2-20	Testing of Balanced Communication Cabling in Accordance with ISO/IEC 11801 – Part 2-20: Patch Cords and Work Area Cords – Blank Detail Specification for Class D Applications
IEC	IEC 62067	Power Cables with Extruded Insulation and Their Accessories for Rated Voltages Above 150 kV (Um = 170 kV) Up To 500 kV (Um = 550 kV) – Test Methods and Requirements
IEC	IEC/TR 62095	Electric Cables – Calculations for Current Ratings – Finite Element Method
IEC	IEC/TS 62100	Cables for Aeronautical Ground Lighting Primary Circuits
IEC	IEC/TR 62125	Environmental Statement Specific to IEC TC 20 – Electric Cables
IEC	IEC/TR 62153-4-0	Metallic Communication Cable Test Methods – Part 4-0: Electromagnetic Compatibility (EMC) – Relationship between Surface Transfer Impedance and Screening Attenuation, Recommended Limits
IEC	IEC 62329-2	Heat-shrinkable Moulded Shapes – Part 2: Methods of Test
IEC	IEC/TS 62395-2	Electrical Resistance Trace Heating Systems for Industrial and Commercial Applications – Part 2: Application Guide for System Design, Installation and Maintenance
IEC	IEC 62440	Electric Cables with a Rated Voltage not Exceeding 450/750 V – Guide to Use
IEC	IEC/TR 62482	Electrical Installations in Ships – Electromagnetic Compatibility – Optimizing of Cable Installations on Ships – Testing Method of Routing Distance
IEC	IEC 62491	Industrial Systems, Installations and Equipment and Industrial Products – Labelling of Cables and Cores

also extended to manufacturing companies, to aid in the development and improvement of equipment used by the utility companies. Today, the AEIC membership is composed primarily of utility and industry chief executives that can direct their company policies. AEIC utility cable standards are presented in Table 10.7.

Insulated Cable Engineers Association (ICEA) was established in 1925. Its purpose was for the development of standards for electrical cable for power, control, and telecommunications

TABLE 10.7 AEIC cable standards

Developer	Standard No.	Title
AEIC	AEIC CS1	Specifications for Impregnated Paper-Insulated Metallic-Sheathed Cable, Solid Type
AEIC	AEIC CS2	Specification for Impregnated Paper and Laminated Paper Polypropylene Insulated Cable, High Pressure Pipe Type
AEIC	AEIC CS3	Specifications for Impregnated-Paper-Insulated Metallic Sheathed Cable, Low Pressure Gas-Filled Type
AEIC	AEIC CS4	Specifications for Impregnated-Paper-Insulated Low and Medium Pressure Self-Contained Liquid Filled Cable
AEIC	AEIC CS5	Specifications for Thermoplastic and Crosslinked Polyethylene Insulated Shielded Power Cables Rated 5 through 69 kV
AEIC	AEIC CS6	Specifications for Ethylene Propylene Rubber Insulated Shielded Power Cables Rated 5 through 69 kV
AEIC	AEIC CS7	Specifications for Crosslinked Polyethylene Insulated Shielded Power Cables Rated 46 through 138 kV
AEIC	AEIC CS8	Specification for Extruded Dielectric Shielded Power Cables Rated 5 through 46 kV
AEIC	AEIC CS9	Specification for Extruded Insulation Power Cables and Their Accessories Rated Above 46 kV through 345 kVAC
AEIC	AEIC CS31	Specifications for Electrically Insulating Pipe Filling Liquids for High-Pressure Pipe-Type Cable
AEIC	AEIC CG1	Guide for Establishing the Maximum Operating Temperatures of Impregnated Paper and Laminated Paper Polypropylene Insulated Cables
AEIC	AEIC CG3	Guide for Installation of Pipe-Type Cable Systems
AEIC	AEIC CG4	Guide for Installation of Extruded Dielectric Insulated Power Cable Systems Rated 69 kV Through 138 kV
AEIC	AEIC CG5	Underground Extruded Power Cable Pulling Guide
AEIC	AEIC CG6	Guide for Establishing the Maximum Operating Temperatures of Extruded Dielectric Insulated Shielded Power Cables
AEIC	AEIC CG7	Guide for Replacement and Life Extension of Extruded Dielectric 5–35 kV Underground Distribution Cables
AEIC	AEIC CG8	Guide for Electric Utility Quality Assurance Program for Extruded Dielectric Power Cables
AEIC	AEIC CG9	Guide for Installing, Operating, and Maintaining Lead Covered Cable Systems Rated 5 kV through 46 kV
AEIC	AEIC CG10	Guide for Developing Specifications for Extruded Power Cables Rated 5 through 46 kV
AEIC	AEIC CG11	Guide for Reduced Diameter Extruded Dielectric Shielded Power Cables Rated 5 through 46 kV
AEIC	AEIC CG12	Guide for Minimizing the Cost of Extruded Dielectric Shielded Power Cables Rated 5 through 46 kV

applications. Its original name was the Insulated Power Cable Engineers Association (IPCEA). ICEA's members are engineers who are sponsored by more than 30 of North America's cable manufacturers. The organization is broken into self-autonomous Sections, including the Power; Control & Instrumentation; Portable; and Communications cables. Their power and control cable standards are listed in Table 10.8.

TABLE 10.8 ICEA power and control cable standards

Developer	Standard No.	Title
ICEA	ICEA P-32-382	Short-Circuit Characteristics of Insulated Cable
ICEA	ICEA P-45-482	Short-Circuit Performance of Metallic Shields and Sheaths
ICEA	ICEA P-50-431	Aluminum Conductors, Bare and Weather Resistant
ICEA	ICEA P-51-432	Copper Conductors, Bare and Weather Resistant
ICEA	ICEA P-52-469	Copper-Covered Steel and Copper Composite Conductors
ICEA	ICEA P-53-426 NEMA WC 50-1976	Ampacities, Including Effect of Shield Losses for Single Conductor Solid Dielectric Power Cable 15 kV through 69 kV
ICEA/NEMA	ICEA P-54-440 NEMA WC51	Ampacities of Cables in Open-Top Trays
ICEA	ICEA P-56-520	Tables of Physical and Electrical Properties of Conductors Used on Overhead Lines
ICEA	ANSI/ICEA P-79-561	Guide for Selecting Aerial Cable Messengers and Lashing Wires
ICEA/NEMA	ICEA P-81-570 NEMA WC 3	ICEA Standards Publication for Direct Burial 600 Volt Cable with Ruggedized Extruded Insulation
ICEA/NEMA	ICEA S-19-81 NEMA WC 3	ICEA/NEMA Standards Publication for Rubber-Insulated Wire and Cable for the Transmission and Distribution of Electrical Energy
ICEA	ANSI/ICEA S-56-434	ICEA/ANSI Standards Publication for Polyolefin-Insulated Communications Cables for Outdoor Use
ICEA	ICEA S-58-679	Standard for Control Cable Conductor Identification
ICEA/NEMA	ICEA S-61-402 NEMA WC 5	ICEA/NEMA Standards Publication for Thermoplastic-Insulated Wire and Cable for the Transmission and Distribution of Electrical Energy
ICEA/NEMA	ICEA S-66-524 NEMA WC 7	ICEA/NEMA Standards Publication for Cross-Linked-Thermosetting Polyethylene-Insulated Wire and Cable for the Transmission and Distribution of Electrical Energy
ICEA/NEMA	ICEA S-67-401 NEMA WC 21	Steel Armor and Associated Coverings for Impregnated-Paper-Insulated Cables
ICEA/NEMA	ICEA S-68-516 NEMA WC 8	ICEA/NEMA Standards Publication for Ethylene-Propylene-Rubber-Insulated Wire and Cable for the Transmission and Distribution of Electrical Energy
ICEA	ANSI/ICEA S-70-547	Weather Resistant Polyethylene Covered Conductors
ICEA/NEMA	ICEA S-73-532 NEMA WC 57	Standard for Control, Thermocouple Extension, and Instrumentation Cables

TABLE 10.8 ICEA power and control cable standards—cont'd

Developer	Standard No.	Title
ICEA/NEMA	ANSI/ICEA S-75-381 NEMA WC58	Portable and Power Feeder Cables for Use in Mines and Similar Applications
ICEA	ANSI/ICEA S-76-474	Neutral Supported Power Cable Assemblies with Weather-Resistant Extruded Insulation Rated 600 Volts
ICEA	ANSI/ICEA S-81-570	600 Volt Rated Cables of Ruggedized Design for Direct Burial Installations as Single Conductors or Assemblies of Single Conductors
ICEA	ANSI/ICEA S-86-634	ICEA/ANSI Standards Publication for Buried Distribution and Service Wire, Filled Polyolefin-Insulated, Copper Conductor
ICEA	ICEA S-89-648	Standard for Aerial Service Wire Technical Requirements
ICEA/NEMA	ANSI/ICEA S-93-639 NEMA WC74	Shielded Power Cables 5,000–46,000 V
ICEA	ANSI/ICEA S-94-649	Concentric Neutral Cables Rated 5 through 46 kV
ICEA/NEMA	ICEA S-95-658 NEMA WC70	Non-Shielded Power Cables Rated 2000 V or less
ICEA/NWM	ICEA S-96-659 NEMA WC71	Non-Shielded Power Cables Rated 2001–5000 V
ICEA	ANSI/ICEA S-97-682	Utility Shielded Power Cables Rated 5 through 46 kV
ICEA	ICEA S-100-685	Thermoplastic Insulated and Jacketed Telecommunications Station Wire for Indoor/Outdoor Use Technical Requirements
ICEA	ANSI/ICEA S-105-692	600 Volt Single Layer Thermoset Insulated Utility Underground Distribution Cables
ICEA	ANSI/ICEA S-108-720	Extruded Insulation Power Cables Rated Above 46 through 345 kV

The Electric Power Research Institute (EPRI) is a private research institute which conducts high-voltage testing for utility companies. It maintains an extensive high-voltage laboratory and is responsible for the development of overhead and underground transmission line reference book standards which are accepted industry wide. A list of their publications is presented in Table 10.9.

TABLE 10.9 EPRI cable reference and guideline books

Developer	Standard No.	Title
EPRI	Red Book	*EPRI AC Transmission Line Reference Book: 200kV and Above*
EPRI	Green Book	*EPRI Underground Transmission Systems Reference Book*
EPRI	Orange Book	*EPRI Transmission Line Reference Book: Wind-Induced Conductor Motion*
EPRI	Yellow Book	*EPRI Overhead Transmission Inspection and Assessment Guidelines*
EPRI	Blue Book	*EPRI Transmission Line Reference Book: 115–345 kV Compact Line Design*
EPRI	Teal Book	*Best Practices and Life Extension Guidelines for Substations*

There are several other North American standards-producing organizations, with standards covering several areas of cable utilization and manufacturing. They include: ASTM International; National Electrical Manufacturers Association (NEMA); the Instrumentation, Systems, and Automation Society (ISA); the Institute of Electrical and Electronics Engineers (IEEE); Canadian Standards Association (CSA); and Underwriters' Laboratories (UL). Table 10.10 contains General Wire and Cable Standards.

TABLE 10.10 General wire and cable standards

Developer	Standard No.	Title
ASTM	ASTM B99	Standard Specification for Copper–Silicon Alloy Wire for General Applications
ASTM	ASTM B99M	Standard Specification for Copper–Silicon Alloy Wire for General Applications
ASTM	ASTM B258	Standard Specification for Standard Nominal Diameters and Cross-Sectional Areas of AWG Sizes of Solid Round Wires Used as Electrical Conductors
ASTM	ASTM B354	Terminology Relating to Uninsulated Metallic Electrical Conductors
ASTM	ASTM B500	Standard Specification for Metallic Coated Stranded Steel Core for Use in Overhead Electrical Conductors
ASTM	ASTM B500M	Standard Specification for Metallic Coated Stranded Steel Core for Use in Overhead Electrical Conductors
ASTM	ASTM B682	Standard Specification for Standard Metric Sizes of Electrical Conductors
ASTM	ASTM B830	Standard Specification for Uniform Test Methods and Frequency
ASTM	ASTM D866	Specification for Styrene-Butadiene (SBR) Synthetic Rubber Jacket for Wire and Cable
ASTM	ASTN D1047	Specification for Polyvinyl Chloride (PVC) Jacket for Wire and Cable
ASTM	ASTM D1248	Standard Specification for Polyethylene Plastics Extrusion Materials For Wire and Cable
ASTM	ASTM D1351	Specification for Polyethylene Insulation for Wire and Cable
ASTM	ASTM D1352	Specification for Ozone-Resisting Butyl Rubber Insulation for Wire and Cable
ASTM	ASTM D1523	Method for Synthetic Rubber Insulation for Wire and Cable, 90°C Operation
ASTM	ASTM D1679	Specification for Synthetic Rubber Heat and Moisture-Resisting Insulation for Wire and Cable, 75°C Operation
ASTM	ASTM F1835	Standard Guide for Cable Splicing Installations
ASTM	ASTM D2219	Specification for Polyvinyl Chloride (PVC) Insulation for Wire and Cable, 60°C Operation
ASTM	ASTM D2220	Specification for Polyvinyl Chloride (PVC) Insulation for Wire and Cable, 75°C Operation
ASTM	ANSI/ASTM D2308	Standard Specification for Thermoplastic Polyethylene Jacket for Electrical Wire and Cable
ASTM	ASTM D2526	Specification for Ozone-Resisting Silicone Rubber Insulation for Wire and Cable
ASTM	ASTM D2655	Specification for Crosslinked Polyethylene Insulation for Wire and Cable Rated 0–2000 V

TABLE 10.10 **General wire and cable standards—cont'd**

Developer	Standard No.	Title
ASTM	ASTM D2656	Specification for Crosslinked Polyethylene Insulation for Wire and Cable Rated 2001–35000 V
ASTM	ASTM D2768	Specification for General-Purpose Ethylene-Propylene Rubber Jacket for Wire and Cable
ASTM	ASTM D2770	Specification for Ozone-Resisting Ethylene-Propylene Rubber Integral Insulation and Jacket for Wire and Cable
ASTM	ASTM D2802	Specification for Ozone-Resistant Ethylene-Propylene Rubber Insulation for Wire and Cable
ASTM	ASTM D3004	Specification for Extruded Thermosetting and Thermoplastic Semi-Conducting Conductor and Insulation Shields
ASTM	ASTM D3485	Specification for Smooth-Wall Coilable Polyethylene (PE) Conduit (Duct) for Preassembled Wire and Cable
ASTM	ASTM D3554	Standard Specification for Track-Resistant Black Thermoplastic High-Density Polyethylene Insulation for Wire and Cable, 75°C Operation
ASTM	ASTM D3555	Specification for Track-Resistant Black Crosslinked Thermosetting Polyethylene Insulation for Wire and Cable
ASTM	ASTM D4244	Specification for Ozone-Resistant Thermoplastic Elastomer Insulation for Wire and Cable, 90°C Dry – 75°C Wet Operation
ASTM	ASTM D4245	Standard Specification for Ozone-Resistant Thermoplastic Elastomer Insulation for Wire and Cable, 90°C Dry/75°C Wet Operation
ASTM	ASTM D4246	Specification for Ozone-Resistant Thermoplastic Elastomer Insulation for Wire and Cable, 90°C Operation
ASTM	ASTM D4247	Specification for General-Purpose Black Heavy-Duty and Black Extra-Heavy Duty Polychloroprene Jackets for Wire and Cable
ASTM	ASTM D4313	Specification for General Purpose Heavy-Duty and Extra-Heavy-Duty Crosslinked Chlorinated Polyethylene Jackets for Wire and Cable
ASTM	ASTM D4314	Specification for General Purpose Heavy-Duty and Extra-Heavy-Duty Crosslinked Chlorosulfonated Polyethylene Jackets for Wire and Cable
ASTM	ASTM D4363	Specification for Thermoplastic Chlorinated Polyethylene Jacket for Wire and Cable
ASTM	ASTM D4388	Standard Specification for Nonmetallic Semi-Conducting and Electrically Insulating Rubber Tapes
ASTM	ASTM D6096	Standard Specification for Poly(Vinyl Chloride) Insulation for Wire and Cable, 90°C Operation
ASTM	ASTM D6585	Standard Specification for Unsintered Polytetrafluoroethylene (PTFE) Extruded Film or Tape
ASTM	ASTM F855	Standard Specifications for Temporary Protective Grounds to be Used on De-energized Electric Power Lines and Equipment
ASTM	ASTM F1835	Standard Guide for Cable Splicing Installations
ASTM	ASTM F1837M	Standard Specification for Heat-Shrink Cable Entry Seals (Metric)

(Continued)

TABLE 10.10   General wire and cable standards—cont'd

Developer	Standard No.	Title
ASTM	ANSI/ASTM F1883	Standard Practice for Selection of Wire and Cable Size in AWG or Metric Units
CSA	CSA C57	Electric Power Connectors for Use in Overhead Line Conductors
CSA	CSA C83	Communication and Power Line Hardware
EPRI	EPRI EL-3014	Optimization of the Design of Metallic Shield-Concentric Conductors of extruded Dielectric Cables Under Fault Conditions
EPRI	EPRI EL-5757	Thermal Overload Characteristics of Extruded Dielectric Cables
ICEA	ICEA P-60-573	Guide for Tapes, Braids, Wraps and Serving Specifications
ICEA	ICEA P-79-561	Aerial Cable Messengers and Lashing Wires
ICEA	ICEA S-19-81 NEMA WC3	Rubber-Insulated Wire and Cable
ICEA	ICEA S-58-679	Standard for Control Cable Conductor Identification
ICEA	ICEA S-61-402 NEMA WC5	Thermoplastic-Insulated Wire and Cable
ICEA	ICEA S-65-375	Varnished-Cloth-Insulated Wire and Cable for the Transmission and Distribution of Electrical Energy
ICEA	ICEA S-66-524 NEMA WC7	Cross-Linked-Thermosetting-Polyethylene Insulated Wire and Cable
ICEA	ICEA S-67-401	Steel Armor and Associated Coverings for Impregnate-Paper Insulated Cables
ICEA	ICEA S-68-516 NEMA WC8	Ethylene-Propylene-Rubber Insulated Wire and Cable
ICEA	ICEA S-70-547	Weather-Resistant Polyolefin-Covered Wire and Cable
ICEA	ICEA S-73-532 NEMA WC57	Control Cables
ICEA	ICEA S-75-381 NEMA WC58	Portable and Power Feeder Cables for Use in Mines and Similar Applications
ICEA	ICEA S-76-474	Neutral-Supported Power Cable Assemblies with Weather-Resistant Extruded Insulation, 600 Volts
ICEA	ICEA S-94-649	Concentric Neutral Cables Rated 5 through 46 kV
ICEA	ICEA S-103-701	Riser Cables Technical Requirements
IEEE	IEEE 590	IEEE Cable Plowing Guide
IEEE	IEEE 1018	IEEE Recommended Practice for Specifying Electric Submersible Pump Cable – Ethylene-Propylene Rubber Insulation
IEEE	IEEE 1019	IEEE Recommended Practice for Specifying Electric Submersible Pump Cable – Polypropylene Insulation
IEEE	ANSI/IEEE 1333	IEEE Guide for Installation of Cable Using the Guided Boring Method
NEMA	ANSI/NEMA WC 67	Standard for Uninsulated Conductors Used in Electrical and Electronic Applications
NEMA	ANSI/NEMA WC 27500	Standard for Aerospace and Industrial Cable
UL	UL 183	Standard for Manufactured Wiring Systems

TABLE 10.10   General wire and cable standards—cont'd

Developer	Standard No.	Title
UL	UL 814	Standard for Gas-Tube-Sign and Ignition Cable
UL	UL 817	Standard for Cord Sets and Power-Supply Cords
UL	UL 1063	Standard for Machine-Tool Wires and Cables
UL	UL 1084	Standard for Hoistway Cables
UL	UL 1263	Standard for Irrigation Cables
UL	UL 1588	Standard for Roof and Gutter De-Icing Cable Units
UL	UL 1807	Standard for Fire Resistant Cable Coating Materials
UL	UL 2029	Standard for Gas/Vapor-Blocked Cable Classified for Use in Class 1 Hazardous (Classified) Locations

General Wire and Cable Standards provide guidance in conductor insulation; conductor standard sizing; insulation standards at different cable operating temperatures; buried cable installation recommendations; and various cable use application standards. Table 10.11 contains materials for cable insulations and jackets.

Table 10.12 on cable testing and data standards contains cable testing requirements; ampacities; cable electrical and physical characteristics; fire test requirements; cable termination testing; etc. Testing is an integral part of electrical equipment and wiring installation safety. When preparing electrical specifications, it is important to consider including some testing requirements by which the cable will be made and by which it will be

TABLE 10.11   Electrical conductor insulation and cable jacket materials

Conductor insulation material	Cable jacket material
Polyethylene Insulation	Styrene-Butadiene (SBR) Synthetic Rubber Jackets
Butyl Rubber Insulation	Polyvinyl Chloride (PVC) Jackets
Synthetic Rubber Insulation	Thermoplastic Polyethylene Jackets
Polyvinyl Chloride (PVC) Insulation	Ethylene-Propylene Rubber Jackets
Silicone Rubber Insulation	Polychloroprene Jackets
Crosslinked Polyethylene Insulation	Crosslinked Chlorinated Polyethylene Jackets
Ethylene-Propylene Rubber Insulation	Crosslinked Chlorosulfonated Polyethylene Jackets
Thermoplastic High-Density Polyethylene Insulation	Thermoplastic Chlorinated Polyethylene Jackets
Crosslinked Thermosetting Polyethylene Insulation	Polyolefin Jacket
Thermoplastic Elastomer Insulation	
Varnished-Cloth-Insulation	
Impregnated-Paper Insulation	

TABLE 10.12 Cable testing and data standards

Developer	Standard No.	Title
ASTM	ASTM B193	Standard Test Method for Resistivity of Electrical Conductor Materials
ASTM	ASTM B263	Standard Test Method for Determination of Cross-Sectional Area of Stranded Conductors
ASTM	ASTM D149	Standard Test Method for Dielectric Breakdown Voltage and Dielectric Strength of Solid Electrical Insulating Materials at Commercial Power Frequencies
ASTM	ASTM D470	Method of Testing Crosslinked Insulations and Jackets for Wire and Cable
ASTM	ASTM D2633	Standard Test Methods for Thermoplastic Insulations and Jackets for Wire and Cable
ASTM	ASTM D4325	Standard Test Methods for Nonmetallic Semi-Conducting and Electrically Insulating Rubber Tapes
ASTM	ASTM D4496	Test Method of DC Resistance or Conductance of Moderately Conductive Materials
ASTM	ASTM D4565	Standard Test Methods for Physical and Environmental Performance Properties of Insulations and Jackets for Telecommunications Wire and Cable
ASTM	ASTM D4566	Standard Test Methods for Electrical Performance Properties of Insulations and Jackets for Telecommunications Wire and Cable
ASTM	ASTM D4568	Standard Test Methods for Evaluating Compatibility Between Cable Filling and Flooding Compounds and Polyolefin Wire and Cable Materials
ASTM	ASTM D5537	Standard Test Method for Heat Release, Flame Spread, Smoke Obscuration, and Mass Loss Testing of Insulating Materials Contained in Electrical or Optical Fiber Cables When Burning in a Vertical Cable Tray Configuration
ASTM	ASTM D6095	Standard Test Method for Longitudinal Measurement of Volume Resistivity for Extruded Crosslinked and Thermoplastic Semiconducting Conductor and Insulation Shielding Materials
ASTM	ASTM D6097	Standard Test Method for Relative Resistance to Vented Water-Tree Growth in Solid Dielectric Insulating Materials
ASTM	ASTM F780	Standard Test Method for Measuring the Insulation Resistance of Mineral-Insulated, Metal-Sheathed Thermocouples and Thermocouple Cable at Room Temperature
CSA	CSA 22.2 No. 0.3-M	Test Methods for Electrical Wires and Cables
CSA	CSA 22.5 No. 249	Standard Tests for Determining Compatibility of Cable-Pulling Lubricants with Wire and Cable
EPRI	EPRI TR-101245	Effect of DC Testing on Extruded Cross-Linked Polyethylene Insulated Cables
EPRI	EPRI TR-101245-V2	Effect of DC Testing on Extruded Cross-Linked Polyethylene Insulated Cables – Phase II
ICEA	ICEA P-32-382	Short-Circuit Characteristics of Insulated Cable
ICEA	ICEA P-34-359	AC/DC Resistance Ratios @ 60 Hz

TABLE 10.12 Cable testing and data standards—cont'd

Developer	Standard No.	Title
ICEA	ICEA P-43-457	Conductor Resistance and Ampacities @ 400 and 800 Hz
ICEA	ICEA P-45-482	Short-Circuit Performance of Metallic Shields and Sheaths
ICEA/IEEE	IEEE S-135 ICEA P-46-426	IEEE-IPCEA Power Cable Ampacities
ICEA	ICEA P-48-426	Paper Cable Ampacities at AEIC Temperatures
ICEA	ICEA P-51-432	Tables of Physical and Electrical Properties of Conductors Used on Overhead Lines
ICEA	ICEA P-52-469	Tables of Physical and Electrical Properties of Conductors Used on Overhead Lines
ICEA/NEMA	NEMA WC50 ICEA P-53-426	Ampacities, 15–69 kV 1/c Power Cable Including Effect of Shield Losses (Solid Dielectric)
ICEA/NEMA	NEMA WC51 ICEA P-54-440	Ampacities of Cables in Open-Top Trays
ICEA	P-56-520	Cable Tray Fire Test Report (Round Robin Project)
ICEA	P-57-653	Guide for the Implementation of Metric Units in ICEA Publications
ICEA	ICEA T-22-294	Test Procedures for Extended Time-Testing of Wire and Cable Insulations for Service in Wet Locations
ICEA	ICEA T-24-380	Guide For Partial-Discharge Test Procedure
ICEA	ICEA T-25-425	Guide for Establishing Stability of Volume Resistivity for Conducting Polymeric Compounds of Power Cables
ICEA/NEMA	NEMA WC54 ICEA T-26-465	Guide for Frequency of Sampling Extruded. Dielectric Cables (2001)
ICEA/NEMA	NEMA WC53 ICEA T-27-581	Test Methods for Extruded Dielectric Cables
ICEA	ICEA T-28-562	Test Method for Measurement of Hot Creep of Polymeric Insulation
ICEA	ICEA T-29-520	Vertical Cable Tray Flame Tests @ 210,000 Btu
ICEA	ICEA T-30-520	Vertical Cable Tray Flame Tests @ 70,000 Btu
ICEA	ANSI/ICEA T-31-610	Test Method for Conducting a Longitudinal Water Penetration Resistance Tests on Blocked Conductors
ICEA	ICEA T-32-645	Guide for Establishing Compatibility of Sealed Conductor Filler Compounds with Conductor Stress Control Materials
ICEA	ICEA T-33-655	Low Smoke, Halogen-Free Polymeric Jackets
ICEA	ANSI/ICEAT-34-664	Test Method for Conducting Longitudinal Water Penetration Resistance Tests on Longitudinal Blocked Cables
IEEE	ANSI/IEEE 48	IEEE Standard Test Procedures and Requirements for Alternating-Current Cable Terminations 2.5 kV through 765 kV
IEEE	IEEE 383	IEEE Standard for Type Test of Class IE Electric Cables, Field Splices, and Connections for Nuclear Power Generating Stations
IEEE	IEEE Std 385	IEEE Standard Power Cable Ampacity Tables – Part1 thru 5

(Continued)

TABLE 10.12 Cable testing and data standards—cont'd

Developer	Standard No.	Title
IEEE	IEEE 400	IEEE Guide for Making High-Direct-Voltage Tests on Power Cable Systems in the Field
IEEE	IEEE 400.1™	Guide for Field Testing of Laminated Dielectric, Shielded Power Cable Systems Rated 5 kV and Above with High Direct Current Voltage
IEEE	ANSI/IEEE 400.2	IEEE Guide for Field Testing of Shielded Power Cable Using Very Low Frequency (VLF)
IEEE	ANSI/IEEE 400.3	IEEE Guide for Partial Discharge Testing of Shielded Power Cable Systems in a Field Environment
IEEE	ANSI/IEEE 532	IEEE Guide for Selecting and Testing Jackets for Power, Instrumentation, and Control Cables
IEEE	IEEE 575	IEEE Guide for the Application of Sheath-Bonding Methods for Single-Conductor Cables and the Calculation of Induced Voltages and Currents in Cable Sheaths
IEEE	ANSI/IEEE 634	IEEE Standard Cable-Penetration Fire Stop Qualification Test
IEEE	IEEE 644	IEEE Standard Procedures for Measurement of Power Frequency Electric and Magnetic Fields from AC Power Lines
IEEE	IEEE 738	IEEE Standard for Calculation of Bare Overhead Conductor Temperature and Ampacity Under Steady-State Conditions
IEEE	IEEE Std 835	IEEE Standard Power Cable Ampacity Tables
IEEE	ANSI/IEEE 848	IEEE Standard Procedure for the Determination of the Ampacity Derating of Fire-Protected Cables
IEEE	ANSI/IEEE 1017	IEEE Recommended Practice for Field Testing Electric Submersible Pump Cable
IEEE	IEEE 1142	IEEE Guide for the Design, Testing, and Application of Moisture-Impervious, Solid Dielectric, 5–35 kV Power Cable Using Metal-Plastic Laminates
IEEE	ANSI/IEEE 1202	IEEE Standard for Flame Testing of Cables for Use in Cable Tray in Industrial and Commercial Occupancies
IEEE	ANSIIEEE 1210	IEEE Standard Tests for Determining Compatibility of Cable-Pulling Lubricants with Wire and Cable
IEEE	IEEE 1368™	Guide for Aeolian Vibration Field Measurements of Overhead Conductors
IEEE	ANSI/IEEE 1407™	IEEE Guide for Accelerated Aging Tests for Medium-Voltage (5 kV–35 kV) Extruded Electric Power Cables Using Water-Filled Tanks
IEEE	IEEE C135.61™	IEEE Standard for the Testing of Overhead Transmission and Distribution Line Hardware
IEEE	IEEE Std 1493	IEEE Guide for the Evaluation of Solvents Used for Cleaning Electrical Cables and Accessories
IEEE	ANSI/IEEE 1511	IEEE Guide for Investigating and Analyzing Power Cable, Joint, and Termination Failures on Systems Rated 5 kV through 46 kV
NEMA	NEMA WC 50	Ampacities, including Effect of Shield Losses for Single Conductor Solid Dielectric Power Cable 15 kV–69 kV

TABLE 10.12 Cable testing and data standards—cont'd

Developer	Standard No.	Title
NEMA	NEMA WC 51	Ampacities of Cables Installed in Cable Trays
NEMA/ICEA	ANSI/NEMA WC 53 ANSI/ICEA T-27-581	Standard Test Methods for Extruded Dielectric Power, Control, Instrumentation, and Portable Cables for Test
NEMA	NEMA WC 56	3.0 kHz Insulation Continuity Proof Testing of Wire and Cable
NEMA	ANSI/NEMA WC 61	Transfer Impedance Testing
NFPA	NFPA 72®	National Fire Alarm Code®
NFPA	ANSI/NFPA 262	Standard Method of Test for Flame Travel and Smoke of Wires and Cables for Use in Air-Handling Spaces, 2007 Edition
UL	UL 1581	Reference Standard for Electrical Wires, Cables, and Flexible Cords
UL	UL 1660	Standard Test for Flame Propagation Height of Electrical and Optical-Fiber Cable Installed Vertically in Shafts
UL	UL 1712	Standard Tests for Ampacity of Insulated Electrical Conductors Installed in the Fire Protective System
UL	UL 2023	Standard Test Method for Flame and Smoke Characteristics of Nonmetallic Wiring Systems (Raceway and Conductors) for Environmental Air-Handling Spaces
UL	UL 2196	Standard for Tests for Fire Resistive Cables

tested after installation is complete or being completed, and on a periodic basis after installation and operation.

A major issue when specifying cable insulation testing is the desirability to not damage the cable insulation. Cable insulation test may use either AC or DC voltage potentials in high potential (hi-pot) testing. The likelihood of cable insulation testing damage is greater for AC hi-pot testing than DC. In fact, cable insulation can withstand a DC potential equal to the basic impulse insulation level (BIL) rating of the system on which it is located, for a sustained period [8].

Cable manufacturers test all cable prior to shipment. Manufacturers normally use an AC voltage applied for 5 minutes. Testing requires the application of the voltage to a conductor and then checking for leakage current flow to a ground return point. Nonshielded cables require being immersed in a water bath to complete this testing. Shielded wire can use the shield as the ground return. A DC potential may also be used in this testing. Manufacturers specify the test voltage type and level and their testing is only on a Pass/Fail basis. They may use criteria set by UL, ICEA, or AEIC. Testing on cables rated above 2 kV may also include checking for corona [9].

Cable insulation tests and periodic maintenance testing will be required after the cable is initially installed. The first step in cable field testing is to establish the test criteria and voltage application scheme. The main objective of the testing is to determine the condition of the cable, identify any potential problem areas before the cable fails, and to not accidentally damage the cable insulation.

Standards normally accepted for cable testing include those developed by ASTM, AEIC, ICEA, and IEEE. Reference Table 10.7 for the list of AEIC cable standards. IEEE 400, *IEEE Guide for Field Testing and Evaluation of the Insulation of Shielded Power Cable Systems – Description* is a major cable testing standard. Its test voltages are much higher than those of AEIC or ICEA testing standards [10]. The higher IEEE 400 test voltages are designed to eliminate or reduce cable failures during normal operation by overstressing the cable while it is de-energized during a planned shutdown period.

Testing requires removal of the cable connections to all equipment and cable terminations, isolating the conductor to be tested from any physical contact with anything other than the test instrument. All other conductors in the cable, not being tested, are grounded. The conductor to be tested has a series of step-voltages applied for a 5 to 15 minute period. During that time, a test micro ammeter is monitored for leakage current. After each test period, the test voltage is increased to the next step level, after a one minute rest period. This process is repeated until the test is completed [11].

The monitored cable test current consists of three components:

Cable capacitance charging current;

Cable dielectric absorption characteristic current;

Insulation leakage current

The capacitance charging current will flow until the cable capacitance is fully charged and then it will cease if the test voltage is DC.

During the initial voltage application, the test ammeter should be monitored for signs of cable insulation failure. This will occur when the monitored current increases slowly at first and then at a rapid pace. This current avalanche indicates a pending cable insulation failure and the cable test should be terminated before cable failure occurs.

Complete or near total insulation failure is not an indication that an avalanche event occurred. The damage expected from an avalanche event would include a sudden flashover event. That would occur after insulation failure from the current avalanche, with an accompanying increase in the potential difference at the cable failure point. Once that potential difference exceeds the air gap spark potential level required, an electric arc event with accompanying flashover will be created to ground or a lower potential point.

Step testing allows recording of a one (1) minute absorption stabilization current at the end of each step voltage test. This data can be used to calculate the insulation leakage current resistance, allowing each step voltage increase to have its leakage current resistance checked and compared to the other step data. If a calculated leakage resistance decreases by 50% or more, testing should be discontinued because the conductor insulation is near failure [12].

The ratio of the leakage current after one (1) minute to the leakage current after 5 minutes is called the *Polarization Index (PI)*. This value should be between 1.25 and 2 or greater for good conductors. Values less than 1 are considered failing. Values between 1 and 1.25 are marginal. Periodic cable insulation testing and recording of the PI data will allow tracking of the cable's ability to safely deliver electrical power and planned scheduling of the cable's replacement before a catastrophic failure event occurs.

Communications Cable

Electrical and electronic communications may consist of the following mechanisms including but not limited to closed-circuit television, computer links, data-processing cables, facsimile, fiber-optic cable, microwave, network-powered broadband communications systems, public-address paging, telegraph, telemetry, telephone, satellite link, etc. There are electrical and electronic cables involved with some of those systems such as coaxial cable, fiber-optic cables, twisted pair, multi-pair shielded cable, power-limited cable, etc.

Articles in the NEC® Chapter 8, *Communications Systems* covering communication systems include Article 800, *Communications Circuits*, Article 810, *Radio and Television Equipment*, Article 820, *Community Antenna Television and Radio Distribution Systems*, and Article 830, *Network-Powered Broadband Communications Systems*. Although Article 770, Optical Fiber Cables and Raceway is not included in NEC Chapter 8, it will be included in this examination since it does have communications and instrumentation applications. Fiber-optic cables are used for data and signal transmission applications and telephone systems.

Article 725, *Class 1, Class 2, and Class 3 Remote-Control, Signaling, and Power-Limited Circuits* covers remote-controlled, signaling, and power-limited circuits. Class 2 and Class 3 remote-control, signaling, and power-limited circuits can be purchased in a single *composite cable* in a single jacket containing both Class 2 or 3 cables and communications circuits; nonconductive and conductive optical fiber cables; CATV and radio distribution cables; or low-power, network-powered broadband communications cables. In those situations the Class 2 or 3 cables should be listed to be installed with the other cable types.

Optical fiber cables as covered by the NEC Article 770 are for use in buildings. They can be grouped into three categories [13]:

> *Nonconductive Cables (OFN)* – containing neither metallic members nor any electrically conductive materials.

> *Conductive Cables (OFC)* – containing non-current carrying conductive members. Metallic members might include strength members, vapor barriers, and armor or sheath.

Composite Cables – containing optical fibers and electrical current-carrying conductors. It may also contain non-current carrying conductive members such as those included in conductive cables above.

Optical fiber cables are listed in accordance with the following designations [14]:

Types OFNP and OFCP – non-conductive and conductive optical fiber cables listed as suitable for installation in ducts, plenums, and environmental air spaces. These cables require adequate fire-resistant and low-smoke producing characteristics.

Types OFNR and OFCR – non-conductive and conductive riser optical fiber cables listed as suitable for installation in vertical runs in shafts or routing between floors. These cables require adequate fire-resistant characteristics to prevent transmission of a fire between floors.

Types OFNG and OFCG – non-conductive and conductive general-purpose optical fiber cables listed as suitable for general purpose use only. They are not suitable for use in plenum or riser applications. They are required to be listed as having resistance to the spread of fire. Reference CSA C22.2 No. 0.3-M-2001, *Test Methods for Electrical Wires and Cables* or ANSI/UL 1581-2000, *Standard for Safety for Vertical-Tray Fire-Propagation and Smoke-Release Tests for Electrical and Optical-Fiber Cables* for a definition for *resistance to the spread of fire*.

Types OFN and OFC – non-conductive and conductive general-purpose optical fiber cables listed as suitable for general purpose use only. They are not suitable for use in plenum, environmental air space, or riser applications. They are required to be listed as having resistance to the spread of fire. Reference CSA C22.2 No. 0.3-M-2001 or ANSI/UL 1581-2000 for a definition for *resistance to the spread of fire*.

Communications cables can be used in overhead routing applications, underground routing applications, and inside building applications. Outdoor communications cables, both overhead and underground applications fall under the auspices of IEEE C2, *National Electrical Safety Code*® on matters of installation. The indoor installation of communications cables is under the auspices of NFPA 70, *National Electrical Code*®. For the purposes of this chapter, the indoor communications cable aspects will be examined.

Communications cables carry the following designators [15]:

Type CM – communications cable listed for general-purpose communications use. It is not listed of use in risers or plenums. But shall be listed as *resistant to the spread of fire*.

Type CMG – general-purpose communications cable listed for general-purpose communications use. It is not listed for use in risers or plenums, but is listed to exhibit resistant to *spread of flame* characteristics.

Type CMP – communications cable listed for use in ducts, plenums, and environmental air spaces. It is also required to exhibit fire-resistant and low-smoke characteristics.

Type CMR – communications cable listed for use in vertical runs in shafts and between floors. It is also required to exhibit fire-resistant characteristics preventing the transmission of fire between floors.

Type CMX – limited use communication cable listed for use in dwellings and raceways. It is also required to be listed for being *resistant to flame spread*. Reference ANSI/UL 1581 for a definition for *resistant to flame spread*.

Type CMUC Undercarpet Wire and Cable – communications cable listed for undercarpet use. It is also required to be listed as *resistant to flame spread*.

Multipurpose (MP) Cables – cables containing communications type cables as well as either optical fiber or coaxial cables. Multipurpose cable designators include *MPP*, *MPR*, *MPG*, and *MP* for plenum, riser, and general-purpose uses respectively.

Communications Circuit Integrity (CI) Cable – include communication cables designed for survivability of critical circuits under specified fire conditions and duration. Cables meeting that designation have an additional classification suffix *"CI"*. For example *CM-CI, CMP-CI, CMR-CI,* and *CMX-CI*.

Hybrid Power and Communications Cable – cables containing both power non-metallic sheathed cables (*NM* or *NM-B*) and communications cables (*CM*) rated a minimum of 600 Volts. The hybrid cable is also required to be listed as *resistant to flame spread*.

Community Antenna Television and Radio Distribution Systems "covers coaxial cable distribution of radio frequency signals typically employed in community antenna television (CATV) systems" [16]. CATV cables have the following designators:

Type CATV – community antenna television cable is listed for general purpose CATV use. It is not listed for plenum or riser installation. It is also required to be listed as *resistant to the spread of fire*.

Type CATVP – community antenna television cable is listed for installation in ducts, plenums, and environmental air spaces. It is also required to be listed as *resistant to the spread of fire*.

Type CATVR – community antenna television cable is listed for installation in vertical runs in shafts or between floors. It is also required to be listed as possessing fire-resistant characteristics to prevent the spread of fire between floors.

Type CATVX – limited-use community antenna television cable is listed for installation in dwellings and for use in raceway. It is also required to be listed as *resistant to flame spread*.

Network-Powered Broadband Communications Systems

provide any combination of voice, audio, video, data, and interactive services through a network interface unit. [17]

Cable designators for this system include the following [18]:

1. Type BMU, Type BM, and Type BMR cables
   a. Network-powered broadband communications medium power underground cable, *Type BMU*. These cables shall be provided with an external jacket and shall be listed for outdoor, underground service.
   b. Network-powered broadband communications medium power cable, *Type BM*. This cable shall be listed for general-purpose use. It is not listed for plenum or riser installation and is also required to be listed as *resistant to the spread of fire*.
   c. Network-powered broadband communications medium power riser cable, *Type BMR*. This cable shall be composed of factory-assembled multiple purpose cables including a jacketed coaxial cable; a combination of a coaxial cable and multiple individual conductors in a single jacket; or a combination of an optical fiber cable with multiple individual conductors in a single jacket. All individual conductors shall have a minimum voltage rating of 300 V. This cable shall be listed for installation in vertical runs in shafts or between floors. It is also required to be listed as possessing fire-resistant characteristics to prevent the spread of fire between floors.
   d. *Type BMU*, *Type BM*, and *Type BMR* cables shall be fed from a power source having a maximum circuit voltage V_{max} of 150 and a maximum power-limitation of volt-amperes VA_{max} of 250.
2. Type BLU, Type BLX, Type BL, Type BLR and Type BLP Cables [19]
   a. Network-powered broadband communications low-power underground cable, *Type BLU*. These cables shall be provided with an external jacket and shall be listed for outdoor, underground service. Type BLU jacketed cables shall be listed for outdoor underground use.
   b. Limited use network-powered broadband communications low-power cable, *Type BLX* shall be listed for outdoor use, dwelling use, and shall be suitable for raceway installation. The cable shall also have a listing as being resistant to flame spread.
   c. Network-powered broadband communications low-power cable, *Type BL*. This cable shall be listed for general-purpose use. It is not listed for plenum or riser installation and is also required to be listed as *resistant to the spread of fire*.
   d. Network-powered broadband communications low-power riser cable, *Type BLR*. This cable shall be listed for installation in vertical runs in shafts or between floors. It is also required to be listed as possessing fire-resistant characteristics to prevent the spread of fire between floors.
   e. Network-powered broadband communications low-power plenum cable, *Type BLP* and shall be suitable for use in ducts, plenums, and environmental air spaces. It shall be

listed as fire-resistant and shall have low smoke-producing characteristics. It shall be factory-assembled and shall consist of jacketed coaxial cable; coaxial cable with individual multiple conductors in a common jacket; or an optical fiber cable with individual multiple conductors in a common jacket. The multiple conductors' insulation shall have a minimum voltage rating of 600 V.

Ethernet Cabling

Ethernet cabling consists of three basic cable groups. They include:

Twisted pair cable

Coaxial cable

Optical fiber cable

Each of these cable types serves specific characteristics regarding the system it supports.

(1) *Twisted pair cable* consists of insulated copper wire pairs, twisted together. This reduces noise susceptibility and crosstalk. *Unshielded twisted pair (UTP)* cable contains, as its name describes, no electrical shielding. It is normally composed of four pairs of wires which are enclosed in a common sheath. UTP generally is rated as having a 100Ω electrical impedance in accordance with ANSI/TIA/EIA-568-B.2, *100-Ohm Balanced Twisted-Pair Cabling Standard*. Not all Ethernet Physical Layers utilize the entire four-pair cables. Only two twisted pairs are used by 10Base-T, 100Base-TX, and 100Base-T2. All four pairs are required by 100Base-T4 and 1000Base-T. Descriptions of the various UTP cables are presented in Table 10.13.

Screened twisted pair (SCTP) is another category of twisted pair cables. It consists of four pairs of twisted cables, surrounded by aluminum foil or a braided screen. This is installed to aid in minimizing EMI radiation. It also reduces the affect of outside noise. *Shielded twisted pair (STP)* cabling normally refers to 150Ω impedance twisted pairs, each wrapped in a foil shield and all pairs are enclosed in an overall outer braided wire shield. This product is normally associated with IBM Cabling System specifications.

(2) *Coaxial cable* is the second major group of Ethernet cables. It is a communication transmission cable, having a copper-coated steel center conductor which is surrounded by an insulating spacer. The insulation is surrounded by another conductor consisting of a braid, foil, or a combination of both. An insulating layer and outer protective jacket complete the cable construction. There are several types of coaxial cable used with Ethernet. They include *Thicknet*, *Thinnet*, *CATV*, and *Twinax*.

TABLE 10.13   Unshielded twisted pair (UTP) cable

Cable category	Ethernet physical layers	Specification	Cable description
Category 1 (CAT 1)	Not suitable for Ethernet	N/A	N/A
Category 2 (CAT 2)	Not suitable for Ethernet	N/A	N/A
Category 3 (CAT 3)	10Base-T 100Base-T4 100-Base-T2	ANSI/EIA/TIA 568	100Ω impedance frequency ≤ 16 MHz
Category 4 (CAT 4)	10Base-T 100Base-T4 100-Base-T2	ANSI/EIA/TIA 568	100Ω impedance frequency ≤ 20 MHz
Category 5 (CAT 5)	10Base-T 100Base-T4 100-Base-T2 100Base-TX	ANSI/EIA/TIA 568	100Ω impedance frequency ≤ 100 MHz
Category 5e (Enhanced CAT 5)	10Base-T 100Base-T4 100-Base-T2 100Base-TX 1000Base-T	ANSI/EIA/TIA 568	100Ω impedance frequency ≤ 100 MHz 1 Gbps
Category 6 (CAT 6)	100Base-T Gigabit Ethernet	ANSI/EIA/TIA 568	100Ω impedance frequency ≤ 250 MHz 10 Gbps
Category 6A	1000Base-T 10GBase-T	ANSI/EIA/TIA 568	$100\ \Omega$ impedance frequency ≤ 500 MHz 10 Gbps
Category 7 (CAT 7)		ANSI/EIA/TIA 568-B.2	100Ω impedance frequency ≤ 600 MHz 10 Gbps

Thicknet coax is 50Ω coaxial cable used on Ethernet 10Base5 networks. It retains it name from its 10 mm "thickness". Thicknet can support a 10 Mb/s transmission rate and has a maximum supported length of 500 meters. Its outer jacket is normally yellow with black bands at 2.5 meter intervals. Those bands mark valid transceiver location points and it can support a maximum of 100 stations per cable segment length.

Thinnet coax is also 50Ω coaxial cable, but is used on Ethernet 10Base2 networks. It retains its name from the fact that it has a 5 mm "thickness" or half that of its Thicknet counterpart. This coaxial cable has a maximum supported length of 185 meters and can support a maximum of 30 stations per cable segment length.

CATV coax is 75Ω coaxial cable, commonly used for cable television signals. It is also utilized on Ethernet 10Broad36 networks. *Twinax* or *Twinaxial* communication transmission cable consists of two center conductors. The conductors are individually covered with insulating spacers. The entire assembly is then surrounded by a tubular outer conductor of braid, foil, or a combination of both. An insulation cover and protective outer layer complete the cable's construction. It is rated as 75Ω coaxial cable and can be used with a 1000Base-CX media system. It has support segments 25 meters long, but is capable of handling a 1.25 Gbaud signal transmission rate.

(3) The third group is *optical fiber cable*. This cabling contains a thin glass fiber that supports the transmission of optical signals. It requires devices to convert electrical signals into optical signals and then to reconvert them back to electrical signals at the other end of the optical fiber cable. The following Ethernet media systems can be transmitted over optical fibers. They include:

> FOIRL
>
> 10Base-FL
>
> 10Base-FB
>
> 10Base-FP
>
> 100Base-FX
>
> 1000Base-LX
>
> 1000Base-SX

There are two primary types of optical fiber cable. They include *single-mode fiber* and *multi-mode fiber.*

Because of its approximately 10 micron optical size, *single-mode fiber (SMF)* can only transmit a single mode of light. Because of the small fiber size, it presents difficulties in coupling a light source with the fiber. Five thousand (5000) meter segment lengths of this cable can support data transmission rates of 1 Gbps.

Multi-mode fiber (MMF) cable has a large optical fiber core, with core diameters between 50 and 100 microns. This is substantially larger than SMF cable. The larger size of its core allows easy coupling with light emitting sources. This cable type can support segment lengths of 2000 meters for 10 and 100 Mbps Ethernet. For 1 Gbps Ethernet transmission rates, segment lengths of up to 550 meters can be supported.

Communications cables standards are presented in Table 10.14. ICEA and some of the *IEC communications cable* standards are listed in respectively in Tables 10.15 and 10.16.

Optical fiber cable and *optical fiber cable connectors* standards are presented in Table 10.17. This table represents many of the International Electrotechnical Commission's standards on optical fibers. The optical fiber standards were separated from the other IEC communication standards to allow the optical fiber standards to be more easily examined.

TABLE 10.14 Communication cable standards

Developer	Standard No.	Title
ASTM	ASTM D4967	Standard Guide for Selecting Materials to Be Used for Insulation, Jacketing, and Strength Components in Fiber-Optic Cables
ASTM	ASTM D5424	Standard Test Method for Smoke Obscuration of Insulating Materials Contained in Electrical or Optical Fiber Cables When Burning in a Vertical Cable Tray Configuration
ASTM	ASTM D5537	Standard Test Method for Heat Release, Flame Spread, Smoke Obscuration, and Mass Loss Testing of Insulating Materials Contained in Electrical or Optical Fiber Cables When Burning in a Vertical Cable Tray Configuration
ASTM	ASTM D4730	Standard Specification for Flooding Compounds for Telecommunications Wire and Cable
ASTM	ASTM D4967	Guide for Selecting Materials to Be Used for Insulation, Jacketing, and Strength Components in Fiber Optic Cables
CSA	CSA 22.5 No. 214	Communications Cables (Bi-National Standard, with UL 444)
CSA	CSA 22.5 No. 232	Optical Fiber Cables
CSA	CSA 22.5 No. 233	Cords and Cord Sets for Communication Systems
CSA	CSA 22.5 No. 262	Optical Fibre Cable Raceway Systems
CSA	CSA T525	Residential and Light Commercial Telecommunication Wiring Standard
CSA	CSA T527	Commercial Building Grounding/Bonding Requirements for Telecommunications
CSA	CSA T528	Administration Standard for the Telecommunications Infrastructure
CSA	CSA T530	Commercial Building Standard for Telecommunications Pathways and Spaces
IEEE	IEEE 789	IEEE Standard Performance Requirements for Communications and Control Cables for Application in High-Voltage Environments
IEEE	IEEE 802.3ab	1000BASE-T
IEEE	IEEE 802.3ae	10Gb/s Ethernet
IEEE	IEEE 802.3an	IEEE Standard for Information Technology- Telecommunications and Information Exchange Between Systems - Local And Metropolitan Area Networks – Specific Requirements Part 3, Amendment 1: Physical Layer and Management Parameters for 10 Gb/s Operation, Type 10GBASE-T
IEEE	IEEE 802.3ak	10GBASE-CX4.
IEEE	IEEE 802.3i	10BASE-T
IEEE	IEEE 802.3j	Fiber Optic Active and Passive Star-Based Segments, Type 10BASE
IEEE	IEEE 802.3u	IEEE Standards for Local and Metropolitan Area Networks: Supplement to Carrier Sense Multiple Access with Collision Detection (CSMA/CD) Access Method and Physical Layer Specifications Media Access Control (MAC) Parameters, Physical Layer, Medium Attachment Units, and Repeater for 100 Mb/s Operation, Type 100BASE-T (Clauses 21-30)
IEEE	IEEE 802.3z	Gigabit Ethernet Standard

TABLE 10.14 Communication cable standards—cont'd

Developer	Standard No.	Title
IEEE	ANSI/IEEE 802.5	Token Ring Access Method and Physical Layer Specifications
IEEE	IEEE 1138	IEEE Standard Construction of Composite Fiber Optic Overhead Ground Wire (OPGW) for Use on Electric Utility Power Lines
IEEE	IEEE 1222™	Standard for All-Dielectric Self-Supporting Fiber Optic Cable
IEEE	IEEE Std 1428	IEEE Guide for Installation Methods for Fiber-Optic Cables in Electric Power Generating Stations and in Industrial Facilities
IEEE	IEEE 1590™	Recommended Practice for the Electrical Protection of Optical Fiber Communication Facilities Serving, or Connected to, Electrical Supply Locations
IEEE	IEEE Std 1594	IEEE Standard for Helically Applied Fiber Optic Cable Systems (Wrap Cable) for Use on Overhead Utility Lines
NECA/BICSI	ANSI/NECA/BICSI 568	Standard for Installing Commercial Building Telecommunications Cabling
NEMA	NEMA WC 63.1	Performance Standard for Twisted Pair Premise Voice and Data Communications Cables
NEMA	NEMA WC 63.2	Performance Standard for Coaxial Premise Data Communications Cable
NEMA	ANSI/NEMA WC 63.2	Performance Standard for Coaxial Premise Data Communications Cables
NFPA	NFPA 70®, Article 770	Optical Fiber Cables and Raceways
TIA/EIA	ANSI/TIA/EIA-568-B.1	Commercial Building Telecommunications Cabling Standard Part 1: General Requirements
TIA/EIA	ANSI/TIA/EIA-568-B.2	Commercial Building Telecommunications Cabling Standard Part 2: Balanced Twisted Pair Cabling Components
TIA/EIA	ANSI/TIA/EIA-568-B.3	Optical Fiber Cabling Components Standard
TIA	ANSI/TIA-568-C.0	Generic Telecommunications Cabling for Customer Premises Standard
TIA/EIA	ANSI/TIA/EIA-569-B	Commercial Building Standard for Telecommunications Pathways and Spaces
TIA/EIA	ANSI/TIA/EIA-570-A	Residential and Light Commercial Telecommunication Wiring Standard
TIA/EIA	ANSI/TIA/EIA-606-A	Administration Standard for the Telecommunications Infrastructure of Commercial Buildings
TIA/EIA	ANSI/J-STD-607-A	Commercial Building Grounding/Bonding Requirements for Telecommunications
TIA/EIA	TIA/EIA-758	Customer-owned Outside Plant Telecommunications Cabling Standard
TIA/EIA	ANSI/TIA/EIA-942	Telecommunications Infrastructure Standard for Data Centers
TIA	ANSI/TIA -1005	Telecommunications Infrastructure Standard for Industrial Premises
UL	UL 444	Communications Cables
UL	UL 910	Standard for Test for Flame-Propagation and Smoke-Density Values for Electrical and Optical-Fiber Cables Used in Spaces Transporting Environmental Air

(Continued)

TABLE 10.14 Communication cable standards—cont'd

Developer	Standard No.	Title
UL	UL 1651	Optical Fiber Cable
UL	UL 1666	Standard Test for Flame Propagation Height of Electrical and Optical-Fiber Cables Installed Vertically in Shafts
UL	UL 1685	Standard for Vertical-Tray Fire-Propagation and Smoke-Release Test for Electrical and Optical-Fiber Cables
UL	UL 1690	Standard for Data Processing (DP) Cables
UL	ANSI/UL 2024	Standard for Optical Fiber and Communication Cable Raceway

TABLE 10.15 ICEA communications cable standards

Developer	Standard No.	Title
ICEA	ICEA P-47-434	Pressurization Characteristics, PE Communication Cable
ICEA	ANSI/ICEA S-56-434	Polyolefin-Insulated Communications Cables for Outdoor Use
ICEA	ICEA P-61-694	Coding Guide for Copper Outside Plant and Riser Cable
ICEA	ANSI/ICEA P-79-561	Guide for Selecting Aerial Cable Messengers and Lashing Wires
ICEA	ICES S-76-474	ICEA/ANSI Standards Publication for Outside Plant Communications Cables, Specifying Metric Wire Sizes
ICEA	ANSI/ICEA S-77-528	Outside Plant Communications Cables, Specifying Metric Wire Sizes
ICEA	ANSI/ICEA S-80-576	Category 1 and 2 Individually Unshielded Twisted Pair Indoor Cables for Use in Communications Wiring Systems
ICEA	ANSI/ICEA S-83-596	Fiber Optic Premises Distribution Cable
ICEA	ANSI/ICEA S-84-608	Telecommunications Cable, Filled Polyolefin-Insulated Copper Conductor
ICEA	ANSI/ICEA S-85-625	Air-core, Polyolefin-Insulated, Copper Conductor Telecommunications Cable
ICEA	ANSI/ICEA S-86-634	Buried Distribution and Service Wire, Filled Polyolefin-Insulated, Copper Conductor
ICEA	ANSI/ICEA S-87-640	Fiber Optic Outside Plant Communications Cable
ICEA	ANSI/ICEA S-88-626	Telephone Cordage and Cord Sets
ICEA	ANSI/ICEA S-89-648	Telecommunications Aerial Service Wire
ICEA	ANSI/ICEA S-90-661	Category 3, 5, and 5e Individually Unshielded Twisted Pair Indoor Cable for Use in General Purpose and LAN Communication Wiring Systems
ICEA	ANSI/ICEA S-91-674	Coaxial and Coaxial/Twisted Pair Composite Buried Service Wires
ICEA	ANSI/ICEA S-92-675	Coaxial and Coaxial/Twisted Pair Composite Aerial Service Wires
ICEA	ANSI/ICEA S-98-688	Broadband Twisted Pair, Telecommunications Cable Air-core, Polyolefin-Insulated Copper Conductors
ICEA	ANSI/ICEA S-99-689	Broadband Twisted Pair Telecommunications Cable Filled, Polyolefin-Insulated Copper Conductors

TABLE 10.15 ICEA communications cable standards—cont'd

Developer	Standard No.	Title
ICEA	ANSI/ICEA S-100-685	TP Telecommunications, Station Wire, Indoor/Outdoor
ICEA	ANSI/ICEA S-101-699	Standard for Category 3 Individually Unshielded Twisted Pair Indoor Cable for Use in General Purpose Non-Lan Telecommunication Wiring Systems Technical Requirements
ICEA	ICEA S-102-700	ICEA Standard for Category 6 Individually Unshielded Twisted Pair Indoor Cables (With or Without an Overall Shield) for Use in Communications Wiring Systems Technical Requirements
ICEA	ICEA S-103-701	Riser Cables Technical Requirements
ICEA	ICEA S-104-696	Standard for Indoor-Outdoor Optical Cable
ICEA	ANSI/ICEA S-106-703	Standard for Broadband Aerial Service Wire – Air-core, Polyolefin, Copper Conductors – Technical Requirements
ICEA	ANSI/ICEA S-107-704	ICEA Standard for Broadband Buried Service Wire-Filled Polyolefin-Insulated, Copper Conductors – Technical Requirements
ICEA	ANSI/ICEA S-109-709	ICEA Standard for Distribution Frame-Wire Technical Requirements
ICEA	ICEA S-110-717	Optical Drop Cables
ICEA	ICEA S-112-718	ICEA Standard for Optical Fiber Cable for Placement in Sewer Environments

TABLE 10.16 IEC communications cables

Developer	Standard No.	Title
IEC	IEC 61196-1-113	Coaxial Communication Cables – Part 1-113: Electrical Test Methods – Test for Attenuation Constant
IEC	IEC 61156-7	Multicore and Symmetrical Pair/Quad Cables for Digital Communications – Part 7: Symmetrical Pair Cables with Transmission Characteristics Up To 1200 mHz – Sectional Specification for Digital and Analog Communication Cables
IEC	IEC 61156-7-1	Multicore and Symmetrical Pair/Quad Cables for Digital Communications – Part 7-1: Symmetrical Pair Cables with Transmission Characteristics Up To 1200 mHz – Blank Detail Specification for Digital and Analog Communication Cables
IEC	IEC 61156-7-2	Multicore and Symmetrical Pair/Quad Cables for Digital Communications – Part 7-2: Symmetrical Pair Cables with Transmission Characteristics Up To 1200 mHz – Capability Approval – Sectional Specification for Digital and Analog Communication Cables
IEC	IEC 61196-1	Coaxial Communication Cables – Part 1: Generic Specification – General, Definitions and Requirements
IEC	IEC 61196-1-1	Coaxial Communication Cables – Part 1-1: Capability Approval for Coaxial Cables
IEC	IEC 61196-1-100	Coaxial Communication Cables – Part 1-100: Electrical Test Methods – General Requirements
IEC	IEC 61196-1-101	Coaxial Communication Cables – Part 1-101: Electrical Test Methods – Test for Conductor DC Resistance of Cable

(Continued)

TABLE 10.16 IEC communications cables—cont'd

Developer	Standard No.	Title
IEC	IEC 61196-1-102	Coaxial Communication Cables – Part 1-102: Electrical Test Methods – Test for Insulation Resistance of Cable Dielectric
IEC	IEC 61196-1-103	Coaxial Communication Cables – Part 1-103: Electrical Test Methods – Test for Capacitance of Cable
IEC	IEC 61196-1-104	Coaxial Communication Cables – Part 1-104: Electrical Test Methods – Test for Capacitance Stability of Cable
IEC	IEC 61196-1-105	Coaxial Communication Cables – Part 1-105: Electrical Test Methods – Test for Withstand Voltage of Cable Dielectric
IEC	IEC 61196-1-106	Coaxial Communication Cables – Part 1-106: Electrical Test Methods – Test for Withstand Voltage of Cable Sheath
IEC	IEC 61196-1-107	Coaxial Communication Cables – Part 1-107: Electrical Test Methods – Test for Cable Microphony Charge Level (Mechanically Induced Noise)
IEC	IEC 61196-1-108	Communication Cables – Part 1-108: Electrical Test Methods – Test for Characteristic Impedance, Phase and Group Delay, Electrical Length and Propagation Velocity
IEC	IEC 61196-1-111	Coaxial Communication Cables – Part 1-111: Electrical Test Methods – Test for Stability of Phase Constant
IEC	IEC 61196-1-112	Coaxial Communication Cables – Part 1-112: Electrical Test Methods – Test for Return Loss (Uniformity of Impedance)
IEC	IEC 61196-1-115	Coaxial Communication Cables – Part 1-115: Electrical Test Methods – Test for Regularity of Impedance (Pulse/Step Function Return Loss)
IEC	IEC 61196-1-122	Coaxial Communication Cables – Part 1-122: Electrical Test Methods – Test for Cross-Talk Between Coaxial Cables
IEC	IEC 61196-1-200	Coaxial Communication Cables – Part 1-200: Environmental Test
IEC	IEC 61196-1-201	Coaxial Communication Cables – Part 1-201: Environmental Test Methods – Test for Cold Bend Performance of Cable
IEC	IEC 61196-1-203	Coaxial Communication Cables – Part 1-203: Environmental Test Methods – Test for Water Penetration of Cable
IEC	IEC 61196-1-205	Coaxial Communication Cables – Part 1-205: Environmental Test Methods – Resistance to Solvents and Contaminating Fluids
IEC	IEC 61196-1-206	Coaxial Communication Cables – Part 1-206: Environmental Test Methods – Climatic Sequence
IEC	IEC 61196-1-208	Coaxial Communication Cables – Part 1-208: Environmental Test Methods – Longitudinal Pneumatic Resistance
IEC	IEC 61196-1-301	Coaxial Communication Cables – Part 1-301: Mechanical Test Methods – Test for Ovality
IEC	IEC 61196-1-302	Coaxial Communication Cables – Part 1-302: Mechanical Test Methods – Test for Eccentricity

TABLE 10.16 IEC communications cables—cont'd

Developer	Standard No.	Title
IEC	IEC 61196-1-308	Coaxial Communication Cables – Part 1-308: Mechanical Test Methods – Test for Tensile Strength and Elongation for Copper-Clad Metals
IEC	IEC 61196-1-310	Coaxial Communication Cables – Part 1-310: Mechanical Test Methods – Test for Torsion Characteristics of Copper-Clad Metals
IEC	IEC 61196-1-313	Coaxial Communication Cables – Part 1-313: Mechanical Test Methods – Adhesion of Dielectric and Sheath
IEC	IEC 61196-1-314	Coaxial Communication Cables – Part 1-314: Mechanical Test Methods – Test for Bending
IEC	IEC 61196-1-316	Coaxial Communication Cables – Part 1-316: Mechanical Test Methods – Test of Maximum Pulling Force of Cable
IEC	IEC 61196-1-317	Coaxial Communication Cables – Part 1-317: Mechanical Test Methods – Test for Crush Resistance of Cable
IEC	IEC 61196-1-318	Coaxial Communication Cables – Part 1-318: Mechanical Test Methods – Heat Performance Tests
IEC	IEC 61196-1-324	Coaxial Communication Cables – Part 1-324: Mechanical Test Methods – Test for Abrasion Resistance of Cable
IEC	IEC 61196-1-325	Coaxial Communication Cables – Part 1-325: Mechanical Test Methods – Aeolian Vibration
IEC	IEC 61196-4	Coaxial Communication Cables – Part 4: Sectional Specification for Radiating Cables
IEC	IEC 61196-5	Coaxial Communication Cables – Part 5: Sectional Specification for CATV Trunk and Distribution Cables
IEC	IEC 61196-5-1	Coaxial Communication Cables – Part 5-1: Blank Detail Specification for CATV Trunk and Distribution Cables
IEC	IEC 61196-6	Coaxial Communication Cables – Part 6: Sectional Specification for CATV Drop Cables
IEC	IEC 61196-6-1	Coaxial Communication Cables – Part 6-1: Blank Detail Specification for CATV Drop Cables
IEC	IEC 61935-1	Testing of Balanced Communication Cabling in Accordance with ISO/IEC 11801 – Part 1: Installed Cabling
IEC	IEC 61935-2	Testing of Balanced Communication Cabling In Accordance with ISO/IEC 11801 – Part 2: Patch Cords and Work Area Cords
IEC	IEC 61935-2-20	Testing of Balanced Communication Cabling in Accordance with ISO/IEC 11801 – Part 2-20: Patch Cords and Work Area Cords – Blank Detail Specification for Class D Applications
IEC	IEC 61950	Cable Management Systems – Specifications for Conduit Fittings and Accessories for Cable Installations for Extra Heavy Duty Electrical Steel Conduit
IEC	IEC 62153-1-1	Metallic Communication Cables Test Methods – Part 1-1: Electrical – Measurement of the Pulse/Step Return Loss in the Frequency Domain Using the Inverse Discrete Fourier Transformation (IDFT)

(Continued)

TABLE 10.16 IEC communications cables—cont'd

Developer	Standard No.	Title
IEC	IEC/TR 62153-4-0	Metallic Communication Cable Test Methods – Part 4-0: Electromagnetic Compatibility (EMC) – Relationship Between Surface Transfer Impedance and Screening Attenuation, Recommended Limits
IEC	IEC/TR 62153-4-1	Metallic Communication Cable Test Methods – Part 4-1: Electromagnetic Compatibility (EMC) – Introduction to Electromagnetic (EMC) Screening Measurements
IEC	IEC 62153-4-2	Metallic Communication Cable Test Methods – Part 4-2: Electromagnetic Compatibility (EMC) – Screening and Coupling Attenuation – Injection Clamp Method
IEC	IEC 62153-4-3	Metallic Communication Cable Test Methods – Part 4-3: Electromagnetic Compatibility (EMC) – Surface Transfer Impedance – Triaxial Method
IEC	IEC 62153-4-4	Metallic Communication Cable Test Methods – Part 4-4: Electromagnetic Compatibility (EMC) – Shielded Screening Attenuation, Test Method for Measuring of the Screening Attenuation as Up To and Above 3 GHz
IEC	IEC 62153-4-5	Metallic Communication Cables Test Methods – Part 4-5: Electromagnetic Compatibility (EMC) – Coupling or Screening Attenuation – Absorbing Clamp Method
IEC	IEC 62153-4-6	Metallic Communication Cable Test Methods – Part 4-6: Electromagnetic Compatibility (EMC) – Surface Transfer Impedance – Line Injection Method
IEC	IEC 62153-4-7	Metallic Communication Cable Test Methods – Part 4-7: Electromagnetic Compatibility (EMC) – Test Method for Measuring the Transfer Impedance and the Screening – or the Coupling Attenuation – Tube in Tube Method
IEC	IEC 62153-4-8	Metallic Communication Cable Test Methods – Part 4-8: Electromagnetic Compatibility (EMC) – Capacitive Coupling Admittance
IEC	IEC 62153-4-9	Metallic Communication Cable Test Methods – Part 4-9: Electromagnetic Compatibility (EMC) – Coupling Attenuation of Screened Balanced Cables, Triaxial Method
IEC	IEC 62153-4-10	Metallic Communication Cable Test Methods – Part 4-10: Electromagnetic Compatibility (EMC) – Shielded Screening Attenuation Test Method for Measuring the Screening Effectiveness of Feed-Through and Electromagnetic Gaskets Double Coaxial Method
IEC	IEC/TR 62222	Fire Performance of Communication Cables Installed in Buildings
IEC/ISO	ISO/IEC 11801	Generic Cabling for Customer Preemies
IEC/ISO	ISO/IEC 14165-114	ISO/IEC Application Standard-A Full Duplex Ethernet Physical Layer Specification for 1,000 Mbit/s Operating Over Balanced Channels Class F (Category 7 Twisted Pair Cabling)

TABLE 10.17   IEC optical fiber cable and cable connector standards

Developer	Standard No.	Title
IEC	Project IEC 61202-1	Fibre Optic Interconnecting Devices and Passive Components – Fibre Optic Isolators – Part 1: Generic Specification
IEC	Project IEC 61280-2-3	Fibre Optic Communication Subsystem Test Procedures – Part 2-3: Digital Systems – Jitter and Wander Measurements
IEC	Project IEC 61280-4-1	Fibre-Optic Communication Subsystem Test Procedures – Part 4-1: Installed Cable Plant – Multimode Attenuation Measurement
IEC	Project IEC 61754-15	Fibre Optic Interconnecting Devices and Passive Components – Fibre Optic Connector Interfaces – Part 15: Type LSH Connector Family
IEC	Project IEC 61754-24-11	Fibre Optic Interconnecting Devices and Passive Components – Fibre Optic Connector Interfaces – Part 24-11: Type SC-RJ Connectors with Protective Housings Based on IEC 61076-3-117
IEC	IEC 60874-1	Connectors for Optical Fibres and Cables – Part 1: Generic Specification
IEC	IEC 60874-1-1	Connectors for Optical Fibres and Cables – Part 1-1: Blank Detail Specification
IEC	IEC 60874-10-1	Connectors for Optical Fibres and Cables – Part 10-1: Detail Specification for Fibre Optic Connector Type BFOC/2,5 Terminated to Multimode Fibre Type A1
IEC	IEC 60874-10-2	Connectors for Optical Fibres and Cables – Part 10-2: Detail Specification for Fibre Optic Connector Type BFOC/2,5 Terminated to Single-Mode Fibre Type B1
IEC	IEC 60874-10-3	Connectors for Optical Fibres and Cables – Part 10-3: Detail Specification for Fibre Optic Adaptor Type BFOC/2,5 for Single and Multimode Fibre
IEC	IEC 60874-14-1	Connectors for Optical Fibres and Cables – Part 14-1: Detail Specification for Fibre Optic Connector Type SC/PC Standard Terminated to Multimode Fibre Type A1a, A1b
IEC	IEC 60874-14-2	Connectors for Optical Fibres and Cables – Part 14-2: Detail Specification for Fibre Optic Connector Type SC/PC Tuned Terminated to Single-Mode Fibre Type B1
IEC	IEC 60874-14-3	Connectors for Optical Fibres and Cables – Part 14-3: Detail Specification for Fibre Optic Adaptor (Simplex) Type SC for Single-Mode Fibre
IEC	IEC 60874-14-4	Connectors for Optical Fibres and Cables – Part 14-4: Detail Specification for Fibre Optic Adaptor (Simplex) Type SC for Multi-Mode Fibre
IEC	IEC 60874-14-5	Connectors for Optical Fibres and Cables – Part 14-5: Detail Specification for Fibre Optic Connector Type SC-PC Untuned Terminated to Single-Mode Fibre Type B1
IEC	IEC 60874-14-6	Connectors for Optical Fibres and Cables – Part 14-6: Detail Specification for Fibre Optic Connector – Type SC-APC 9° Untuned Terminated to Single-Mode Fibre Type B1
IEC	IEC 60874-14-7	Connectors for Optical Fibres and Cables – Part 14-7: Detail Specification for Fibre Optic Connector Type SC-APC 9° Tuned Terminated to Single-Mode Fibre Type B1
IEC	IEC 60874-14-9	Connectors for Optical Fibres and Cables – Part 14-9: Fibre Optic Connector Type SC-APC Tuned 8° Terminated on Single Mode Fibre Type B1 – Detail Specification

(Continued)

TABLE 10.17 IEC optical fiber cable and cable connector standards—cont'd

Developer	Standard No.	Title
IEC	IEC 60874-14-10	Connectors for Optical Fibres and Cables – Part 14-10: Fibre Optic Pigtail Or Patch Cord Connector Type SC-APC Untuned 8° Terminated on Single Mode Fibre Type B1 – Detail Specification
IEC	IEC 60874-17	Connectors for Optical Fibres and Cables – Part 17: Sectional Specification for Fibre Optic Connector – Type F-05 (Friction Lock
IEC	IEC 60874-19-1	Fibre Optic Interconnecting Devices and Passive Components – Connectors for Optical Fibres and Cables – Part 19-1: Fibre Optic Patch Cord Connector Type SC-PC (Floating Duplex) Standard Terminated on Multimode Fibre Type A1a, A1b – Detail Specification
IEC	IEC 60874-19-2	Connectors for Optical Fibres and Cables – Part 19-2: Fibre Optic Adaptor (Duplex) Type SC for Single-Mode Fibre Connectors – Detail Specification
IEC	IEC 61073-1	Fibre Optic Interconnecting Devices and Passive Components – Mechanical Splices and Fusion Splice Protectors for Optical Fibres and Cables – Part 1: Generic Specification

Instrumentation Cable

Some instrumentation cable standards are presented in Tables 10.18 and 10.19. One type of instrumentation cable includes those covered in NEC Article 725, Class 1, Class 2, and Class 3 Remote-Control, Signaling, and Power-Limited Circuits. There are four main cable designations or classes associated with that Article including:

1. *Circuit integrity (CI) cable* – is utilized to assure survivability and operation under fire conditions for a remote control, signaling, or power-limited system critical circuit.

2. *Class 1 Power-limited circuit cable (CL1)* – is the wiring between the load side of an overcurrent protection device and the equipment to which it connects. Source voltage on this circuit is limited to not more than 30 Volts and power is limited to 1000 Volt-Amperes (VA).

   Class 1 Remote-control and signaling circuits must not exceed a voltage source of 600 Volts. Its output power is not limited.

   a. Conductor insulation is required to be suitably rated for 600 Volts. Conductors larger than 16 WG are required to comply with NEC Article 310. Conductors "… 18 AWG and 16 AWG shall be Type FFH-2, KF-2, KFF-2, PAF, PAFF, PF, PFF, PGF, PGFF, PTF,PTFF, RFH-2, RFHH-2, RFHH-3, SF-2, SFF=2, TF, TFFN, TFN, ZF, OR ZFF" [20]

3. *Class 2 Power-limited circuit cable (CL2)* – is the segment of the wiring between the load side of a Class 2 power source and the connected load. This class deals primarily

TABLE 10.18 Some instrumentation cable standards

Developer	Standard No.	Title
CSA	CSA 22.5 No. 208	Fire Alarm and Signal Cable
CSA	CSA 22.5 No. 239	Control and Instrumentation Cables
ICEA	ICEA S-73-532 ANSI/NEMA WC 57	Standard for Control, Thermocouple Extension, and Instrumentation Cables
ICEA	ICEA S-82-552	Instrumentation Cables and Thermocouple Wire
ISA	ISA 50.2	Fieldbus Standard for Use in Industrial Control Systems
IEEE	ANSI/IEEE 1143	IEEE Guide on Shielding Practice for Low-Voltage Cables
NEMA	NEMA HP 100-HP 100.4	High Temperature Instrumentation and Control Cables
NEMA	ANSI/ICEA T-27-581/NEMA WC 53	Standard Test Methods for Extruded Dielectric Power, Control, Instrumentation, and Portable Cables for Test
NEMA	NEMA WC 57	Standard for Control, Thermocouple Extension, and Instrumentation Cables
NEMA	NEMA WC 67	Standard for Uninsulated Conductors Used in Electrical and Electronic Applications
NFPA	NFPA 70, Article 727	Instrumentation Tray Cable: Type ITC
NFPA	NFPA 70, Article 760.81	Listing and Marking of Non-Power-Limited Fire Alarm (NPLFA) Cables
NFPA	NFPA 70, Article 760.82	Listing and Marking of Power-Limited Fire Alarm Cables (PLFA) Cables and Insulated Continuous Line-Type Fire Detectors
UL	UL 13	Power-Limited Circuit Cables
UL	UL 1424	Cables for Power-Limited Fire-Alarm Circuits
UL	UL 1425	Cables for Non-Power-Limited Fire-Alarm Circuits
UL	UL 1690	Data-Processing Cable
UL	UL 2250	Instrumentation Tray Cable

with the elimination of a fire initiation event with the circuit power limitations as well as acceptable electrical shock protection.

4. *Class 3 Power-limited circuit cable (CL3)* – is the segment of the wiring between the load side of a Class 3 power source and the connected load. This class deals primarily with the elimination of a fire initiation event with the circuit power limitations. Because the current and voltage limitations are higher than a Class 2 circuit, additional protection must be provided for electrical shock hazard potential.

Class 1, Class 2, and Class 3 power-limited cables have additional designators for special use cables. Class 2 and Class 3 power-limited cables may also be listed for use in plenums and as riser cables. Designations for plenum use are CL2P and CL3P. Riser designators are CL2R and CL3R. Designations for residential power-limited cable are CL2X and CL3X.

TABLE 10.19 Some IEC instrumentation cables

Developer	Standard No.	Title
IEC	IEC 60092-350	Electrical Installations in Ships – Part 350: General Construction and Test Methods of Power, Control and Instrumentation Cables for Shipboard and Offshore Applications
IEC	IEC 60092-351	Electrical Installations in Ships – Part 351: Insulating Materials for Shipboard and Offshore Units, Power, Control, Instrumentation, Telecommunication and Data Cables
IEC	IEC 60092-376	Electrical Installations in Ships – Part 376: Cables for Control and Instrumentation Circuits 150/250 V (300 V)
IEC	IEC 60092-504	Electrical Installations in Ships – Part 504: Special Features – Control and Instrumentation
IEC	IEC 60331-23	Tests for Electric Cables Under Fire Conditions - Circuit Integrity – Part 23: Procedures and Requirements – Electric Data Cables
IEC	IEC 60331-25	Tests for Electric Cables Under Fire Conditions - Circuit Integrity – Part 25: Procedures and Requirements - Optical Fibre Cables
IEC	IEC 61892-4	Mobile and Fixed Offshore Units – Electrical Installations – Part 4: Cables

Power-Limited Tray Cable (PLTC)

Power-limited tray cable is defined in Underwriters Laboratories UL13, *UL Standard for Safety Power-Limited Circuit Cables.* That category covers a variety of cables, principally for Class 3 and Class 2 circuits described in the National Electrical Code® Article 725, *Class 1, Class 2, and Class 3 Remote-Control, Signaling, and Power-Limited Circuits* and in other parts of that document. Type PLTC is not listed for either plenum or riser use, but is listed for Class 3 and Class 2 circuits generally and in cable tray use specifically.

PLTC may be shielded or unshielded. It may also contain one or more optical-fiber members and under that situation would carry an identifying cable designator suffix of "-OF". Its overall jacket is required to be "sunlight resistant" and pass a 720-hour sunlight resistance test. PLTC may also be listed for direct burial use and must comply with a 1000-lbf crushing test. If the cable has an armor, metal braid or metal sheath covering, it must also have an outer jacket over the metal covering.

PLTC also may have various smoke and fire resistance requirements; the degree of those depends upon its application, use, and Class designation. Its overall outer jacket is a "gas/vapor tight continuous sheath" [21] This cable is acceptable for use in Class I, Division 2 hazardous areas. When that application occurs, sealing of that cable must comply with the requirements of NEC® Article 501.15(E).

Instrumentation Tray Cable (ITC)

Instrumentation tray cable is cable that is listed for installation in cable tray for use with instrumentation and control circuits which operate at 150 Volts or less and 5 amperes or less [22]. NEC Article 727.2 defines ITC as:

> A Factory assembly of two or more insulated conductors, with or without a grounding conductor(s), enclosed in a nonmetallic sheath. [23]

ITC cable may also be enclosed with smooth metallic sheath, continuous corrugated metallic sheath, or interlocking armor, over its non-metallic sheath. When utilized without an external metallic sheath, ITC cable has a requirement that only 15 meters (50 feet) can be routed outside of a cable tray. There are other requirements for its use, which are outlined in NEC® Article 727.

Fire Alarm Cable

Installation of *fire alarm systems* is covered in NEC® Article 760. Additional fire alarm requirements are also covered by NFPA 72®, *National Fire Alarm Code®*. NFPA 1221, *Standard for the Installation, Maintenance, and Use of Emergency Services Communications Systems* provides installation, maintenance, and use requirements for all public fire service communications facilities.

Fire alarm circuits are classified in only two ways, either non-power-limited or power-limited. *Non-power-limited fire alarm (NPLFA)* circuits are those supplied with power sources that must comply with NEC Article 760.21 and must have an output voltage of 600 V or less. These circuits cannot be supplied through either ground-fault circuit interrupters or arc-fault circuit interrupters.

Non-power-limited fire alarm conductor insulation must be [24]:

> Type KF-2, KFF-2, PAFF-2, PTFF-2, PF, PFF, PGF, PGFF, RFH-2, RFHH-2, RFHH-3, SF-2, SFF-2, TF, TFF, TFN, TFFN, ZF, or ZFF. Conductors with other types and thickness of insulation shall be permitted if listed for non-power-limited fire alarm use.

Non-power-limited fire alarm cable when installed in ducts and plenums must be listed as NPLFP. Similarly, NPLF cable installed in risers must be listed as NPLFR.

Fire alarm circuit integrity (CI) cable is installed for continued use of critical circuits under fire conditions. When it is critical that NPLF, NPLFP, or NPLFR cables be provided with 2-hour circuit integrity-rated cables, Types NPLF-CI, NPLFP-CI, and NPLFR-CI listed cables must be installed.

Power-Limited Fire Alarm (PLFA) Cable

Power-limited fire alarm (PLFA) circuits are those that are supplied by power complying with NEC® Article 760, Section III. PLFA cables shall be fed from any of the following sources [25]:

1. A listed PLFA or Class 3 transformer
2. A listed PLFA or Class 3 power supply
3. Listed equipment marked to identify the PLFA power source

PFLA cables are required to be copper and have a voltage rating of 300 or more Volts. The following are PLFA cables:

FPLP – power-limited fire alarm plenum cables are required to be listed and suitable for installation in ducts, plenums, and other environmental air spaces. They are required to be listed with fire-resistant and low smoke- producing characteristics.

FPLR – power-limited fire alarm riser cables are required to be listed and suitable for installation in shafts or run from floor to floor. They are also required to be listed with fire-resistant characteristics to prevent fire spread between floors.

FPL – power-limited fire alarm cables are required to be listed and suitable for general-purpose fire alarm use. They are not suitable as replacements for FPLP and FPLR applications. Those cables are also required to be resistant to fire spread.

Fire alarm circuit integrity *(CI)* cables are used where survivability of critical circuits is required under fire conditions. Cables listed for this service would be designated as *FPL-CI, FPLP-CI,* or *FPLR-CI.*

Coaxial cables with 30% electrical conductivity copper-covered steel center conductor wire may be installed if listed as Type *FPLP, FPLR,* or *FPL.*

Ethernet and Optical Fiber Cables

Ethernet and optical fiber cables can be used for instrumentation purposes. They are also suitable for communications use. For specific information and standards associated with those cables, reference the "Communication Cables" section of this chapter for information on those cables.

Temperature Detector Cables

Temperature detectors play an important role in instrumentation practices by monitoring temperatures in critical areas or sections in mechanical and electrical equipment, in chemical and petroleum processes, in turbines and reciprocating engines, in transformers, motors, and

TABLE 10.20 Thermocouple elements

Common Thermocouple Elements	
First Metallic Component	**Second Metallic Component**
Platinum	90% platinum and 10% rhodium
Platinum	87% platinum and 13% rhodium
Chromel P*	*Alumel
Iron*	*Constantan
Chromel P	Constantan
Copper	Constantan

*Most commonly used thermocouple elements.

generators. Electrically operated temperature detectors consist of *resistance thermometer device* (RTD) elements, light detectors, pressure-gage thermometers, and thermocouple devices. RTD elements consist of a coil of platinum wire wrapped around a mica frame and enclosed in a glass tube. The device is connected to a thermometer bridge for measuring coil resistance. Light detectors determine high temperatures by sensing the wavelength of the radiation being emitted by a heat source. Pressure-gage detectors utilize a metal bulb connected to a pressure gage through a capillary tube. The vapor pressure of the fluid in the bulb is proportional to the temperature range for which it is monitoring. Fluid-expansion liquids or gases are also utilized.

A thermocouple is a junction of dissimilar metals that produces a voltage dependent on the temperature of the junction. Since the generated voltages are low, specialized cables are required to extend the distance from the device to a receiver. The extension cable material must be matched to the thermocouple material Thermocouples are primarily used for high temperature applications.

Thermocouple elements consist of two dissimilar metals, with different coefficients of expansion. Some common thermocouples include those listed in Table 10.20.

The EMF values recorded from the thermocouple devices are measured in milliVolts. Some of the temperature detector standards which have been developed are presented in Tables 10.21 and 10.22.

Electrical Raceway, Conduit, and Cable Tray

Raceway is defined by the National Electrical Code® as:

> An enclosed channel of metal or nonmetallic materials designed expressly for holding wires, cables, or busbars with additional functions as permitted in this Code. Raceways include, but

TABLE 10.21 Temperature detector standards

Developer	Standard No.	Title
ASTM	ASTM E207	Standard Test Method for Thermal EMF Test of Single Thermoelement Materials by Comparison with a Reference Thermoelement of Similar EMF-Temperature Properties
ASTM	ASTM E220	Standard Test Method for Calibration of Thermocouples by Comparison Techniques
ASTM	ASTM E230	Standard Specification and Temperature-Electromotive Force (EMF) Tables for Standardized Thermocouples
ASTM	ASTM E235	Standard Specification for Thermocouples, Sheathed, Type K and Type N, for Nuclear or for Other High-Reliability Applications
ASTM	ASTM E377	Standard Practice for Internal Temperature Measurements in Low-Conductivity Materials
ASTM	ASTM E452	Standard Test Method for Calibration of Refractory Metal Thermocouples Using a Radiation Thermometer
ASTM	ASTM E459	Standard Test Method for Measuring Heat Transfer Rate Using a Thin-Skin Calorimeter
ASTM	ASTM E574-06	Standard Specification for Duplex, Base Metal Thermocouple Wire with Glass Fiber or Silica Fiber Insulation
ASTM	ASTM E585	Standard Specification for Compacted Mineral-Insulated, Metal-Sheathed, Base Metal Thermocouple Cable
ASTM	ASTM E585M	Standard Specification for Compacted Mineral-Insulated, Metal-Sheathed, Base Metal Thermocouple Cable
ASTM	ASTM E601	Standard Test Method for Measuring Electromotive Force (EMF) Stability of Base-Metal Thermoelement Materials with Time in Air
ASTM	ASTM E608	Standard Specification for Mineral-Insulated, Metal-Sheathed Base Metal Thermocouples
ASTM	ASTM E608M	Standard Specification for Mineral-Insulated, Metal-Sheathed Base Metal Thermocouples
ASTM	ASTM E633	Standard Guide for Use of Thermocouples in Creep and Stress-Rupture Testing to 1800 °F (1000 °C) in Air
ASTM	ASTM E696	Standard Specification for Tungsten-Rhenium Alloy Thermocouple Wire
ASTM	ASTM E780	Standard Test Method for Measuring the Insulation Resistance of Mineral-Insulated, Metal-Sheathed Thermocouples and Thermocouple Cable at Room Temperature
ASTM	ASTM E839	Standard Test Methods for Sheathed Thermocouples and Sheathed Thermocouple Material
ASTM	ASTM E988	Standard Temperature-Electromotive Force (EMF) Tables for Tungsten-Rhenium Thermocouples
ASTM	ASTM E1652	Standard Specification for Magnesium Oxide and Aluminum Oxide Powder and Crushable Insulators Used in the Manufacture of Metal-Sheathed Platinum Resistance Thermometers, Base Metal Thermocouples, and Noble Metal Thermocouples

TABLE 10.21 Temperature detector standards—cont'd

Developer	Standard No.	Title
ASTM	ASTM E1684	Standard Specification for Miniature Thermocouple Connectors
ASTM	ASTM F882	Standard Performance and Safety Specification for Cryosurgical Medical Instruments
ASTM	ASTM WK21141	Compacted, Mineral-Insulated, Metal-Sheathed Cable used in Industrial Resistance Thermometers
ICEA/NEMA	ICEA S-82-552 NEMA WC 55	ICEA/NEMA Standards Publication for Instrumentation Cables and Thermocouple Wire
ISA	ISA MC96.1	Temperature Measurement Thermocouples
SAE	SAE ARP 690	Standard Exposed Junction Thermocouple for Controlled Conduction Errors in Measurement of Air of Exhaust Gas Temperature (Reaffirmed: Mar 1984 Feb 1992)
SAE	SAE AS 428	Exhaust Gas Temperature Instruments
SAE	SAE AS 5419	Cable, Thermocouple Extension, Shielded and Unshielded
SAE	SAE AS 8005A	Minimum Performance Standard Temperature Instruments
NEMA	NEMA WC 57	Standard for Control, Thermocouple Extension, and Instrumentation Cables

are not limited to, rigid metal conduit, rigid nonmetallic conduit, intermediate metal conduit, liquidtight flexible conduit, flexible metallic tubing, flexible metal conduit, electrical non-metallic tubing, electrical metallic tubing, underfloor raceways, cellular concrete floor raceways, cellular metal floor raceways, surface raceways, wireways, and busways. [26]

It is important to note here that cable tray is not considered raceway since it is a support system for several wiring methods. NEC Table 392.3(A) lists wiring methods which can be installed in cable tray. Reference Table 10.23 for a list of the cable and raceway suitable for installation in cable tray. It should be noted here that that National Electrical Code did have some exceptions and minimum criteria that must be met before some of the cable and raceway can be installed

TABLE 10.22 IEC Thermocouple Standards

Developer	Standard No.	Title
IEC	IEC 60584-1	Thermocouples – Part 1: Reference Tables
IEC	IEC 60584-2	Thermocouples. – Part 2: Tolerances
IEC	IEC 60584-3	Thermocouples – Part 3: Extension and Compensating Cables – Tolerances and Identification System
IEC	IEC 61515	Mineral Insulated Thermocouple Cables and Thermocouples
IEC	IEC 62460	Temperature - Electromotive Force (EMF) Tables for Pure-Element Thermocouple Combinations

TABLE 10.23 **Cable and raceway suitable for installation in cable tray**

Power and control cables	Communication cables	Instrumentation cables	Raceway
Armored cables	Armored cables	Armored cables	CATV raceways
Medium-voltage cables	CATV cables	Class 2 and Class 3 cables	Communications raceway
Metal-clad cables	Metal-clad cables		Electrical metallic tubing
Mineral-insulated, metal-sheathed cables	Multipurpose and communications cables	Instrumentation tray cables	
		Metal-clad cables	Electrical nonmetallic tubing
Multiconductor service-entrance cables	Network-powered broadband communication cable	Mineral insulated cables	Flexible metal conduit
		Non-power-limited fire alarm cables	Flexible metallic tubing
	Optical fiber cables		
Multiconductor underground feeder and branch-circuit cables		Optical fiber cables	Intermediate metal conduit
		Power-limited fire alarm cables	Liquidtight flexible metal conduit
		Power-limited tray cables	Liquidtight flexible nonmetallic conduit
Nonmetallic-sheathed cables			
			Optical fiber raceways
Power and control tray cables			Polyvinyl chloride (PVC) conduit
			Reinforced thermosetting resin (RTRC) conduit
			Rigid metal conduit
			Rigid nonmetallic conduit
			Signaling raceway

in a cable tray. It is recommended that the latest edition of NFPA 70 be referenced for specific applications.

One additional consideration must be reviewed before any cable is installed in areas that have been classified as Class I, II, or III; Groups A, B, C, D, E, F, or G; Division 1 or 2; or Class I; Zone 0, 1 or 2 electrically hazardous. Cable, raceway, and cable tray installed in those locations must be approved for that specific service.

TABLE 10.24 NEC basic conduit types

NEC Article No.	Conduit type
342	Intermediate Metal Conduit: Type IMC
344	Rigid Metal Conduit: Type RMC
348	Flexible Metal Conduit: Type FMC
350	Liquidtight Flexible Metal Conduit: Type LFMC
352	Rigid Nonmetallic Conduit: Type RNC
353	High Density Polyethylene Conduit: Type HDPE
354	Nonmetallic Underground Conduit with Conductors: Type NUCC
356	Liquidtight Flexible Nonmetallic Conduit: Type LFNC
358	Electrical Metallic Tubing: Type EMT
360	Flexible Metallic Tubing: Type FMT
362	Electrical Nonmetallic Tubing: Type ENT

The National Electrical Code® places certain restrictions when using the cable and raceway listed in Table 10.23 [27]. All of the cable and raceway listed are suitable for installation in cable tray; however, any restrictions placed by the NEC Article for each cable and raceway type would be necessitated for its installation in cable tray.

The 11 types of conduit noted in Table 10.24 are the most commonly used in power and control, instrumentation and communication applications. However, the National Electrical Code will only allow their use in certain applications or situations and in certain electrical hazardous area classifications. Also of significant importance in using conduit is the number of conductors which can be safely installed in the conduit and the affect that number of conductors has on the ampacity of the cable or conductors. Ampacity derating can involve ambient and conductor operating temperatures, conductor temperature rating compatibility with the equipment terminal temperature rating on which it is terminated, or the installation of four or more current carrying conductors in a conduit.

Conduit selection is based on a number of factors. Some include:

Area classification

Application

Corrosion

Mechanical protection

Environmental factors

Resistance to fire and the spread of flame

Material and installation costs

Underground service

Maintenance considerations

Table 10.25 lists many of the codes, standards, and recommended practices for electrical conduit, tubing, and duct systems. As can be seen from that table, there are many standards generating organizations that tender documents for various aspects of each of those systems.

Table 10.26 lists some of the raceway standards used in North America and the cable trunking and ducting systems used in IEC standards. The IEC cable trunking and ducting systems are surface or underfloor channel raceway similar to NEC Article 386 Surface Meal Raceway; Article 388 Surface Nonmetallic Raceways; and Article 390 Underfloor Raceways. The IEC standard responsible for that raceway is IEC 61084. Reference Table 10.26 for many of the IEC cable trunking and ducting standards.

Many of the most common cable tray standards used in North America are presented in Table 10.27. That table includes ampacity table for cable tray use, cable tray testing, and specific cable tray type standards. Reference NEC Article 392 for cable tray installation. There are several cable tray types, including ladder type, ventilated trough, ventilated channel, solid bottom, etc. Tray may be either of metallic or nonmetallic construction. Partitions and walls may be required depending on the cable installed.

Cable Support and Restraint Systems

Requirements for cable *supports* and *restraints* [28] are addressed by either national wiring standards or industry-specific standards. In the United States, the relevant national wiring standard is the NFPA 70, *National Electrical Code*. Examples of industry-specific standards are American Bureau of Shipping, *ABS Steel Vessel Rules*; Title 46 – Shipping, Chapter II—Maritime Administration, Department of Transportation, Subchapter J—Miscellaneous; and IEEE RP45, *Recommended Practice for Electric Installations on Shipboard*.

While the NEC does not define *support,* it specifies support requirements based on the type of cable or wire. Webster's *New World Dictionary of American English*, Third College Edition, © 1988, defines *support* as: *"to carry or bear the weight of; keep from falling, slipping or sinking"*. This definition suggests that horizontal cable and wiring runs in cable tray or raceway are supported by the cable tray rungs or raceway. In vertical runs, however, the cable and wiring support must be provided by other means (e.g. cable cleats, clamps, ties, etc) in order to keep the cable or wiring from falling or slipping. The NEC also does not define

TABLE 10.25 Electrical Conduit, Tubing, and Duct Standards

Developer	Standard No.	Title
ASME	ANSI/ASME B.1.20.1	Pipe Threads, General Purpose (Inch)
ASTM	ASTM A239	Standard Practice for Locating the Thinnest Spot in a Zinc (Galvanized) Coating on Iron or Steel Articles
ASTM	ASTM C875	Standard Specification for Asbestos-Cement Conduit
ASTM	ASTM D3485	Standard Specification for Smooth-Wall Coilable Polyethylene (PE) Conduit (Duct) for Preassembled Wire and Cable
ASTM	ASTM D6070	Standard Test Methods for Physical Properties of Smooth-Wall, Coilable, Polyethylene (PE) Conduit (Duct) for Preassembled Wire and Cable
ASTM	ASTM F512	Standard Specification for Smooth-Wall Poly(Vinyl Chloride) (PVC) Conduit and Fittings for Underground Installation
ASTM	ASTM F2160	Standard Specification for Solid Wall High Density Polyethylene (HDPE) Conduit Based on Controlled Outside Diameter (OD)
CSA	CAN/CSA-C22.2 No. 45.1	Electrical Rigid Metal Conduit – Steel (Tri-National Standard, with UL 6 and NMX- J-534-ANCE-2007)
CSA	CAN/CSA-C22.2 No. 83.1-07	Electrical Metallic Tubing – Steel (Tri-National Standard, with UL 797 and NMX-J-536-ANCE-2007)
CSA	CAN/CSA-C22.2 No. 227.2.1	Liquid-Tight Flexible Nonmetallic Conduit (Bi-National Standard, with UL 1660)
IEC	IEC 60423	Conduit Systems for Cable Management – Outside Diameters of Conduits for Electrical Installations and Threads for Conduits and Fittings
IEC	IEC 60981	Extra Heavy-Duty Electrical Rigid Steel Conduits
IEC	IEC 61386-1	Conduit Systems for Cable Management – Part 1: General Requirements
IEC	IEC 61386-21	Conduit Systems for Cable Management – Part 21: Particular Requirements – Rigid Conduit Systems
IEC	IEC 61386-22	Conduit Systems for Cable Management – Part 22: Particular Requirements – Pliable Conduit Systems
IEC	IEC 61386-23	Conduit Systems for Cable Management – Part 23: Particular Requirements – Flexible Conduit Systems
IEC	IEC 61386-24	Conduit Systems for Cable Management – Part 24: Particular Requirements – Conduit Systems Buried Underground
IEC	IEC 61950	Cable management systems – Specifications for Conduit Fittings and Accessories for Cable Installations for Extra Heavy Duty Electrical Steel Conduit
NECA	NECA 101	Standard for Installing Steel Conduits (Rigid, IMC, EMT)
NEMA	ANSI C80.1	American National Standard for Electric Rigid Steel Conduit (ERSC)
NEMA	ANSI C80.3	American National Standard for Steel Electrical Metallic Tubing (EMT)
NEMA	ANSI C80.5	American National Standard for Electrical Rigid Aluminum Conduit (ERAC)
NEMA	ANSI C80.6	American National Standard for Intermediate Metal Conduit (EIMC)

(Continued)

TABLE 10.25 Electrical Conduit, Tubing, and Duct Standards—cont'd

Developer	Standard No.	Title
NEMA	ANSI/NEMA FB 1	Fittings, Cast Metal Boxes, and Conduit Bodies for Conduit, Electrical Metallic Tubing and Cable
NEMA	NEMA FB 2.10	Selection and Installation Guidelines for Fittings for Use with Non-Flexible Metallic Conduit or Tubing (Rigid Metal Conduit, Intermediate Metal Conduit, and Electrical Metallic Tubing)
NEMA	NEMA FB 2.20	Selection and Installation Guidelines for Fittings for Use with Flexible Electrical Conduit and Cable
NEMA	ANSI/NEMA OS 1	Sheet-Steel Outlet Boxes, Device Boxes, Covers and Box Supports
NEMA	ANSI/NEMA OS 2	Nonmetallic Outlet Boxes, Device Boxes, Covers and Box Supports
NEMA	NEMA RN 1	Polyvinyl Chloride (PVC) Externally Coated Galvanized Rigid Steel Conduit and Intermediate Metal Conduit
NEMA	NEMA RN 2	Packaging of Master Bundles for Steel Rigid Conduit, Intermediate Metal Conduit (Imc), And Electrical Metallic Tubing
NEMA	NEMA RN 3	Product Identification Numbers for Metallic Tubular Conduit Products for Use with Bar Coding and Electric Data Interchange Applications
NEMA	NEMA TC 2	Electrical Polyvinyl Chloride (PVC) Conduit
NEMA	NEMA TC 3	Polyvinyl Chloride (PVC) Fittings for Use with Rigid PVC Conduit and Tubing
NEMA	NEMA TC 7	Smooth-Wall Coilable Electrical Polyethylene Conduit
NEMA	NEMA TC 9	Fittings for Polyvinyl Chloride (PVC) Plastic Utilities Duct for Underground Installation
NEMA	NEMA TC 13	Electrical Nonmetallic Tubing (ENT)
NEMA	NEMA TC 14	Reinforced Thermosetting Resin Conduit (RTRC) and Fittings
NEMA	NEMA TC 6 & 8	Polyvinyl Chloride (PVC) Plastic Utilities for Underground Installations
NEMA	NEMA TCB 2	NEMA Guidelines for the Selection and Installation of Underground Nonmetallic Duct
NEMA	NEMA TCB 3	User's Manual for the Installation of Underground Corrugated Coilable Plastic Utility Duct (CCD)
NFPA	NFPA 70, Article 342	Intermediate Metal Conduit: Type IMC
NFPA	NFPA 70, Atricle 344	Rigid Metal Conduit: Type RMC
NFPA	NFPA 70, Article 348	Flexible Metal Conduit: Type FMC
NFPA	NFPA 70, Article 350	Liquidtight Flexible Metal Conduit: Type LFMC
NFPA	NFPA 70, Article 352	Rigid Polyvinyl Chloride Conduit: Type PVC
NFPA	NFPA 70, Article 353	High Density Polyethylene Conduit: Type HDPE
NFPA	NFPA 70, Article 354	Nonmetallic Underground Conduit with Conductors: Type NUCC

TABLE 10.25 Electrical Conduit, Tubing, and Duct Standards—cont'd

Developer	Standard No.	Title
NFPA	NFPA 70, Article 355	Reinforced Thermosetting Resin Conduit: Type RTRC
NFPA	NFPA 70, Article 356	Liquidtight Flexible Nonmetallic Conduit: Type LFNC
NFPA	NFPA 70, Article 358	Electrical Metallic Tubing: Type EMT
NFPA	NFPA 70, Article 360	Flexible Metallic Tubing: Type FMT
NFPA	NFPA 70, Article 362	Electrical Nonmetallic Tubing: Type ENT
UL	ANSI/UL 1	Flexible Metal Conduit
UL	ANSI/UL 6	Electrical Rigid Metal Conduit – Steel
UL	UL 6A	Standard for Electrical Rigid Metal Conduit – Aluminum and Stainless Steel
UL	UL 360	Standard for Liquid-Tight Flexible Steel Conduit
UL	UL 514A	Metallic Outlet Boxes
UL	UL 514B	Conduit, Tubing, and Cable Fittings
UL	UL 514C	Standard for Nonmetallic Outlet Boxes, Flush-Device Boxes, and Covers
UL	UL 651	Standard for Schedule 40 and 80 Rigid PVC Conduit and Fittings
UL	UL 651A	Type EB and A Rigid PVC Conduit and HDPE Conduit
UL	ANSI/UL 651B	Standard for Continuous Length HDPE Conduit
UL	ANSI/UL 797	Electrical Metallic Tubing – Steel
UL	ANSI/UL 797A	Electrical Metallic Tubing – Aluminum
UL	ANSI/ UL 1242	Electrical Metal Intermediate Conduit – Steel
UL	ANSI/ UL 1653	Electrical Nonmetallic Tubing
UL	UL 1660	Liquid-Tight Flexible Nonmetallic Conduit
UL	ANSI/ UL 1660	Standard for Safety for Liquid-Tight Flexible Nonmetallic Conduit
UL	UL 1684	Reinforced Thermosetting Resin Conduit (RTRC) and Fittings
UL	ANSI/ UL 1684A	Supplemental Requirements for Extra Heavy Wall Reinforced Thermosetting Resin Conduit (RTRC) and Fittings
UL	ANSI/ UL 1990	Standard for Safety for Nonmetallic Underground Conduit with Conductors
UL	ANSI/ UL 2239	Standard for Hardware for the Support of Conduit, Tubing and Cable

restraint; however, the same dictionary defines *restrain* as: *"to hold back; keep under control; limit; restrict; prevent or suppress and control some action"*.

Cable cleats, clamps, ties, etc can provide restraint if they are capable of controlling or limiting the movement of the cables or wires during all foreseeable conditions. Otherwise, these products would only provide support and wire management purposes. Since it is reasonably

TABLE 10.26 Raceway standards

Developer	Standard No.	Title
CSA	CSA C22.2 No. 62	Surface Raceway Systems
CSA/UL	CSA C22.2 No. 62.1	Nonmetallic Surface Raceways and Fittings (Bi-National standard, with UL 5A)
CSA	CSA C22.2 No. 79	Cellular Metal and Cellular Concrete Floor Raceways and Fittings
CSA	CSA C22.2 No. 80	Underfloor Raceways and Fittings
CSA	CSA CAN/CSA-C22.2 No. 262	Optical Fibre Cable Raceway Systems
IEEE	ANSI/IEEE 628	IEEE Standard Criteria for the Design, Installation and Qualification of Raceway Systems for Class 1E Circuits for Nuclear Power Generating Stations
NFPA	NFPA 70 Article 372	Cellular Concrete Floor Raceways
NFPA	NFPA 70 Article 374	Cellular Metal Floor Raceways
NFPA	NFPA 70 Article 376	Metal Wireways
NFPA	NFPA 70 Article 378	Nonmetallic Wireways
NFPA	NFPA 70 Article 384	Strut-Type Channel Raceway
NFPA	NFPA 70 Article 386	Surface Metal Raceways
NFPA	NFPA 70 Article 388	Surface Nonmetallic Raceways
NFPA	NFPA 70 Article 390	Underfloor Raceways
NECA	ANSI/ NECA 111	Standard for Installing Nonmetallic Raceways (RNC, ENT, LFNC)
UL	ANSI/ UL 5	Standard for Surface Metal Raceways and Fittings
UL	ANSI/ UL 5A	Standard for Safety for Non-Metallic Surface Raceways and Fittings
UL	ANSI/ UL 5B	Standard for Strut-Type Channel Raceways and Fittings
UL	ANSI/ UL 5C	Standard for Surface Raceways and Fittings for Use with Data, Signal, and Control Circuits
UL	ANSI/ UL 209	Standard for Cellular Metal Floor Raceways and Fittings
UL	ANSI/UL 870	Standard for Wireways, Auxiliary Gutters, and Associated Fittings
UL	ANSI/UL 884	Standard for Safety for Underfloor Raceways and Fittings
ULC	ULC/ORD-C2024	Standard Method of Fire Tests for Optical Fibre Cable Raceway
IEC	IEC 61084-1	Cable trunking and ducting systems for electrical installations – Part 1: General requirements
IEC	IEC 61084-2-1	Cable trunking and ducting systems for electrical installations – Part 2: Particular requirements – Section 1: Cable trunking and ducting systems intended for mounting on walls or ceilings
IEC	IEC 61084-2-2	Cable trunking and ducting systems for electrical installations – Part 2-2: Particular requirements – Cable trunking systems and cable ducting systems intended for underfloor and flushfloor installations
IEC	IEC 61084-2-4	Cable trunking and ducting systems for electrical installations – Part 2: Particular requirements – Section 4: Service poles

TABLE 10.27 Cable Tray Standards

Developer	Standard No.	Title
ASTM	ASTM D5424	Standard Test Method for Smoke Obscuration of Insulating Materials Contained in Electrical or Optical Fiber Cables When Burning in a Vertical Cable Tray Configuration
ASTM	ASTM D5537	Standard Test Method for Heat Release, Flame Spread, Smoke Obscuration, and Mass Loss Testing of Insulating Materials Contained in Electrical or Optical Fiber Cables When Burning in a Vertical Cable Tray Configuration
ICEA	P-54-440	Ampacities of Cables in Open-Top Cable Trays (USE NEMA WC 51)
ICEA	P-56-520	Cable Tray Flame Tests – a Round-Robin Study
ICEA	T-29-520	Conducting Vertical Cable Tray Flame Tests with Theoretical Heat Input Rate of 210,000 BTU/Hour
ICEA	T-30-520	Conducting Vertical Cable Tray Flame Tests with Theoretical Heat Input Rate of 70,000 BTU/Hour
IEC	IEC 61537	Cable Management – Cable Tray Systems and Cable Ladder Systems
NECA	NECA/NEMA 105	Standard for Installing Metal Cable Tray Systems (ANSI)
NEMA	NEMA FG-1	Nonmetallic Cable Tray Systems
NEMA	ANSI/ NEMA VE 1	Metal Cable Tray Systems
NEMA	ANSI/ NEMA VE 2	Metal Cable Tray Installation Guidelines
NEMA	NEMA WC 51	Ampacities of Cables Installed in Cable Trays
UL	UL 568	Nonmetallic Cable Tray Systems

foreseeable that cables and wires are subject to overcurrent situations (as evidenced by the requirements for overcurrent protective devices), it is incumbent on the design engineer and Authority Having Jurisdiction (AHJ) to ensure that cable and wiring systems include adequate means for both support and restraint.

The forces generated between adjacent cables during short-circuit events may be severe, especially in circuits where the available three-phase fault current duty exceeds 10 kA RMS-Symmetrical. The NEC provides a performance specification for adequate restraint in Article 392.8 (D) by requiring single conductor cables to be:

> securely bound in circuit groups to prevent excessive movement due to fault-current magnetic forces. [29]

Throughout the NEC, the use of subjective language empowers the AHJ to subjectively interpret the Code. In the case of Article 392.8 (D), the NEC requires the AHJ to evaluate the integrity of the "securely bound circuit groups" to determine whether a cable or wiring system may be subject to excessive movement. However, the NEC neither defines excessive movement, nor offers adequate criteria for such evaluation. If cables and their raceways or cable tray are damaged by movement (e.g. due to fault-current magnetic forces), one can reasonably infer with a degree of certainty that the movement was excessive. Other detrimental movement causes which may affect cable and wiring installations include seismic activity and vibration.

A design engineer should have a strong working knowledge of the relevant standards for the supporting and restraining devices in order to design a cable and wiring system that is adequately supported and restrained. Two such standards are UL 1565, *Positioning Devices* (the standard that applies to cable ties) and UL 2239, *Hardware for the Support of Conduit, Tubing and Cable.* Both UL 1565 and UL 2239 provide product testing criteria. However, these are static tests and do not address the dynamic performance requirements due to fault-current magnetic forces (i.e. electromechanical short-circuit forces). One electrical standard provides the necessary tools to evaluate supporting and restraining devices for cable and wiring systems, as well as defining a product that is specifically intended to provide such functionality. That standard is IEC 61914:2009, Edition 1, *Cable Cleats for Electrical Installations.*

IEC 61914 provides the following definitions:

- *Retention:* limiting the lateral and/or axial movement of the cable

- *Securing:* fixing to or from a mounting surface or another product

- *Cable cleat:* device designed to provide securing of cables when installed at intervals along the length of cables.

In addition to classifications according to lateral and/or axial retention, IEC 61914 includes two categories for resistance to electromechanical short-circuit forces:

- §6.4.3, withstanding one short circuit test – Category-1

- §6.4.4, withstanding more than one short circuit test – Category-2

These ratings are based on dynamic cable restraint testing under short-circuit conditions.

To successfully pass the *Category-1* test for resistance to electromechanical short-circuit forces, cable cleats are subjected to a single short-circuit and must remain undamaged while restraining and protecting the cables from cuts or damage to the outer sheath. The *Category-2* test requires cable cleats to successfully pass the test for resistance to electromechanical short-circuit forces for more than one short circuit. In those tests cable cleats are subjected to a second short-circuit test with the requirement that the cleats must remain undamaged while restraining and protecting the cables from cuts or damage to the outer sheath. The cables must then pass a voltage withstand test to validate the integrity of the conductor insulation. Category-2 is the highest rating for a cable restraint device. However, when comparing products, it is vitally important for the design engineer to obtain the manufacturer testing data, including the test cable sizes to independently verify the cable cleat was tested at force levels at least as severe as the intended application.

IEC 61914 defines the following formulas for calculating the forces between cabled conductors during short-circuits:

- For a single-phase or DC short-circuit, the maximum force on a conductor can be calculated by:

$$F_s = 0.2 \cdot i_1 \cdot i_2 / S \tag{Eq. 10.4}$$

- For a three-phase short-circuit with cables in a trefoil configuration, the maximum force on a conductor can be calculated by:

$$F_t = 0.17 \cdot i_p^2 / S \tag{Eq. 10.5}$$

- For a three-phase short-circuit with the conductors in flat configuration, the forces on the two outer conductors are always directed outwards from the central conductor. The force on the central conductor is oscillating. The maximum force on the outer conductors in flat formation can be calculated by:

$$F_{fo} = 0.16 \cdot i_p^2 / S \tag{Eq. 10.6}$$

The maximum force on the middle conductors in flat formation can be calculated by:

$$F_{fm} = 0.17 \cdot i_p^2 / S \tag{Eq. 10.7}$$

Where:

F_s = maximum force on the cable conductors in flat formation for a single-phase or DC short-circuit [N/m],

F_{fo} = maximum force on the outer cable conductors in flat formation for a three-phase short-circuit [N/m],

F_{fm} = maximum force on the middle cable conductor in flat formation for a three-phase short-circuit [N/m],

F_t = maximum force on the cable conductor in a trefoil configuration for a three-phase short-circuit [N/m],

i_p = the 1st cycle asymmetrical peak short-circuit current magnitude [kA],

i_1 = the 1st cycle asymmetrical peak short-circuit current magnitude in cable 1 [kA],

i_2 = the 1st cycle asymmetrical peak short-circuit current magnitude in cable 2 [kA],

S = center to center distance between two adjacent conductors [m].

Figure 10.1: Trefoil cable cleat (reprinted with permission from kVA Strategies, LLC)

If the three-phase RMS symmetrical short circuit fault duty is known, the first-cycle asymmetrical peak short-circuit current magnitude may be estimated by applying the following calculation:

$$i_p = i_{rms} \cdot 2.2$$

An example of a cable cleat is depicted in Figure 10.1.

A summary of the cable tie, cleat, and support standards is presented in Table 10.28.

TABLE 10.28   Cable ties, cable cleats, and cable supports

Developer	Standard No.	Title
API	API RP 14F	Electrical Systems for Offshore Facilities, Paragraph 6.7.1.4
CSA	CSA 22.2 No. 18.5	Positioning Devices (Bi-National standard, with UL 1565)
IEC	IEC 61914	Cable Cleats for Electrical Installations
IEC	IEC 62275	Cable Management Systems – Cable Ties for Electrical Installations
NFPA	NFPA 70, Article 392.8 (D)	National Electrical Code, Cable Installation – Connected in Parallel
UL	UL 1565	Positioning Devices
UL	UL 2239	Hardware for the Support of Conduit, Tubing, and Cable

References

1. Pender, Harold and Del Mar, William A., Eds, *Electrical Engineers' Hanbook – Electric Power*; 4th Edition, June, 1967, page 14-201. John Wiley & Sons, Inc.; New York.

2. Hilado, Carlos J., *Flammability Handbook for Plastics*; 4th Edition, 1990, page 6, Section 1.4. Technomic Publishing Co, Inc.; Lancaster, PA.

3. NFPA 70, *National Electrical Code*®; 2008, Article 310.13 FPN. National Fire Protection Association; Quincy, MA.

4. Ibid., Article 100.

5. IEEE 141-1993(R1999), *IEEE Recommended Practice for Electric Power Distribution for Industrial Plants*; 1993, Section 12.2.3, page 556. Institute of Electrical and Electronics Engineers; New York.

6. Ibid., page 558, Paragraph a(4).

7. Ibid., page 559, Table 12–3.

8. Ibid., Section 12.11.2, page 608.

9. Ibid., Section 12.11.3, page 608.

10. Ibid., Section 12.11.4, page 610.

11. Ibid., Section 12.11.4, page 610.

12. Ibid., Section 12.11.4, page 611.

13. NFPA 70, *National Electrical Code*®; 2008, Article 770.9. National Fire Protection Association; Quincy, MA.

14. Ibid., 770.179.

15. Ibid., Article 800.179.

16. Ibid., Article 820.1.

17. Ibid., Article 830.1.

18. Ibid., Article 830.179(A)(1).

19. Ibid., Article 830.179(A)(2).

20. Ibid., Article 725.27(B).

21. UL 13, *UL Standard for Safety Power-Limited Circuit Cables*; July 23, 2007, UL Document Information – Scope. Underwriters Laboratories, Inc.; Northbrook, IL.

22. NFPA 70, *National Electrical Code*®; 2008, Article 727.1. National Fire Protection Association; Quincy, MA.

23. Ibid., Article 725.2.

24. Ibid., Article 760.27(B).

25. Ibid., Article 760.41.

26. Ibid., Article 100.

27. Ibid., Table 392.3(A).

28. The information in this section was provided by Charles A. Darnell, PE, CFEI of KVA Strategies, L.L.C.

29. NFPA 70, *National Electrical Code*®; 2008, Article 392.8(D). National Fire Protection Association; Quincy, MA.

CHAPTER 11

Transformers, Capacitors, and Reactors

Transformers

A transformer can be defined as an electromagnetic device comprised of two or more windings (coils) coupled by a mutual magnetic field. The coils consist of a *primary winding* and a *secondary winding*. The primary winding is normally connected to an Alternating-Voltage power source, which creates an alternating magnetic flux linking both windings.

In an *ideal transformer,* the transform voltage is directly proportional to the ratio of the primary and secondary winding turns (N_1 = number of primary winding turns and N_2 = number of secondary winding turns) and can be represented by

$$V_1 = \frac{N_1}{N_2} V_2$$ (Eq. 11.1)

Its current is inversely proportional to the turns ratio.

$$I_1 = -\frac{N_2}{N_1} I_2$$ (Eq. 11.2)

The transformed impedance is proportional to the square of the turn's ratio as:

$$\frac{V_1}{I_1} = \left(\frac{N_1}{N_2}\right)^2 Z_2$$ (Eq. 11.3)

where Z_2 is the complex impedance of the secondary windings.

An *ideal transformer* is one in which:

All flux is linked only through the transformer core linking both windings;

Both winding resistances are negligible;

Core losses are considered negligible; and

The core permeability is high.

The transformer windings are normally wound on a core of iron or other ferromagnetic material, such as silicon steel, compressed powdered permalloy, or other similar types of materials. The core material selection is based on the application. The magnetic flux linking both windings produces mutual inductance between each transformer winding and leakage or self inductance at each winding.

Transformer terminal voltages (v_1 and v_2) are represented by the following equations:

$$v_1 = r_1 i_1 + L_{11}\frac{di_1}{dt} + L_{12}\frac{di_2}{dt} \qquad \text{(Eq. 11.4 [1])}$$

$$v_2 = r_2 i_2 + L_{22}\frac{di_2}{dt} + L_{12}\frac{di_1}{d_t} \qquad \text{(Eq. 11.5 [2])}$$

Where:

v_1 = primary winding voltage

v_2 = secondary winding voltage

r_1 = primary winding resistance

r_2 = secondary winding resistance

L_{11} and L_{22} = primary and secondary windings self-inductances

L_{12} = primary and secondary windings mutual inductance

i_1 and i_2 = primary and secondary windings currents

Equations 11.4 and 11.5 illustrate the interrelationship of primary and secondary currents, as well as mutual and self-inductances of those windings have in transforming voltages.

Transformer Classifications

There are a significant number of transformers in use for a variety of applications. For the purposes of this chapter, the discussion on transformers will be limited to those covered in NFPA 70®, *National Electrical Code*®, Article 450 Transformers and Transformer Vaults (Including Secondary Ties) and Instrument Transformers.

IEEE 141, *IEEE Recommended Practice for Electric Power Distribution for Industrial Plants* classifies [3] power transformers by:

1. Distribution and Power

    a. Distribution transformers – 3 to 500 kVA

    b. Power transformers – above 500 kVA

2. Insulation

   a. Dry-type transformers

      (1) Open-wound

      (2) Cast coil

      (3) Vacuum pressure impregnation

      (4) Encapsulated and vacuum pressure encapsulated

   b. Liquid-insulated – dielectric coolant

      (1) Mineral oil

      (2) Nonflammable liquid

      (3) Low-flammable liquid

      (4) Biodegradable

   c. Combination of liquid-, vapor-, and gas-filled

3. Substation or Unit Substation

   a. Primary substation transformer – secondary windings rated $\geq$1000 V

   b. Secondary substation transformer – secondary windings rated <1000 V

A *Substation* transformer is a power transformer with termination equipment for cable and overhead lines. A *Unit Substation* is integrally bus connected with enclosed bus to its primary and/or secondary windings, usually on the same base skid with the transformer. A substation transformer can be bus connected to the primary and/or secondary switchgear, but commonly through open bus.

As a minimum, transformers require the following specification information:

1. Power Rating: Kilovolt-amperes or megavolt-amperes

2. Phase: Single-phase or three-phase

3. Frequency: Unit of measurement in Hertz

4. Voltage Rating: Primary and secondary

5. Impedance (Base rating %Z)

6. Winding Connections: Single, two or three phase; zigzag, auto-transformer, T; Delta or wye

7. Voltage Taps: Load-changing or non-load changing, $\pm$% tap range, automatic or manual tap changer

8. Basic Impulse Level (BIL)

9. Temperature Rise Rating

10. Ambient Temperature Rating

11. Service: Indoor or outdoor

12. Insulation Type: Dry liquid-immersed, gas-filled

13. Termination Type and Location

14. Sound Level Requirements

15. Cooling Requirements: Self-cooling or forced-cooling w/fans, oil-water heat exchanger or forced-air cooling

16. Surge Arrester: High-voltage and/or low-voltage terminations

17. Alarm Devices: Pressure-vacuum, pressure-relief, liquid-level, temperature, rapid pressure rise

18. Gauges: Liquid-level, thermometer, pressure-vacuum, pressure-relief, hot-spot indicator, shipping shock indicator

19. Winding Material: Copper or aluminum

20. Integrally Mounted Bushing Current Transformer

21. Electrostatic Shields

22. Grounding Connections and Pads

Voltage and Power Ratings [4]

Transformer ratings are given in kilovolt-amperes or megavolt-amperes, at a specified winding temperature rise and will also list its rating for forced-cooling if provided. The temperature rise is established by resistance test. Liquid-filled power transformers have ratings based on a 65°C winding temperature rise with an average ambient temperature of 30°C to 40°C maximum, over a 24-hour period. When dual temperature rises are indicated, such as 55°C/65°C rise, this indicates the transformer has a 100% load rating with a 55°C winding temperature rise and typically a 112% rating with a 65°C. Higher percentage overload temperature ratings may be available from some manufacturers.

Dry-type transformers are available in three general insulation classes, including Class H (220°C), Class F (185°C), and Class B (150°C). Temperature rises associated with those insulation classes include:

1. 150°C Temperature Rise for Class H Insulation

2. 115°C Temperature Rise for Class F and H Insulation

3. 80°C Temperature Rise for Class B, F, and H Insulation

Transformer winding hot spot allowances are provided for all three classes.

Transformer winding temperature rise (by resistance) is measured with an average ambient temperature of 30°C over a 24-hour period with a maximum ambient temperature of 40°C. Longer life dry-type transformers have lower temperature rises of 80°C to 115°C. These transformers are typically capable of 15% and 30% overload operation respectively. IEEE 141 reports [5] most dry-type transformers 30kVA and larger are provided with a 220°C insulation system.

In 2005, the United States Department of Energy established a minimum efficiency standard for low-voltage dry-type distribution transformers. This was done as a means of improving the power losses incurred from those transformers in the distribution of electricity. The energy standards adopted were established by the National Electrical Manufacturers Association (NEMA) in conjunction with the U.S. Department of Energy (DOE). NEMA Standard TP-1-2002, *Guide for Determining Energy Efficiency for Distribution Transformers*, Table 4-2, Efficiency Levels for Low-Voltage Dry-Type Distribution Transformers were adopted by DOE for any transformers manufactured on or after January 1, 2007. The efficiency levels are noted in Table 11.1. Reference Chapter 6 for DOE's 2010 distribution transfer efficiency requirements.

DOE limited their regulation requirements for 2007 to low-voltage dry-type distribution transformers. DOE defined distribution transformers as follows:

§ 431.192 Definitions concerning distribution transformers. Distribution transformer means a transformer that –

(1) Has an input voltage of 34.5 kiloVolts or less;
(2) Has an output voltage of 600 Volts or less; and
(3) Is rated for operation at a frequency of 60 Hertz; however, the term "distribution transformer" does not include –
    (i) A transformer with multiple voltage taps, the highest of which equals at least 20 percent more than the lowest;
    (ii) A transformer that is designed to be used in a special purpose application and is unlikely to be used in general purpose applications, such as a drive transformer, rectifier

TABLE 11.1 DOE 2007 efficiency levels for low-voltage, dry-type distribution transformers

Single-phase efficiency		Three-phase efficiency	
Transformer kVA	Low voltage	Transformer kVA	Low voltage
15	97.7	15	97.0
25	98.0	30	97.5
37.5	98.2	45	97.7
50	98.3	75	98.0
75	98.5	112.5	98.2
100	98.6	150	98.3
167	98.7	225	98.5
250	98.8	300	98.6
333	98.9	500	98.7
500		750	98.8
667		1000	98.9
833		1500	
		2000	
		2500	

transformer, auto-transformer, Uninterruptible Power System transformer, impedance transformer, regulating transformer, sealed and non-ventilating transformer, machine tool transformer, welding transformer, grounding transformer, or testing transformer; or

(iii) Any transformer not listed in paragraph (3)(ii) of this definition that is excluded by the Secretary by rule because –

  (A) The transformer is designed for a special application;
  (B) The transformer is unlikely to be used in general purpose applications; and
  (C) The application of standards to the transformer would not result in significant energy savings.

Low-voltage dry-type distribution transformer means a distribution transformer that –

(1) Has an input voltage of 600 volts or less;
(2) Is air-cooled; and
(3) Does not use oil as a coolant. [6]

DOE adopted NEMA Standard TP-1 energy efficiency standards in 2007 for low-voltage dry-type distribution transformers, single-phase and three-phase. For 2010, it also adopted energy efficiency standards for liquid-immersed single-phase liquid-filled distribution transformers 10 kVA to 833 kVA and three-phase liquid-filled distribution transformers 15 kVA and above. The 2007 DOE mandate only covered low-voltage, dry type distribution transformers with input voltages 600 Volts or less. This was the only classification of distribution which the DOE Energy Conservation Standards now mandated in 2007. The medium-voltage, dry type distribution transformers to 95 kV and the liquid-immersed dry type distribution transformers will be covered by 10 CFR Part 431 on January 1, 2010.

Transformer Tests

A number of tests are required to physically determine the electrical characteristics of power and distribution transformers. Many of those tests are indicated below [7]:

1. Resistance

2. Ratio

3. Polarity and voltage vector relations

4. No-load loss and exciting current

5. Impedance loss and impedance voltage

6. Temperature tests (heat run)

7. Dielectric tests

Procedures for transformer testing are contained in IEEE C57.12.90, IEEE standard test code for liquid-immersed distribution, power, and regulating transformers and IEEE guide for short-circuit testing of distribution and power transformers; and C57.12.90, IEEE standard test code for dry type distribution and power transformers.

Resistance Test [8]

Transformer windings $I^2 R$ loss can be calculated with the establishment of the winding resistance values. This testing also allows establishment of data on the winding temperature rise at the end of the test. Testing can be conducted with the transformer oil either in or out. In dry situations, transformer thermocouples or thermometers are placed in contact with the windings. In devices with cooling oil, the temperature indicating instruments need only to be placed as near as possible to the coils, but in contact with the oil surrounding those windings.

Coil resistance can be measured by either instrument bridge equipment or by the drop of potential method. Test current is normally kept at or below 15% of the transformer full load current. Winding inductance and capacitance will determine the length of time required to reach steady state current values.

Winding Turns Ratio Test [9]

There are three basic types of construction normally used when winding transformer coils. They include helical coils, disk coils of multiple layers, and disk coils of only one turn per layer. Coil layers are separated by either oil ducts in liquid-filled transformers or by paper, treated cloth, or other insulating materials. Transformer windings are wrapped around a core of ferromagnetic material. Transformers are normally classified either core type or shell type, which describes the construction of the core.

Transformers are normally supplied with multiple winding tap terminations, allowing turns ratio field adjustment to obtain a desired output voltage. Once a transformer has been constructed, it is necessary to verify that the winding taps were constructed to the specified requirements. This necessitates that each tap setting turns ratio be tested. There are three basic methods for winding ratio testing, including:

Voltage application and measurement

Comparison with a known standard transformer

Resistance potentiometer method

Voltage application and measurement involves the application of a known voltage, at or below transformer rated voltage and frequency and the simultaneous measurement of voltage across both windings being tested. At least four tests should be conducted, varying in voltage amplitude in $\pm$ 10% increments. Comparison testing involves parallel operation of both transformers with a suitable alternating current supply. Electronic test equipment is available to conduct this testing and contains an internal calibrated reference transformer. The resistance potentiometer method uses a resistance potentiometer connected across the transformer high-voltage windings with an applied alternating-current supply. High-voltage and low-voltage like-polarity leads are connected on one side. The other low-voltage lead is connected to the potentiometer through a AC null indicator, adjusting the potentiometer until the null indicator zeros. The potentiometer resistance ratios equal transformer turns ratio.

Polarity and Voltage Vector Diagram Tests [10]

There are three methods used to determine if a transformer has been constructed with the correct specified polarity and voltage vector relations. They include:

AC Method: Connection of one high-voltage lead to an adjacent low-voltage lead, energizing the primary winding, and measuring the difference in potential between the other high-voltage and low-voltage leads

Comparison to a known identical standard transformer

DC Method: Use of the inductive kick method

The inductive kick method involves the application of a direct current (DC) source across the transformer's high-voltage windings, along with a high-voltage DC voltmeter. With the circuit energized, each voltmeter lead is transferred to their respective low-voltage bushing, observing the direction of voltmeter polarity change. Commercial test equipment is available to conduct these tests.

No-Load Loss and Exciting Current Tests

Primary and secondary current will flow when a voltage source is placed across the primary windings of a transformer and a load is placed across the secondary winding. The primary current is composed of a load component and an exciting component. The load component will be required to counteract the magnetomotive force (mmf) created by the secondary winding current. The exciting component is composed of two segments. The first is the core-loss component resulting from the hysteresis and eddy-current losses in the transformer core. The second is the magnetizing current component. The exciting current will also be present when a voltage source is placed across the primary winding with the secondary winding open-circuited. The exciting current can be as much as 5% of the primary winding full load current, with rated voltage and frequency applied to that winding.

The no-load loss and exciting current tests for single-phase transformers are conducted by placing a voltage source across the transformer primary winding equal to its rated voltage and frequency. The secondary winding is open-circuited. An ammeter is placed in series with the primary winding and a voltmeter and wattmeter are placed across that winding. The resulting primary winding current flow is the exciting current. The input power measured is approximately equal to the core-loss. The no-load loss can be better approximated using the following relationship [11]:

$$\text{Sine-Wave No-Load Loss} = \frac{\text{Measured No} - \text{Load Loss}}{0.8 + 0.2k^2} \qquad \text{(Eq. 11.6)}$$

Where k is the rms Test Voltage/Rated Voltage, 0.8 is the per-unit hysteresis loss and 0.2 is the per-unit eddy-current loss for iron. If the core is not iron, then (0.2) loss constant should be changed for the loss constant of core material being used.

Impedance Loss Tests

The test circuit for these tests consists of shorting the transformer secondary winding for single-phase transformers. A voltage source at rated frequency is applied to the primary windings and is varied until full load current is developed in the primary winding. That voltage is defined as the *impedance voltage (IV)*. Winding temperature is recorded in °C immediately before and after the tests. The average of those two temperatures will be used as the winding temperature. The impedance voltage is corrected to the standard temperature of 75°C using the relationship [12]:

$$\text{Adjusted IV} = \text{IV@}T_1 \times \frac{(234.5 + T_2)}{(234.5 + T_1)} \qquad \text{(Eq. 11.7)}$$

where T_1 in °C is the average measured winding temperature and T_2 is the adjusted temperature in °C.

An ammeter in series with the primary coil is used to record the primary current. A voltmeter and wattmeter are placed across the primary winding. The impedance wattage loss measured on the wattmeter is the copper loss and should be corrected similarly as that in Eq. 11.7. The percentage impedance (%Z) is equal to:

$$\%Z = 100 \times \frac{(\text{Impedance Voltage @}T_1)}{(\text{Winding Rated Voltage})} \qquad \text{(Eq. 11.8)}$$

where T_1 is the average winding test temperature. This impedance should also be adjusted to75°C. The percentage resistance (%R) is

$$\%R = 100 \times \frac{(\text{Measured Resistance Loss in Kilowatts})}{(\text{Transformer Rated KVA})} \qquad \text{(Eq. 11.9)}$$

The percentage reactance (%X) is

$$\%\text{X} = \sqrt{(\text{Percentage Impedance})^2 - (\text{Percentage Resistance})^2} \qquad \text{(Eq. 11.10)}$$

Three-phase transformers utilize the average test values for current and voltage for the three phases. The following relationships describe the percentage impedance (%Z) [13]:

$$Z_1 = \frac{1}{2}(Z_{12} + Z_{13} - Z_{23}) \qquad \text{(Eq. 11.11)}$$

$$Z_2 = \frac{1}{2}(Z_{23} + Z_{12} - Z_{13}) \qquad \text{(Eq. 11.12)}$$

$$Z_3 = \frac{1}{2}(Z_{13} + Z_{23} - Z_{12}) \qquad \text{(Eq. 11.13)}$$

where Z_{12} is the impedance with windings 1 and 2 short-circuited with circuit 3 open-circuited, Z_{13} is the impedance with windings 1 and 3 short-circuited with circuit 2 open-circuited, and Z_{23} is the impedance with windings 2 and 3 short-circuited with circuit 1 open-circuited.

Temperature Tests (Heat Run)

Temperature testing consists of a setup similar to that used for impedance testing. One secondary winding is short circuited on a three-phase transformer and sufficient voltage at rated frequency is applied to the primary winding to develop wattage losses equal to the sum of the no-load excitation loss plus the full load copper loss. That test current is held for one hour or until the transformer oil temperature stabilizes. This oil temperature is referred to as the *ultimate oil temperature*. The temperature is recorded before the test begins and after the oil temperature increases and remains constant. The difference between these temperatures is the *transformer oil temperature rise*.

The test continues with the primary voltage adjusted so that the primary current equals full load current at rated frequency. The test is conducted for one hour at which time the oil temperature is recorded and the winding resistance is measured. The winding resistance should also be measured at ambient temperature. The copper temperature must be calculated using the following relationship:

$$T_2 = (234.5 + T_1) \times \frac{\text{Winding Resistance } @T_2}{(\text{Winding Resistance } @T_1) - 234.5} \qquad \text{(Eq. 11.14)}$$

The difference between this copper temperature and the simultaneous corresponding oil temperature is added to the ultimate oil rise previously determined to obtain the copper rise by resistance. [14]

The winding hot resistance must be measured within 4 minutes after the transformer is shutdown. If this is not accomplished, correction factors may be used.

Dielectric Tests [15]

Dielectric testing is conducted to verify the insulation levels of a transformer. Three common dielectric tests conducted on transformers include:

Low-frequency applied potential tests

Low-frequency induced potential tests

Impulse tests

The details of those tests can be found in ANSI/IEEE C57.12.90, *IEEE Standard Test Code for Liquid-Immersed Distribution, Power, and Regulating Transformers* and *IEEE Guide for Short-Circuit Testing of Distribution and Power Transformers*.

Reactors

The 2008 *National Electrical Code® Handbook* describes the use of reactors in the explanation section after Article 470.1. It notes there that:

> Reactors are installed in a circuit to introduce inductance for motor starting, combined with a capacitor to make a filter, controlling the current, and paralleling transformers. Current-limiting reactors are installed to limit the amount of current that can flow in a circuit when a short circuit occurs. Reactors can be divided into two classes: those with iron cores and those with no magnetic materials in the windings. Either type may be air cooled or oil immersed. [16]

Other reactor uses include generator grounding schemes [17] which are typically connected to a wye-configured generator neutral. The rating of a neutral grounding reactor is its thermal current rating. The reactor should be rated to carry the generator rms ground fault current for its rated time under standard conditions without overheating. The neutral reactor's current rating is equal [18] to the rms symmetrical current calculated using the generator's transient reactance to represent positive-sequence reactance and the system's negative-sequence reactance and zero-sequence reactance. The reactor's reactance can be calculated using

$$X_{\mathrm{N}} = \frac{X_1 + X_0}{3}$$ (Eq. 11.15 [19])

where X_1 is the generator positive-sequence reactance, X_0 is the generator zero-sequence reactance, and X_{N} is the Reactor reactance.

Reactors can be used to reduce short-circuit current levels with generators by tying the generator load bus to a synchronizing bus through reactors. This can allow the lowering of the interrupting duty on circuit breakers. This can be helpful in situations where additional generation capacity is added to existing bus, increasing the available fault current. Reactors have also been used in series with generator outputs to limit fault current. However, this should be carefully examined where low power-factor loads are encountered.

Transformer and Reactor Standards

Many North American codes, standards and recommended practices for transformers and reactors are presented in Table 11.2. These documents cover a variety of topics associated with that equipment, including insulating materials; dielectric fluids and their handling; switching and protection schemes; bar coding; pad-mounted, pole-mounted, and underground equipment; instrument transformers; dry-type and liquid-filled equipment; apparatus bushings; testing; guides for loading; unit substations; determination of power losses; etc.

TABLE 11.2 Distribution, power, instrument transformers, and reactors standards

Developer	Standard No.	Title
IEEE	IEEE Std 1™	Temperature/Evaluation of Electrical Insulation
IEEE	IEEE Std 62™	IEEE Guide for Diagnostic Field Testing of Electric Power Apparatus – Part 1: Oil Filled Power Transformers, Regulators, and Reactors
IEEE	IEEE Std 98™	Test/Evaluation of Insulating Materials
IEEE	IEEE Std 99™	Test/Evaluation of Insulation Systems
IEEE	IEEE Std 259™	IEEE Standard Test Procedure for Evaluation of Systems of Insulation for Specialty Transformers
IEEE	IEEE Std 315™	Graphic Symbols for Diagrams
IEEE	IEEE Std 637™	IEEE Guide for the Reclamation of Insulating Oil and Criteria for Its Use
IEEE	IEEE Std 638™	IEEE Standard for Qualification of Class IE Transformers for Nuclear Power Generating Stations
IEEE	IEEE 799	IEEE Guide for Handling and Disposal of Transformer Grade Insulating Liquids Containing PCBs
IEEE	IEEE 1158	IEEE Recommended Practice for Determination of Power Losses in High-Voltage Direct-Current (HVDC) Converter Stations – Description
IEEE	IEEE Std 1276	IEEE Trial-Use Guide for the Application of High-Temperature Insulation Materials in Liquid-Immersed Power Transformers
IEEE	IEEE Std 1277	IEEE Trial-Use Standard General Requirements and Test Code for Dry-Type and Oil-Immersed Smoothing Reactors for DC Power Transmission
IEEE	IEEE Std 1312™	IEEE Standard Preferred Voltage Ratings for Alternating-Current Electrical Systems and Equipment Operating at Voltages Above 230 kV Nominal
IEEE	IEEE Std 1313.1™	IEEE Standard for Insulation Coordination – Definitions, Principles, and Rules
IEEE	IEEE Std 1313.2™	IEEE Guide for the Application of Insulation Coordination
IEEE	IEEE Std 1388™	IEEE Standard for the Electronic Reporting of Transformer Test Data
IEEE	IEEE Std 1538™	IEEE Guide for Determination of Maximum Winding Temperature Rise in Liquid-Filled Transformers
IEEE	IEEE C37.015	IEEE Application Guide for Shunt Reactor Switching
IEEE	ANSI/IEEE C37.109	IEEE Guide for the Protection of Shunt Reactors
IEEE	IEEE C57.113	IEEE Guide for Partial Discharge Measurement in Liquid-Filled Power Transformers and Shunt Reactors
IEEE	ANSI/IEEE C57.12.00	IEEE Standard General Requirements for Liquid-Immersed Distribution, Power, and Regulating Transformers
IEEE	ANSI/IEEE C57.12.01	IEEE Standard General Requirements for Dry-Type Distribution and Power Transformers Including Those with Solid-Cast and/or Resin Encapsulated Windings

(Continued)

TABLE 11.2 Distribution, power, instrument transformers, and reactors standards—cont'd

Developer	Standard No.	Title
IEEE	ANSI C57.12.10	American National Standard for Transformers – 230 kV and Below 833/958 through 8333/10 417 kVA, Single-Phase, and 750/862 through 60 000/80 000/100 000 kVA, Three-Phase without Load Tap Changing; and 3750/4687 through 60 000/80 000/100 000 kVA with Load Tap Changing – Safety Requirements
IEEE	ANSI C57.12.20	American National Standard for Transformers Standard For Overhead Type Distribution Transformers, 500 kVA and Smaller: High Voltage, 34500 Volts and Below; Low Voltage, 7970/13800Y Volts and Below
IEEE	ANSI C57.12.21	American National Standard for Transformers – Pad-Mounted, Compartmental-Type, Self-Cooled, Single-Phase Distribution Transformers with High-Voltage Bushings; High Voltage, 34 500 GRYD/19920 Volts and Below; Low Voltage, 240/120 Volts; 167 kVA and Smaller
IEEE	ANSI C57.12.22	American National Standard for Transformers – Pad-Mounted, Compartmental-Type, Self-Cooled, Three-Phase Distribution Transformers with High-Voltage Bushings, 2500 kVA and Smaller: High Voltage, 34 500GrdY/19 920 Volts and Below; Low Voltage, 480 Volts and Below – Requirements
IEEE	ANSI C57.12.23	IEEE Standard for Transformers – Underground-Type, Self-Cooled, Single-Phase Distribution Transformers with Separable, Insulated, High-Voltage Connectors; High Voltage (24 940 GrdY/14 400 V and Below) and Low Voltage (240/120 V, 167 kVA and Smaller)
IEEE	ANSI C57.12.24	American National Standard for Transformers Underground-Type Three-Phase Distribution Transformers, 2500 kVA and Smaller; High Voltage, 34 500 GrdY/19 920 Volts and Below; Low Voltage, 480 Volts and Below – Requirements
IEEE	ANSI C57.12.25	American National Standard for Transformers Pad-Mounted, Compartmental-Type, Self-Cooled, Single- Phase Distribution Transformers with Separable Insulated High-Voltage Connectors; High Voltage, 34 500 Grd Y/ 19 920 Volts and Below; Low Voltage, 240/120 Volts; 167 kVA and Smaller Requirements
IEEE	ANSI C57.12.26	IEEE Standard for Pad-Mounted, Compartmental-Type, Self-Cooled, Three-Phase Distribution Transformers for Use with Separable Insulated High-Voltage Connectors (34 500 Grd Y/19 920 V and Below; 2500 kVA and Smaller)
IEEE	IEEE C57.12.28	IEEE Standard for Pad-Mounted Equipment Enclosure Integrity
IEEE	ANSL C57.12.29	American National Standard for Switchgear and Transformers – Pad-Mounted Equipment – Enclosure Integrity for Coastal Environments
IEEE	ANSI C57.12.31	American National Standard Pole-Mounted Equipment – Enclosure Integrity
IEEE	ANSI C57.12.32	American National Standard Submersible Equipment – Enclosure Integrity
IEEE	IEEE C57.12.34	IEEE Standard Requirements for Pad-Mounted , Compartmental-Type, Self-Cooled, Three-Phase Distribution Transformers, 2500 kVA and Smaller-High Voltage: 34 500 GrdY/19 920 Volts and Below; Low Voltage: 480 Volts and Below

TABLE 11.2 Distribution, power, instrument transformers, and reactors standards—cont'd

Developer	Standard No.	Title
IEEE	IEEE C57.12.35	IEEE Standard for Bar Coding for Distribution Transformers
IEEE	IEEE C57.12.36	IEEE Standard Requirements for Liquid-Immersed Distribution Substation Transformers
IEEE	IEEE C57.12.37	IEEE Standard for the Electronic Reporting of Distribution Transformer Test Data
IEEE	ANSI C57.12.40	American National Standard for Secondary Network Transformers Subway and Vault Types (Liquid Immersed) – Requirements
IEEE	IEEE C57.12.44	IEEE Standard Requirements for Secondary Network Protectors
IEEE	ANSI C57.12.50	American National Standard Requirements for Ventilated Dry-Type Distribution Transformers, 1 to 500 kVA, Single-Phase, and 15 to 500 kVA, Three-Phase, with High-Voltage 601 to 34 500 Volts, Low-Voltage 120 to 600 Volts
IEEE	ANSI C57.12.51	American National Standard Requirements for Ventilated Dry-Type Power Transformers, 501 kVA and Larger, Three-Phase, with High-Voltage 601 to 34 500 Volts, Low-Voltage 208Y/120 to 4160 Volts
IEEE	ANSI/IEEE C57.12.52	American National Standard Requirements for Sealed Dry-Type Power Transformers, 501 kVA and Larger, Three-Phase, with High-Voltage 601 to 34 500 Volts, Low-Voltage 208Y/120 to 4160 Volts
IEEE	ANSI C57.12.55	American National Standard for Transformers Dry-Type Transformers Used in Unit Installations, Including Unit Substations Conformance Standard
IEEE	ANSI C57.12.56	IEEE Standard Test Procedure for Thermal Evaluation of Insulation Systems for Ventilated Dry-Type Power and Distribution Transformers
IEEE	ANSI C57.12.57	American National Standard for Transformers – Ventilated Dry-Type Network Transformers 2500 kVA and Below, Three-Phase, with High-Voltage 34 500 Volts and Below, Low-Voltage 216Y/125 and 480Y/277 Volts – Requirements
IEEE	IEEE C57.12.58	IEEE Guide for Conducting a Transient Voltage Analysis of a Dry-Type Transformer Coil
IEEE	IEEE C57.12.59	Standard for Dry-Type Transformer Through-Fault Current Duration
IEEE	IEEE C57.12.60	IEEE Guide for Test Procedures for Thermal Evaluation of Insulation Systems for Solid-Cast and Resin-Encapsulated Power and Distribution Transformers
IEEE	ANSI C57.12.70	American National Standard Terminal Markings and Connections for Distribution and Power Transformers
IEEE	ANSI/IEEE C57.12.80	IEEE Standard Terminology for Power and Distribution Transformers
IEEE	ANSI/IEEE C57.12.90	IEEE Standard Test Code for Liquid-Immersed Distribution, Power, and Regulating Transformers and IEEE Guide for Short-Circuit Testing of Distribution and Power Transformers
IEEE	ANSI/IEEE C57.12.91	IEEE Standard Test Code for Dry-Type Distribution and Power Transformers

(*Continued*)

TABLE 11.2 Distribution, power, instrument transformers, and reactors standards—cont'd

Developer	Standard No.	Title
IEEE	IEEE C57.13	IEEE Standard Requirements for Instrument Transformers
IEEE	IEEE C57.13.1	IEEE Guide for Field Testing of Instrument Transformers
IEEE	IEEE C57.13.2	IEEE Standard Conformance Test Procedure for Instrument Transformers
IEEE	IEEE C57.13.3	IEEE Guide for the Grounding of Instrument Transformer Secondary Circuits and Cases
IEEE	IEEE C57.13.5	Trial-Use Standard of Performance and Test Requirements for Instrument Transformers of a Nominal System Voltage of 115 kV and Above
IEEE	IEEE C57.13.6	IEEE Standard for High Accuracy Instrument Transformers
IEEE	IEEE C57.15	IEEE Standard Requirements, Terminology, and Test Code for Step-Voltage and Induction-Voltage Regulators
IEEE	IEEE C57.16	IEEE Standard Requirements, Terminology, and Test Code for Dry-Type Air-Core Series-Connected Reactors
IEEE	IEEE C57.18.10	IEEE Standard Practices and Requirements for Semiconductor Power Rectifier Transformers
IEEE	ANSI/IEEE C57.19.00	IEEE Standard General Requirements and Test Procedure for Outdoor Power Apparatus Bushings
IEEE	ANSI/IEEE C57.19.01	IEEE Standard Performance Characteristics and Dimensions for Outdoor Apparatus Bushings
IEEE	IEEE C57.19.03	IEEE Standard Requirements, Terminology, and Test Code for Bushings for DC Applications
IEEE	IEEE C57.19.100	IEEE Guide for Application of Power Apparatus Bushings
IEEE	IEEE C57.19.21	IEEE Standard Requirements, Terminology, and Test Code for Shunt Reactors Rated Over 500 kVA
IEEE	IEEE C57.91	IEEE Guide for Loading Mineral-Oil Immersed Transformers
IEEE	IEEE C57.93	IEEE Guide for Installation of Liquid-immersed Power Transformers
IEEE	ANSI/IEEE C57.94	IEEE Recommended Practice for Installation, Application, Operation, and Maintenance of Dry-Type General Purpose Distribution and Power Transformers
IEEE	ANSI/IEEE C57.96	IEEE Guide for Loading Dry Type Distribution and Power Transformers
IEEE	IEEE C57.98	IEEE Guide for Transformer Impulse Tests
IEEE	ANSI/IEEE C57.100	IEEE Standard Test Procedure for Thermal Evaluation of Liquid-Immersed Distribution and Power Transformers
IEEE	IEEE C57.104	IEEE Guide for the Interpretation of Gases Generated in Oil-Immersed Transformers
IEEE	IEEE C57.105	IEEE Guide for Application of Transformer Connections in Three-Phase Distribution Systems
IEEE	IEEE C57.106	IEEE Guide for Acceptance and Maintenance of Insulating Oil in Equipment
IEEE	IEEE C57.109	IEEE Guide for Liquid-Immersed Transformer Through-Fault-Current Duration

TABLE 11.2 Distribution, power, instrument transformers, and reactors standards—cont'd

Developer	Standard No.	Title
IEEE	ANSI/IEEE C57.110	IEEE Recommended Practice for Establishing Transformer Capability When Supplying Non-sinusoidal Load Currents
IEEE	IEEE C57.111	IEEE Guide for Acceptance of Silicone Insulating Fluid and Its Maintenance in Transformers
IEEE	IEEE C57.113	IEEE Guide for Partial Discharge Measurement in Liquid-Filled Power Transformers and Shunt Reactors
IEEE	IEEE C57.116	IEEE Guide for Transformers Directly Connected to Generators
IEEE	ANSI/IEEE C57.117	IEEE Guide for Reporting Failure Data for Power Transformers and Shunt Reactors on Electric Utility Power Systems
IEEE	IEEE C57.119	Recommended Practice for Performing Temperature Rise Tests on Oil Immersed Power Transformers at Loads Beyond Nameplate Ratings
IEEE	IEEE C57.120	IEEE Loss Evaluation Guide for Power Transformers and Reactors
IEEE	IEEE C57.121	IEEE Guide for Acceptance and Maintenance of Less Flammable Hydrocarbon Fluid in Transformers
IEEE	ANSI/IEEE C57.12.123	Guide for Transformer Loss Measurement
IEEE	IEEE C57.124	IEEE Recommended Practice for the Detection of Partial Discharge and the Measurement of Apparent Charge in Dry-Type Transformers
IEEE	IEEE C57.125	IEEE Guide for Failure Investigation, Documentation, and Analysis for Power Transformers and Shunt Reactors
IEEE	IEEE C57.127	IEEE Guide for the Detection and Location of Acoustic Emissions for Partial Discharges in Oil-Immersed Power Transformers and Reactors
IEEE	IEEE C57.129	IEEE Standard for General Requirements and Test Code for Oil-Immersed HVDC Converter Transformers
IEEE	IEEE C57.131	IEEE Standard Requirements for Load Tap Changers
IEEE	IEEE C57.134	Guide for Determination of Hottest Spot Temperature in Dry Type Transformers
IEEE	IEEE C57.135	IEC/IEEE Guide for the Application, Specification, and Testing of Phase-shifting Transformers
IEEE	IEEE C57.136	Guide for Sound Abatement and Determination for Liquid-Immersed Power Transformers and Shunt Reactors Rated over 500 kVA
IEEE	IEEE C57.138	IEEE Recommended Practice for Routine Impulse Test for Distribution Transformers
IEEE	IEEE C57.140	Guide for the Evaluation and Reconditioning of Liquid Immersed Power Transformers
IEEE	IEEE C57.144	IEEE Guide for Metric Conversion of Transformer Standards
IEEE	IEEE C57.146	IEEE Guide for the Interpretation of Gases Generated in Silicone-Immersed Transformers
IEEE	IEEE C57.147	IEEE Guide for Acceptance and Maintenance of Natural Ester Fluids in Transformers
NEMA	NEMA TP 1	Guide for Determining Energy Efficiency for Distribution Transformers

(Continued)

TABLE 11.2 Distribution, power, instrument transformers, and reactors standards—cont'd

Developer	Standard No.	Title
NEMA	NEMA TP 2	Standard Test Method for Measuring the Energy Consumption of Distribution Transformers
NEMA	NEMA TR 1	Transformers, Regulators and Reactors
NEMA	NEMA ST 20	Dry Type Transformers for General Applications
NFPA	NFPA 70®	National Electrical Code; Article 450 Transformers and Transformer Vaults
NFPA	NFPA 70®	National Electrical Code; Article 470 Resistors and Reactors
UL	UL 1062	Standard for Unit Substations
UL	UL 1446	Systems of Insulating Materials – General
UL	UL 1561	Standard for Dry-Type General Purpose and Power Transformers
UL	UL 1562	Transformers, Distribution, Dry-Type – over 600 Volts
UL	UL 5085-1	Low-Voltage Transformers – Part 1: General Requirements
UL	UL 5085-2	Low-Transformers – Part 2: General Purpose Transformers
UL	UL 5085-3	Low-Voltage Transformers – Part 3: Class 2 and Class 3 Transformers
CSA	CSA C9	Dry-Type Transformers
CSA	CSA CAN3-C13	Instrument Transformers
CSA	CSA C50	Mineral Insulating Oil, Electrical, for Transformers and Switches
CSA	CAN/CSA-C88	Power Transformers and Reactors
CSA	CAN/CSA-C88.1	Power Transformer and Reactor Bushings
CSA	CSA C199	Three-Phase Network Distribution Transformers
CSA	CSA C227.3	Low-profile, Single-phase, Pad-mounted Distribution Transformers with Separable Insulated High-voltage Connectors
CSA	CSA C227.4	Three-Phase, Pad-mounted Distribution Transformers with Separable Insulated High-Voltage Connector
CSA	CSA C227.5	Three-Phase Live-Front Pad-Mounted Distribution Transformers
CSA	CSA C301.1	Single-Phase Submersible Distribution Transformers
CSA	CSA C301.2	Three-Phase Submersible Distribution Transformers
CSA	CAN/CSA-C60044-1	Instrument Transformers – Part 1: Current Transformers (Adopted CEI/IEC 60044-1:1996+A1:2000+A2:2002, edition 1.2, 2003-02)
CSA	CAN/CSA-C60044-2	Instrument Transformers – Part 2: Inductive Voltage Transformers (Adopted CEI/IEC 60044-2:1997+A1:2000+A2:2002, edition 1.2, 2003-02)
CSA	CAN/CSA-C60044-3	Instrument Transformers – Part 3: Combined Transformers (Adopted CEI/IEC 60044-3:2002, second edition, 2002-12)
CSA	CAN/CSA-C60044-5	Instrument Transformers – Part 5: Capacitor Voltage Transformers (Adopted CEI/IEC 60044-5:2004, first edition, 2004-04)
CSA	CAN/CSA-C60044-6	Instrument Transformers – Part 6: Requirements for Protective Current Transformers for Transient Performance (Adopted CEI/IEC 44-6:1992, first edition, 1992-03)

TABLE 11.2 Distribution, power, instrument transformers, and reactors standards—cont'd

Developer	Standard No.	Title
CSA	AN/CSA-C60044-7	Instrument Transformers – Part 7: Electronic Voltage Transformers (Adopted CEI/IEC 60044-7:1999, first edition, 1999-12)
CSA	CAN/CSA-C60044-8	Instrument Transformers – Part 8: Electronic Current Transformers (Adopted IEC 60044-8:2002, first edition, 2002-07)
CSA	CAN/CSA-E61558-1	Safety of Power Transformers, Power Supply Units and Similar – Part 1: General Requirements and Tests (Adopted CEI/IEC 61558-1:1997 + A1:1998, edition 1.1, 1998-07, with Canadian deviations)
CSA	CAN/CSA-E61558-2-1	Safety of Power Transformers, Power Supply Units and Similar – Part 2: Particular Requirements for Separating Transformers for General Use (Adopted CEI/IEC 61558-2-1:1997, first edition, 1997-02)
CSA	CAN/CSA-E61558-2-2	Safety of Power Transformers, Power Supply Units and Similar – Part 2-2: Particular Requirements for Control Transformers (Adopted CEI/IEC 61558-2-2:1997, first edition, 1997-10)
CSA	CAN/CSA-E61558-2-4	Safety of Power Transformers, Power Supply Units and Similar – Part 2: Particular Requirements for Isolating Transformers for General Use (Adopted CEI/IEC 61558-2-4:1997, first edition 1997-02)
CSA	CAN/CSA-E61558-2-5	Safety of Power Transformers, Power Supply Units and Similar – Part 2-5: Particular Requirements for Shaver Transformers and Shaver Supply Units (Adopted CEI/IEC 61558-2-5:1997, first edition, 1997-12, with Canadian deviations)
CSA	CAN/CSA-E61558-2-6	Safety of Power Transformers, Power Supply Units and Similar – Part 2: Particular Requirements for Safety Isolating Transformers for General Use (Adopted CEI/IEC 61558-2-6:1997, first edition, 1997-02)
CSA	CAN/CSA-E61558-2-13	Safety of Power Transformers, Power Supply Units and Similar Devices – Part 2-13: Particular Requirements for Auto-Transformers for General Use (Adopted CEI/IEC 61558-2-13:1999, first edition, 1999-10, with Canadian deviations)
CSA	CAN/CSA-C22.2 NO. 47	Air-Cooled Transformers (Dry Type)
CSA	CSA C22.2 NO. 66.1	Low-Voltage Transformers – Part 1: General Requirements (Binational Standard with UL 5085-1)
CSA	CSA C22.2 NO. 66.2	Low-Voltage Transformers – Part 2: General Purpose Transformers (Bi-National standard, with UL 5085-2)
CSA	CSA C22.2 NO. 66.3	Low-Voltage Transformers – Part 3: Class 2 and Class 3 Transformers (Bi-National standard, with UL 5085-3)
CSA	CSA C22.2 NO. 180	Series Isolating Transformers for Airport Lighting
CSA	CSA CAN/CSA-E742	Isolating Transformers and Safety Isolating Transformers – Requirements (Adopted IEC 742:1983, first edition, including Amendment 1:1992, with Canadian Deviations)
FM Global	FM 3990	Approval Standard for Less or Nonflammable Liquid-Insulated Transformers
FM Global	FM 6930	Approval Standard for Flammability Classification of Industrial Fluids
FM Global	FM 6933	Approval Standard for Less Flammable Transformer Fluids
FM Global	FM 6934	Approval Standard for Nonflammable Transformer Fluids

NEMA Standard TR 1, *Transformers, Regulators, and Reactors* provides lists of ANSI, IEEE, and NEMA transformer and reactor standards by device type including those shown in Table 11.3:

Table 11.4 contains a list of many of the International Electrotechnical Commission (IEC) transformer and reactor standards. It contains standards for power and instrument transformers; reactors; testing; equipment insulating bushings; dry and liquid-immersed equipment; tap-changers; core specifications; markings; liquid dielectric material standards; application guides; etc.

Power Capacitors

Power capacitors can be used in motor starting applications as well as for power factor improvement. Power factor is defined as the cosine φ of the phase displacement angle by which current leads or lags voltage in a circuit. Power factor is also defined as the ratio of active power in kW to apparent power in kVA. Power factor can be described as leading or lagging. Capacitors and overexcited synchronous motors supply reactive power (kvar) that has a leading power factor. Inductors and motors supply lagging power factors. Both capacitors and motors/reactors are considered kilovar (kvar) generators; however, their respective leading and lagging power factors cause them to arithmetically cancel out their kvar contributions.

To better understand the concept of power factor improvement, refer to the relationship of real and apparent power in Equation 11.16.

$$\text{kVA} = \sqrt{(\text{kW})^2 + (\text{kvar})^2} \qquad \text{(Eq. 11.16)}$$

TABLE 11.3 Transformer and reactor standards

Device Type	NEMA TR 1 Part No.
Power transformers	Part 1
Distribution transformers	Part 2
Secondary network transformers	Part 3
Dry-type transformers	Part 4
Unit substation transformers	Part 5
Transmission and distribution voltage regulators	Part 8
Current-limiting reactors	Part 9
Arc furnace transformers	Part 10
Shunt reactors	Part 11
Underground-type three-phase distribution transformers	Part 12

TABLE 11.4 IEC transformers and reactors

Developer	Standard No.	Title
IEC	IEC 60044-1	Instrument Transformers – Part 1: Current Transformers
IEC	IEC 60044-2	Instrument Transformers – Part 2 : Inductive Voltage Transformers
IEC	IEC 60044-3	Instrument Transformers – Part 3: Combined Transformers
IEC	IEC 60044-5	Instrument Transformers – Part 5: Capacitor Voltage Transformers
IEC	IEC 60044-6	Instrument Transformers – Part 6: Requirements for Protective Current Transformers for Transient Performance
IEC	IEC 60044-7	Instrument Transformers – Part 7: Electronic Voltage Transformers
IEC	IEC 60044-8	Instrument Transformers – Part 8: Electronic Current Transformers
IEC	IEC 60050-321	International Electrotechnical Vocabulary. Chapter 321: Instrument Transformers
IEC	IEC 60050-421	International Electrotechnical Vocabulary. Chapter 421: Power Transformers and Reactors
IEC	IEC 60076-1	Power Transformers – Part 1: General
IEC	IEC 60076-2	Power Transformers – Part 2: Temperature Rise
IEC	IEC 60076-3	Power Transformers – Part 3: Insulation Levels, Dielectric Tests and External Clearances in Air
IEC	IEC 60076-4	Power Transformers – Part 4: Guide to the Lightning Impulse and Switching Impulse Testing – Power Transformers and Reactors
IEC	IEC 60076-5	Power Transformers – Part 5: Ability to Withstand Short Circuit
IEC	IEC 60076-6	Power Transformers – Part 6: Reactors
IEC	IEC 60076-7	Power Transformers – Part 7: Loading Guide for Oil-Immersed Power Transformers
IEC	IEC 60076-8	Power Transformers – Part 8: Application Guide
IEC	IEC 60076-10	Power Transformers – Part 10: Determination of Sound Levels
IEC	IEC 60076-10-1	Power Transformers – Part 10-1: Determination of Sound Levels – Application Guide
IEC	IEC 60076-11	Power Transformers – Part 11: Dry-Type Transformers
IEC	IEC 60076-12	Power Transformers – Part 12: Loading Guide for Dry-Type Power Transformers
IEC	IEC 60076-13	Power Transformers – Part 13: Self-Protected Liquid-Filled Transformers
IEC	IEC 60076-14	Power Transformers – Part 14: Design and Application of Liquid-Immersed Power Transformers Using High-Temperature Insulation Materials
IEC	IEC 60076-15	Power Transformers – Part 15: Gas-Filled Power Transformers
IEC	IEC 60092-303	Electrical Installations in Ships. Part 303: Equipment – Transformers for Power and Lighting
IEC	IEC 60137	Insulated Bushings for Alternating Voltages Above 1000 V
IEC	IEC 60146-1-3	Semiconductor Convertors - General Requirements and Line Commutated Convertors – Part 1-3: Transformers and Reactors
IEC	IEC 60214-1	Tap-Changers – Part 1: Performance Requirements and Test Methods

(Continued)

TABLE 11.4 IEC transformers and reactors—cont'd

Developer	Standard No.	Title
IEC	IEC 60214-2	Tap-Changers – Part 2: Application Guide
IEC	IEC 60247	Insulating Liquids – Measurement of Relative Permittivity, Dielectric Dissipation Factor (Tan D) and DC Resistivity
IEC	IEC 60296	Fluids for Electrotechnical Applications – Unused Mineral Insulating Oils for Transformers and Switchgear
IEC	IEC 60310	Railway Applications – Traction Transformers and Inductors on Board Rolling Stock
IEC	IEC 60401-3	Terms and Nomenclature for Cores Made of Magnetically Soft Ferrites – Part 3: Guidelines on the Format of Data Appearing in Manufacturers' Catalogues of Transformer and Inductor Cores
IEC	IEC 60422	Mineral Insulating Oils in Electrical Equipment – Supervision and Maintenance Guidance
IEC	IEC 60445	Basic and Safety Principles for Man-Machine Interface, Marking and Identification – Identification of Equipment Terminals and Conductor Terminations
IEC	IEC 60450	Measurement of the Average Viscometric Degree of Polymerization of New and Aged Cellulosic Electrically Insulating Materials
IEC	IEC 60567	Oil-Filled Electrical Equipment – Sampling of Gases and of Oil for Analysis of Free and Dissolved Gases – Guidance
IEC	IEC 60588-1	Askarels for Transformers and Capacitors – Part 1: General
IEC	IEC 60588-2	Askarels for Transformers and Capacitors – Part 2: Test Methods
IEC	IEC 60588-3	Askarels for Transformers and Capacitors – Part 3: Specifications for New Askarels
IEC	IEC 60588-4	Askarels for Transformers and Capacitors – Part 4: Guide for Maintenance of Transformer Askarels in Equipment
IEC	IEC 60588-5	Askarels for Transformers and Capacitors – Part 5: Screening Test for Compatibility of Materials and Transformer Askarels
IEC	IEC 60588-6	Askarels for Transformers and Capacitors – Part 6: Screening Test for Effects of Materials on Capacitor Askarels
IEC	IEC 60599	Mineral Oil-Impregnated Electrical Equipment in Service – Guide to the Interpretation of Dissolved and Free Gases Analysis
IEC	IEC/TR 60616	Terminal and Tapping Markings for Power Transformers
IEC	IEC 60618	Inductive Voltage Dividers
IEC	IEC 60647	Dimensions for Magnetic Oxide Cores Intended for Use in Power Supplies (EC-Cores)
IEC	IEC 60740-1	Laminations for Transformers and Inductors – Part 1: Mechanical and Electrical Characteristics
IEC	IEC/TR 60787	Application Guide for the Selection of High-Voltage Current-Limiting Fuse-Links for Transformer Circuits
IEC	IEC 60836	Specifications for Unused Silicone Insulating Liquids for Electrotechnical Purposes
IEC	IEC 60851-4	Winding Wires – Test Methods – Part 4: Chemical Properties

TABLE 11.4 IEC transformers and reactors—cont'd

Developer	Standard No.	Title
IEC	IEC 60944	Guide for the Maintenance of Silicone Transformer Liquids
IEC	IEC 60989	Separating Transformers, Autotransformers, Variable Transformers and Reactors.
IEC	IEC 61000-2-12	Electromagnetic Compatibility (EMC) – Part 2-12: Environment – Compatibility Levels for Low-Frequency Conducted Disturbances and Signalling in Public Medium-Voltage Power Supply Systems
IEC	IEC 61000-3-13	Electromagnetic Compatibility (EMC) – Part 3-13: Limits – Assessment of Emission Limits for the Connection of Unbalanced Installations to MV, HV and EHV Power Systems
IEC	IEC 61099	Specifications for Unused Synthetic Organic Esters for Electrical Purposes
IEC	IEC 61181	Mineral Oil-Filled Electrical Equipment – Application of Dissolved Gas Analysis (DGA) to Factory Tests on Electrical Equipment
IEC	IEC 61203	Synthetic Organic Esters for Electrical Purposes – Guide for Maintenance of Transformer Esters in Equipment
IEC	IEC 61378-1	Convertor Transformers – Part 1: Transformers for Industrial Applications
IEC	IEC 61378-2	Convertor Transformers – Part 2: Transformers for HVDC Applications
IEC	IEC 61378-3	Converter Transformers – Part 3: Application Guide
IEC	IEC/TS 61463	Bushings – Seismic Qualification
IEC	IEC 61558-1	Safety of Power Transformers, Power Supplies, Reactors and Similar Products – Part 1: General Requirements and Tests
IEC	IEC 61558-2-1	Safety of Power Transformers, Power Supplies, Reactors and Similar Products – Part 2-1: Particular Requirements and Tests for Separating Transformers and Power Supplies Incorporating Separating Transformers for General Applications
IEC	IEC 61558-2-4	Safety of Transformers, Reactors, Power Supply Units and Similar Products for Supply Voltages Up To 1,100 V – Part 2-4: Particular Requirements and Tests for Isolating Transformers and Power Supply Units Incorporating Isolating Transformers
IEC	IEC 61558-2-6	Safety of Transformers, Reactors, Power Supply Units and Similar Products for Supply Voltages Up To 1,100 V – Part 2-6: Particular Requirements and Tests for Safety Isolating Transformers and Power Supply Units Incorporating Safety Isolating Transformers
IEC	IEC 61558-2-13	Safety of Transformers, Reactors, Power Supply Units and Similar Products for Supply Voltages Up To 1,100 V – Part 2-13: Particular Requirements and Tests for Auto Transformers and Power Supply Units Incorporating Auto Transformers
IEC	IEC 61596	Magnetic Oxide EP-Cores and Associated Parts for Use in Inductors and Transformers – Dimensions
IEC	IEC 61620	Insulating Liquids – Determination of the Dielectric Dissipation Factor by Measurement of the Conductance and Capacitance – Test Method

(Continued)

TABLE 11.4 IEC transformers and reactors—cont'd

Developer	Standard No.	Title
IEC	IEC 61639	Direct Connection Between Power Transformers and Gas-Insulated Metal-Enclosed Switchgear for Rated Voltages of 72.5 kV and Above
IEC	IEC 61869-1	Instrument Transformers – Part 1: General Requirements
IEC	IEC 61936-1	Power Installations Exceeding 1 kV AC – Part 1: Common Rules
IEC	IEC 62032	Guide for the Application, Specification, and Testing of Phase-Shifting Transformers
IEC	IEC 62041	Power Transformers, Power Supply Units, Reactors and Similar Products – EMC Requirements
IEC	IEC 62044-1	Cores Made of Soft Magnetic Materials – Measuring Methods – Part 1: Generic Specification
IEC	IEC 62044-2	Cores Made of Soft Magnetic Materials – Measuring Methods – Part 2: Magnetic Properties at Low Excitation Level
IEC	IEC 62044-3	Cores Made of Soft Magnetic Materials – Measuring Methods – Part 3: Magnetic Properties at High Excitation Level
IEC	IEC 62199	Bushings for DC Application

By decreasing the kvar reactive power component through power factor correction techniques, the apparent power (kVA) requirements will become lower and the operating current will decrease. This reduction in system power requirements creates the *release of capacity*, allowing additional loads to be installed on a distribution system.

The use of capacitors for power factor correction will also have the tendency to improve or raise a circuit's voltage. The reason for that voltage improvement can be see in Eq. 11.17 for voltage drop [20]:

$$\Delta V \cong RI \, \text{Cos} \, \varphi \pm XI \, \text{Sin} \, \varphi \qquad \text{(Eq. 11.17)}$$

where resistance (R) and reactance (X) are in ohms and current (I) is in amperes. Typically, the value of the reactive component $X \, \text{Sin} \, \varphi$ is substantially larger that $R \, \text{Cos} \, \varphi$. φ is the power factor angle. ($+$) is utilized when the power factor is lagging and ($-$) is used when the circuit power factor is leading. Reducing a circuit's voltage drop will increase the system voltage.

IEEE 141 provides a mathematical relationship to determine capacitor voltage drop improvement [21].

$$\%\Delta V = \frac{(\text{Capacitor kvar}) \times (\%\text{Transformer Impedance})}{\text{Transformer kVA}} \qquad \text{(Eq. 11.18)}$$

The capacitor installation scheme will determine how the voltage will be affected by the presence of a capacitor bank. If the capacitor is permanently installed on a bus, it will provide a permanent boost in voltage. If the capacitor is switched, voltage will increase when it is turned on and decrease when it is turned off.

Capacitors can also be installed on the load side of a motor starter for induction motors with poor power factors. However, caution should be exercised in that practice. Squirrel-cage induction motors can have power factors of 80–90% at full load operation. That power factor level can drop off depending on the percentage load at which it operates. However, the motor reactive power does not change substantially between no load and full load. If the motor is substantially large enough and does not operate at or near full load, then power factor capacitor installation might be examined more closely.

There are several considerations which must be investigated before installing power factor capacitors directly to a motor starter. Capacitors connected to the load side of a motor starter will begin to discharge when they are switched off. This may have material affect on the motor time constant. A motor time constant is the amount of time that the motor must remain off before safely reconnecting the motor on line. NFPA 70, Article 460.28(A) Means to Reduce the Residual Voltage requires that the capacitor must be discharged to 50 Volts or less within 5 minutes of disconnection from the power supply. This may become critical with motors with high-inertial drive applications and with fast reclosing switching [22].

Application of incorrectly sized capacitors to motors can result in excess voltage levels during switching. Prudent practice would require consultation with the motor manufacturer before any capacitor is added to a motor circuit. IEEE 141 recommends [23] that motor-capacitor applications should be avoided or involve detailed technical investigations conducted when considering the use of capacitors on the load side of motor starters in the following applications:

Reversing or plugging motor;

Restarting motors that are still turning after being turned off;

Crane or elevator motors in which the load may drive the motor and multi-speed motor applications;

Wye-delta connected, open-transition reduced-voltage starters, where the capacitor should be connected to the line side of the starter.

Power capacitor standards are presented in Table 11.5. It includes many of the CSA, NEMA, UL, ASTM, and IEEE standards.

Table 11.6 contains some of the IEC codes, standards, and recommended practices involved with power capacitors. The subjects involved with these standards include switches; capacitive voltage transformers; enclosures; testing; protection.

TABLE 11.5 Power capacitor standards

Developer	Standard No.	Title
CSA	CSA C22.2 No. 190	Capacitors for Power Factor Correction
CSA	CAN/CSA 60044-5	Instrument Transformers – Part 5: Capacitor Voltage Transformers
CSA	CAN/CSA 60871-1	Shunt Capacitors for AC Power Systems Having a Rated Voltage Above 1000 V – Part 1: General – Performance, Testing and Rating - Safety Requirements – Guide for Installation and Operation
CSA	CAN/CSA 60871-2	Shunt Capacitors for AC Power Systems Having a Rated Voltage Above 1000 V – Part 2: Endurance Testing
ASTM	ASTM D2296	Standard Specification for Continuity of Quality of Electrical Insulating Polybutene Oil for Capacitors
ASTM	ASTM D3809	Standard Test Methods for Synthetic Dielectric Fluids for Capacitors
ASTM	ASTM D831	Standard Test Method for Gas Content of Cable and Capacitor Oils
IEEE	ANSI/IEEE 18	Shunt Power Capacitors
IEEE	ANSI/IEEE 21	General Requirements and Test Procedures for Outdoor Apparatus Bushings – Part 1
IEEE	ANSI/IEEE 100	Dictionary of Electrical and Electronic Terms
IEEE	IEEE Std. 824	IEEE Standard for Series Capacitor Banks in Power Systems
IEEE	IEEE 1036	IEEE Guide for Application of Shunt Power Capacitors
IEEE	ANSI/IEEE 1534	Recommended Practice for Specifying Thyristor Controlled Series Capacitors
IEEE	IEEE 1726	Guide for the Functional Specification of Fixed Transmission Series Capacitor Banks for Transmission System Applications
IEEE	ANSI/IEEE C37.012	IEEE Application Guide for Capacitance Current Switching for AC High-Voltage Circuit Breakers
IEEE	ANSI C37.30	Definitions and Requirements for High-Voltage Air Switches, Insulators and Bus Supports
IEEE	IEEE C37.66	IEEE Standard Requirements for Capacitor Switches for AC Systems (1 kV to 38 kV)
IEEE	IEEE C37.99	IEEE Guide for the Protection of Shunt Capacitor Banks
IEEE	IEEE C57.12.30	Standard for Pole-Mounted Equipment – Enclosure Integrity for Coastal Environments
IEEE	IEEE C57.12.31	Standard for Pole-Mounted Equipment – Enclosure Integrity
NEMA	NEMA/ANSI C93.1	Power-Line Carrier Coupling Capacitors and Coupling Capacitor Voltage Transformers (CCVT)
NEMA	NEMA CP 1	Shunt Capacitors
NFPA	NFPA 70®	National Electrical Code; Article 460, Capacitors
UL	UL 810	Standard for Safety for Capacitors

TABLE 11.6 IEC power capacitors

Developer	Standard No.	Title
IEC	IEC 60044-5	Instrument Transformers – Part 5: Capacitor Voltage Transformers
IEC	IEC 60050-436	International Electrotechnical Vocabulary. Chapter 436: Power Capacitors
IEC	IEC 60051-5	Direct Acting Indicating Analogue Electrical Measuring Instruments and Their Accessories – Part 5: Special Requirements for Phase Meters, Power Factor Meters and Synchroscopes
IEC	IEC 60110-1	Power Capacitors for Induction Heating Installations – Part 1: General
IEC	IEC 60110-2	Power Capacitors for Induction Heating Installations – Part 2: Ageing Test, Destruction Test, and Requirements for Disconnecting Internal Fuses
IEC	IEC 60143-1	Series Capacitors for Power Systems – Part 1: General
IEC	IEC 60143-2	Series Capacitors for Power Systems – Part 2: Protective Equipment for Series Capacitor Banks
IEC	IEC 60143-3	Series Capacitors for Power Systems – Part 3: Internal Fuses
IEC	IEC 60252-1	AC Motor Capacitors – Part 1: General – Performance, Testing and Rating – Safety Requirements – Guide for Installation and Operation
IEC	IEC 60252-2	AC Motor Capacitors – Part 2: Motor Start Capacitors
IEC	IEC 60358	Coupling Capacitors and Capacitor Dividers
IEC	IEC 60481	Coupling Devices for Power Line Carrier Systems
IEC	IEC 60549	High-Voltage Fuses for the External Protection of Shunt Power Capacitors
IEC	IEC 60567	Oil-Filled Electrical Equipment – Sampling of Gases and of Oil for Analysis of Free and Dissolved Gases – Guidance
IEC	IEC 60588-1	Askarels for Transformers and Capacitors – Part 1: General
IEC	IEC 60588-2	Askarels for Transformers and Capacitors – Part 2: Test Methods
IEC	IEC 60588-3	Askarels for Transformers and Capacitors – Part 3: Specifications for New Askarels
IEC	IEC 60588-4	Askarels for Transformers and Capacitors – Part 4: Guide for Maintenance of Transformer Askarels in Equipment
IEC	IEC 60588-5	Askarels for Transformers and Capacitors – Part 5: Screening Test for Compatibility of Materials and Transformer Askarels
IEC	IEC 60588-6	Askarels for Transformers and Capacitors – Part 6: Screening Test for Effects of Materials on Capacitor Askarels
IEC	IEC 60831-1	Amendment 1 – Shunt Power Capacitors of the Self-Healing Type for AC Systems Having a Rated Voltage Up To and Including 1000 V – Part 1: General – Performance, Testing and Rating – Safety Requirements – Guide for Installation and Operation
IEC	IEC 60831-2	Shunt Power Capacitors of the Self-Healing Type for AC Systems Having a Rated Voltage Up To and Including 1000 V – Part 2: Ageing Test, Self-Healing Test and Destruction Test
IEC	IEC 60871-1	Shunt Capacitors for AC Power Systems Having a Rated Voltage Above 1000 V – Part 1: General

(*Continued*)

TABLE 11.6 IEC power capacitors—cont'd

Developer	Standard No.	Title
IEC	IEC/TS 60871-2	Shunt Capacitors for AC Power Systems Having a Rated Voltage Above 1000 V – Part 2: Endurance Testing
IEC	IEC/TS 60871-3	Shunt Capacitors for AC Power Systems Having a Rated Voltage Above 1000 V – Part 3: Protection of Shunt Capacitors and Shunt Capacitor Banks
IEC	IEC 60871-4	Shunt Capacitors for AC Power Systems Having a Rated Voltage Above 1000 V – Part 4: Internal Fuses
IEC	IEC 60931-1	Shunt Power Capacitors of the Non-Self-Healing Type for AC Systems Having a Rated Voltage Up To and Including 1000 V – Part 1: General – Performance, Testing and Rating – Safety Requirements – Guide for Installation and Operation
IEC	IEC 60931-2	Shunt Power Capacitors of the Non-Self-Healing Type for AC Systems Having a Rated Voltage Up To and Including 1000 V – Part 2: Ageing Test and Destruction Test
IEC	IEC 60931-3	Shunt Capacitors of the Non-Self-Healing Type for AC Power Systems Having a Rated Voltage Up To and Including 1000 V – Part 3: Internal Fuses
IEC	IEC 61642	Industrial AC Networks Affected by Harmonics – Application of Filters and Shunt Capacitors
IEC	IEC 61921	Power Capacitors – Low-Voltage Power Factor Correction Banks
IEC	IEC 61936-1	Power Installations Exceeding 1 kV AC – Part 1: Common Rules
IEC	IEC 61954	Power Electronics for Electrical Transmission and Distribution Systems – Testing of Thyristor Valves for Static VAR Compensators

If a capacitor is installed on the load side of a motor starter with overload relays, consideration may be required for downsizing the overload relays if the capacitor lowers the motor full load current. Also, if capacitors are connected to a bus or line through switching devices, NFPA 70, Article 460.24 requires that switching devices rated over 600 Volts shall be rated to carry continuous current of not less that 135% of the capacitor rated current. If the switching device is not rated as a load-interrupting device, then it is required to be either interlocked with a load-interrupting device or be provided with a caution sign in accordance with NFPA 70, Article 490.22 to prevent switching the capacitor under load. Reference ANSI/IEEE C37.012, *IEEE Application Guide for Capacitance Current Switching for AC High-Voltage Circuit Breakers* for the requirements in switching medium-voltage capacitors.

The use of capacitors with harmonic-producing loads should be investigated to prevent the creation of a harmonic resonance condition. The technical investigation into the circuit's resonance frequency and the harmonic components existing in the circuit are crucial in preventing overcurrent and overvoltage conditions. The study may find that a reactance will be required to be placed in series with the capacitor to prevent a harmonic resonance condition.

References

1. Fitzgerald, A.E. and Kingsley, Charles, Jr., *Electric Machinery – The Dynamics and Statics of Electromechanical Energy Conversion*; Second Edition, 1961, page 26, Eq. 1-68. McGraw-Hill Book Company, Inc. New York.
2. Ibid., page 26, Eq. 1-69.
3. IEEE 141-1993 (R1999), *IEEE Recommended Practice for Electric Power Distribution for Industrial Plants*; 1999, page 503; Institute of Electrical and Electronic Engineers; New York.
4. Ibid., page 508.
5. Ibid.
6. Energy Policy Act of 2005, 10 CFR Ch. II (1-1-06) Edition; Subpart K-Distribution Transformers; 70 FR 60416, October 18, 2005; Part 431.192.
7. Pender, Harold and Del Mar, William A., Eds., *Electrical Engineers' Handbook – Electric Power*; 4th Edition, June, 1967, page 10-53. John Wiley & Sons, Inc.; New York.
8. Ibid., pages 10-44 and 10–53.
9. Ibid., page 10-53.
10. Ibid., pages 10-53 and 10-54.
11. Ibid., page 10-54, Equation (10).
12. Ibid., page 10-55.
13. Fitzgerald, A.E. and Kingsley, Charles, Jr., *Electric Machinery – The Dynamics and Statics of Electromechanical Energy Conversion*; 2nd Edition, 1961, page 375. McGraw–Hill Book Company, Inc.; New York.
14. Pender, Harold and Del Mar, William A., Eds., *Electrical Engineers' Handbook – Electric Power*; 4th Edition, June, 1967, page 10-56. John Wiley & Sons, Inc.; New York.
15. Ibid., page 10-57.
16. Earley, Mark W., Sargent, Jeffrey S., Sheehan, Joseph V., and Buss, E. William, *NEC®* *2008 Handbook: NFPA 70: National Electrical Code*; 2008, Article 470.1, page 632. National Fire Protection Association; Quincy, MA.
17. Beeman, Donald, Ed., *Industrial Power Systems Handbook*; 1955, page 382. McGraw-Hill Book Company, Inc., New York.
18. Beeman, Donald, Ed., *Industrial Power Systems Handbook*; 1955, page 382. McGraw-Hill Book Company, Inc., New York.
19. Ibid., page 382, Equation (6.6).
20. IEEE 141-1993(R1999), *IEEE Recommended Practice for Electric Power Distribution for Industrial Plants*; 1999, page 339. Institute of Electrical and Electronic Engineers; New York.
21. Ibid., page 400.
22. Ibid., Section 8.9.2, page 414.
23. Ibid., Section 8.9.2, page 415.

CHAPTER 12

Electrical Transmission and Distribution Systems

Electric utilities recognize the use of IEEE C2, *National Electrical Safety Code* (NESC) for design and operation of their low-, medium-, high-, and extra-high voltage transmission and distribution systems, and electrical substations. That standard is also applicable for distribution and substation systems in industrial facilities. It is sometimes mistakenly confused by individuals outside of the utility industry with NFPA 70®, *National Electrical Code®*; however, they are completely different documents.

The NESC defines its *Scope* in Article 011 as covering:

> supply and communication lines, equipment, and associated work practices employed by a public or private electric supply, communications, railway, or similar utility in the exercise of its function as a utility. They cover similar systems under the control of qualified persons, such as those associated with an industrial complex or utility interactive system.

> NESC rules do not cover installations in mines, ships, railway rolling equipment, aircraft, or automotive equipment, or utilization wiring except as covered in Parts 1 and 3. For building utilization wiring requirements, see the National Electrical Code, ANSI/NFPA 70 … [1]

The NESC is not a design specification. It does provide minimum clearance distances for energized conductors above pedestrian accessible areas; water ways and bodies of water; roadways; structures; billboards; railways; etc. It also provides minimum clearance and separation distances between conductors, both overhead and on poles and crossarms, as well as environmental loading criteria for those poles and conductors. The Code provides minimum clearance distances from live parts in substations.

The NESC provides some guidance for the minimum safety criteria for electrical transmission and distribution systems, for both overhead lines and underground cables. It notes in Section 012.A General Rules that:

> All electric supply and communication lines and equipment shall be designed, constructed, operated, and maintained to meet the requirements of these rules.

The US Department of Labor, Occupational Safety and Health Administration (OSHA) 29 CFR 1910, *Safety and Health Regulations for General Industry*, Section

1910.269 Appendix E recognizes the NESC as a national consensus standard that may be followed

> … in complying with the more performance-oriented requirements of OSHA's final rule.

Part 2 of the NESC establishes minimum loading criteria for wind and ice loading on overhead conductors and structures. Rule 250 in that document establishes overload factors for structures, crossarms, guys, foundations, and anchors. NESC Rule 260 establishes strength requirements and safety factors for metal, prestressed-concrete, reinforced-concrete, and wood structures. The Rule also establishes minimum strength requirements for insulators. Strength factors for crossarms and open supply and static conductors are also established.

Part 3 of the NESC deals with underground electric supply and communication lines. Cable routing guidelines are established, including separation recommendations from other buried lines, piping, sewer lines, foundations, etc. Minimum burial depths are also addressed. That part also addresses employee safety considerations; including minimum approach distances for live alternating current live line work; and other electrical safe work practices.

Power Distribution System Design Considerations

IEEE C2, *National Electrical Safety Code*® is primarily a support standard used to assist in the safety design elements of electrical transmission and distribution systems, with voltage levels up to 814 kV. The NESC provides ground, structure, and roadway clearance distances which are necessary to assure public safety in and around electrical transmission and distribution systems.

CAN/CSA C22.1 Canadian Electrical Code Part III – *Electricity Distribution and Transmission Systems* is the Canadian standard that applies to electric supply, communication lines and equipment, located entirely outside of supply/substations. Table 12.1 indicates the areas covered by this CSA Code.

The Institute of Electrical and Electronic Engineers *Color Book Series* provides some guidance and recommendations in the design of industrial electrical distribution systems. Table 12.2 lists most of those IEEE standards. The IEEE 602, *IEEE Recommended Practice for Electric Systems in Health Care Facilities (White Book)* was not included in Table 12.2.

Each IEEE Color Book discusses general power distribution design considerations in a variety of applications. For example, IEEE Standard 141, *Red Book,* reviews the following chapter topics for industrial plant distribution as noted in its Table of Contents [2]:

> *Overview* – provides general information on electrical engineering issues, responsibilities; sources of technical references; design, safety, and maintenance issues, and other power plant considerations.

TABLE 12.1 Canadian codes covering electricity distribution and transmission systems

Code No.	Title
CSA C22.3 No. 1	Overhead Systems
CSA C22.3 No. 3	Electrical Coordination
CSA C22.3 No. 4	Control of Electrochemical Corrosion of Underground Metallic Structures
CSA C22.3 No. 5.1	Recommended Practices for Electrical Protection – Electric Contact Between Overhead Supply and Communication Lines
CSA C22.3 No. 6	Principles and Practices of Electrical Coordination Between Pipelines and Electric Supply Lines
CSA C22.3 No. 7	Underground Systems
CSA C22.3 No. 8	Railway Electrification Guidelines
CSA C22.3 No. 9	Interconnection of Distributed Resources and Electricity Supply Systems
CSA C22.3 No. 60826	Design Criteria of Overhead Transmission Lines
CSA C22.3 No. 61936-1	Power Installations Exceeding 1 kV AC – Part 1: Common Rules (Adopted CEI/IEC 61936-1:2002, first edition, 2002-10, with Canadian deviations)

System Planning – deals with safety and reliability design, as well as distribution design topics.

Voltage Considerations – covers voltage classes, control, selection, voltage drop, effects of voltage variation on different loads, phase-voltage imbalance, voltage dips and flicker, harmonics, and voltage drop calculations.

Fault Calculations – considers fault current sources and calculation procedures.

Application and Coordination of System Protective Devices – discusses protection devices, techniques, principles, requirements, and testing.

Surge Voltage Protection – explores surge voltage characteristics, insulation withstand properties, and surge arresters.

Grounding – describes grounding techniques and methods.

Power Factor and Related Considerations – analyzes fundamentals, system power loses, capacitor use, transients, resonances and harmonics.

Harmonics in Power Systems – discusses voltage harmonics topics.

Power Switching, Transformation, and Motor-Control Apparatus – reviews switchgear, transformers, substations, and motor-control equipment.

Instruments and Meters – presents a discussion on a variety of instruments, meters, and auxiliary devices.

TABLE 12.2 IEEE Color Books for industrial/commercial power distribution systems

Book Color	Standard No.	Title
IEEE Red Book™	IEEE 141	IEEE Recommended Practice for Electric Power Distribution for Industrial Power Plants
IEEE Green Book™:	IEEE 142	IEEE Recommended Practice for Grounding of Industrial and Commercial Power Systems
IEEE Gray Book™	IEEE 241	IEEE Recommended Practice for Electric Power Systems in Commercial Buildings
IEEE Buff Book™	IEEE 242	IEEE Recommended Practice for Protection and Coordination of Industrial and Commercial Power Systems
IEEE Brown Book™	IEEE 399	IEEE Recommended Practice for Industrial and Commercial Power Systems Analysis
IEEE Orange Book™	IEEE 446	IEEE Recommended Practice for Emergency and Standby Power Systems for Industrial and Commercial Applications
IEEE Gold Book™	IEEE 493	IEEE Recommended Practice for the Design of Reliable Industrial and Commercial Power Systems
IEEE Violet Book™	IEEE 551	Recommended Practice for Calculating Short-Circuit Currents in Industrial and Commercial Power Systems
IEEE Bronze Book™	IEEE 739	IEEE Recommended Practice for Energy Conservation and Cost Effective Planning in Industrial Facilities
IEEE Yellow Book™	IEEE 902	IEEE Guide for Maintenance, Operation and Safety of Industrial and Commercial Power Systems
IEEE Blue Book™	IEEE 1015	IEEE Recommended Practice for Applying Low-Voltage Circuit Breakers Used in Industrial and Commercial Power Systems
IEEE Emerald Book™	IEEE 1100	IEEE Recommended Practice for Powering and Grounding Electronic Equipment

Cable Systems – reviews cable characteristics, ratings, installation, connections and terminations, splices, testing, and fault locating.

Busways – examines busway uses, types, selection, installation, and testing.

Electrical Conservation through Energy Management – Review energy saving methods.

Industrial Substations: Plant–Utility Interface Considerations – discusses planning, design, construction, and operation issues with electrical industrial substations.

Cost Estimating of Industrial Power Systems – describes industrial power system considerations and cost estimates.

Power Generation Considerations

Selection of economically feasible, reliable, and readily available energy sources is essential for the commercial generation of electrical energy. An energy source must be provided to produce the mechanical, chemical, or electrical energy necessary to reliably drive generators or directly produce electricity. The selection of a variety of energy sources can help guarantee generation system reliability and the production of electricity. Some available electrical generator energy driver sources include:

Hydroelectric energy conversion utilizing the potential and kinetic energy of water stored in reservoirs or water falls.

Hydrokinetic electric energy conversion, including wave action, tidal flow, river flow, etc.

Energy conversion from the burning of hydrocarbon fuels, such natural gas, diesel, fuel oil, etc. for powering steam generation equipment.

Energy conversion from geothermal sources.

Energy conversion from the burning of carbon based fuels, such as coal, for powering steam generation equipment.

Energy conversion from the burning of waste products powering steam generation equipment.

Energy conversion from the recovery of process waste heat energy powering steam generation equipment.

The recovery of wind driven kinetic energy as a mechanical energy source.

The use of nuclear fuel sources creating steam for power generation equipment.

The conversion of solar energy sources to electrical energy.

Electrochemical energy conversion to electrical energy using fuel cells.

The above-noted energy sources are presently available, with some more cost efficient and capable of commercially producing larger amounts of electricity than others. Other considerations in selecting and developing energy conversion sources include governmental regulatory and environmental considerations, costs of transportation and storage, consistency of availability, costs of processing and converting the fuel, byproducts produced, etc.

The US Department of Energy, Energy Information Administration maintains data on the consumption of fuels used in utility electricity generation. The information developed for the first 11 months of 2008 is presented in Table 12.3. Coal, natural gas, and nuclear fuels account for almost 87% of the electricity generated during that time period.

Utility generated electrical power can be supplemented on their distribution system by the use of *Distributed Energy Resources (DER)* as small as 3 kW in size. Those sources could be privately or utility generated electrical power, used as an alternative to or an enhancement of the traditional utility electric power system. Where allowed by state/municipal utility commissions,

TABLE 12.3 Total United States electric power industry
generation fuel source summary, January – November, 2008

Fuel description	Megawatt-hour %
Coal	47.02
Petroleum liquids	0.72
Petroleum coke	0.31
Natural gas	20.82
Other gases	0.33
Nuclear	18.83
Conventional hydroelectric	6.25
Other renewables	2.70
Wood and wood-derived fuels	0.89
Other biomass	0.37
Geothermal	0.32
Solar thermal and photovoltaic	0.02
Wind	1.05
Hydroelectric pumped storage	0.13
Other energy sources	0.24

Source: US Department of Energy – Energy Information Administration

excess generated Distributed Energy Resources can be sold to an electrical utility. The United State Department of Energy (DOE) defines Distributed Energy Resources as:

> small, modular, energy generation and storage technologies that provide electric capacity or energy where you need it. Typically producing less than 10 megawatts (MW) of power, DER systems can usually be sized to meet your particular needs and installed on site. DER systems may be either connected to the local electric power grid or isolated from the grid in stand-alone applications.

> DER technologies include wind turbines, photovoltaics (PV), fuel cells, microturbines, reciprocating engines, combustion turbines, cogeneration, and energy storage systems. [3]

> More than 12 million DG units are installed across the United States today, with a total capacity over 200 GW. In 2003, these units generated approximately 250,000 GWh. Over 99% of these units are small emergency reciprocating engine generators or photovoltaic systems, installed with inverters that do not feed electricity directly into the distribution grid. However ... this large number of smaller machines represents a relatively small fraction of the total installed capacity (Energy Information Administration 2005). [4]

Table 12.4 lists a number of codes, standards, and recommended practices associated with Distributed Resources. States and state utility regulatory bodies have developed their own rules and regulations regarding Distributed Resources sale of electricity back to utility

TABLE 12.4 **Distributed resource standards**

Developer	Standard No.	Title
IEEE	IEEE 446	IEEE Recommended Practice for Emergency and Standby Power Systems for Industrial and Commercial Applications
IEEE	ANSI/IEEE Std 928	IEEE Recommended Criteria for Terrestrial Photovoltaic Power Systems – Description
IEEE	ANSI/IEEE Std 929	IEEE Recommended Practice for Utility Interface of Residential and Intermediate Photovoltaic (PV) Systems – Description
IEEE	IEEE 937	IEEE Recommended Practice for Installation and Maintenance of Lead-Acid Batteries for Photovoltaic (PV) Systems
IEEE	IEEE Std 946	IEEE Recommended Practice for the Design of DC Auxiliary Power Systems for Generating Stations – Description
IEEE	IEEE 1013	IEEE Recommended Practice for Sizing Lead-Acid Batteries for Stand-Alone Photovoltaic (PV) Systems
IEEE	IEEE Std 1020	IEEE Guide for Control of Small Hydroelectric Power Plants – Description
IEEE	IEEE 1248	IEEE Guide for the Commissioning of Electrical Systems in Hydroelectric Power
IEEE	IEEE Std 1361™	IEEE Guide for Selection, Charging, Test, and Evaluation of Lead-Acid Batteries Used in Stand-Alone Photovoltaic (PV) Systems – Description
IEEE	IEEE Std 1526™	IEEE Recommended Practice for Testing the Performance of Stand-Alone Photovoltaic Systems – Description
IEEE	ANSI/IEEE 1547™	Standard for Interconnecting Distributed Resources with Electric Power Systems
IEEE	IEEE 1547.1™	Standard for Conformance Test Procedures for Equipment Interconnecting Distributed Resources with Electric Power Systems
IEEE	IEEE 1547.2™	Application Guide for IEEE 1547 Standard for Interconnecting Distributed Resources with Electric Power Systems
IEEE	IEEE 1547.3™	Guide for Monitoring, Information Exchange, and Control of Distributed Resources Interconnected with Electric Power Systems
IEEE	IEEE 1547.4™	Draft Guide for Design, Operation, and Integration of Distributed Resource Island Systems with Electric Power Systems
IEEE	IEEE 1547.5™	Draft Technical Guidelines for Interconnection of Electric Power Sources Greater Than 10 MVA to the Power Transmission Grid
IEEE	IEEE 1547.6™	Draft Recommended Practice for Interconnecting Distributed Resources with Electric Power Systems Distribution Secondary Networks
IEEE	IEEE 1547.7™	Draft Guide to Conducting Distribution Impact Studies for Distributed Resource Interconnection
IEEE	IEEE 1561	IEEE Guide for Optimizing the Performance and Life of Lead-Acid Batteries in Remote Hybrid Power Systems
IEEE	IEEE 1562™	Guide for Array and Battery Sizing in Stand-Alone Photovoltaic (PV) Systems

(*Continued*)

TABLE 12.4 **Distributed resource standards—cont'd**

Developer	Standard No.	Title
IEEE	IEEE 1661™	IEEE Guide for Test and Evaluation of Lead-Acid Batteries Used in Photovoltaic (PV) Hybrid Power Systems
IEEE	IEEE 1589	Standard for Conformance Tests Procedures for Equipment Interconnecting Distributed Resources with Electric Power Systems (Draft Standard)
NFPA	NFPA 37	Standard for the Installation and Use of Stationary Combustion Engines and Gas Turbines
NFPA	NFPA 70® Article 705	Interconnected Electric Power Production Sources
NFPA	NFPA 111	Standard on Stored Electrical Energy Emergency and Standby Power Systems
UL	UL 1741	Standard for Inverters, Converters, Controllers and Interconnection System Equipment for Use with Distributed Energy Resources
UL	UL 2200	Standard for Safety for Stationary Engine Generator Assemblies

companies. Local utility companies also have requirements for connection of Distributed Resource systems into their distribution network. IEEE 1547™ and IEEE 1547.1 through 1547.7 form the standards basis for design and connection of Distributed Resource systems.

Table 12.5 presents many of the codes, standards, and recommended practices associated with a <u>wind turbine</u> Distributed Resource. Some private developers and utility companies have designed and installed large commercially productive wind turbine farms. Smaller, residential-sized wind turbines are commercially available. The American Wind Energy Association (AWEA) is an organization that was formed to promote the growth of wind turbine electrical energy generation. AWEA reports that wind turbine generation is one of the fastest growing forms of electrical power generation in the United States. The leading states in wind turbine generation capacity installed in the United States as of December 31, 2007 include [5]:

1. Texas 4,356 MW

2. California 2,439 MW

3. Minnesota 1,299 MW

4. Iowa 1,273 MW

5. Washington (state) 1,163 MW

Fuel cells are electrochemical devices that convert energy stored in a fuel source directly into electrical energy. Direct Current (DC) electrical energy is produced from those devices. Table 12.6 list some of the codes, standards, and recommended practices associated with fuel cells.

TABLE 12.5 Wind turbine system standards

Developer	Standard No.	Title
AGMA	ANSI/AGMA/ AWEA 6006-A03	Design and Specification of Gearboxes for Wind Turbines
ASME	ANSI/ASME PTC 42	Wind Turbines
CSA	CAN/CSA C61400-1	Wind Turbines – Part 1: Design Requirements (Adopted IEC 61400-1 with Canadian deviations)
CSA	CAN/CSA C61400-11	Wind Turbine Generator Systems – Part 11: Acoustic Noise Measurement Techniques (Adopted IEC 61400-11+A1)
CSA	CAN/CSA C61400-12-1	Wind Turbines – Part 12-1: Power Performance Measurements of Electricity Producing Wind Turbines (Adopted IEC 61400-12-1)
CSA	CAN/CSA C61400-24	Wind Turbine Generator Systems – Part 24: Lightning Protection (Adopted IEC/TR 61400-24 W/Canadian Deviations)
IEC	IEC 60050-415	International Electrotechnical Vocabulary – Part 415: Wind Turbine generator systems
IEC	IEC 61400-1	Wind Turbines – Part 1: Design Requirements
IEC	IEC 61400-2	Wind Turbines – Part 2: Design Requirements for Small Wind Turbines
IEC	IEC 61400-3	Wind Turbines – Part 3: Design Requirements for Offshore Wind Turbines
IEC	IEC 61400-11	Wind Turbine Generator Systems – Part 11: Acoustic Noise Measurement Techniques
IEC	IEC 61400-12-1	Wind Turbines – Part 12-1: Power Performance Measurements of Electricity Producing Wind Turbines
IEC	IEC/TS 61400-13	Wind Turbine Generator systems – Part 13: Measurement of Mechanical Loads
IEC	IEC 61400-14	Wind turbines – Part 14: Declaration of Apparent Sound Power Level and Tonality Values
IEC	IEC 61400-21	Wind Turbines – Part 21: Measurement and Assessment of Power Quality Characteristics of Grid Connected Wind Turbines
IEC	IEC 61400-23	Wind Turbine Generator Systems – Part 23: Full-scale Structural Testing of Rotor Blades
IEC	IEC 61400-24	Wind Turbine Generator Systems – Part 24: Lightning Protection
IEC	IEC 61400-25-1	Wind Turbines – Part 25-1: Communications for Monitoring and Control of Wind Power Plants – Overall Description of Principles and Models
IEC	IEC 61400-25-2	Wind Turbines – Part 25-2: Communications for Monitoring and Control of Wind Power Plants – Information Models
IEC	IEC 61400-25-3	Wind Turbines – Part 25-3: Communications for Monitoring and Control of Wind Power Plants – Information Exchange Models
IEC	IEC 61400-25-4	Wind Turbines – Part 25-4: Communications for Monitoring and Control of Wind Power Plants – Mapping to Communication Profile

(*Continued*)

TABLE 12.5 Wind turbine system standards—cont'd

Developer	Standard No.	Title
IEC	IEC 61400-25-5	Wind turbines – Part 25-5: Communications for Monitoring and Control of Wind Power Plants – Conformance Testing
IEC	IEC 61400-SER-02	Wind Turbine Generator Systems – All Parts
ISO	ISO 8068	Lubricants, Industrial Oils and Related Products (class L) – Family T (Turbines) – Specification for Lubricating Oils for Turbines
ISO	ISO 81400-4	Wind Turbines – Part 4: Design and Specification of Gearboxes
NFPA	NFPA 780	Standard for the Installation of Lightning Protection Systems

TABLE 12.6 Fuel Cell Standards

Developer	Standard No.	Title
ASME	ANSI/ASME PTC 50	Fuel Cell Power Systems Performance
CSA America, Inc.	ANSI/CSA America FC 1	Stationary Fuel Cell Power Systems
CSA	CAN/CSA-C22.2 NO. 62282-2	Fuel Cell Technologies – Part 2: Fuel Cell Modules (Adopted CEI/IEC 62282-2:2004, with Canadian deviations)
IEC	IEC/TS 62282-1	Fuel Cell Technologies – Part 1: Terminology
IEC	IEC 62282-2	Fuel Cell Technologies – Part 2: Fuel Cell Modules
IEC	IEC 62282-3-1	Fuel Cell Technologies – Part 3-1: Stationary Fuel Cell Power Systems – Safety
IEC	IEC 62282-3-2	Fuel Cell Technologies – Part 3-2: Stationary Fuel Cell Power Systems – Performance Test Methods
IEC	IEC 62282-3-3	Fuel Cell Technologies – Part 3-3: Stationary Fuel Cell Power Systems – Installation
NFPA	NFPA 853	Standard for the Installation of Stationary Fuel Cell Power Plants
UL	UL 2075	Standard for Gas and Vapor Detectors and Sensors
UL	UL 2262	PEM Type Fuel Cell Power Plants/Modules

A fuel cell consists of a *fuel processor* section that converts the fuel into a useable form. A hydrogen fuel source would not require a fuel processor; however, it may require filtration to remove impurities from the hydrogen.

Hydrogen-rich hydrocarbon fuels will require a *reformer* to convert those fuels into gaseous hydrogen and carbon compounds. Those compounds are then treated in a *reactor* to remove impurities. High temperature process fuel cells, such as molten carbonate and solid oxide fuel cells have their fuels *internally reformed* inside the fuel cell stacks; however, those fuels may also require filtration. Waste heat recovery can be used in fuel cells to produce steam and generate additional electrical energy with the use of steam turbines.

Fuel cell stacks are the devices that actually chemically convert the fuel into direct current (DC) electricity. Substantial numbers of fuel cell stacks, much like batteries in a flashlight, must be electrically connected to generate a sufficient amount of electrical energy to operate electrical equipment. Also, since the energy produced by those devices is DC, electrical "inverters" and "power conditioning" equipment must also be employed to produce alternating current (AC) electricity. Voltage amplitude, frequency, and harmonic content must be controlled during that power conditioning and DC to AC conversion process.

The DOE [6] lists the following available categories of fuel cells:

Polymer Electrolyte Membrane (PEM) Fuel Cell

Direct Methanol Fuel Cell

Alkaline Fuel Cell

Phosphoric Acid Fuel Cell

Molten Carbonate Fuel Cell

Solid Oxide Fuel Cell

Generative Fuel Cell

The US Department of Energy, *Hydrogen, Fuel Cell & Infrastructure Technologies Program* has developed a characteristics and usage comparison chart for five different fuel cell sources. That chart is presented in Table 12.7. A review of the chart provides some indication of the fuel cell technologies which are most useful as utility Distributed Energy Resources. The chart was based on the most commonly available fuel cell technologies in 2007.

Electrical Transmission Systems

Electrical transmission systems involve the movement of electrical energy from its generation points to distribution substations for local distribution. To accomplish that task economically, the generated electricity voltage is stepped up at generation transmission substations, to transmission levels. A major concern with the transmission of electrical energy is to minimize energy losses due to the impedance associated with the electrical power lines. Transmission voltage levels can typically be above 100 kV, and commonly within extra-high voltage (EHV) levels to voltage levels less than 1000 kV. Transmission systems in the western United States and Canada with extra-long routes are also utilizing transmission voltage levels in the ultra-high voltage (UHV) range above 1000 kV [8].

The United States is divided into three separate power grids, as can be seen in Figure 12.1. A power grid is a mechanism for the bulk transmission of electrical energy. It facilitates the movement of electrical energy over large distances, through a variety of routes. The power

TABLE 12.7 Comparison of Fuel Cell Technologies Department of Energy – Hydrogen Program [7]

Fuel cell type	Common electrolyte	Operating temperature	System output	Electrical efficiency	Combined heat and power (CHP) efficiency	Applications	Advantages
Polymer electrolyte membrane (PEM)*	Solid organic polymer poly-perfluorosulfonic acid	50–1000°C 122–212°F	<1kW–250 kW	53–58% (transportation) 25–35% (stationary)	70–90% (low-grade waste heat)	• Backup power • Portable power • Small distributed generation • Transportation • Specialty vehicles	• Solid electrolyte reduces corrosion and electrolyte management problems • Low temperature • Quick start-up
Alkaline (AFC)	Aqueous solution of potassium hydroxide soaked in a matrix	90–1000°C 194–2120°F	10 kW–100 kW	60%	>80% (low-grade waste heat)	• Military • Space	• Cathodic reaction faster in alkaline electrolyte, leads to higher performance • Can use a variety of catalysts
Phosphoric acid (PAFC)	Liquid phosphoric acid soaked in a matrix	150–2000°C 302–3920°F	50 kW–1 MW (250 kW module typical)	>40%	>85%	• Distributed generation	• Higher overall efficiency with CHP • Increased tolerance to impurities in hydrogen

	Electrolyte	Power	Temperature	Efficiency	Applications	Advantages
Molten carbonate (MCFC)	Liquid solution of lithium, sodium, and/or potassium carbonates, soaked in a matrix	<1 kW–1 MW (205 kW module typical)	600–7000°C 1112–12920°F	45–47%	• Electric utility • Large distributed generation	• High efficiency • Fuel flexibility • Can use a variety of catalysts • Suitability for CHP
Solid oxide (SOFC)	Yttria stabilized zirconia	<1 kW–3 MW	600–10000°C 1202–18320°F	35–43%	• Auxiliary power • Electric utility • Large distributed generation	• High efficiency • Fuel flexibility • Can use a variety of catalysts • Solid electrolyte reduces electrolyte management problems • Suitability for CHP • Hybrid/GT cycle

*Direct methanol fuel cells (DMFC) are a subset of PEM typically used for small portable power applications with a size range of about a subwatt to 100 W and operating at 60–90°C.

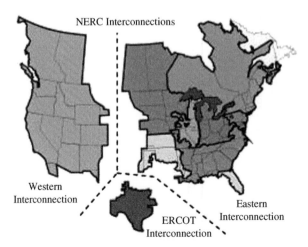

Figure 12.1: North America Electric Reliability Corporation electrical transmission grids (courtesy of US Department of Energy, Washington, DC) [8]

grid allows the transfer of electrical energy from areas with spare generation capacity to areas whose peak energy demand exceeds its own generation capacity. The three power grids illustrated in Figure 12.1 are interconnected, but are not simultaneously linked together.

> The Federal Energy Regulatory Commission (FERC) for jurisdictional utilities regulates interstate wholesale trade and the associated transmission interconnections. Federally owned utilities, and State and municipally owned utilities are not regulated by FERC, but must follow Federal regulations if they wish to buy and sell electricity in the wholesale market or use the transmission facilities of jurisdictional utilities. [9]

FERC is an independent regulatory agency within the United States Department of Energy. The responsibilities of the Federal Energy Regulatory Commission include [10]:

- Regulates the transmission and sale of natural gas for resale in interstate commerce;

- Regulates the transmission of oil by pipeline in interstate commerce;

- Regulates the transmission and wholesale sales of electricity in interstate commerce;

- Licenses and inspects private, municipal, and state hydroelectric projects;

- Approves the siting and abandonment of interstate natural gas pipelines and storage facilities, and ensures the safe operation and reliability of proposed and operating LNG terminals;

- Ensures the reliability of high-voltage interstate transmission system;

- Monitors and investigates energy markets;

- Uses civil penalties and other means against energy organizations and individuals who violate FERC rules in the energy markets;

- Oversees environmental matters related to natural gas and hydroelectricity projects and major electricity policy initiatives; and

- Administers accounting and financial reporting regulations and conduct of regulated companies.

Regulatory areas that are not under the purview of FERC include [11]:

- Regulation of retail electricity and natural gas sales to consumers;

- Approval for the physical construction of electric generation, transmission, or distribution facilities; except for hydropower and certain electric transmission facilities located in National interest electric transmission corridors;

- Regulation of activities of the municipal power systems, federal power marketing agencies like the *Tennessee Valley Authority*, and most rural electric cooperatives;

- Regulation of nuclear power plants by the *Nuclear Regulatory Commission*;

- Issuance of State Water Quality Certificates;

- Oversight for the construction of oil pipelines;

- Abandonment of service as related to oil facilities;

- Mergers and acquisitions as related to oil companies;

- Responsibility for *pipeline safety* or for pipeline transportation on or across the Outer Continental Shelf;

- Regulation of local distribution pipelines of natural gas; and the

- Development and operation of natural gas vehicles.

The North America Electric Reliability Corporation (NERC) was established in 1981 by the electric utilities as an informal, voluntary organization consisting of 12 regional councils, to formulate the coordination of the transmission of bulk electrical energy on the transmission systems in the United States and Canada. That organization was originally founded in 1968 as the National Electric Reliability Council (NERC). The name change was implemented to recognize Canadian membership; however, its original name acronym was maintained.

The United Sates Energy Policy Act of 2005 was legislated by the United States Congress. Part of that Act established a self-regulatory Electrical Reliability Organization (ERO) to assure electrical transmission system reliability. The Federal Energy Regulatory Commission

(FERC) was the Federal agency designated with regulatory jurisdiction over the owners, operators, and users of the electric bulk power system. FERC was granted the enforcement authority to assure compliance with the ERO created bulk power system reliability standards.

FERC adopted the North American Electric Reliability Corporation (NERC) as their designated ERO in 2006. FERC noted in a news release dated April 17, 2008 [12] that

> FERC designated NERC as the ERO under section 215 of the Federal Power Act a new provision added by the Energy Policy Act of 2005 to establish a system of mandatory, enforceable reliability standards under the Commission's oversight. NERC has submitted 107 proposed Reliability Standards to FERC. In Order No. 693, issued March 16, 2007, FERC approved 83 of these standards. FERC also directed NERC to develop modifications to 56 of the 83 approved standards.

FERC also approved the use of some eight reliability standards from the Western Electricity Coordinating Council (WECC) and eight Critical Infrastructure Protection (CIP) cyber-security related standards. The Department of Homeland Security (DHS) was the Federal agency assigned the responsibility for creation of the National Infrastructure Protection Plan (NIPP). DHS designated the US Department of Energy (DOE) as its Sector-Specific Agency (SSA) for the energy sector. DOE was assigned the task of coordinating the establishment of an Energy Sector-Specific Plan (SSP) to be included in DHS's overall National Infrastructure Protection Plan. Both FERC and NERC groups assisted the DOE in the development of its CIP planning. The FERC reliability standards for the generation and transmission of electrical energy are included in Table 12.8.

The Electric Power Research Institute (EPRI) is an independent, nonprofit collaborative R&D energy and environmental research institute founded in 1973. The Institute was formed by companies from all sectors of the United States electric utility industry. Those collaborative companies are responsible for the generation and delivery of electrical energy to more than 90% of the United States. EPRI does not have regulatory authority to establish industry wide codes, standards, and recommended practices; however, all organizing members have adopted use of its publications, some of which are listed in Table 12.9.

There were approximately 161,100 miles of electrical power transmission lines in the United States in 2002. Figure 12.2 illustrates the lengths of those lines by transmission voltage level, based on data developed by the US Department of Energy. Of the total transmission line mileage, approximately 3300 miles utilize direct current (DC) transmission voltage and 157,800 miles utilize alternating current (AC) transmission voltage. DC transmission lines offer more system stability than their AC line counterpart; however, their costs are only competitive over long transmission line distances. The majority of the DC transmission lines are in the western and central United States. DC transmission lines voltages of 1000 kVDC are operational in Canada and China.

TABLE 12.8 FERC reliability standards for electrical power transmission and generation

Developer	Standard No.	Title
NERC	BAL-001-0	Real Power Balancing Control Performance
NERC	BAL-002-0	Disturbance Control Performance
NERC	BAL-003-0	Frequency Response and Bias
NERC	BAL-004-0	Time Error Correction
NERC	BAL-005-0	Automatic Generation Control
NERC	BAL-006-1	Inadvertent Interchange
NERC	CIP-001-1	Sabotage Reporting
NERC	COM-001-1	Telecommunications
NERC	COM-002-2	Communications and Coordination
NERC	EOP-001-0	Emergency Operations Planning
NERC	EOP-002-2	Capacity and Energy Emergencies
NERC	EOP-003-1	Load Shedding Plans
NERC	EOP-004-1	Disturbance Reporting
NERC	EOP-005-1	System Restoration Plans
NERC	EOP-006-1	Reliability Coordination – System Restoration
NERC	EOP-007-0*	Establish, Maintain, and Document a Regional Blackstart Capability Plan
NERC	EOP-008-0	Plans for Loss of Control Center Functionality
NERC	EOP-009-0	Documentation of Blackstart Generating Unit Test Results
NERC	FAC-001-0	Facility Connection Requirements
NERC	FAC-010-2	System Operating Limits Methodology for the Planning Horizon
NERC	FAC-002-0	Coordination of Plans for New Facilities
NERC	FAC-003-1	Transmission Vegetation Management Program
NERC	FAC-008-1	Facility Ratings Methodology
NERC	FAC-009-1	Establish and Communicate Facility Ratings
NERC	FAC-010-1	System Operating Limits Methodology for the Planning Horizon
NERC	FAC-011-1	System Operating Limits Methodology for the Operations Horizon
NERC	FAC-011-2	System Operating Limits Methodology for the Operations Horizon
NERC	FAC-012-1	Transfer Capabilities Methodology
NERC	FAC-013-1	Establish and Communicate Transfer Capabilities
NERC	FAC-014-1	Establish and Communicate System Operating Limits
NERC	FAC-014-2	Establish and Communicate System Operating Limits
NERC	INT-001-3	Interchange Transaction Tagging
NERC	INT-003-2	Interchange Transaction Implementation
NERC	INT-004-2	Interchange Transaction Modifications
NERC	INT-005-2	Interchange Authority Distributes Arranged Interchange
NERC	INT-006-2	Response to Interchange Authority

(*Continued*)

TABLE 12.8 FERC reliability standards for electrical power transmission and generation—cont'd

Developer	Standard No.	Title
NERC	INT-007-1	Interchange Confirmation
NERC	INT-008-2	Interchange Authority Distributes Status
NERC	INT-009-1	Implementation of Interchange
NERC	INT-010-1	Interchange Coordination Exceptions
NERC	IRO-001-1	Reliability Coordination – Responsibilities and Authorities
NERC	IRO-002-1	Reliability Coordination – Facilities
NERC	IRO-003-2	Reliability Coordination – Wide Area View
NERC	IRO-004-1	Reliability Coordination – Operations Planning
NERC	IRO-005-1	Reliability Coordination – Current Day Operations
NERC	IRO-006-3	Reliability Coordination – Transmission Loading Relief
NERC	IRO-014-1	Procedures, Processes, or Plans to Support Coordination Between Reliability Coordinators
NERC	IRO-015-1	Notifications and Information Exchange Between Reliability Coordinators
NERC	IRO-016-1	Coordination of Real-time Activities Between Reliability Coordinators
NERC	MOD-001-0[*]	Documentation of TTC and ATC Calculation Methodologies
NERC	MOD-002-0[*]	Review of TTC and ATC Calculations and Results
NERC	MOD-003-0[*]	Procedure for Input on TTC and ATC Methodologies and Values
NERC	MOD-004-0[*]	Documentation of Regional CBM Methodologies
NERC	MOD-005-0[*]	Procedure for Verifying CBM Values
NERC	MOD-006-0	Procedures for Use of CBM Values
NERC	MOD-007-0	Documentation of the Use of CBM
NERC	MOD-008-0[*]	Documentation and Content of Each Regional TRM Methodology
NERC	MOD-009-0[*]	Procedure for Verifying TRM Values
NERC	MOD-010-0	Steady-State Data for Transmission System Modeling and Simulation
NERC	MOD-011-0[*]	Regional Steady-State Data Requirements and Reporting Procedures
NERC	MOD-012-0	Dynamics Data for Transmission System Modeling and Simulation
NERC	MOD-013-1[*]	RRO Dynamics Data Requirements and Reporting Procedures
NERC	MOD-014-0[*]	Development of Interconnection-Specific Steady State System Models
NERC	MOD-015-0[*]	Development of Interconnection-Specific Dynamics System Models
NERC	MOD-016-1	Actual and Forecast Demands, Net Energy for Load, Controllable DSM
NERC	MOD-017-0	Aggregated Actual and Forecast Demands and Net Energy for Load
NERC	MOD-018-0	Reports of Actual and Forecast Demand Data
NERC	MOD-019-0	Forecasts of Interruptible Demands and DCLM Data
NERC	MOD-020-0	Providing Interruptible Demands and DCLM Data
NERC	MOD-021-0	Accounting Methodology for Effects of Controllable DSM in Forecasts
NERC	MOD-024-1[*]	Verification of Generator Gross and Net Real Power Capability
NERC	MOD-025-1[*]	Verification of Generator Gross and Net Reactive Power Capability
NERC	NUC-001-1	Nuclear Plant Interface Coordination

TABLE 12.8 FERC reliability standards for electrical power transmission and generation—cont'd

Developer	Standard No.	Title
NERC	PER-001-0	Operating Personnel Responsibility and Authority
NERC	PER-002-0	Operating Personnel Training
NERC	PER-003-0	Operating Personnel Credentials
NERC	PER-004-1	Reliability Coordination – Staffing
NERC	PRC-001-1	System Protection Coordination
NERC	PRC-002-1	Define and Document Disturbance Monitoring Equipment Requirements
NERC	PRC-003-1*	Regional Requirements for Analysis of Misoperations of Transmission and Generation Protection Systems
NERC	PRC-004-1	Analysis and Mitigation of Transmission and Generation Protection System Misoperations
NERC	PRC-005-1	Transmission and Generation Protection System Maintenance and Testing
NERC	PRC-006-0*	Development and Documentation of Regional UFLS Programs
NERC	PRC-007-0	Assuring Consistency with Regional UFLS Program
NERC	PRC-008-0	Underfrequency Load Shedding Equipment Maintenance Programs
NERC	PRC-009-0	UFLS Performance Following an Underfrequency Event
NERC	PRC-010-0	Assessment of the Design and Effectiveness of UVLS Program
NERC	PRC-011-0	UVLS System Maintenance and Testing
NERC	PRC-012-0*	Special Protection System Review Procedure
NERC	PRC-013-0*	Special Protection System Database
NERC	PRC-014-0*	Special Protection System Assessment
NERC	PRC-015-0	Special Protection System Data and Documentation
NERC	PRC-016-0	Special Protection System Misoperations
NERC	PRC-017-0	Special Protection System Maintenance and Testing
NERC	PRC-018-1	Disturbance Monitoring Equipment Installation and Data Reporting
NERC	PRC-020-1*	Under-Voltage Load Shedding Program Database
NERC	PRC-021-1	Under-Voltage Load Shedding Program Data
NERC	PRC-022-1	Under-Voltage Load Shedding Program Performance
NERC	TOP-001-1	Reliability Responsibilities and Authorities
NERC	TOP-002-2	Normal Operations Planning
NERC	TOP-003-0	Planned Outage Coordination
NERC	TOP-004-1	Transmission Operations
NERC	TOP-005-1	Operational Reliability Information
NERC	TOP-006-1	Monitoring System Conditions
NERC	TOP-007-0	Reporting SOL and IROL Violations
NERC	TOP-008-1	Response to Transmission Limit Violations
NERC	TPL-001-0	System Performance Under Normal Conditions
NERC	TPL-002-0	System Performance Following Loss of a Single BES Element

(*Continued*)

TABLE 12.8 FERC reliability standards for electrical power transmission and generation—cont'd

Developer	Standard No.	Title
NERC	TPL-003-0	System Performance Following Loss of Two or More BES Elements
NERC	TPL-004-0	System Performance Following Extreme BES Events
NERC	TPL-005-0*	Regional and Interregional Self-Assessment Reliability Reports
NERC	TPL-006-0*	Assessment Data from Regional Reliability Organizations
NERC	VAR-001-1	Voltage and Reactive Control
NERC	VAR-002-1	Generator Operations for Maintaining Network Voltage Schedules
WECC	STD-002-0	Operating Reserves
WECC	IRO-STD-006-0	Qualified Path Unscheduled Flow Relief
WECC	PRC-STD-001-1	Certification of Protective Relay Applications and Settings
WECC	PRC-STD-003-1	Protective Relay and Remedial Action Scheme Misoperation
WECC	PRC-STD-005-1	Transmission Maintenance
WECC	TOP-STD-007-0	Operating Transfer Capability
WECC	VAR-STD-002a-1	Automatic Voltage Regulators
WECC	VAR-STD-002b-1	Power System Stabilizers
CIP	CIP-002-1	Critical Cyber Asset Identification
CIP	CIP-003-1	Security Management Controls
CIP	CIP-004-1	Personnel and Training
CIP	CIP-005-1	Electronic Security Perimeter(s)
CIP	CIP-006-1	Physical Security of Critical Cyber Assets
CIP	CIP-007-1	Systems Security Management
CIP	CIP-008-1	Incident Reporting and Response Planning
CIP	CIP-009-1	Recovery Plans for Critical Cyber Assets

*Standard approval pending FERC adoption.

Some typical transmission line voltage categories in the United States include [14]:

High voltage (HV) AC:	69 kV, 115 kV, 138 kV, 161 kV, 230 kV
Extra-high voltage (EHV) AC:	345 kV, 500 kV, 765 kV
Ultra-high voltage (UHV) AC:	1100 kV, 1500 kV, 2225 kV kV
Direct-current high voltage (DC HV):	$\pm$125 kV, $\pm$200 kV, $\pm$250 kV, $\pm$400 kV, $\pm$500 kV

DC transmission lines require that the generated AC voltage be converted or rectified from three-phase, 50/60 Hz to DC. Once the electrical energy reaches its distribution point, it must be re-converted or inverted back to three-phase, 50/60 Hz AC. Those conversion processes are the principal reason for the increased costs for DC transmission lines installation. An AC transmission line will require three line conductors and larger support structures than the two lines required for DC transmission. There is less line voltage drop

TABLE 12.9   Electric Power Research Institute Power Transmission Books

Developer	Standard No.	Title
EPRI	Red Book	EPRI AC Transmission Line Reference Book: 200 kV and Above
EPRI	Green Book	EPRI Underground Transmission Systems Reference Book
EPRI	Orange Book	EPRI Transmission Line Reference Book: Wind-Induced Conductor Motion
EPRI	Yellow Book	EPRI Overhead Transmission Inspection and Assessment Guidelines
EPRI	Blue Book	EPRI Transmission Line Reference Book: 115-345 kV Compact Line Design
EPRI	Teal Book	Best Practices and Life Extension Guidelines for Substations

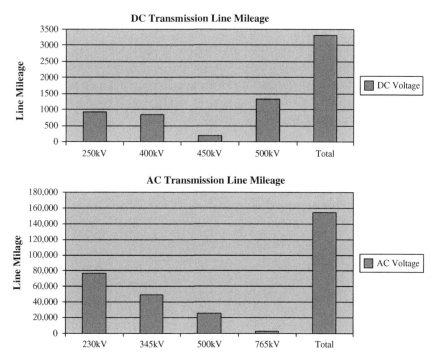

Figure 12.2: United States 2002 electrical transmission line mileage (courtesy of US Department of Energy, Washington, DC) [13]

associated with DC transmission lines because the DC lines do not have the inductive reactance line impedance associated with AC transmission lines. Both AC and DC transmission line conductors have associated line resistance which creates conductor voltage drops with corresponding power losses. It has been estimated that as high as 10% of the electrical energy generated, transmitted, and distributed could be lost in some line conductor installations.

Electrical Distribution Systems

Electrical energy is distributed to customers through a utility's *electrical distribution network*. That network consists of electrical distribution substations which step-down the transmission line voltage levels between 69 kV and 765 kV to distribution voltage levels, usually 35 kV or less. Typical distribution voltages range from 34,500Y/19,920 V to 4,160Y/2400 V. Distribution networks can consist of overhead electrical lines, as well as underground cable systems. Voltages at utility customer delivery points may require further reduction or stepping-down, either by utility transformers or customer owned and operated transformers.

Transmission and Distribution Systems Considerations

There are many codes, standards, and recommended practices which are applicable to the design and construction of electrical transmission and distribution systems. The first area of consideration would be the overhead and underground lines associated with those systems. Table 12.10 illustrates some of the codes, standards, and recommended practices associated with electrical transmission and distribution lines.

Each electrical utility company, electric cooperative and municipal utility entity has their own engineering and design standards for transmission and distribution systems. The US Department of Agriculture, Rural Utilities Service is the United States Federal agency charged with administering loan programs for electrification and telephone service in rural areas. It has established a number of codes, standards, and recommended practices for the transmission and distribution of electrical energy. They are presented in Table 12.11.

Many of the Canadian codes, standards, and recommended practices used for electrical utility transmission and distribution systems are presented in Table 12.12. The CSA Standards utilize technical committees for development. A consensus process is followed, utilizing the principles of inclusive participation. The committees are composed of a membership with diverse interest and the principles of transparency are maintained. Substantial agreement is required among committee members. Any standards approval must be based on membership approval of more than a simple majority.

Overhead Line Support Structures

Four most common means of supporting electrical transmission and distribution overhead lines include wood poles, steel poles and structures, concrete poles, and composite/fiberglass poles. Each of the support structure types will be briefly examined, along with some applicable standards.

TABLE 12.10   Partial list of codes, standards, and recommended practices associated with electrical transmission and distribution systems

Developer	Standard No.	Title
IEC	IEC CISPR 18-2	Radio Interference Characteristics of Overhead Power Lines on High-Voltage Equipment – Part 2: Methods of Measurement and Procedure for Determining Limits
IEC	IEC 60050-603	International Electrotechnical Vocabulary – Part 603: Generation, Transmission and Distribution of Electricity – Power System Planning and Management
IEC	IEC 60050-614	International Electrotechnical Vocabulary – Part 614: Electric Power System – Power System and Planning Operation
IEC	IEC 60050-615	International Electrotechnical Vocabulary – Part 601: Generation, Transmission and Distribution of Electricity – General
IEC	IEC 60099-8	Surge Arresters – Part 8: Metal-Oxide Surge Arresters with External Series Gap (EGLA) for Overhead Transmission and Distribution Lines of AC Systems Above 1 kV
IEC	IEC 60353	Line Traps for AC Power Systems
IEC	IEC 60826	Loading and Strength of Overhead Transmission Lines
IEC	IEC 61936-1	Power Installations Exceeding 1 kV AC – Part 1: Common Rules
IEC	IEC/TS 61640	Rigid High-Voltage, Gas-Insulated Transmission Lines for Rated Voltage of 72.5 kV and Above
IEC	IEC/TR 61911	Live Working – Guidelines for Installation of Distribution Line Conductors – Stringing Equipment and Accessory Items
IEEE	IEEE 524	IEEE Guide to Grounding During the Installation of Overhead Transmission Line Conductors
IEEE	IEEE 656	IEEE Standard for the Measurement of Audible Noise from Overhead Transmission Lines
IEEE	IEEE 738	IEEE Standard for Calculating the Current-Temperature of Bare Overhead Conductors
IEEE	IEEE 1410	IEEE Guide for Improving the Lightning Performance of Electric Power Overhead Distribution Lines
IEEE	IEEE 1048™	IEEE Guide for Protective Grounding of Power Lines
IEEE	IEEE 1441	IEEE Guide for Inspection of Overhead Transmission Line Construction
IEEE	IEEE C2	National Electrical Safety Code
IEEE	IEEE C37.114	IEEE Guide for Determining Fault Location on AC Transmission and Distribution Lines

Wood Poles

The codes, standards, and recommended practices associated with wood poles are presented in Table 12.13. Wood poles can consist of a single pole cut, shaped, and chemically treated, from a variety of trees. They can also be constructed of a number or interconnected wood poles (H-frame, etc.) and of laminated wood products for high strength and specialty application service.

TABLE 12.11 USDA rural utilities service electrical transmission and distribution standards

Developer	Standard No.	Title
USDA/RUS	1724D-101A	Electric System Long-Range Planning Guide
USDA/RUS	1724D-101B	System Planning Guide – Construction Work Plans
USDA/RUS	1724D-103	System Planning Guide, System Mapping Guide
USDA/RUS	1724D-104	Engineering Economics Computer Workbook Procedure – also available: Economic Analysis Worksheet
USDA/RUS	1724D-106	Considerations for Replacing Storm-Damaged Conductors
USDA/RUS	1724D-112	The Application of Capacitors on Rural Electric Systems
USDA/RUS	1724E-104	Reduced Size Neutral Conductors for Overhead Rural Distribution Lines
USDA/RUS	1724E-150	Unguyed Distribution Poles – Strength Requirements
USDA/RUS	1724E-151	Mechanical Loading on Distribution Crossarms
USDA/RUS	1724E-152	The Mechanics of Overhead Distribution Line Conductors
USDA/RUS	1724E153	Electric Distribution Line Guys and Anchors
USDA/RUS	1724E-200	Design Manual for High-Voltage Transmission Lines
USDA/RUS	1724E-202	An Overview of Transmission System Studies
USDA/RUS	1724E-203	Guide for Upgrading Transmission Lines
USDA/RUS	1724E-204	Guide Specifications for Steel Single Pole and H-Frame Structures
USDA/RUS	1724E-205	Design Guide: Embedment Depths for Concrete and Steel Poles
USDA/RUS	1724E-206	Guide Specification for Spun, Prestressed Concrete Poles and Concrete Pole Structures
USDA/RUS	1724E-214	Guide Specification for Standard Class Steel Transmission Poles
USDA/RUS	1724E-216	Guide Specification for Standard Class Spun, Prestressed Concrete Transmission Poles
USDA/RUS	1724E-220	Procurement and Application Guide for Non-Ceramic Composite Insulators, Voltage Class 34.5 kV and Above
USDA/RUS	1724E-224	Electric Transmission Guide Specifications and Drawings for Steel Pole Construction – 34.5 to 230 kV
USDA/RUS	1724E-226	Electric Transmission Guide Specifications and Drawings for Concrete Pole Construction – 34.5 to 230 kV
USDA/RUS	1724E-300	Design Guide for Rural Substations
USDA/RUS	1724E-301	Guide for the Evaluation of Large Power Transformer Losses
USDA/RUS	1724E-302	Design Guide for Oil Spill Prevention and Control at Substations
USDA/RUS	1724E-400	Building Plans and Specifications
USDA/RUS	1726-601	Electric System Construction Policies and Procedures – Interpretations
USDA/RUS	1726A-125	Joint Use Agreements with CATV Companies
USDA/RUS	1726C-115	Checking Sag in a Conductor Using the Return Wave Method
USDA/RUS	1726I-602	Attachments to Electric Program Standard Contract Forms
USDA/RUS	1728F-700	Specification for Wood Poles, Stubs and Anchor Logs (Incorporated by reference – §1728.97)

TABLE 12.11 USDA rural utilities service electrical transmission and distribution standards—cont'd

Developer	Standard No.	Title
USDA/RUS	1728F-800	Construction Assembly Unit Numbers and Standard Format
USDA/RUS	1728F-803	Specifications and Drawings for 24.9/14.4 kV Line Construction (Incorporated by reference – §1728.97)
USDA/RUS	1728F-804	Specifications and Drawings for 12.5/7.2 kV Line Construction (Incorporated by reference – §1728.97)
USDA/RUS	1728F-806	Specifications and Drawings for Underground Electric Distribution (Incorporated by reference – §1728.97)
USDA/RUS	1728F-810	Electric Transmission Specifications and Drawings, 34.5 kV through 69 kV (Incorporated by reference – §1728.97)
USDA/RUS	1728F-811	Electric Transmission Specifications and Drawings, 115kV through 230 kV (Incorporated by reference – §1728.97)
USDA/RUS	1728H-701	Specifications for Wood Crossarms, Transmission Timbers, and Pole Keys (Codified – §1728.201)
USDA/RUS	1728H-702	Specifications for Quality Control and Inspection of Timber Products (Codified – §1728.202)
USDA/RUS	1730B-2	Guide for Electric System Emergency Restoration Plan
USDA/RUS	1730B-121	Pole Inspection and Maintenance
USDA/RUS	1730-1	Electric System Operation and Maintenance (O&M)

There are several varieties of trees which have been used for utility pole service. The most common of those include:

- Southern Yellow Pine

- Red (Norway) Pine

- Lodgepole Pine

- Sugar Pine

- Jack Pine

- Ponderosa (Western)Pine

- Western White Pine

- Northern White Pine

- Eastern Larch

- Western Larch

- Southern White Cedar

TABLE 12.12 Canadian transmission and distribution codes, standards, and recommended practices

Developer	Standard No.	Title
CAN/CSA	CAN/CSA C22.3 No. 1	Overhead Systems
CAN/CSA	CAN/CSA C22.3 No. 3	Electrical Coordination
CAN/CSA	CAN/CSA C22.3 No. 4	Control of Electrochemical Corrosion of Underground Metallic Structures
CAN/CSA	CAN/CSA C22.3 No. 5.1	Recommended Practices for Electrical Protection – Electric Contact Between Overhead Supply and Communication Lines
CAN/CSA	CAN/CSA C22.3 No. 6-M	Principles and Practices of Electrical Coordination Between Pipelines and Electric Supply Lines
CAN/CSA	CAN/CSA C22.3 No. 7	Underground Systems
CAN/CSA	CAN/CSA C22.3 No. 8-M	Railway Electrification Guidelines
CAN/CSA	CAN/CSA C22.3 No. 9	Interconnection of Distributed Resources and Electricity Supply Systems
CAN/CSA	CAN/CSA-C22.3 NO. 60826	Design Criteria for Overhead Transmission Lines (adopted CEI/IEC 60826)
CAN/CSA	CAN/CSA-C22.3 NO. 61936-1	Power Installations Exceeding 1 kV AC – Part 1: Common Rules (adopted CEI/IEC 61936-1)
CAN/CSA	C83	Communication and Power Line Hardware
CAN/CSA	CAN/CSA-C156.1-M	Ceramic and Glass Station Post Insulators
CAN/CSA	CAN/CSA-C156.3-M	Test Methods for Station Post Insulators
CAN/CSA	CAN/CSA-C233.1	Gapless Metal Oxide Surge Arresters for Alternating Current Systems
CAN/CSA	CAN3-C235	Preferred Voltage Levels for AC Systems, 0 to 50 000 V
CAN/CSA	CAN/CSA-C411.1-M	AC Suspension Insulators
CAN/CSA	CAN/CSA-C1325	Insulators for Overhead Lines with a Nominal Voltage Above 1000 V – Ceramic or Glass Insulator Units for DC Systems – Definitions, Test Methods and Acceptance Criteria (adopted CEI/IEC 1325)
CAN/CSA	CAN/CSA-C62155	Hollow Pressurized and Unpressurized Ceramic and Glass Insulators for Use in Electrical Equipment with Rated Voltages Greater than 1000 V

- Alaska Yellow Cedar

- Sitka Spruce

- White Fir

- Douglas Fir

- Eastern Hemlock

- Western Hemlock

- Redwood

TABLE 12.13 Wood transmission and distribution pole standards

Developer	Standard No.	Title
ASTM	ASTM D9	Standard Terminology Relating to Wood and Wood-Based Products
ASTM	ASTM D1036	Standard Test Methods of Static Tests of Wood Poles
ASTM	ASTM D2555	Standard Methods for Establishing Clear Wood Strength Values
ASTM	ASTM D1036	Standard Test Methods of Static Tests of Wood Poles
ATIS	ANSI O5.1	Wood Poles – Specifications and Dimensions
ATIS	ANSI O5.2	Wood Products – Structural Glued Laminated Timber for Utility Structures
ATIS	ANSI O5.3	Solid Sawn-Wood Crossarms and Braces – Specifications and Dimensions
AWAP	AWAP A1	Standard Methods for Analysis of Creosote and Oil-Type Preservatives
AWAP	AWAP A2	Standard Methods for Analysis of Waterborne Preservatives and Fire-Retardant Formulations
AWAP	AWAP A3	Standard Methods for Determining Penetration of Preservatives and Fire Retardants
AWAP	AWAP A5	Standard Methods for Analysis of Oil-Borne Preservatives
AWAP	AWAP A6	Method for the Determination of Oil-Type Preservatives and Water in Wood
AWAP	AWAP A7	Standard for Wet Ashing Procedures for Preparing Wood for Chemical Analysis
AWAP	AWAP A9	Standard Method for Analysis of Treated Wood and Treating Solutions by X-ray Spectroscopy
AWAP	AWAP A11	Standard Method for Analysis of Treated Wood and Treating Solutions by X-ray Spectroscopy
AWAP	AWAP A12	Wood Densities for Preservative Retention Calculations by Standards A2, A9 and A11
AWAP	AWAP A15	Referee Methods
AWAP	AWAP A41	Standard Method for Determination of Naphthenic Acid in Copper Naphthenate Wood and Treating Solutions by Gas Chromatography
AWAP	AWAP C1	All Timber Products – Preservative Treatment by Pressure Processes
AWAP	AWAP C4	Poles – Preservative Treatment by Pressure Processes
AWAP	AWAP C8	Western Red Cedar and Alaska Yellow Cedar Poles – Preservative Treatment by the Full-Length Thermal Process
AWAP	AWAP E10	Standard Method of Testing Wood Preservatives by Laboratory Soil-Block Cultures
AWAP	AWAP E23	Accelerated Method of Evaluating Wood Preservatives in Soil Contact
AWPA	AWPA M1	Standard for the Purchase of Treated Wood Products
AWPA	AWPA M2	Standard for Inspection of Treated Wood Products
AWPA	AWPA M3	Standard Quality Control Procedures for Wood Preserving Plants
AWPA	AWPA M4	Standard for the Care of Preservative-Treated Wood Products
AWPA	AWPA M5	Glossary of Terms Used in Wood Preservation
AWPA	AWPA M9	An Overview of the ISO 9000 Quality Standards for Consumers and Producers of Treated Wood Products

(Continued)

TABLE 12.13 Wood transmission and distribution pole standards—cont'd

Developer	Standard No.	Title
AWPA	AWPA M13	A Guideline for the Physical Inspection of Poles in Service
AWAP	AWAP P1/P13	Standard for Creosote Preservative
AWAP	AWAP P5	Standard for Waterborne Preservatives
AWAP	AWAP P8	Standard for Oil-Borne Preservatives
AWAP	AWAP P9	Standards for Solvents and Formulations for Organic Preservative Systems
AWAP	AWAP P25	Standard for Inorganic Boron (SBX)
AWAP	AWAP T1	Use Category System: Processing and Treatment Standard
AWAP	AWAP U1	Use Category System: User Specification for Treated Wood
SPIB	SPIB 1003	Standard Grading Rules
SPIB	SPIB 1018	Special Product Rules for Structural, Industrial and Railroad Freight Car Lumber
SPIB	SPIB 1027	Kiln Drying Southern Pine
WCLIB	WCLIB Standard 17	Grading Rules for West Coast Lumber

Wood poles are subject to decay from fungi and invasion by termites, woodpeckers, and insects. Several wood preservative products have been developed to extend the useful life of wood utility poles. They include:

1. Oil-type preservatives

   - Copper naphthenate

   - Creosote and creosote solutions

   - Pentachlorophenol

2. Water-borne preservatives

   - Ammoniacal copper zinc arsenate (ACZA)

   - Ammoniacal copper quat (ACQ)

   - Chromated copper arsenate (CCA)

   - Copper azole

Borate, another water-based wood preservative, does not precipitate into the wood substrate like other wood preservatives. It will readily leach from the wood if exposed to rain or wet soil. That product is not recommended as a utility pole preservative.

Wood pole size material specifications, classification, and engineering properties are established in ANSI O5.1, *American National Standard for Wood Products – Specifications and Dimensions*. Pole Classification (Class) designations are based on physical dimension values for minimum circumference in inches at the pole tip (top), as well as the minimum circumference in inches at 6 feet (6'0") from the butt end. ANSI wood pole distribution pole Classes are Class 10 to Class 1 and transmission Classes H1 through H5. Distribution Class 10 has the smallest tip circumference and lowest fiber stress value while Class 1 has the largest tip circumference and the greatest fiber stress value.

ANSI O5.1 establishes the requirements related to pole species acceptance criteria, manufacturing requirements, length and Class, and code marking. It lists some 25 native species for the United States as having the physical characteristics suitable for use as utility poles. Data are provided on each species, including fiber stress (lb/in^2), pole sizes, and shape values. Wolfe and Moody [15] report that of the 25 ANSI O5.1 listed species, not all are now commonly used. Six of the species repeatedly account for 90% of the wood poles used in the United States. Those species include Loblolly, Longleaf, Shortleaf, and Slash Pines, as well as Douglas Fir and Western Red Cedar.

Steel, Concrete, Fiber-Reinforced Polymer Poles

Concrete, fiber-reinforced polymer, and steel utility poles are also used to support power transmission and distribution overhead lines. These poles have some advantages that can make them competitive with their wood pole counterparts. The alternative poles offer protection from decay, insect damage, and avian damage. Steel and concrete poles offer protection from fire damage. All three substitutes have a longer lifespan than wood. Some can offer increased conductor span lengths, requiring fewer poles. Fiberglass composite poles can offer energy absorption characteristics which can be desirable from a vehicle impact standpoint. A list of many of the codes, standards, and recommended practices for concrete, polymer and steel utility poles is presented in Table 12.14. Some of the ASTM standards listed are also available in metric dimensions. Those documents have the identical ASTM standards number as noted in Table 12.14 with an (M) designation at the end.

Transmission and Distribution Hardware and Equipment

Overhead transmission and distribution conductors require substantial hardware for support and attachment to poles and structures, maintenance of conductor spacing requirements, and for prevention of the development of flashover paths between energized conductors and ground. Additional hardware may be required for protection of overhead lines from lightning and voltage surges. Transmission and distribution system protective equipment and switches permit clearance of electrical short circuit and overcurrent events; transmission and distribution circuit and equipment isolation or switching.

Table 12.15 lists a number of American codes, standards, and recommend practices for transmission and distribution hardware. Line insulators provide critical conductor support and protection. They can be mounted either horizontally or vertically, depending upon the

TABLE 12.14 Steel, concrete, fiber-reinforced polymer utility pole standards

Developer	Standard No.	Title
ASCE	ASCE 7	Minimum Design Loads for Building and Other Structures
ASCE	ASCE 10	Design of Latticed Steel Transmission Structures
ASCE	ASCE/SEI 48	Design of Steel Transmission Pole Structures
ASCE	ASCE 104	Recommended Practice for Fiber-Reinforced Polymer Products for Overhead Utility Line Structures
ASCE	ASCE 596	Guide for the Design and Use of Concrete Poles
ASTM	ASTM A36	Standard Specification for Carbon Structural Steel
ASTM	ASTM A53	Standard Specification for Pipe, Steel, Black and Hot-Dipped, Zinc-Coated, Welded and Seamless
ASTM	ASTM A123	Standard Specification for Zinc (Hot-Dip Galvanized) Coatings on Iron and Steel Products
ASTM	ASTM A143	Standard Practice for Safeguarding Against Embrittlement of Hot-Dip Galvanized Structural Steel Products and Procedure for Detecting Embrittlement
ASTM	ASTM A153	Standard Specification for Zinc Coating (Hot-Dip) on Iron and Steel Hardware
ASTM	ASTM A325	Standard Specification for Structural Bolts, Steel, Heat Treated, 120/105 ksi Minimum Tensile Strength
ASTM	ASTM A370	Standard Test Methods and Definitions for Mechanical Testing of Steel Products
ASTM	ASTM A376	Standard Practice for Measuring Coating Thickness by Magnetic-Field or Eddy-Current (Electromagnetic) Test Methods
ASTM	ASTM A384	Standard Practice for Safeguarding Against Warpage and Distortion During Hot-Dip Galvanizing of Steel Assemblies
ASTM	ASTM A500	Standard Specification for Cold-Formed Welded and Seamless Carbon Steel Structural Tubing in Rounds and Shapes
ASTM	ASTM A501	Standard Specification for Hot-Formed Welded and Seamless Carbon Steel Structural Tubing
ASTM	ASTM A563	Standard Specification for Carbon and Alloy Steel Nuts
ASTM	ASTM A572	Standard Specification for High-Strength Low-Alloy Columbium-Vanadium Structural Steel
ASTM	ASTM A588	Standard Specification for High-Strength Low-Alloy Structural Steel, up to 50 ksi 345 MPa Minimum Yield Point, with Atmospheric Corrosion Resistance
ASTM	ASTM A780	Repair of Damaged and Uncoated Areas of Hot-Dip Galvanized Coatings
ASTM	ASTM A871	Standard Specification for High-Strength Low-Alloy Structural Steel Plate With Atmospheric Corrosion Resistance
ASTM	ASTM C935	Standard Specification for General Requirements for Prestressed Concrete Poles Statically Cast
ASTM	ASTM C1089	Standard Specification for Spun Cast Prestressed Concrete Poles
ASTM	ASTM D4923	Standard Specification for Reinforced Thermosetting Plastic Poles

TABLE 12.14 Steel, concrete, fiber-reinforced polymer utility pole standards—cont'd

Developer	Standard No.	Title
IEEE	IEEE 691	IEEE Guide for Transmission Structure Foundation Design and Testing
IEEE	IEEE 1025	IEEE Guide to the Assembly and Erection of Concrete Pole Structures
NEMA	NEMAANSI C136.6	Roadway and Area Lighting Equipment – Metal Heads and Reflector Assemblies – Mechanical and Optical Interchangeability
NEMA	ANSI C136.20	American National Standard for Roadway and Area Lighting Equipment – Fiber-Reinforced Composite (FRC) Lighting Poles
NEMA	ANSI C136.36B	American National Standard for Roadway and Area Lighting Equipment – Concrete Lighting Poles

TABLE 12.15 Transmission and distribution insulators and hardware

Developer	Standard No.	Title
NEMA	NEMA/ANSI C29.1	Test Methods for Electrical Power Insulators
NEMA	NEMA/ANSI C29.2	For Insulators, Wet-Process Porcelain and Toughened Glass – Suspension Type
NEMA	NEMA/ANSI C29.3	Wet-Process Porcelain Insulators-(Spool Type)
NEMA	NEMA/ANSI C29.4	Wet-Process Porcelain Insulators – Strain Type
NEMA	NEMA/ANSI C29.5	Wet Process Porcelain Insulators (Low- and Medium-Voltage Pin Type)
NEMA	NEMA/ANSI C29.6	Wet-Process Porcelain Insulators-High-Voltage Pin Type
NEMA	NEMA/ANSI C29.7	Porcelain Insulators-High-Voltage Line-Post Type
NEMA	NEMA/ANSI C29.8	Apparatus, Cap And Pin Type (Wet Process Porcelain
NEMA	NEMA/ANSI C29.9	Wet-Process Porcelain Insulators (Apparatus, Post-Type)
NEMA	NEMA/ANSI C29.10	Wet-Process Porcelain Insulators – Indoor Apparatus Type
NEMA	NEMA/ANSI C29.11	Composite Suspension Insulators for Overhead Transmission Lines – Tests
NEMA	NEMA/ANSI C29.12	Insulators – Composites – Suspension Type
NEMA	NEMA/ANSI C29.13	Insulators – Composite Distribution Dead End Type
NEMA	NEMA/ANSI C29.17	For Insulators – Composite-Line Post Type
NEMA	NEMA/ANSI C29.18	Insulators Composite – Distribution Line Post Type
NEMA	NEMA HV 2	Application Guide for Ceramic Suspension Insulators
IEEE	IEEE C135.61	IEEE Standard for Testing of Overhead Transmission and Distribution Line Hardware
IEEE	IEEE 386™	Standard for Separable Insulated Connector Systems for Power Distribution Systems Above 600 V
IEEE	IEEE 957	IEEE Guide for Cleaning Insulators
IEEE	IEEE 1283	IEEE Guide for Determining the Effects of High-Temperature Operation on Conductors, Connectors, and Accessories

(Continued)

TABLE 12.15 Transmission and distribution insulators and hardware—cont'd

Developer	Standard No.	Title
IEEE	IEEE C135.1	IEEE Standard for Zinc-Coated Steel Bolts and Nuts for Overhead Line Construction
IEEE	IEEE C135.2	IEEE Standard for Threaded Zinc-Coated Ferrous Strand-Eye Anchor Rods and Nuts for Overhead Line Construction
IEEE	IEEE C135.3	IEEE Standard for Zinc-Coated Ferrous Lag Screws for Overhead Line Construction
IEEE	IEEE C135.20	IEEE Standard for Zinc-Coated Ferrous Insulator Clevises for Overhead Line Construction
IEEE	ANSI C135.22	American National Standard for Zinc-Coated Ferrous Pole-top Insulator Pins with Lead Threads for Overhead Line Construction
IEEE	ANSI C135.30	American National Standard for Zinc-Coated Ferrous Ground Rods for Overhead or Underground Line Construction
IEEE	ANSI C135.31	American National Standard for Zinc-Coated Ferrous Single and Double Upset Spool Insulator Bolts for Overhead Line Construction
IEEE	ANSI C135.35	American National Standard for Zinc-Coated Ferrous Cable Racks and Cable Rack Hooks for Underground Line Construction
IEEE	ANSI C135.38	American National Standard for Zinc-Coated Ferrous Washerhead Bolts and Washer Nuts
IEEE	IEEE C135.61	Testing of Overhead Transmission and Distribution Line Hardware
IEEE	IEEE C135.62	Standard for Zinc-Coated Forged Anchor Shackles
IEEE	IEEE C135.63	IEEE Standard for Shoulder Live Line Extension Links for Overhead Line Construction
IEEE	IEEE C135.64	Guide for Design Testing of Bolted Deadend Strain Clamps
ASTM	ASTM D1049	Standard Specification for Rubber Insulating Covers
ASTM	ASTM D116	Standard Test Methods for Vitrified Ceramic Materials for Electrical Applications
ASTM	ASTM F855	Standard Specifications for Temporary Protective Grounds to Be Used on De-energized Electric Power Lines and Equipment

application, with the most common mounting scheme on pole crossarms. Line conductors can be attached to line pin and post insulators using mechanical clamps on specifically designed insulators or through the use of tie wires, attaching the conductor to either top or side grooves of an insulator. Tie wires do not support a line conductor, but only prevent the line conductor from slipping out of the insulator groove, placing the line conductor support weight on the insulator. Some applications require that a uninsulated overhead conductor be reinforced with armor rods at an insulator attachment. The armor rods increase conductor support, reduce vibration, and prevent conductor chafing and flashover.

The main types of line conductor insulators in electrical distribution systems include line pin and post insulators, suspension insulators, dead-end insulators, and spool insulators.

Transmission and distribution system operating voltage governs the insulator design requirements. Insulator geometry plays a major role in establishing the distance required to prevent leakage current flow and flashover. The insulator surface leakage current path can be affected by surface moisture and dirt deposits. Insulating surface shapes are designed to conform to line conductor electrostatic field flow lines. Insulator material thickness must be capable of preventing puncture of the insulating material by overvoltages created by a combination of the normal line voltage and any induced or generated line transient voltages.

Electrical parameters that define overhead line insulators include dry flashover voltage, leakage distance in inches, dry arcing distance in inches, and wet arcing distance in inches. Dry flashover voltage is the voltage required to initiate a flashover event during dry conditions. An insulator's geometry determines its effective physical surface distance path between the line conductor attachment point and the insulator mounting pin or supporting crossarm. That distance must exceed the distances associated with wet and dry arcing distances, dry flashover, etc.

Leakage distance is the shortest distance between the insulator cap and pin or the conductor and pin and is measured along the insulator surface. Dry arcing distance is the shortest distance between the pin and conductor or pin and insulator cap. Its path is measured entirely through air. The wet arcing distance is the shortest distance between the cap or conductor and pin measured over the insulator cap surface and through the air across insulator ridges.

The International Electrotechnical Commission (IEC) has developed a substantial number of standards for transmission and distribution systems. A number of those standards are listed in Table 12.16. Also included in that table are some standards for the distribution of electrical energy on electric railway systems.

Line conductor insulators and associated hardware, reclosers, circuit breakers, fused cutouts, surge arresters, switches, etc., are all circuit protection and/or disconnect devices associated with electrical transmission and distribution systems. Codes, standards, and recommended practices for those devices are listed in Tables 12.17 and 12.18 for IEC and IEEE standards.

IEEE 141, *IEEE Recommended Practice for Electrical Power Distribution for Industrial Plants* provides guidance in the selection of surge arresters. There are several factors which must be considered when selecting a surge arrester for a specific application. Those considerations include [16]:

Arrester Maximum Continuous Operating Voltage (MCOV)

Lightning and Switching Impulse (Protective Characteristics)

Temporary Overvoltage (TOV) and Switching Surge (Durability)

Operating Service Conditions (Capacitor Switching, Substation, etc.)

Arrester Pressure-Relief Requirements (Fault Current Withstand)

TABLE 12.16 IEC transmission and distribution standards

Developer	Standard No.	Title
IEC	IEC 60038	IEC Standard Voltages
IEC	IEC 60050-466	International Electrotechnical Vocabulary. Chapter 466: Overhead Lines
IEC	IEC 60050-601	International Electrotechnical Vocabulary. Chapter 601: Generation, Transmission and Distribution of Electricity – General
IEC	IEC 60050-602	International Electrotechnical Vocabulary. Chapter 602: Generation, Transmission and Distribution of Electricity – Generation
IEC	IEC 60050-603	International Electrotechnical Vocabulary. Chapter 603: Generation, Transmission and Distribution of Electricity – Power Systems Planning and Management
IEC	IEC 60050-604	International Electrotechnical Vocabulary. Chapter 604: Generation, Transmission and Distribution of Electricity – Operation
IEC	IEC 60050-605	International Electrotechnical Vocabulary. Chapter 605: Generation, Transmission and Distribution of Electricity – Substations
IEC	IEC 60050-614	International Electrotechnical Vocabulary – Part 614: Electric Power System – Power System and Planning Operation
IEC	IEC 60050-615	International Electrotechnical Vocabulary – Part 601: Generation, Transmission and Distribution of Electricity – General
IEC	IEC 60077-4	Railway Applications – Electric Equipment for Rolling Stock – Part 4: Electrotechnical Components – Rules for AC Circuit-Breakers
IEC	IEC 60099-8	Surge Arresters – Part 8: Metal-Oxide Surge Arresters with External Series Gap (EGLA) for Overhead Transmission and Distribution Lines of AC Systems above 1 kV
IEC	IEC 60353	Line Traps for AC Power Systems
IEC	IEC 60358	Coupling Capacitors and Capacitor Dividers
IEC	IEC 60652	Loading Tests on Overhead Line Structures
IEC	IEC 60826	Design Criteria of Overhead Transmission Lines
IEC	IEC 60909-0	Short-Circuit Currents in Three-Phase AC Systems – Part 0: Calculation of Currents
IEC	IEC 60909-1	Short-Circuit Currents in Three-Phase AC Systems – Part 1: Factors for the Calculation of Short-Circuit Currents According to IEC 60909-0
IEC	IEC 60909-2	Short-Circuit Currents in Three-Phase AC Systems – Part 2: Data of Electrical Equipment for Short-Circuit Current Calculations
IEC	IEC 60909-3	Short-Circuit Currents in Three-Phase AC Systems – Part 3: Currents during Two Separate Simultaneous Line-to-Earth Short Circuits and Partial Short-Circuit Currents Flowing through Earth
IEC	IEC 60913	Electric Traction Overhead Lines
IEC	IEC/TS 61000-6-5	Electromagnetic Compatibility (EMC) – Part 6-5: Generic Standards – Immunity for Power Station and Substation Environments
IEC	IEC 61284	Overhead Lines – Requirements and Tests for Fittings

TABLE 12.16 IEC transmission and distribution standards—cont'd

Developer	Standard No.	Title
IEC	IEC/TR 61334-1-4	Distribution Automation using Distribution Line Carrier Systems – Part 1: General Considerations – Section 4: Identification of Data Transmission Parameters Concerning Medium and Low-Voltage Distribution Mains
IEC	IEC 61334-5-1	Distribution Automation Using Distribution Line Carrier Systems – Part 5-1: Lower Layer Profiles – The Spread Frequency Shift Keying (S-FSK) Profile
IEC	IEC 61334-5-5	Distribution Automation Using Distribution Line Carrier Systems – Part 5-5: Lower Layer Profiles – Spread Spectrum – Fast Frequency Hopping (SS-FFH) Profile
IEC	IEC/TS 61394	Overhead Lines – Characteristics of Greases for Aluminium, Aluminium Alloy and Steel Bare Conductors
IEC	IEC 61395	Overhead Electrical Conductors – Creep Test Procedures for Stranded Conductors
IEC	IEC/TS 61640	Rigid High-Voltage, Gas-Insulated Transmission Lines for Rated Voltage of 72. kV and above
IEC	IEC 61773	Overhead Lines – Testing of Foundations for Structures
IEC	IEC 61854	Overhead Lines – Requirements and Tests for Spacers
IEC	IEC 61865	Overhead Lines – Calculation of the Electrical Component of Distance between Live Parts and Obstacles – Method of Calculation
IEC	IEC 61897	Overhead Lines – Requirements and Tests for Stockbridge Type Aeolian Vibration Dampers
IEC	IEC 61936-1	Power Installations Exceeding 1 kV AC – Part 1: Common Rules
IEC	IEC 61850-7-2	Communication Networks and Systems in Substations – Part 7-2: Basic Communication Structure for Substation and Feeder Equipment – Abstract Communication Service Interface (ACSI)
IEC	IEC 61954	Power Electronics for Electrical Transmission and Distribution Systems – Testing of Thyristor Valves for Static VAR Compensators
IEC	IEC 62004	Thermal-Resistant Aluminium Alloy Wire for Overhead Line Conductor Maintenance Result Date: 2011
IEC	IEC 62539	Guide for the Statistical Analysis of Electrical Insulation Breakdown Data

Line voltage and neutral grounding configuration (ungrounded, resistance/reactance grounded, solidly grounded) require consideration when selecting the device voltage rating. Surge arresters must be rated at 100% of the system line-to-line operating voltage on ungrounded systems. That is required because the full line-to-line voltage can be impressed on a surge arrester in the event of a line-to-ground fault occurrence on a single line conductor. IEEE 141, Tables 6-5 and 6-6 [17] provide metal oxide arrester characteristics for station- and intermediate-class MOVs (metal oxide varistors), as well as distribution-class and riser pole MOVs.

TABLE 12.17 IEC transmission, distribution, and station insulators

Developer	Standard No.	Title
IEC	IEC 60050-471	International Electrotechnical Vocabulary – Part 471: Insulators
IEC	IEC 60120	Dimensions of Ball and Socket Couplings of String Insulator Units
IEC	IEC 60168	Tests on Indoor and Outdoor Post Insulators of Ceramic Material or Glass for Systems with Nominal Voltages Greater Than 1000 V
IEC	IEC 60273	Characteristic of Indoor and Outdoor Post Insulators for Systems with Nominal Voltages Greater Than 1000 V
IEC	IEC 60305	Insulators for Overhead Lines with a Nominal Voltage Above 1000 V – Ceramic or Glass Insulator Units for AC Systems – Characteristics of Insulator Units of the Cap and Pin Type
IEC	IEC 60383-1	Insulators for Overhead Lines with a Nominal Voltage Above 1000 V – Part 1: Ceramic or Glass Insulator Units for AC Systems – Definitions, Test Methods and Acceptance Criteria
IEC	IEC 60383-2	Insulators for Overhead Lines with a Nominal Voltage Above 1000 V – Part 2: Insulator Strings and Insulator Sets for AC Systems – Definitions, Test Methods and Acceptance Criteria
IEC	IEC 60433	Insulators for Overhead Lines with a Nominal Voltage Above 1000 V – Ceramic Insulators for AC Systems – Characteristics of Insulator Units of the Long Rod Type
IEC	IEC 60437	Radio Interference Test on High-Voltage Insulators
IEC	IEC 60471	Dimensions of Clevis and Tongue Couplings of String Insulator Units
IEC	IEC 60507	Artificial Pollution Tests on High-Voltage Insulators to Be Used on AC Systems
IEC	IEC/TR 60575	Thermal-Mechanical Performance Test and Mechanical Performance Test on String Insulator Units
IEC	IEC 60660	Insulators – Tests on Indoor Post Insulators of Organic Material for Systems with Nominal Voltages Greater Than 1000 V Up To but Not Including 300 kV
IEC	IEC 60720	Characteristics of Line Post Insulators
IEC	IEC/TR 60797	Residual Strength of String Insulator Units of Glass or Ceramic Material for Overhead Lines After Mechanical Damage of the Dielectric
IEC	IEC/TS 60815-1	Selection and Dimensioning of High-Voltage Insulators Intended for Use in Polluted Conditions – Part 1: Definitions, Information and General Principles
IEC	IEC/TS 60815-2	Selection and Dimensioning of High-Voltage Insulators Intended for Use in Polluted Conditions – Part 2: Ceramic and Glass Insulators for AC Systems
IEC	IEC/TS 60815-3	Selection and Dimensioning of High-Voltage Insulators Intended for Use in Polluted Conditions – Part 3: Polymer Insulators for AC Systems
IEC	IEC 61109	Insulators for Overhead Lines – Composite Suspension and Tension Insulators for AC Systems with a Nominal Voltage Greater Than 1000 V – Definitions, Test Methods and Acceptance Criteria
IEC	IEC 61211	Insulators of Ceramic Material or Glass for Overhead Lines with a Nominal Voltage Greater Than 1000 V – Impulse Puncture Testing in Air
IEC	IEC/TS 61245	Artificial Pollution Tests on High-Voltage Insulators to Be Used on DC Systems

TABLE 12.17 IEC transmission, distribution, and station insulators—cont'd

Developer	Standard No.	Title
IEC	IEC 61325	Insulators for Overhead Lines with a Nominal Voltage Above 1000 V – Ceramic or Glass Insulator Units for DC Systems – Definitions, Test Methods and Acceptance Criteria
IEC	IEC 61462	Composite Hollow Insulators – Pressurized and Unpressurized Insulators for Use in Electrical Equipment with Rated Voltage Greater Than 1000 V – Definitions, Test Methods, Acceptance Criteria and Design Recommendations
IEC	IEC 61466-1	Composite String Insulator Units for Overhead Lines with a Nominal Voltage Greater Than 1000 V – Part 1: Standard Strength Classes and End Fittings
IEC	IEC 61466-2	Composite String Insulator Units for Overhead Lines with a Nominal Voltage Greater Than 1000 V – Part 2: Dimensional and Electrical Characteristics
IEC	IEC 61467	Insulators for Overhead Lines – Insulator Strings and Sets for Lines with a Nominal Voltage Greater Than 1000 V – AC Power Arc Tests
IEC	IEC 61952	Insulators for Overhead Lines – Composite Line Post Insulators for AC Systems with a Nominal Voltage Greater Than 1000 V – Definitions, Test Methods and Acceptance Criteria
IEC	IEC/TR 62039	Selection Guide for Polymeric Materials for Outdoor Use Under HV Stress
IEC	IEC 62073	Guidance on the Measurement of Wettability of Insulator Surfaces
IEC	IEC 62155	Hollow Pressurized and Unpressurized Ceramic and Glass Insulators for Use in Electrical Equipment with Rated Voltages Greater Than 1000 V
IEC	IEC 62217	Polymeric Insulators for Indoor and Outdoor Use with a Nominal Voltage > 1000 V – General Definitions, Test Methods and Acceptance Criteria
IEC	IEC 62231	Composite Station Post Insulators for Substations with AC Voltages Greater Than 1000 V Up To 245 kV – Definitions, Test Methods and Acceptance Criteria
IEC	IEC/TS 62371	Characteristics of Hollow Pressurised and Unpressurised Ceramic and Glass Insulators for Use in Electrical Equipment with Rated Voltages Greater Than 1000 V

TABLE 12.18 Distribution class switches and circuit breakers

Developer	Standard no.	Title
IEEE	IEEE C37.30	IEEE Standard Requirements for High-Voltage Switches
IEEE	ANSI C37.32	American National Standard High-Voltage Air Disconnect Switches, Interrupter Switches, Fault Initiating Switches, Grounding Switches, Bus Supports and Accessories Control Voltage Ranges – Schedules of Preferred Ratings, Construction Guidelines and Specifications
IEEE	IEEE C37.34	IEEE Standard Test Code for High-Voltage Air Switches
IEEE	IEEE C37.35	IEEE Guide for the Application, Installation, Operation, and Maintenance of High-voltage Air Disconnecting and Load Interrupter Switches

(Continued)

TABLE 12.18 Distribution class switches and circuit breakers—cont'd

Developer	Standard no.	Title
IEEE	IEEE C37.36B	IEEE Guide to Current Interruption with Horn-Gap Air Switches
IEEE	IEEE C37.37	IEEE Guide to Current Interruption with Horn-Gap Air Switches
IEEE	IEEE C37.40	IEEE Standard Service Conditions and Definitions for High-Voltage Fuses, Distribution Enclosed Single-Pole Air Switches, Fuse Disconnecting Switches, and Accessories
IEEE	IEEE C37.40a	IEEE Standard Service Conditions and Definitions for External Fuses for Shunt Capacitors. Supplement to IEEE Std C37.40
IEEE	IEEE C37.41	IEEE Standard Design for High-Voltage Fuses, Distribution Enclosed Single-Pole Air Switches, Fuse Disconnecting Switches, and Accessories
IEEE	ANSI C37.42	American National Standard Specification for High-Voltage Expulsion Type Distribution Class Fuses, Cutouts, Fuse Disconnecting Switches and Fuse Links
IEEE	IEEE C37.43	IEEE Standard Specifications for High-Voltage Expulsion, Current-Limiting, and Combination-Type Distribution and Power Class External Fuses, with Rated Voltages from 1 kV through 38 kV, Used for the Protection of Shunt Capacitors
IEEE	IEEE C37.45	IEEE Standard Specifications for High-Voltage Distribution Class Enclosed Single-Pole Air Switches with Rated Voltages from 1 kV through 8.3 kV
IEEE	ANSI C37.46	American National Standard for High-Voltage Distribution Class Enclosed Single-Pole Air Switches with Rated Voltages from 1 kV to 8.3 kV
IEEE	ANSI C37.47	American National Standard for High-Voltage Current-Limiting Type Distribution Class Fuses and Fuse Disconnecting Switches
IEEE	IEEE C37.48	IEEE Guide for Application, Operation, and Maintenance of High-Voltage Fuses, Distribution Enclosed Single-Pole Air Switches, Fuse Disconnecting Switches, and Accessories
IEEE	IEEE C37.48.1	IEEE Guide for the Operation, Classification, Application and Coordination of Current-Limiting Fuses with Rated Voltages 1-38 kV
IEEE	IEEE C37.60	Standard Requirements for Overhead, Pad-Mounted, Dry Vault, and Submersible Automatic Circuit Reclosers and Fault Interrupters for Alternating Current Systems up to 38 kV
IEEE	IEEE C37.73	IEEE Standard Requirements for Pad-Mounted Fused Switchgear
IEEE	IEEE C37.74	Standard Requirements for Subsurface, Vault, and Pad-Mounted Load-Interrupter Switchgear and Fused Load-Interrupter Switchgear for Alternating Current Systems up to 38 kV
IEEE	IEEE C37.104	Guide for Automatic Reclosing of Line Circuit Breakers for AC Distribution and Transmission Lines
IEEE	IEEE C57.12.28	IEEE Standard for Pad-Mounted Equipment Enclosure Integrity
IEEE	IEEE C57.12.29	IEEE Standard for Pad-Mounted Equipment-Enclosure Integrity for Coastal Environments
IEEE	IEEE P1247	Standard for Interrupter Switches for Alternating Current, Rated Above 1000 Volts

Four basic classes of surge arresters include Station Class, Intermediate Class, Distribution Class (heavy duty and normal duty), and Secondary [18]. Table 12.19 lists some common surge arrester standards. Surge arresters are characterized in Table 12.20 by the arrester discharge-current withstand values by which they must be tested. The table provides an indication of the level of service in which each is capable of operating [19].

TABLE 12.19 Surge arrester standards

Developer	Standard No.	Title
IEC	IEC 60099-1	Surge Arresters – Part 1: Non-Linear Resistor Type Gapped Arresters for AC Systems
IEC	IEC/TR 60099-3	Surge Arresters – Part 3: Artificial Pollution Testing of Surge Arresters
IEC	IEC 60099-4	Surge Arresters – Part 4: Metal-Oxide Surge Arresters without Gaps for AC Systems
IEC	IEC 60099-5	Surge Arresters – Part 5: Selection and Application Recommendations
IEC	IEC 60099-6	Surge Arresters – Part 6: Surge Arresters Containing Both Series and Parallel Gapped Structures – Rated 52 kV and Less
IEC	IEC 61643-1	Low-Voltage Surge Protective Devices – Part 1: Surge Protective Devices Connected to Low-Voltage Power Distribution Systems – Requirements and Tests
IEC	IEC 61643-12	Low-Voltage Surge Protective Devices – Part 12: Surge Protective Devices Connected to Low-Voltage Power Distribution Systems – Selection and Application Principles
IEC	IEC-61643-21	Low-Voltage Surge Protective Devices – Part 21: Surge Protective Devices Connected to Telecommunications and Signalling Networks – Performance Requirements and Testing Methods
IEC	IEC 61643-311	Components for Low-Voltage Surge Protective Devices – Part 311: Specification for Gas Discharge Tubes (GDT)
IEC	IEC 61643-321	Components for Low-Voltage Surge Protective Devices – Part 321: Specifications for Avalanche Breakdown Diode (ABD)
IEC	IEC 61643-331	Components for Low-Voltage Surge Protective Devices – Part 331: Specification for Metal Oxide Varistors (MOV)
IEC	IEC 61643-341	Components for Low-Voltage Surge Protective Devices – Part 341: Specification for Thyristor Surge Suppressors (TSS)
IEC	IEC 61992-5	Railway Applications – Fixed Installations – DC Switchgear – Part 5: Surge Arresters and Low-Voltage Limiters for Specific Use in DC Systems
IEC	IEC 61992-7	Surge Arresters – Part 7: Glossary of Terms and Definitions from IEC Publications 60099-1, 60099-4, 60099-6, 61643-1, 61643-12, 61643-21, 61643-311, 61643-321, 61643-331 And 61643-341
IEEE	IEEE 141	IEEE Recommended Practice for Electric Power Distribution for Industrial Plants
IEEE	IEEE 1299	Guide for the Connection of Surge Arresters to Protect Insulated, Shielded Electric Power Cable Systems

(Continued)

TABLE 12.19 Surge arrester standards—cont'd

Developer	Standard No.	Title
IEEE	IEEE C62.1™	IEEE Standard for Gapped Silicon-Carbide Surge Arresters for AC Power Circuits
IEEE	IEEE C62.2™	IEEE Guide for the Application of Gapped Silicon-Carbide Surge Arresters for Alternating Current Systems
IEEE	IEEE C62.11™	Standard for Metal-Oxide Surge Arresters for AC Power Circuits (> 1 kV)
IEEE	IEEE C62.21™	IEEE Guide for the Application of Surge Voltage Protective Equipment on AC Rotating Machinery 1000 Volts and Greater
IEEE	IEEE C62.22™	IEEE Guide for Application of Metal-Oxide Surge Arresters for Alternating-Current Systems
IEEE	IEEE C62.22.1	IEEE Guide for the Connection of Surge Arresters to Protect Insulated, Shielded Electric Power Cable Systems
IEEE	IEEE C62.23™	IEEE Application Guide for Surge Protection of Electric Generating Plants
IEEE	IEEE C62.31	IEEE Standard Test Specifications for Gas-Tube Surge-Protective Devices
IEEE	IEEE C62.32	IEEE Standard Test Methods for Low-Voltage Air Gap Surge-Protective Device Components (Excluding Valve and Expulsion Types)
IEEE	IEEE C62.33	IEEE Standard Test Specifications for Varistor Surge-Protective Devices
IEEE	IEEE C62.34	Standard for Performance of Low-Voltage (1000 V rms or Less, Frequency between 48 Hz and 62 Hz) Surge Protective Devices (Secondary Arresters)
IEEE	IEEE C62.35	IEEE Standard Test Specifications for Avalanche Junction Semiconductor Surge Protective Devices
IEEE	IEEE C62.36	IEEE Standard Test Methods for Surge Protectors Used in Low-voltage data, Communications, and Signaling Circuits
IEEE	IEEE C62.37.1	IEEE Guide for the Application of Thyristor Surge Protective Devices
IEEE	IEEE C62.41.1™	Guide on the Surge Environment in Low-Voltage (1000 V and Less) AC Power Circuits
IEEE	IEEE C62.41.2™	Recommended Practice on Characterization of Surges in Low-Voltage (1000 V and Less) AC Power Circuits
IEEE	IEEE C62.42	IEEE Guide for the Application of Component Surge-Protective Devices for Use in Low-Voltage [Equal to or Less than 1000 V (ac) or 1200 V (dc)] Circuits
IEEE	IEEE C62.43	IEEE Draft Guide for the Application of Surge Protectors Used in Low-Voltage (Equal to or Less than 1000 V rms or 1200 VDC) Data, Communications, and Signaling Circuits
IEEE	IEEE C62.45™	Recommended Practice on Surge Testing for Equipment Connected to Low-Voltage (1000 V and Less) AC Power Circuits
IEEE	IEEE C62.48™	IEEE Guide on Interactions Between Power System Disturbances and Surge-Protective Devices

TABLE 12.19   Surge arrester standards—cont'd

Developer	Standard No.	Title
NEMA	NEMA/ANSI C62.61	Gas Tube Surge Arresters on Wire Line Telephone Circuits
IEEE	IEEE C62.62™	IEEE Standard Test Specifications for Surge Protective Devices for Low-Voltage AC Power Circuits
IEEE	IEEE C62.64	IEEE Standard Specifications for Surge Protectors Used in Low-Voltage Data, Communications, and Signaling Circuits
IEEE	IEEE C62.72™	IEEE Guide for the Application of Surge-Protective Devices for Low-Voltage (1000 V or Less) AC Power Circuits
NEMA	NEMA LA 1	Surge Arresters
NEMA	NEMA LS 1	Low-Voltage Surge Protection Devices
NETA	NETA ATS	NETA Acceptance Testing Specifications for Electrical Power Distribution Equipment and Systems
UL	UL 1449	Standard for Surge Protective Devices

TABLE 12.20   Arrester class differences – discharge-current withstand test [19]

Arrester discharge-current withstand test description	Station Class	Intermediate Class	Distribution Normal duty	Distribution Heavy duty	Secondary Class
High current, short duration test					
65 kA crest, (4—6)/(10–15) μs wave shape	✔	✔	✔		
100 kA crest, (4-6)/(10–15) μs wave shape				✔	
10 kA crest, (4-6)/(10–15) μs wave shape					✔
Low current, long duration test					
150–200 mile line capacitance discharge	✔				
100 mile line capacitance discharge		✔			
Rectangular shaped wave 250 A min. surge current, 2000 μs duration				✔	
Rectangular shaped wave 75 A min. surge current, 2000 μs duration			✔		

Arrester location is an important design consideration. Ideally, the device location should be at the terminals of the equipment being protected. However, the sum of the arrester lead length plus the conductor length from the arrester lead junction should be kept to a total of 10 feet (10') or less. ANSI C62.2, *IEEE Guide for the Application of Gapped Silicon-Carbide Surge Arresters for Alternating Current Systems* should be consulted regarding that requirement.

Consideration of employee safety during overhead live line maintenance and repair is mandated by governmental regulations. The International Electrotechnical Commission and the US Department of Labor, Occupational Safety and Health Administration (OSHA) address those concerns. Codes, standards, and recommended practices reflecting live line work practices are presented in Tables 12.21 and 12.22. Both provide a variety of standards including those covering: safe work practices; procedures; equipment maintenance and testing; personnel protective equipment (PPE); and service equipment.

TABLE 12.21   IEC live working standards

Developer	Standard No.	Title
IEC	IEC 60050-651	International Electrotechnical Vocabulary – Part 651: Live Working
IEC	IEC 60743	Live Working – Terminology for Tools, Equipment and Devices
IEC	IEC 60832	Insulating Poles (Insulating Sticks) and Universal Tool Attachments (Fittings) for Live Working
IEC	IEC 60855	Insulating Foam-Filled Tubes and Solid Rods for Live Working
IEC	IEC 60895	Live Working – Conductive Clothing for Use at Nominal Voltage Up To 800 kV AC and +/− 600 kV DC
IEC	IEC 60900	Live Working – Hand Tools for Use Up To 1000 VAC and 1500 VDC
IEC	IEC 60903	Live Working – Gloves of Insulating Material
IEC	IEC 60984	Sleeves of Insulating Material for Live Working
IEC	IEC 61057	Aerial Devices with Insulating Boom Used for Live Working
IEC	IEC 61219	Live Working – Earthing or Earthing and Short-Circuiting Equipment Using Lances as a Short-Circuiting Device – Lance Earthing
IEC	IEC 61229	Rigid Protective Covers for Live Working on AC Installations
IEC	IEC 61230	Live Working – Portable Equipment for Earthing or Earthing and Short-Circuiting
IEC	IEC 61235	Live Working – Insulating Hollow Tubes for Electrical Purposes
IEC	IEC 61236	Saddles, Pole Clamps (Stick Clamps) and Accessories for Live Working
IEC	IEC 61243-1	Live Working – Voltage Detectors – Part 1: Capacitive Type to Be Used for Voltages Exceeding 1 kV AC
IEC	IEC 61243-2	Live Working – Voltage Detectors – Part 2: Resistive Type to Be Used for Voltages of 1 kV to 36 kV AC
IEC	IEC 61243-3	Live Working – Voltage Detectors – Part 3: Two-Pole Low-Voltage Type
IEC	IEC 61243-5	Live Working – Voltage Detectors – Part 5: Voltage Detecting Systems (VDS)

TABLE 12.21 IEC live working standards—cont'd

Developer	Standard No.	Title
IEC	IEC 61318	Live Working – Conformity Assessment Applicable to Tools, Devices and Equipment
IEC	IEC/TR 61328	Live Working – Guidelines for the Installation of Transmission Line Conductors and Earthwires – Stringing Equipment and Accessory Items
IEC	IEC 61472	Live Working – Minimum Approach Distances for AC Systems in the Voltage Range 72.5 kV to 800 kV – a Method of Calculation
IEC	IEC 61477	Live Working – Minimum Requirements for the Utilization of Tools, Devices and Equipment
IEC	IEC 61478	Live Working – Ladders of Insulating Material
IEC	IEC 61479	Live Working – Flexible Conductor Covers (Line Hoses) of Insulating Material
IEC	IEC 61481	Live Working – Portable Phase Comparators for Use on Voltages from 1 kV to 36 kV AC
IEC	IEC 61482-1	Live Working – Flame-Resistant Materials for Clothing for Thermal Protection of Workers – Thermal Hazards of an Electric Arc – Part 1: Test Methods
IEC	IEC 61482-1-2	Live Working – Protective Clothing Against the Thermal Hazards of an Electric Arc – Part 1-2: Test Methods – Method 2: Determination of Arc Protection Class of Material and Clothing by Using a Constrained and Directed Arc (Box Test)
IEC	IEC/TS 61813	Live Working – Care, Maintenance and In-Service Testing of Aerial Devices with Insulating Booms
IEC	IEC/TR 61911	Live Working – Guidelines for Installation of Distribution Line Conductors – Stringing Equipment and Accessory Items
IEC	IEC 61936-1	Power Installations Exceeding 1 kV AC – Part 1: Common Rules
IEC	IEC 62192	Live Working – Insulating Ropes
IEC	IEC 62193	Live Working – Telescopic Sticks and Telescopic Measuring Sticks
IEC	IEC 62237	Live Working – Insulating Hoses with Fittings for Use with Hydraulic Tools and Equipment
IEC	IEC/TR 62263	Live Working – Guidelines for the Installation and Maintenance of Optical Fibre Cables on Overhead Power Lines

Electrical Substations

Electrical substations are used in both transmission and distribution systems. Transmission substations are responsible for stepping up generated power to transmission voltage levels. Substations are also used to step-down transmission voltage levels to distribution system voltage levels of 23 kV or 34.5 kV. The distribution voltages may require further level reduction at a *secondary distribution substation*. Common secondary distribution voltages are 4160 V, 5000 V, 12,470 V, 13,200 V, and 13,800 V.

Substation voltage control involves the regulation of the substation voltage amplitude differences, between no-load and full-load conditions, to specific ± system voltage swing

TABLE 12.22 29 CFR 1910.269 Appendix E, *Referenced Documents*

Developer	Standard No.	Title
ASME	ANSI/ASME B20.1	Safety Standard for Conveyors and Related Equipment
ASTM	ASTM D120	Specification for Rubber Insulating Gloves
ASTM	ASTM D149	Test Method for Dielectric Breakdown Voltage and Dielectric Strength of Solid Electrical Insulating Materials at Commercial Power Frequencies
ASTM	ASTM D178	Specification for Rubber Insulating Matting
ASTM	ASTM D1048	Specification for Rubber Insulating Blankets
ASTM	ASTM D1049	Specification for Rubber Insulating Covers
ASTM	ASTM D1050	Specification for Rubber Insulating Line Hose
ASTM	ASTM D1051	Specification for Rubber Insulating Sleeves
ASTM	ASTM F478	Specification for In-Service Care of Insulating Line Hose and Covers
ASTM	ASTM F479	Specification for In-Service Care of Insulating Blankets
ASTM	ASTM F496	Specification for In-Service Care of Insulating Gloves and Sleeves
ASTM	ASTM F711	Specification for Fiberglass-Reinforced Plastic (FRP) Rod and Tube Used in Live Line Tools
ASTM	ASTM F712	Test Methods for Electrically Insulating Plastic Guard Equipment for Protection of Workers
ASTM	ASTM F819	Definitions of Terms Relating to Electrical Protective Equipment for Workers
ASTM	ASTM F855	Specifications for Temporary Grounding Systems to Be Used on De-Energized Electric Power Lines and Equipment
ASTM	ASTM F887	Specifications for Personal Climbing Equipment
ASTM	ASTM F914	Test Method for Acoustic Emission for Insulated Aerial Personnel Devices
ASTM	ASTM F968	Specification for Electrically Insulating Plastic Guard Equipment for Protection of Workers
ASTM	ASTM F1116	Test Method for Determining Dielectric Strength of Overshoe Footwear
ASTM	ASTM F1117	Specification for Dielectric Overshoe Footwear
ASTM	ASTN F1236	Guide for Visual Inspection of Electrical Protective Rubber Products
ASTM	ASTM F1505	Standard Specification for Insulated and Insulating Hand Tools
ASTM	ASTM F1506	Standard Performance Specification for Textile Materials for Wearing Apparel for Use by Electrical Workers Exposed to Momentary Electric Arc and Related Thermal Hazards
IEEE	ANSI C2	National Electrical Safety Code
IEEE	ANSI/IEEE Std. 4	IEEE Standard Techniques for High-Voltage Testing
IEEE	IEEE 62	IEEE Guide for Field Testing Power Apparatus Insulation
IEEE	ANSI/IEEE 100	IEEE Standard Dictionary of Electrical and Electronic Terms
IEEE	IEEE 524	IEEE Guide to the Installation of Overhead Transmission Line Conductors
IEEE	ANSI/IEEE 516	IEEE Guide for Maintenance Methods on Energized Power-Lines
IEEE	ANSI/IEEE 935	IEEE Guide on Terminology for Tools and Equipment to Be Used in Live Line Working

TABLE 12.22   29 CFR 1910.269 Appendix E, Referenced Documents—cont'd

Developer	Standard No.	Title
IEEE	ANSI/IEEE 957	IEEE Guide for Cleaning Insulators
IEEE	ANSI/IEEE 978	IEEE Gide for In-Service Maintenance and Electrical Testing of Live-Line Tools
IEEE	IEEE 1048	IEEE Guide for Protective Grounding of Power Lines
IEEE	IEEE 1067	IEEE Guide for the In-Service Use, Care, Maintenance, and Testing of Conductive Clothing for Use on Voltages up to 765 kV AC
IEEE	ANSI/IEEE Std. 4	IEEE Standard Techniques for High-Voltage Testing
ISA*	ANSI Z133.1	American National Standard for Arboricultural Operations – Pruning, Trimming, Repairing, Maintaining, and Removing Trees, and Cutting Brush – Safety Requirements
SIA	ANSI/SIA A92.2	American National Standard for Vehicle-Mounted Elevating and Rotating Aerial Devices

*ISA = International Society of Arboriculture.

percentages. Substation voltage control can be maintained using several techniques, including substation bus regulation, feeder regulation, or both. Bus voltage regulation can be accomplished through a distribution transformer's secondary winding tap-changing-under-load equipment (load-ratio control) and by induction or step-voltage regulators connected in series with the distribution transformer secondaries. Capacitor bank switching and synchronous condensers can also be employed for voltage regulation. Feeder line voltage regulation can be accomplished using shunt capacitors and step-voltage regulators. Table 12.23 contains codes, standards, and recommended practices for reactors and voltage regulators.

Substations contain a substantial amount of equipment, including bus; circuit breakers; transformers; switching devices; high-voltage switchgear; protection devices and relays; metering; support structures; low-voltage motor control centers, panelboards, and transformers; communications and control equipment; etc. Voltage regulation equipment can also be included in substations.

A *unit substation* is a prefabricated transformer substation, skid-mounted to allow remote fabrication and transportation to any location. This modular type substation is composed of three main sections, including a *primary section*, a *transformer section*, and a *secondary section*. The primary section provides the connection point for one or more incoming high-voltage circuits. Switching and interrupting devices can be included as part of that section. The transformer section contains one or more transformers which can be provided with automatic tap-changing-under-load or tap-changing-not-under load equipment. The secondary section provides connections for one or more secondary connections, each provided with switching or interrupting devices.

TABLE 12.23　Reactors and voltage regulator standards

Developer	Standard No.	Title
IEEE	IEEE C37.109	IEEE Gide for Liquid-Immersed Transformer Through-Fault-Current Duration
IEEE	IEEE C57.12.35	Standard for Bar Coding for Distribution Transformers and Step-Voltage Regulators
IEEE	IEEE C57.15	IEEE Standard Requirements, Terminology, and Test Code for Step-Voltage Regulators
IEEE	IEEE C57.16	IEEE Standard Requirements, Terminology, and Test Code for Dry-Type Air-Core Series-Connected Reactors
IEEE	IEEE C57.21	IEEE Standard Requirements, Terminology, and Test Code for Shunt Reactors Rated Over 500 kVA
IEEE	IEEE C57.113	IEEE Guide for Partial Discharge Measurement in Liquid-Filled Power Transformers and Shunt Reactors
IEEE	IEEE C57.117	IEEE Guide for Reporting Failure Data for Power Transformers and Shunt Reactors on Electric Utility Power Systems
IEEE	IEEE C57.120	IEEE Loss Evaluation Guide for Power Transformers and Reactors
IEEE	IEEE C57.125	IEEE Guide for Failure Investigation, Documentation, and Analysis for Power Transformers and Shunt Reactors
IEEE	IEEE C57.127	IEEE Guide for the Detection and Location of Acoustic Emissions from Partial Discharges in Oil-Immersed Power Transformers and Reactors
IEEE	IEEE PC57.130	IEEE Trial-Use Guide for the Use of Dissolved Gas Analysis During Factory Temperature Rise Tests for the Evaluation of Oil-Immersed Transformers and Reactors
IEEE	IEEE C57.136	IEEE Guide for Sound Level Abatement and Termination for Liquid-Immersed Power Transformers and Shunt Reactors Rated over 500 kVA
NEMA	NEMA TR 1	Transformers, Regulators, and Reactors

IEEE Std 141 [20] lists configurations in which unit substations are normally designed and classified. Those classifications are based on the number of transformers and bus configurations for the unit substations. They include:

Radial

Primary Selective and Primary Loop

Secondary Selective

Secondary Spot Network

Distributed Network

Duplex (Breaker and a Half Scheme)

The International Electrotechnical Commission (IEC) has established a group of Standards for Substation design, installation, and operation. Many of those standards are presented in

Table 12.24, representing extensive communication requirements for substation control and coordination within the utility system. They also contain standards for substation component parts. Table 12.25 contains the IEC standards for high-voltage switchgear and controlgear, as well as prefabricated unit substations. The protective device standards include high-voltage sulfur (sulphur) hexafluoride (SF_6) switchgear; high-voltage circuit breakers;

TABLE 12.24 IEC Substation and Substation Equipment standards

Developer	Standard No.	Title
IEC	IEC CISPR 11	Industrial Scientific and Medical (ISM) Radio-Frequency Equipment – Electromagnetic Disturbance Characteristics – Limits and Methods of Measurement
IEC	IEC Guide 109	Environmental Aspects – Inclusion in Electrotechnical Standards
IEC	IEC Guide 111	Electrical High-Voltage Equipment in High-Voltage Substations – Common Recommendations for Product Standards
IEC	IEC 60027-1	Letter Symbols to Be Used in Electrical Technology – Part 1: General
IEC	IEC 60038	IEC Standard Voltages
IEC	IEC 60050-605	International Electrotechnical Vocabulary. Chapter 605: Generation, Transmission and Distribution of Electricity – Substations
IEC	IEC 60050-616	International Electrotechnical Vocabulary – Part 605: Generation, Transmission and Distribution of Electricity – Substations
IEC	IEC 60059	IEC Standard Current Ratings
IEC	IEC 60060-1	High-Voltage Test Techniques – Part 1: General Definitions and Test Requirements
IEC	IEC 60068 (all parts)	Environmental Testing
IEC	IEC 60068-2-18	Environmental Testing – Part 2-18: Tests – Test R and Guidance: Water
IEC	IEC 60068-2-75	Environmental Testing – Part 2-75: Tests – Test Eh: Hammer Tests
IEC	IEC 60071-1	Insulation Co-Ordination – Part 1: Definitions, Principles and Rules
IEC	IEC 60071-2	Insulation Co-Ordination – Part 2: Application Guide
IEC	IEC 60216 (all parts)	Electric Insulating Materials – Properties of Thermal Endurance
IEC	IEC 60296	Specification for Unused Mineral Insulating Oils for Transformers and Switchgear
IEC	IEC 60417 (all parts)	Graphical Symbols for Use on Equipment
IEC	IEC 60439-5	Low-Voltage Switchgear and Controlgear Assemblies – Part 5: Particular Requirements for Assemblies for Power Distribution in Public Networks
IEC	IEC 60633	Terminology for High-Voltage Direct Current (HVDC) Transmission
IEC	IEC 60660	Insulators – Tests on Indoor Post Insulators of Organic Material for Systems with Nominal Voltages Greater Than 1 000 V Up To But Not Including 300 kV

(Continued)

TABLE 12.24 IEC Substation and Substation Equipment standards—cont'd

Developer	Standard No.	Title
IEC	IEC60664-1	Insulation Coordination for Equipment Within Low-Voltage Systems – Part 1: Principles, Requirements and Tests
IEC	IEC 60695	Fire Hazard Testing
IEC	IEC 60721 (all parts)	Classification of Environmental Conditions
IEC	IEC 60721-1	Classification of Environmental Conditions – Part 1: Environmental Parameters and Their Severities
IEC	IEC 60721-2-2	Classification of Environmental Conditions – Part 2-2: Environmental Conditions Appearing in Nature – Precipitation and Wind
IEC	IEC 60721-2-4	Classification of Environmental Conditions – Part 2-4: Environmental Conditions Appearing in Nature – Solar Radiation and Temperature
IEC	IEC 60721-2-6	Classification of Environmental Conditions – Part 2-6: Environmental Conditions Appearing in Nature – Earthquake Vibration and Shock
IEC	IEC 60867	Insulating Liquids – Specifications for Unused Liquids Based on Synthetic Aromatic Hydrocarbons
IEC	IEC 60870-5-103	Telecontrol Equipment and Systems – Part 5-103: Transmission Protocols – Companion Standard for the Informative Interface of Protection Equipment
IEC	IEC/TS 60870-5-601	Telecontrol Equipment and Systems – Part 5-601: Conformance Test Cases for the IEC 60870-5-101 Companion Standard
IEC	IEC 61000-4-12	Electromagnetic Compatibility (EMC) – Part 4-12: Testing and Measurement Techniques – Ring Wave Immunity Test
IEC	IEC 61000-4-18	Electromagnetic Compatibility (EMC) – Part 4-18: Testing and Measurement Techniques – Damped Oscillatory Wave Immunity Test
IEC	IEC /TS 61000-6-5	Electromagnetic Compatibility (EMC) – Part 6-5: Generic Standards – Immunity for Power Station and Substation Environments
IEC	IEC 61000-6-6	Electromagnetic Compatibility (EMC) – Part 6-6: Generic Standards – HEMP Immunity for Indoor Equipment
IEC	IEC 61100	Classification of Insulating Liquids According of Fire-Point and Net Caloric Value
IEC	IEC 61024	Protection of Structures Against Lightning
IEC	IEC 61330	High-Voltage/Low-Voltage Prefabricated Substations
IEC	IEC 61660-1	Short-Circuit Currents in D.C. Auxiliary Installations in Power Plants and Substations – Part 1: Calculation of Short-Circuit Currents
IEC	IEC 61660-2	Short-Circuit Currents in D.C. Auxiliary Installations in Power Plants and Substations – Part 2: Calculation of Effects
IEC	IEC 61660-3	Short-Circuit Currents in D.C. Auxiliary Installations in Power Plants and Substations – Part 3: Examples of Calculations
IEC	IEC 61850-1	Short-Circuit Currents in D.C. Auxiliary Installations in Power Plants and Substations – Part 3: Examples of Calculations
IEC	IEC 61850-2	Communication Networks and Systems in Substations – Part 2: Glossary
IEC	IEC 61850-3	Communication Networks and Systems in Substations – Part 3: General Requirements

TABLE 12.24 IEC Substation and Substation Equipment standards—cont'd

Developer	Standard No.	Title
IEC	IEC 61850-4	Communication Networks and Systems in Substations – Part 4: System and Project Management
IEC	IEC 61850-5	Communication Networks and Systems in Substations – Part 5: Communication Requirements for Functions and Device Models
IEC	IEC 61850-6	Communication Networks and Systems in Substations – Part 6: Configuration Description Language for Communication in Electrical Substations Related to IEDs
IEC	IEC/TS 60870-5-604	Telecontrol Equipment and Systems – Part 5-604: Conformance Test Cases for the IEC 60870-5-104 Companion Standard
IEC	IEC/TS 60870-6-602	Telecontrol Equipment and Systems – Part 6-602: Telecontrol Protocols Compatible with ISO Standards and ITU-T Recommendations – TASE Transport Profiles
IEC	IEC 61000-2	Electromagnetic Compatibility (EMC) – Part 2-12: Environment – Compatibility Levels for Low-Frequency Conducted Disturbances and Signaling in Public Medium-Voltage Power Supply Systems
IEC	IEC 61850-7-1	Communication Networks and Systems in Substations – Part 7-1: Basic Communication Structure for Substation and Feeder Equipment – Principles and Models
IEC	IEC 61850-7-2	Communication Networks and Systems in Substations – Part 7-2: Basic Communication Structure for Substation and Feeder Equipment – Abstract Communication Service Interface (ACSI)
IEC	IEC 61850-7-3	Communication Networks and Systems in Substations – Part 7-3: Basic Communication Structure for Substation and Feeder Equipment – Common data classes
IEC	IEC 61850-7-4	Communication Networks and Systems in Substations – Part 7-4: Basic Communication Structure for Substation and Feeder Equipment – Compatible Logical Node Classes and Data Classes
IEC	IEC 61850-8-1	Communication Networks and Systems in Substations – Part 8-1: Specific Communication Service Mapping (SCSM) – Mappings to MMS (ISO 9506-1 and ISO 9506-2) and to ISO/IEC 8802-3
IEC	IEC 61850-9-1	Communication Networks and Systems in Substations – Part 9-1: Specific Communication Service Mapping (SCSM) – Sampled Values Over Serial Unidirectional Multi-drop Point to Point Link
IEC	IEC 61850-9-2	Communication Networks and Systems in Substations – Part 9-2: Specific Communication Service Mapping (SCSM) – Sampled Values Over ISO/IEC 8802-3
IEC	IEC 61850-10	Communication Networks and Systems in Substations – Part 10: Conformance Testing
IEC	IEC 61850-80-1	Communication Networks and Systems for Power Utility Automation – Part 80-1: Guideline to Exchanging Information from a CDC-based Data Model Using IEC 60870-5-101 or IEC 60870-5-104
IEC	IEC 61850-90-1	Communication Networks and Systems for Power Utility Automation – Part 90-1: Use of IEC 61850 for the Communication Between Substations

(Continued)

TABLE 12.24 IEC Substation and Substation Equipment standards—cont'd

Developer	Standard No.	Title
IEC	IEC 61850-SER	Communication Networks and Systems in Substations – All Parts
IEC	IEC 61936-1	Power Installations Exceeding 1 kV AC – Part 1: Common Rules
IEC	IEC 61973	High-Voltage Direct Current (HVDC) Substation Audible Noise
IEC	IEC 62313	Railway Applications – Power Supply and Rolling Stock – Technical Criteria for the Coordination Between Power Supply (Substation) and Rolling Stock
IEC	IEC/TS 62371	Characteristics of Hollow Pressurised and Unpressurised Ceramic and Glass Insulators for Use in Electrical Equipment with Rated Voltages Greater Than 1000 V
IEC	IEC 62445-2	Use of IEC 61850 for the Communication Between Control Centers and Substations
IEC	IEC 62590	Railway Applications – Fixed Installations – Electronic Power Converters for Substations

TABLE 12.25 IEC high-voltage switchgear/controlgear standards

Developer	Standard No.	Title
IEC	IEC 60296	Specification for Unused Mineral Insulating Oils for Transformers and Switchgear
IEC	IEC 60298	AC Metal-Enclosed Switchgear and Controlgear for Rated Voltages Above 1 kV and Up To and Including 52 kV
IEC	IEC 60376	Specification of Technical Grade Sulfur Hexafluoride (SF6) for Use in Electrical Equipment
IEC	IEC 60466	AC Insulation-Enclosed Switchgear and Controlgear for Rated Voltages Above 1 kV and Up To and Including 38 kV
IEC	IEC 60480	Guidelines for the Checking and Treatment of Sulfur Hexafluoride (SF6) Taken From Electrical Equipment and Specification for Its Re-Use
IEC	IEC 60517	Gas-Insulated Metal-Enclosed Switchgear for Rated Voltages of 72.5 kV and Above
IEC	IEC 60529	Degrees of Protection Provided by Enclosures (IP Code)
IEC	IEC 60694	Common Specifications for High-Voltage Switchgear and Controlgear Standards
IEC	IEC 60943	Guidance Concerning the Permissible Temperature Rise for Parts of Electrical Equipment, in Particular for Terminals
IEC	IEC 61100	Classification of Insulating Liquids According of Fire-Point and Net Caloric Value
IEC	IEC 61180-1	High-Voltage Test Techniques for Low-Voltage Equipment – Part 1: Definitions, Test and Procedure Requirements
IEC	IEC 61634	High-Voltage Switchgear and Controlgear – Use and Handling of Sulphur Hexafluoride (SF6) in High-Voltage Switchgear and Controlgear

TABLE 12.25 IEC high-voltage switchgear/controlgear standards—cont'd

Developer	Standard No.	Title
IEC	IEC 62271-001	Common Specifications for High-Voltage Switchgear and Controlgear Standards
IEC	IEC 62271-100	High-Voltage Alternating Current Circuit-Breakers
IEC	IEC 62271-101	Synthetic Testing of High-Voltage Alternating Current Circuit-Breakers
IEC	IEC 62271-102	Alternating Current Disconnectors and Earthing Switches
IEC	IEC 62271-103	Switches for Rated Voltages Above 1 kV and Less Than 52 kV
IEC	IEC 62271-104	High-Voltage Switches for Rated Voltages of 52 kV and Above
IEC	IEC 62271-105	Alternating Current Switch-Fuse Combinations
IEC	IEC 62271-106	High-Voltage Alternating Current Contactors and Contactor-Based Motor-Starters
IEC	IEC 62271-107	High-Voltage Alternating Current Switchgear-Fuse Combinations
IEC	IEC 62271-108	Switchgear Having Combined Functions
IEC	IEC 62271-109	Alternating-Current Series Capacitor By-Pass Switches
IEC	IEC 62271-200	AC Metal Enclosed Switchgear and Controlgear for Rated Voltages Above 1 kV and Up To and Including 52 kV
IEC	IEC 62271-201	AC Insulation-Enclosed Switchgear and Controlgear for Rated Voltages Above 1 kV and Up To and Including 38 kV
IEC	IEC 62271-202	High-Voltage/Low-Voltage Prefabricated Substations
IEC	IEC 62271-203	Gas-Insulated Metal-Enclosed Switchgear for Rated Voltages of 72.5 kV and Above
IEC	IEC 62271-204	Rigid High-Voltage, Gas-Insulated Transmission Lines for Rated Voltages of 72.5 kV and Above
IEC	IEC 62271-300	Guide for Seismic Qualification of High-Voltage Alternating Current Circuit-Breakers
IEC	IEC 62271-301	High-Voltage Alternating Current Circuit-Breakers – Inductive Load Switching
IEC	IEC 62271-302	High-Voltage Alternating Current Circuit-Breakers – Guide for Short-Circuit and Switching Test Procedures for Metal-Enclosed and Dead Tank Circuit-Breakers
IEC	IEC 62271-303	High-Voltage Switchgear and Controlgear – Use and Handling of Sulphur Hexafluoride (SF6) in High-Voltage Switchgear and Controlgear
IEC	IEC 62271-304	Additional Requirements for Enclosed Switchgear and Controlgear from 1 kV to 72.5 kV to Be Used in Severe Climatic Conditions
IEC	IEC 62271-305	Cable Connections for Gas-Insulated Metal-Enclosed Switchgear for Rated Voltages of 72.5 kV and Above – Fluid-Filled and Extruded Insulation Cables – Fluid-Filled and Dry Type Cable-Terminations
IEC	IEC 62271-306	Direct Connection Between Power Transformers and Gas-Insulated Metal-Enclosed Switchgear for Rated Voltages of 72.5 kV and Above
IEC	IEC 62271-307	High-Voltage Switchgear and Controlgear – the Use of Electronic and Associated Technologies in Auxiliary Equipment of Switchgear and Controlgear
IEC	IEC 62271-308	Guide for Asymmetrical Short-Circuit Breaking Test Duty T100a

prefabricated, unit substations; metal-enclosed switchgear; low-voltage switchgear; metal-clad switchgear; etc.

North American substation equipment codes, standards, and recommended practices are presented in Table 12.26. Some of those standards were written specifically for substations and unit substations. Many of them could also be applicable for high-voltage control and equipment protection use in other applications. Transformers, reactors, and protective relaying were not included in Table 12.26, but are included elsewhere.

TABLE 12.26   Substation and substation-related equipment standards

Developer	Standard No.	Title
ASTM	ASTM D149-97a	Standard Test Method for Dielectric Breakdown Voltage and Dielectric Strength of Solid Electrical Insulating Materials at Commercial Power Frequencies
ASTM	ASTM D150-98	Standard Test Methods for AC Loss Characteristics and Permittivity (Dielectric Constant) of Solid Electrical Insulation
ASTM	ASTM D256ae1	Standard Test Methods for Determining the Izod Pendulum Impact Resistance of Plastics
ASTM	ASTM D257	Standard Test Methods for DC Resistance or Conductance of Insulating Materials
ASTM	ASTM D648	Standard Test Method for Deflection Temperature of Plastics Under Flexural Load in the Edgewise Position
ASTM	ASTM D790E1	Standard Test Methods for Flexural Properties of Unreinforced and Reinforced Plastics and Electrical Insulating Materials
ASTM	ASTM D2307a	Standard Test Method for Thermal Endurance of Film-Insulated Round Magnet Wire
ASTM	ASTM D5628	Standard Test Method for Impact Resistance of Flat, Rigid Plastic Specimens by Means of a Falling Dart (Tup or Falling Mass)
CAN/CSA	CAN/CSA C22.2 No. 31	Switchgear Assemblies
CAN/CSA	CAN/CSA-C50052	Cast Aluminum Alloy Enclosures for Gas-Filled High-Voltage Switchgear and Controlgear (Adopted CENELEC Standard EN 50052)
CAN/CSA	CAN/CSA-C50064	Wrought Aluminum and Aluminum Alloy Enclosures for Gas-Filled High-Voltage Switchgear and Controlgear (Adopted CENELEC Standard EN 50064
CAN/CSA	CAN/CSA-C50068	Wrought Steel Enclosures for Gas-Filled High-Voltage Switchgear and Controlgear (Adopted CENELEC Standard EN 50068)
CAN/CSA	CAN/CSA-C50069	Welded Composite Enclosures of Cast and Wrought Aluminum Alloys for Gas-Filled High-Voltage Switchgear and Controlgear (Adopted CENELEC Standard EN 50069
CAN/CSA	CAN/CSA-C50089	Cast Resin Partitions for Metal-Enclosed Gas-Filled High-Voltage Switchgear and Controlgear (Adopted CENELEC Standard EN 50089)

TABLE 12.26 Substation and substation-related equipment standards—cont'd

Developer	Standard No.	Title
IEEE	IEEE 4	IEEE Standard Techniques for High-Voltage Testing
IEEE	IEEE 1247	IEEE Standard for Interrupter Switches for Alternating Current, Rated Above 1000 V
IEEE	IEEE 1291	IEEE Guide for Partial Discharge Measurement in Power Switchgear
IEEE	IEEE 1325	IEEE Recommended Practice for Reporting Field Failure Data for Power Circuit Breakers
IEEE	IEEE IEC 62271-111	High-Voltage Switchgear and Controlgear – Part 111: Overhead, Pad-mounted, Dry Vault, and Submersible Automatic Circuit Reclosers and Fault Interrupters for Alternating Current Systems up to 38 kV Corrigendum to C37.60-2003
IEEE	IEEE C37.04	IEEE Standard for Rating Structure for AC High-Voltage Circuit Breakers Rated on a Symmetrical Current Basis
IEEE	IEEE C37.04a	IEEE Standard Rating Structure for AC High-Voltage Circuit Breakers Rated on a Symmetrical Current Basis Amendment 1: Capacitance Current Switching
IEEE	ANSI C37.06	American National Standard for Switchgear – AC High-Voltage Circuit Breakers Rated on a Symmetrical Current Basis – Preferred Ratings and Related Required Capabilities
IEEE	ANSI C37.06.1	American National Standard Trial-Use Guide for High-Voltage Circuit Breakers Rated on a Symmetrical Current Basis – Designated "Definite Purpose for Fast Transient Recovery Voltage Rise Times"
IEEE	ANSI/IEEE C37.09	Test Procedure for AC High-Voltage Circuit Breakers Rated on a Symmetrical Current Basis (includes supplements ANSI/IEEE C37.09c-1983, ANSI/IEEE C37.09e-1983 and ANSI/IEEE C37.09g-1991)
IEEE	IEEE C37.09a	IEEE Standard Test Procedure for AC High-Voltage Circuit Breakers Rated on a Symmetrical Current Basis – Amendment 1: Capacitance Current Switching
IEEE	IEEE C37.010	IEEE Application Guide for AC High-Voltage Circuit Breakers Rated on a Symmetrical Current Basis
IEEE	IEEE C37.10	IEEE Guide for Diagnostics and Failure Investigation of Power Circuit Breakers
IEEE	IEEE C37.10.1	IEEE Guide for the Selection of Monitoring for Circuit Breakers
IEEE	IEEE C37.11	IEEE Standard Requirements for Electrical Control for AC High-Voltage Circuit Breakers Rated on a Symmetrical Current Basis
IEEE	IEEE C37.011	IEEE Application Guide for Transient Recovery Voltage for AC High-Voltage Circuit Breakers Rated on a Symmetrical Current Basis
IEEE	IEEE C37.12.1	IEEE Guide for High-Voltage (>1000 V) Circuit Breaker Instruction Manual Content
IEEE	IEEE C37.012	IEEE Application Guide for Capacitance Current Switching for AC High-Voltage Circuit Breakers Rated on a Symmetrical Current Basis
IEEE	IEEE C37.13	IEEE Standard for Low-Voltage AC Power Circuit Breakers Used in Enclosures

(Continued)

TABLE 12.26 Substation and substation-related equipment standards—cont'd

Developer	Standard No.	Title
IEEE	IEEE C37.13.1	IEEE Standard for Definite-Purpose Switching Devices for Use in Metal-Enclosed Low-Voltage Power Circuit Breaker Switchgear
IEEE	IEEE C37.013	IEEE Standard for AC High-Voltage Generator Circuit Breakers Rated on a Symmetrical Current Basis
IEEE	IEEE C37.013a	IEEE Standard for AC High-Voltage Generator Circuit Breakers Rated on a Symmetrical Current Basis – Amendment 1: Supplement for Use with Generators Rated 10-100 MVA
IEEE	IEEE C37.015	IEEE Application Guide for Shunt Reactor Switching
IEEE	IEEE C37.016	IEEE Standard for AC High-Voltage Circuit Switcher rated 15.5 kV through 245 kV
IEEE	ANSI C37.14	American National Standard IEEE Standard for Low-Voltage DC Power Circuit Breakers Used in Enclosures
IEEE	ANSI C37.16	American National Standard Low-Voltage Power Circuit Breakers and AC Power Circuit Protectors – Preferred Ratings
IEEE	ANSI C37.17	American National Standard for Trip Devices for AC and General Purpose DC Low-Voltage Power Circuit Breakers
IEEE	IEEE C37.18	Standard Field Discharge Circuit Breakers for Rotating Electric Machinery
IEEE	IEEE C37.20.1	IEEE Standard for Metal-Enclosed Low-Voltage Power Circuit Breaker Switchgear
IEEE	IEEE C37.20.1a	IEEE Standard for Metal-Enclosed Low-Voltage Power Circuit Breaker Switchgear Amendment 1: Short-Time and Short-Circuit Withstand Current Tests – Minimum Areas for Multiple Cable Connections
IEEE	IEEE C37.20.1b	IEEE Standard for Metal-Enclosed Low-Voltage Power Circuit Breaker Switchgear Amendment 2: Additional Requirements for Control and Auxiliary Power Wiring in DC Traction Power Switchgear
IEEE	IEEE C37.20.2	IEEE Standard for Metal-Clad Switchgear
IEEE	IEEE C37.20.3	IEEE Standard for Metal-Enclosed Interrupter Switchgear
IEEE	IEEE C37.20.4	IEEE Standard for Indoor AC Switches (1 kV–38 kV) for Use in Metal-Enclosed Switchgear
IEEE	IEEE C37.20.6	IEEE Standard for 4.76 kV to 38 kV Rated Ground and Test Devices Used in Enclosures
IEEE	IEEE C37.20.7	IEEE Guide for Testing Metal-Enclosed Switchgear Rated up to 38 kV for Internal Arcing Faults
IEEE	IEEE C37.21	IEEE Standard for Control Switchboards
IEEE	ANSI C37.22	American National Standard Preferred Ratings and Related Capabilities for Indoor AC Medium-Voltage Switches Used in Metal-Enclosed Switchgear
IEEE	IEEE C37.23	IEEE Standard for Metal-Enclosed Bus
IEEE	IEEE C37.24	IEEE Guide for Evaluating the Effect of Solar Radiation on Outdoor Metal-Enclosed Switchgear
IEEE	IEEE C37.26	IEEE Guide for Methods of Power Factor Measurement for Low-Voltage Inductive Test Circuits

TABLE 12.26 **Substation and substation-related equipment standards—cont'd**

Developer	Standard No.	Title
IEEE	IEEE C37.27	IEEE Standard Application Guide for Low-Voltage AC Non-integrally Fused Power Circuit Breakers (Using Separately Mounted Current-Limiting Fuses)
IEEE	IEEE C37.30	IEEE Standard Requirements for High-Voltage Switches
IEEE	ANSI C37.32	American National Standard for High-Voltage Switches, Bus Supports, and Accessories – Schedules of Preferred Ratings, Construction Guidelines
IEEE	IEEE C37.34	IEEE Standard Test Code for High-Voltage Air Switches
IEEE	IEEE C37.35	IEEE Guide for the Application, Installation, Operation, and Maintenance of High-Voltage Air Disconnecting and Load Interrupter Switches
IEEE	IEEE C37.36b	IEEE Guide to Current Interruption with Horn-Gap Air Switches
IEEE	IEEEC37.37	IEEE Loading Guide for AC High-Voltage Air Switches (in Excess of 1000 V)
IEEE	IEEE C37.40	IEEE Standard Service Conditions and Definitions for High-Voltage Fuses, Distribution Enclosed Single-Pole Air Switches, Fuse Disconnecting Switches, and Accessories
IEEE	IEEE C37.40b	IEEE Standard Service Conditions and Definitions for External Fuses for Shunt Capacitors.
IEEE	IEEE C37.41	IEEE Standard Design Tests for High-Voltage Fuses, Distribution Enclosed Single-Pole Air Switches, Fuse Disconnecting Switches, and Accessories
IEEE	ANSI C37.42	American National Standard Specification for High-Voltage Expulsion Type Distribution Class Fuses, Cutouts, Fuse Disconnecting Switches and Fuse Links
IEEE	IEEE C37.45	IEEE Standard Specifications for High-Voltage Distribution Class Enclosed Single-Pole Air Switches with Rated Voltages from 1 kV through 8.3 kV
IEEE	ANSI C37.46	American National Standard Specifications for Power Fuses and Fuse Disconnecting Switches
IEEE	ANSI C37.47	American National Standard for High-Voltage Current-Limiting Type Distribution Class Fuses and Fuse Disconnecting Switches
IEEE	IEEE C37.48	IEEE Guide for Application, Operation, and Maintenance of High-Voltage Fuses, Distribution Enclosed Single-Pole Air Switches, Fuse Disconnecting Switches, and Accessories
IEEE	IEEE C37.48.1	IEEE Guide for the Operation, Classification, Application, and Coordination of Current-Limiting Fuses with Rated Voltages 1–38 kV
IEEE	ANSI C37.50	American National Standard for Switchgear – Low-Voltage AC Power Circuit Breakers Used in Enclosures – Test Procedures
IEEE	ANSI C37.51	American National Standard for Switchgear – Metal-Enclosed Low-Voltage AC Power-Circuit-Breaker Switchgear Assemblies – Conformance Test

(Continued)

TABLE 12.26 Substation and substation-related equipment standards—cont'd

Developer	Standard No.	Title
IEEE	ANSI C37.52	American National Standard Test Procedures for Low-Voltage AC Power Circuit Protectors Used in Enclosures
IEEE	ANSI C37.53.1	American National Standard High-Voltage Current-Limiting Motor-Starter Fuses – Conference Test Procedures
NEMA	ANSI C37.54	Indoor Alternating Current High-Voltage Circuit Breakers Applied as Removable Elements in Metal-Enclosed Switchgear – Conformance Test Procedures
IEEE	ANSI C37.55	American National Standard for Switchgear – Metal-Clad Switchgear Assemblies – Conformance Test Procedures
IEEE	ANSI C37.57	American National Standard for Switchgear-Metal-Enclosed Interrupter Switchgear Assemblies – Conformance Testing
IEEE	ANSI C37.58	Indoor AC Medium-Voltage Switches for Use in Metal-Enclosed Switchgear – Conformance Test Procedures
IEEE	IEEE C37.59	IEEE Standard Requirements for Conversion of Power Switchgear Equipment
IEEE	IEEE C37.60	IEEE Standard Requirements for Overhead, Pad-Mounted, Dry Vault, and Submersible Automatic Circuit Reclosers and Fault Interrupters for Alternating Current Systems up to 38 kV – Corrigendum 1
IEEE	IEEE C37.63	IEEE Standard Requirements for Overhead, Pad-Mounted, Dry-Vault, and. Submersible Automatic Line. Sectionalizers for AC Systems
IEEE	IEEE C37.66	IEEE Standard Requirements for Capacitor Switches for AC Systems (1 kV to 38 kV)
IEEE	IEEE C37.74	IEEE Standard Requirements for Subsurface, Vault, and Pad-Mounted Load-Interrupter Switchgear and Fused Load-Interrupter Switchgear for Alternating Current Systems Up to 38 kV
IEEE	IEEE C37.81	IEEE Guide for Seismic Qualification of Class 1E Metal-Enclosed Power Switchgear Assemblies
IEEE	IEEE C37.081	IEEE Guide for Synthetic Fault Testing of AC High-Voltage Circuit Breakers Rated on a Symmetrical Current Basis
IEEE	IEEE C37.081a	Supplement to IEEE Guide for Synthetic Fault Testing of AC High-Voltage Circuit Breakers Rated on a Symmetrical Current Basis. 8.3.2: Recovery Voltage for Terminal Faults; Asymmetrical Short-Circuit Current
IEEE	ANSI/IEEE C37.82	IEEE Standard for the Qualification of Switchgear Assemblies for Class 1E Applications in Nuclear Power Generating Stations
IEEE	ANSI/IEEE C37.082	IEEE Standard Methods for the Measurement of Sound Pressure Levels of AC Power Circuit Breakers
IEEE	IEEE C37.083	IEEE Guide for Synthetic Capacitive Current Switching Tests of AC High-Voltage Circuit Breakers
NEMA	ANSI C37.85	American National Standard for Switchgear – Alternating-Current High-Voltage Power Vacuum Interrupters-Safety Requirements for X-radiation Limits
IEEE	IEEE C37.100	IEEE Standard Definitions for Power Switchgear

TABLE 12.26 Substation and substation-related equipment standards—cont'd

Developer	Standard No.	Title
IEEE	IEEE C37.100.1	IEEE Standard of Common Requirements for High-Voltage Power Switchgear Rated Above 1000 V
IEEE	IEEE C37.104	IEEE Guide for Automatic Reclosing of Line Circuit Breakers for AC Distribution and Transmission Lines
IEEE	IEEE C37.119	IEEE Guide for Breaker Failure Protection of Power Circuit Breakers
IEEE	ANSI C37.121	American National Standard for Switchgear – Unit Substations – Requirements
IEEE	IEEE C62.122.	IEEE Guide for Gas-Insulated Substations
IEEE	IEEE C37.123	IEEE Guide to Specifications for Gas-Insulated, Electric Power Substation Equipment
NEMA	NEMA BU 1.1	General Instructions for Proper Handling, Installation, Operation and Maintenance of Busway Rated 600 V or Less
NEMA	ANSI C12.19	Utility Industry End Device Data Tables
NEMA	ANSI C37.50	American National Standard for Switchgear – Low-Voltage AC Power Circuit Breakers Used in Enclosures – Test Procedures
NEMA	ANSI C37.51	American National Standard for Switchgear – Metal-Enclosed Low-Voltage AC Power Circuit Breaker Switchgear Assemblies – Conformance Test Procedures
NEMA	ANSI C37.52	Phase-Shifting Devices Used in Metering, Marking and Arrangement of, Terminals for
NEMA	ANSI C37.54	American National Standard for Indoor AC High-Voltage Circuit Breakers Applied as Removable Elements in Metal-Enclosed Switchgear – Conformance Test Procedures
NEMA	ANSI C37.55	American National Standard for Switchgear – Medium-Voltage Metal-Clad Assemblies – Conformance Test Procedures
NEMA	ANSI C37.57	American National Standard for Switchgear – Metal-Enclosed Interrupter Switchgear Assemblies – Conformance Testing
NEMA	ANSI C37.58	American National Standard for Switchgear – Indoor AC Medium-Voltage Switches for Use in Metal-Enclosed Switchgear – Conformance Test Procedures
IEEE	ANSI C57.12.55	American National Standard for Transformers – Used in Unit Installations, Including Unit Substations – Conformance Standard
NEMA	ANSI/NEMA CC 1	Electric Power Connection for Substations
NEMA	NEMA ICS 2	AC Vacuum-Break Magnetic Controllers Rated 1500 V AC
NEMA	NEMA ICS 2, Part 8	Controllers, Contactors and Overload Relays Rated Not More Than 2000 V AC or 750 V DC – Part 8: Disconnect Devices for Use in Industrial Control Equipment
NEMA	NEMA ICS 2.3	Instructions for the Handling, Installation, Operation and Maintenance of Motor Control Centers Rated Not More than 600 V
NEMA	NEMA ICS 3	Medium-Voltage Controllers Rated 2001 to 7200 V AC

(Continued)

TABLE 12.26 **Substation and substation-related equipment standards—cont'd**

Developer	Standard No.	Title
NEMA	NEMA ICS 3.1	Handling, Storage and Installation Guide for AC General-Purpose Medium-Voltage Contactors and Class E Controllers, 50 and 60 Hz
NEMA	NEMA ICS 10	Part 1: Electromechanical AC Transfer Switch Equipment
NEMA	NEMA ICS 18	Motor Control Centers
NEMA	NEMA KS 1	Enclosed and Miscellaneous Distribution Equipment Switches (600 V Maximum)
NEMA	NEMA LI 1	NEMA Standards for Polyester-Glass Laminates
NEMA	NEMA LI6	Relative Temperature Indices of Industrial Thermosetting Laminates
NEMA	NEMA PB 1	Panelboards
NEMA	ANSI/NEMA PB 1.1	General Instructions for Proper Installation, Operation and Maintenance of Panelboards Rated 600 V or Less
NEMA	NEMA PB 2	Deadfront Distribution Switchboards
NEMA	ANSI/NEMA PB 2.1	General Instructions for Proper Handling, Installation, Operation and Maintenance of Deadfront Distribution Switchboards Rated 600 V or Less
NEMA	NEMA PE 5	Utility Type Battery Chargers
NEMA	NEMA SC-4	AC High-Voltage Circuit Breaker
NEMA	NEMA SG 10	Guide to OSHA and NFPA 70E Safety Regulations when Servicing and Maintaining Medium-Voltage Switchgear and Circuit Breakers Rated above 1000 V
NEMA	NEMA TR 1	Transformers, Regulators, and Reactors
NEMA	NEMA 250	Enclosures for Electrical Equipment (1000 V Maximum)
NEMA	NEMA 260	Safety Labels for Padmounted Switchgear and Transformers Sited in Public Areas
UL	UL 857	Busways

IEEE 141 designates [21] three basic classifications for switchgear, including *open switchgear assemblies*, *enclosed switchgear assemblies*, and *metal-clad switchgear*. Open switchgear does not use an enclosure as structural support. Metal-enclosed switchgear is encased on the top and all sides by a sheet metal enclosure, with its interior only accessible by doors and/or removable panels. That equipment can be designed for either indoor or outdoor use and for either AC or DC service. There are three types of metal-enclosed switchgear, including metal-clad, low-voltage power circuit breaker switchgear, and interrupter switchgear. Metal-clad switchgear is metal-enclosed switchgear; however, it must meet other specific criteria outlined in IEEE 141 [22]:

- Utilize a removable main circuit switching and interrupting device. Those devices must utilize mechanical means to move the device physically between the connected and disconnected positions to the bus. They must also have self-aligning and self-coupling primary and secondary disconnects.

- Grounded metal barriers shall be used to enclose the circuit switching or interrupting devices, bus, instrument transformers, etc. Inner barriers shall be provided as an integral component to prevent the exposure of any of the interrupting devices, assuring that no energized primary circuit components will be exposed when the front access door is opened.
- Grounded metal compartments shall be utilized to enclose all energized parts. While the removable element is in either the test, disconnected, or withdrawn positions, a metal shutter assembly shall be utilized for prevention of contact with primary circuit components.
- Suitable insulating material shall be utilized for covering all primary bus conductors and connections. Insulated barriers can be specified for installation between all phases and between phase and ground.
- Safe and proper operation shall be assured by the use of mechanical interlocks.
- Grounded metal barriers shall be utilized for isolation of all instruments, meters, relays, secondary control equipment, and their wiring.
- The access door utilized for insertion of the main circuit switching or interrupting device can be used for mounting of instruments and relays and shall provide access for the secondary or control section in the switchgear enclosure.

Metal-enclosed power circuit breaker switchgear is rated for 1000 V or less. It is also considered metal-enclosed type switchgear, with the following equipment requirements [23]:

- Fused or non-fused 1000 V or less power circuit breakers are installed in the enclosure;
- Bus and bus connections are non-insulated; however, insulated and isolated bus can be installed;
- Both instrument and control power transformers are utilized;
- Instruments, meters, and protective relays can be installed;
- Control wiring and accessories are utilized;
- Facilities are provided for cable and bus terminations;
- Withdrawing of circuit breakers will automatically engage protective shutters covering the line-side contacts;
- Power circuit breakers can be either drawout type or stationary installed in individual grounded metal compartments;
- Mechanical interlocks are provided on drawout type circuit breakers to ensure personnel safety and the proper operating sequence.

Metal-enclosed interrupter switchgear is considered as power switchgear. It is metal-enclosed and is required to have the following equipment requirements [24]:

- Circuit disconnect switches and/or interrupter switches;
- If needed, power fuses can be installed;

- The bus and bus connections are non-insulated;
- Both instrument and power control transformers can be installed;
- Control wiring and accessories are utilized;
- Switches can be either stationary or removable types;
- Removable switches must be provided with mechanical interlocks to ensure personnel safety and the proper operating sequence.

It should be noted when reviewing the standards in Table 12.26, a committee has been established to develop two new standards, C37.30.1, *Standard Requirements for High-Voltage Air Switches, Switching Devices, and Interrupters* and C37.30.2, *Guide for Wind Loading Evaluation of High-Voltage (>1000 V) Air Break Switches.* That process began in 2008 and has been charged with the task of consolidating the existing standards C37.30, C37.32, C37.33, C37.34, C37.35, C37.36b, C37.37, and P1247 into the new standards C37.30.1 and C37.30.2. Procedures mandate that during that consolidation process, the contents of the existing standards will be reviewed, reaffirmed or revised on their normal review schedule until the new standards are approved and issued. The consolidated standards cover bus supports and equipment, switches, fused switches, and interrupter switches rated above 1000 V AC.

Substation, Transmission, Distribution, and Transformer Grounding

The standards in Table 12.26 cover high-voltage, medium-voltage, and low-voltage distribution switches and protective equipment, all of which are applicable for use in distribution substations. Also included in that table were standards for battery/battery charger systems for protective device operation, bus supports and hardware, circuit breakers, disconnects, and fused disconnects, Transformers, capacitors, and protective relaying were not included in the table; however, they are examined in separate chapters.

Grounding standards for substations; transmission and distribution systems; and transformer neutrals are presented in Table 12.27. There are four basic grounding schemes available for use with transformer neutrals, including *ungrounded, solidly grounded, resistance grounded,* and *reactive grounded.* In addition, it is critical that safe grounding practices be implemented during the design of an AC electrical substation and its maintenance and operation procedures. The standards presented in Table 12.27 list many of those which provide guidance for substation grounding practices.

IEEE 80, *IEEE Guide for Safety in AC Substation Grounding* examines the requirements for proper substation grounding practices. From a personal safety standpoint, the possibility does exist for the creation of voltage potential differences inside a substation during a ground fault or surge arrester operation event. A potential difference can be created between the feet of anyone walking (*step-potential*) in the substation or between substation metal equipment enclosures and support frames (*touch-potential*) under fault conditions. The design of

TABLE 12.27 Substation, transmission, distribution, and transformer grounding

Developer	Standard No.	Title
ASTM	ANSI/ASTM D448	Specifications for Standard Sizes of Course Aggregate for Highway Construction
CSA	CAN/CSA C22.1, Part III	Standard Canadian Electrical Code Part III – Electricity Distribution and Transmission Systems
IEC	IEC 60479-1	Effects of Current on Human Beings and Livestock – Part 1: General Aspects
IEC	IEC 60479-2	Effects of Current on Human Beings and Livestock – Part 2: Special Aspects
IEC	IEC 60479-3	Effects of Current on Human Beings and Livestock – Part 3: Effects of Currents Passing Through the Body of Livestock
IEC	IEC 60479-4	Effects of Current on Human Beings and Livestock – Part 4: Effects of Lightning Strokes on Human Beings and Livestock
IEC	IEC 60479-5	Effects of Current on Human Beings and Livestock – Part 5: Touch Voltage Threshold Values for Physiological Effects
IEC	IEC 60909-3	Short-Circuit Currents in Three-Phase AC Systems – Part 3: Currents During Two Separate Simultaneous Line-to-Earth Short Circuits and Partial Short-Circuit Currents Flowing Through Earth
IEC	IEC/TS 61201	Use of Conventional Touch Voltage Limits – Application Guide
IEC	IEC 61936-1	Power Installations Exceeding 1 kV AC – Part 1: Common Rules
IEEE	IEEE 80	IEEE Guide for Safety in AC Substation Grounding
IEEE	ANSI/IEEE Std. 81	IEEE Guide for Measuring Earth Resistivity, Ground Impedance, and Earth Surface Potentials of a Ground System – Part 1: Normal Measurements
IEEE	ANSI/IEEE Std 142	IEEE Recommended Practice for Grounding of Industrial and Commercial Power Systems
IEEE	IEEE 367	IEEE Guide for the Maximum Electric Power Station Ground Potential Rise and Induced Voltage from a Power Fault
IEEE	IEEE 590	Cable Plowing Guide
IEEE	ANSI/IEEE 837	IEEE Standard for Qualifying Permanent Connections Used in Substation Grounding
IEEE	ANSI/IEEE 1246	Guide for Temporary Protective Grounding Systems Used in Substations
IEEE	IEEE C2	National Electrical Safety Code
IEEE	IEEE C 62.91	IEEE Guide for the Application of Neutral Grounding in Electrical Utility Systems – Part 1: Introduction
IEEE	IEEE C62.92.2	IEEE Guide for the Application of Neutral Grounding in Electrical Utility Systems, Part 2: Grounding of Synchronous Generator Systems
IEEE	IEEE C62.92.3	IEEE Guide for the Application of Neutral Grounding in Electrical Utility Systems – Part 3: Generator Auxiliary Systems
IEEE	IEEE C62.92.4	IEEE Guide for the Application of Neutral Grounding in Electrical Utility Systems – Part 4: Distribution
IEEE	IEEE C62.92.5	IEEE Guide for the Application of Neutral Grounding in Electrical Utility Systems – Part 5: Transmission Systems and Subtransmission Systems
NEMA	ANSI/NEMA GR 1	Grounding Rod Electrodes and Grounding Rod Electrode Couplings
NFPA	NFPA 70®	National Electrical Code®

a substation grounding plan should create a ground grid system which will minimize the differences in ground and equipment enclosures voltage potentials creating an equipotential surface.

A *ground grid system* consists of buried copper conductors and multiple installed ground rods (electrodes) in a specific geometrical arrangement, with all cable intersection nodes electrically and mechanically connected. The design requires the extensive use of equipment enclosure and metal structural support bonding jumpers, electrically connected to the ground grid. It also necessitates electrical bonding connections of any metallic rebar in concrete and metal fencing into the ground grid system. Bonding jumpers are required between fixed fence posts and gates. Connection of a substation's lightning protection system and surge arresters into the ground grid system is instrumental in assuring an equipotential grid system design. The application of highway course aggregate on the ground surface as insulating material is another technique covered in some of the standards listed in Table 12.27. The purpose of the station ground grid design is to develop the lowest possible grid resistance to remote earth, based on the requirements necessitated by the soil resistivity inside the substation. IEEE Standard 80 provides guidance in designing of and detailed calculations for substation ground grids.

IEEE 141-1993, *IEEE Recommended Practice for Electric Power Distribution for Industrial Power Plants* provides a warning regarding interconnection of an industrial plant's grounding system into a substation's ground grid system. It notes:

> The ground grid of the utility substation is often interconnected with the industrial plant grounding system, either intentionally by overhead service or a buried ground wire or unintentionally through cable tray, conduit systems, or bus duct enclosures. As a result of this interconnection, the plant grounding system is elevated to the same potential above remote earth as the substation grid during a high-voltage fault in the substation. Dangerous surface potentials within an industrial plant as well as within the substation also must be prevented. In certain cases, hazardous surface potentials may be eliminated by effectively isolating the substation ground system from the plant ground system. In most cases, integrating the two grids together and suitably analyzing both systems for step and touch potentials have reduced these potentials to acceptable levels. [25]

References

1. IEEE C2-2007, *National Electrical Safety Code*; 2007, Article 011. Institute of Electrical and Electronic Engineers; New York, NY.
2. IEEE 141-1993 (R1999), *IEEE Recommended Practice for Electric Power Distribution for Industrial Power Plants*; 1999, pages vi–xvi. Institute of Electrical and Electronic Engineers; Piscataway, NJ.

3. US Department of Energy, *Using Distributed Energy Resources: A How-To Guide for Federal Facility Managers*; 2002, DOE/GO-101001-1520. National Renewable Energy Laboratory; Washington, DC.

4. US Department of Energy, *The Potential Benefits of Distributed Generation and Rate-Related Issues that May Impede Their Expansion*; February 2007, page 1-7. US DOE; Washington, DC.

5. American Wind Energy Association Website: http://www.awea.org/newsroom/pdf/Fast_Facts.pdf.

6. US Department of Energy, Fuel Cells, Hydrogen, Fuel Cell & Infrastructure Technologies Program; http://www1.eere.energy.gov/hydrogenandfuelcells/fuelcells/fc_types.html.

7. http://www1.eere.energy.gov/hydrogenandfuelcells/fuelcells/pdfs/fc_comparison_chart.pdf.

8. US Department of Energy, *National Transmission Grid Study*, May 2002. US DOE; Washington, DC.

9. Ibid.

10. Federal Energy Regulatory Commission, *What FERC Does*; 17 December, 2008; http://www.ferc.gov/about/ferc-does.asp.

11. Ibid.

12. FERC News Release, April 17, 2008, Docket Nos: RM08-7-000, AD08-6-000 and RM05-30-002.

13. US Department of Energy, *National Transmission Grid Study*, May 2002, page 3. US DOE; Washington, DC.

14. US Department of Labor, OSHA, *Electric Power Generation, Transmission, Distribution eTool – Illustrated Glossary:* Substations; http://www.osha.gov/SLTC/etools/electric_power/illustrated_glossary/substation.html.

15. Wolfe, Ronald and Moody, Russell, *Standard Specifications for Wood Poles*; 1997, page 2. Utility Pole Structures Conference, November 6–7, 1997; Reno/Sparks, Nevada.

16. IEEE 141-1993 (R1999), *IEEE Recommended Practice for Electrical Power Distribution for Industrial Plants*; 1999, page 337. Institute of Electrical and Electronic Engineers; New York.

17. Ibid., pages 333–334.

18. Ibid., page 335.

19. Ibid., page 336, Section 6.4.6.

20. Ibid., page 519.

21. Ibid., page 492, Sections 10.3.2 and 10.3.3.

22. Ibid., page 493, Section 10.3.3.1.

23. Ibid., pages 493–494, Section 10.3.3.2.

24. Ibid., page 494, Section 10.3.3.3.

25. Ibid., page 373.

Type of Products Requiring NRTL Approval

The product descriptions given here in Table A.1 are as per provisions of the General Industry Standards (Part 1910 of Title 29, Code of Federal Regulations – 29 CFR Part 1910).

TABLE A.1 General product descriptions

1	Electrical conductors or equipment (see Listing of Specific References under 1910.303 and 1910.307)
2	Automatic sprinkler systems
3	Fixed extinguishing systems (dry chemical, water spray, foam, or gaseous agents)
4	Fixed extinguishing systems components and agents
5	Fire detection device for automatic actuation of total flooding system
6	Portable fire extinguishers
7	Automatic fire detection devices and equipment
8	Employee alarm systems
9	Self-closing fire doors (openings to inside storage rooms for flammable or combustible liquids)
10	Fire doors [1½ hour (B) rated] (openings, to other parts of a building, of storage rooms for liquefied petroleum gas (LPG))
11	Metal frame of windows in partitions of inside acetylene generator rooms used in oxygen-fuel welding and cutting
12	Heat actuated (closing) devices (dip tanks containing flammable or combustible liquids)
13	Self-closing fire doors (including frames and hardware) used in openings into an exit
14	Flame arresters, check valves, hose (transfer stations), portable tanks and safety cans – (flammable/combustible liquids)
15	Pumps and self-closing faucets (for dispensing Class I liquids)
16	Flexible connectors (piping, valves, fittings) – (flammable liquids)
17	Service station dispensing units (automotive, marine)
18	Mechanical or gravity ventilation systems (automotive service station dispensing area)
19	Automotive service station latch-open devices for dispensing units
20	New commercial and industrial LPG consuming appliances

(Continued)

TABLE A.1 General product descriptions—cont'd

21	Flexible connectors (piping, valves, fittings) – LPG
22	Powered industrial truck LPG conversion equipment
23	LPG storage and handling systems (DOT containers, cylinders)
24	Automatic shut-off devices (portable LPG heaters including salamanders)
25	LPG container assemblies (non-DOT) for interchangeable installation above or under ground
26	Fixed electrostatic apparatus and devices (coating operations)
27	Electrostatic hand spray apparatus and devices
28	Electrostatic fluidized beds and associated equipment
29	Each appurtenance (e.g., pumps, compressors, safety relief devices, liquid-level gauging devices, valves and pressure gages) in storage and handling of anhydrous ammonia
30	Gasoline, LPG, diesel, or electrically powered industrial trucks used in hazardous atmospheres
31	Acetylene apparatus (torches, regulators or pressure-reducing valves, generators [stationary and portable], manifolds)
32	Acetylene generator compressors or booster systems
33	Acetylene piping protective devices
34	Manifolds (fuel gas or oxygen) – separately for each component part or as assembled units
35	Scaffolding and power or manually operated units of single-point adjustable suspension scaffolds
36	Hoisting machine and supports (Stone setters' adjustable multiple-point suspension scaffold)
37	Hoisting machines (two-point suspension scaffolds; Masons' adjustable multiple-point suspension scaffold

Source: OSHA website: http://www.osha.gov/dts/otpca/nrtl/prodcatg.html.

Occupational Safety and Health Administration Occupational Safety and Health Standards – 29 CFR

TABLE B.1 US Department of Labor Occupational Safety and Health Administration: Standards – 29 CFR

PART 24 Procedures for the Handling of Retaliation Complaints under Federal Employee Protection Statutes

PART 70 Production or Disclosure of Information or Materials

PART 70A Protection of Individual Privacy in Records

PART 71 Protection of Individual Privacy and Access to Records under the Privacy Act of 1974

PART 1900 Reserved

PART 1901 Procedures for State Agreements

PART 1902 State Plans for the Development and Enforcement of State Standards

PART 1903 Inspections, Citations, and Proposed Penalties

PART 1904 Recording and Reporting Occupational Injuries and Illness

PART 1905 Rules of Practice

PART 1906 Administration Witness and Documentations in Private Litigation

PART 1908 Consultation Agreements

PART 1910 Occupational Safety and Health Standards

PART 1911 Rules of Procedure for Promulgating, Modifying or Revoking OSHA Standards

PART 1912 Advisory Committees on Standards

PART 1912A National Advisory Committee on OSHA

PART 1913 Rules Concerning OSHA Access to Employee Medical Records

PART 1915 Occupational. Safety and Health Standards for Shipyard Employment

PART 1917 Marine Terminals

PART 1918 Safety and Health Regulations for Longshoring

PART 1919 Gear Certification

PART 1920 Procedure for Variations under Longshoremen's Act

PART 1921 Rules of Practice in Enforcement under Section 41 of Longshoremen's Act

PART 1922 Investigational Hearings under Section 41 of the Longshoremen's and Harbor Workers' Compensation Act

(Continued)

TABLE B.1 US Department of Labor Occupational Safety and Health Administration: Standards – 29 CFR—cont'd

PART 1924 Safety Standards Applicable to Workshops and Rehab. Facilities

PART 1925 Safety and Health Standards for Federal Service Contracts

PART 1926 Safety and Health Regulations for Construction

PART 1927 Reserved

PART 1928 Occupational. Safety and Health Standards for Agriculture

PART 1949 Office of Training and Education, OSHA

PART 1952 Approved State Plans for Enforcement of State Standards

PART 1953 Changes to State Plans

PART 1954 Procedures for the Evaluation and Monitoring of Approved State Plans

PART 1955 Procedures for Withdrawal of Approval of State Plans

PART 1956 Plans for State and Local Government Employees without Approved Plans

PART 1960 Basic Program Elements for Federal Employees OSHA

PART 1975 Coverage of Employees under the Williams–Steiger OSHA 1970

PART 1977 Discrimination against Employees under OSHA Act of 1970

PART 1978 Rules for Implementing Section 405 of the STAA of 1982

PART 1979 Procedures for the Handling of Discrimination Complaints under Section 519 of the Wendell H. Ford Aviation Investment and Reform Act for the 21st Century

PART 1980 Procedures for the Handling of Discrimination Complaints under Section 806 of the Corporate and Criminal Fraud Accountability Act of 2002, Title VIII of the Sarbanes-Oxley Act of 2002

PART 1981 Procedures for the Handling of Discrimination Complaints under Section 6 of the Pipeline Safety Improvement Act of 2002

PART 1990 Identification, Classification, and Regulation of Carcinogens

PART 2200 OSHA Review Commission

PART 2201 Regulations Implementing the Freedom of Information Act.

PART 2202 Rules of Ethics and Conduct of Review Commission Employees

PART 2203 Regulations Implementing the Government in the Sunshine Act

PART 2204 Implementation of the Equal Access to Justice Act

PART 2205 Enforcement of Nondiscrimination on the Basis of Handicap in Programs or Activities Conducted by the Occupational Safety and Health Review Commission

PART 2400 Regulations Implementing the Privacy Act

Source: www.osha.gov/pls/oshaweb/owasrch.search_form?p_doc_type=STANDARDS&p_toc_level=0.

Comparison 29 CFR 1910.269 versus 29 CFR 1910.147 Hazardous Energy Control Requirements

This Appendix compares hazardous energy control requirements in Regulations (Standards – 29 CFR) The Control of Hazardous Energy (Lockout/Tagout) – 1910.147 and Regulations (Standards – 29 CFR) Electric Power Generation, Transmission, and Distribution – 1910.269. This material should be used for reference only. The comparisons presented here should be used to exemplify any differences or similarities of those Regulations. The reader should refer to the latest editions of both standards for any revisions, additions, or deletions from the information presented here.

The following Tables are presented for review. The verbiage in the center column reflects that which is common to both Standards. *{Bracket}* represents specific verbiage included in 29 CFR 1910.269. *[Bracket]* represents specific verbiage included in 29 CFR 1910.147.

Table Description Summary

Table C.1	Energy Control Program
Table C.2	Procedures/Energy Control Procedures
Table C.3	Periodic Inspection
Table C.4	Training/Training and Communication
Table C.5	Tagout Systems
Table C.6	Retraining/Employee Retraining
Table C.7	Protective Materials and Hardware
Table C.8	Energy Isolation/Notification of Employees
Table C.9	Lockout/Tagout Application/Application of Control
Table C.10	Release from Lockout/Tagout
Table C.11	Additional Requirements
Table C.12	Group Lockout or Tagout/Outside Personnel (Contractors, etc.)

TABLE C.1 Comparison 1910.269 v. 1910.147 Requirements: Energy Control Program

29 CFR 1910.269	{General} [Energy Control Program]	29 CFR 1910.147
1910.269(d)(2)(i)	The employer shall establish a program consisting of energy control procedures, employee training, and periodic inspections to ensure that, before any employee performs any servicing or maintenance on a machine or equipment where the unexpected energizing, start up, or release of stored energy could occur and cause injury, the machine or equipment is isolated from the energy source and rendered inoperative.	1910.147(c)(1)
1910.269(d)(2)(ii)	The employer's energy control program under paragraph (d)(2) of this section shall meet the following requirements:	N/A
1910.269(d)(2)(ii)(A)	If an energy isolating device is not capable of being locked out, the employer's *[energy control]* program *[under paragraph (c)(1) of this section shall utilize]*{shall use}a tagout system.	1910.147(c)(2)(i)
1910.269(d)(2)(ii)(B)	If an energy isolating device is capable of being locked out, the employer's {program}[energy control program under paragraph (c)(1) of this section shall utilize] {shall use} lockout, unless the employer can demonstrate that the {use}[utilization] of a tagout system will provide full employee protection as {follows:}[set forth in paragraph (c)(3) of this section.]	1910.147(c)(2)(ii)
1910.269(d)(2)(ii)(B)(1)	When a tagout device is used on an energy isolating device which is capable of being locked out, the tagout device shall be attached at the same location that the lockout device would have been attached, and the employer shall demonstrate that the tagout program will provide a level of safety equivalent to that obtained by using a lockout program.	1910.147(c)(3)(i)
1910.269(d)(2)(ii)(B)(2)	In demonstrating that a level of safety is achieved in the tagout program *[which is]* equivalent to the level of safety obtained by the {use} [using] of a lockout program, the employer shall demonstrate full compliance with all tagout-related provisions of this standard together with such additional elements as are necessary to provide the equivalent safety available from the use of a lockout device. Additional means to be considered as part of the demonstration of full employee protection shall include the implementation of additional safety measures such as the removal of an isolating circuit element, blocking of a controlling switch, opening of an extra disconnecting device, or the removal of a valve handle to reduce the likelihood of inadvertent {energizing} [energization].	1910.147(c)(3)(ii)
1910.269(d)(2)(ii)(C)	After {November 1, 1994} [January 2, 1990], whenever replacement or major repair, renovation or modification of a machine or equipment is performed, and whenever new machines or equipment are installed, energy isolating devices for such machine or equipment shall be designed to accept a lockout device.	1910.147(c)(3)(iii)

[Brackets] indicate verbiage included in 1910.147
{Brackets} indicate verbiage included in 1910.269.

TABLE C.2 Comparison of 1910.269 v. 1910.147 Requirements: Procedures/Energy Control Procedures

29 CFR 1910.269	{Procedures} [Energy Control Procedures]	29 CFR 1910.147
1910.269(d)(2)(iii)	Procedures shall be developed, documented and *{used}* *[utilized]* for the control of potentially hazardous energy *[when employees are engaged in the activities]* covered by *{paragraph (d) of}* this section.[a]	1910.147(c)(4)(i)
1910.269(d)(2)(iv)	The procedure shall clearly and specifically outline the scope, purpose, *{responsibility,}* authorization, rules, and techniques to be *[utilized for]{applied to}* the control of hazardous energy, and the measures to enforce compliance including, but not limited to, the following:	1910.147(c)(4)(ii)
1910.269(d)(2)(iv)(A)	A specific statement of the intended use of *{this}* *[the]* procedure;	1910.147(c)(4)(ii)(A)
1910.269(d)(2)(iv)(B)	Specific procedural steps for shutting down, isolating, blocking and securing machines or equipment to control hazardous energy;	1910.147(c)(4)(ii)(B)
1910.269(d)(2)(iv)(C)	Specific procedural steps for the placement, removal and transfer of lockout devices or tagout devices and the responsibility for them; and	1910.147(c)(4)(ii)(C)
1910.269(d)(2)(iv)(D)	Specific requirements for testing a machine or equipment to determine and verify the effectiveness of lockout devices, tagout devices, and other energy control measures.	1910.147(c)(4)(ii)(D)

{Bracket} indicates verbiage included in 1910.269.

[Bracket] indicates verbiage included in 1910.147

[a]1910.147 Exception: The employer need not document the required procedure for a particular machine or equipment, when all of the following elements exist: (1) The machine or equipment has no potential for stored or residual energy or reaccumulation of stored energy after shut down which could endanger employees; (2) the machine or equipment has a single energy source which can be readily identified and isolated; (3) the isolation and locking out of that energy source will completely deenergize and deactivate the machine or equipment; (4) the machine or equipment is isolated from that energy source and locked out during servicing or maintenance; (5) a single lockout device will achieve a locked-out condition; (6) the lockout device is under the exclusive control of the authorized employee performing the servicing or maintenance; (7) the servicing or maintenance does not create hazards for other employees; and (8) the employer, in utilizing this exception, has had no accidents involving the unexpected activation or reenergization of the machine or equipment during servicing or maintenance.

TABLE C.3   Comparison of 1910.269 v. 1910.147 Requirements: Periodic Inspection

29 CFR 1910.269	{Periodic Inspection} [Periodic Inspection]	29 CFR 1910.147
1910.269(d)(2)(v)	The employer shall conduct a periodic inspection of the energy control procedure at least annually to ensure that the procedure and the *{provisions of paragraph (d) of this section}* *[requirements of this standard]* are being followed.	1910.147(c)(6)(i)
1910.269(d)(2)(v)(A)	The periodic inspection shall be performed by an authorized employee *{who is not using}* *[other than the one(s) using]* the energy control procedure being inspected.	1910.147(c)(6)(i)(A)
1910.269(d)(2)(v)(B)	The periodic inspection shall be *{designed to identify and}* *[conducted to]* correct any deviations or inadequacies.	1910.147(c)(6)(i)(B)
1910.269(d)(2)(v)(C)	*{If lockout is used for energy control,}* *[Where lockout is used for energy control,]* the periodic inspection shall include a review, between the inspector and each authorized employee, of that employee's responsibilities under the energy control procedure being inspected.	1910.147(c)(6)(i)(C)
1910.269(d)(2)(v)(D)	Where tagout is used for energy control, the periodic inspection shall include a review, between the inspector and each authorized and affected employee, of that employee's responsibilities under the energy control procedure being inspected, and the elements set forth in paragraph *{(d)(2)(vii)}* *[(c)(7)(ii)]* of this section.	1910.147(c)(6)(i)(D)
1910.269(d)(2)(v)(E)	The employer shall certify that the *[periodic]* inspections *{required by paragraph (d)(2)(v) of this section}* have been *{accomplished}* *[performed]*. The certification shall identify the machine or equipment on which the energy control procedure was being {used} [utilized], the date of the inspection, the employees included in the inspection, and the person performing the inspection.	1910.147(c)(6)(i)(E)

{Bracket} indicates verbiage included in 1910.269.
[Bracket] indicates verbiage included in 1910.147

TABLE C.4   Comparison of 1910.269 v. 1910.147 Requirements: Training/Training and Communication

29 CFR 1910.269	{Training} [Training and Communication]	29 CFR 1910.147
1910.269(d)(2)(vi)	The employer shall provide training to ensure that the purpose and function of the energy control program are understood by employees and that the knowledge and skills required for the safe application, usage, and removal of energy controls are acquired by employees. The training shall include the following:	1910.147(c)(7)(i)
1910.269(d)(2)(vi)(A)	Each authorized employee shall receive training in the recognition of applicable hazardous energy sources, the type and magnitude of energy available in the workplace, and in the methods and means necessary for energy isolation and control.	1910.147(c)(7)(i)(A)
1910.269(d)(2)(vi)(B)	Each affected employee shall be instructed in the purpose and use of the energy control procedure.	1910.147(c)(7)(i)(B)
1910.269(d)(2)(vi)(C)	All other employees whose work operations are or may be in an area where energy control procedures may be {used} [utilized] shall be instructed about the procedure{s} and about the prohibition relating to attempts to restart or reenergize machines or equipment that are locked out or tagged out.	1910.147(c)(7)(i)(C)

{Bracket} indicates verbiage included in 1910.269.
[Bracket] indicates verbiage included in 1910.147.

TABLE C.5   Comparison of 1910.269 v. 1910.147 Requirements: Tagout Systems

29 CFR 1910.269	{Tagout Systems} [Tagout Systems]	29 CFR 1910.147
1910.269(d)(2)(vii)(A)	When tagout systems are used, employees shall also be trained in the following limitations of tags:	1910.147(c)(7)(ii)
1910.269(d)(2)(vii)(B)	Tags are essentially warning devices affixed to energy isolating devices and do not provide the physical restraint on those devices that is provided by a lock.	1910.147(c)(7)(ii)(A)
1910.269(d)(2)(vii)(C)	When a tag is attached to an energy isolating means, it is not to be removed without authorization of the authorized person responsible for it, and it is never to be bypassed, ignored, or otherwise defeated.	1910.147(c)(7)(ii)(B)
1910.269(d)(2)(vii)(D)	Tags must be legible and understandable by all authorized employees, affected employees, and all other employees whose work operations are or may be in the area, in order to be effective.	1910.147(c)(7)(ii)(C)
1910.269(d)(2)(vii)(E)	Tags and their means of attachment must be made of materials which will withstand the environmental conditions encountered in the workplace.	1910.147(c)(7)(ii)(D)
1910.269(d)(2)(vii)(F)	Tags may evoke a false sense of security, and their meaning needs to be understood as part of the overall energy control program.	1910.147(c)(7)(ii)(E)
1910.269(d)(2)(vii)(G)	Tags must be securely attached to energy isolating devices so that they cannot be inadvertently or accidentally detached during use.	1910.147(c)(7)(ii)(F)

{Bracket} indicates verbiage included in 1910.269.
[Bracket] indicates verbiage included in 1910.147

TABLE C.6   Comparison of 1910.269 v. 1910.147 Requirements: Retraining/Employee Retraining

29 CFR 1910.269	{Retraining} [Employee retraining]	29 CFR 1910.147
1910.269(d)(2)(viii)	*{Retraining shall be provided by the employer as follows:}* *[Employee retraining.]*	1910.147(c)(7)(iii)
1910.269(d)(2)(viii)(A)	Retraining shall be provided for all authorized and affected employees *{whenever}* *[when]* there is a change in their job assignments, a change in machines, equipment, or processes that present a new hazard or whenever there is a change in the energy control procedures.	1910.147(c)(7)(iii)(A)
1910.269(d)(2)(viii)(B)	Retraining shall also be conducted whenever a periodic inspection under paragraph *{(d)(2)(v)}* *[(c)(6)]* of this section reveals, or whenever the employer has reason to believe, that there are deviations from or inadequacies in an employee's knowledge or use of the energy control procedures.	1910.147(c)(7)(iii)(B)
1910.269(d)(2)(viii)(C)	The retraining shall reestablish employee proficiency and shall introduce new or revised control methods and procedures, as necessary.	1910.147(c)(7)(iii)(C)
1910.269(d)(2)(ix)	The employer shall certify that employee training has been accomplished and is being kept up to date. The certification shall contain each employee's name and dates of training.	1910.147(c)(7)(iv)

{Bracket} indicates verbiage included in 1910.269.
[Bracket] indicates verbiage included in 1910.147

TABLE C.7 Comparison of 1910.269 v. 1910.147 Requirements: Protective Materials and Hardware

29 CFR 1910.269	{Protective materials and hardware} [Protective materials and hardware]	29 CFR 1910.147
1910.269(d)(3)	Protective materials and hardware.	1910.147(c)(5)
1910.269(d)(3)(i)	Locks, tags, chains, wedges, key blocks, adapter pins, self-locking fasteners, or other hardware shall be provided by the employer for isolating, securing, or blocking of machines or equipment from energy sources.	1910.147(c)(5)(i)
1910.269(d)(3)(ii)	Lockout devices and tagout device{s} [(s)] shall be singularly identified; {shall} [may] be the only devices used for controlling energy; may not be used for other purposes; and shall meet the following requirements:	1910.147(c)(5)(ii)
1910.269(d)(3)(ii)(A)	Lockout devices and tagout devices shall be capable of withstanding the environment to which they are exposed for the maximum period of time that exposure is expected.	1910.147(c)(5)(ii)(A)(1)
1910.269(d)(3)(ii)(A)(1)	Tagout devices shall be constructed and printed so that exposure to weather conditions or wet and damp locations will not cause the tag to deteriorate or the message on the tag to become illegible.	1910.147(c)(5)(ii)(A)(2)
1910.269(d)(3)(ii)(A)(2)	Tagout devices shall {be so constructed as} not {to} deteriorate when used in corrosive environments{.} [, such as areas where acid and alkali chemicals are handled and stored.]	1910.147(c)(5)(ii)(A)(3)
1910.269(d)(3)(ii)(B)	Lockout devices and tagout devices shall be standardized within the facility in at least one of the following criteria: color, shape, [or] size{. Additionally}, [; and additionally], in the case of tagout devices, print and format shall be standardized.	1910.147(c)(5)(ii)(B)
1910.269(d)(3)(ii)(C)	Lockout devices shall be substantial enough to prevent removal without the use of excessive force or unusual techniques, such as with the use of bolt cutters or metal cutting tools.	1910.147(c)(5)(ii)(C)(1)
1910.269(d)(3)(ii)(D)	Tagout devices, including their means of attachment, shall be substantial enough to prevent inadvertent or accidental removal. Tagout device attachment means shall be of a non-reusable type, attachable by hand, self-locking, and non-releasable with a minimum unlocking strength of no less than 50 pounds and shall have the general design and basic characteristics of being at least equivalent to a one-piece, all-environment-tolerant nylon cable tie.	1910.147(c)(5)(ii)(C)(2)
1910.269(d)(3)(ii)(E)	{Each lockout} [Lockout] device[s] or tagout device[s] shall {include provisions for the identification} [indicate the identity] of the employee applying the device[s].	1910.147(c)(5)(ii)(D)

TABLE C.7 Comparison of 1910.269 v. 1910.147 Requirements: Protective Materials and Hardware—cont'd

29 CFR 1910.269	{Protective materials and hardware} [Protective materials and hardware]	29 CFR 1910.147
1910.147(d)(3)(ii)(F)	Tagout devices shall warn against hazardous conditions if the machine or equipment is energized and shall include a legend such as the following: {Do Not Start, Do Not Open, Do Not Close, Do Not Energize, Do Not Operate.} [Do Not Start, Do Not Open, Do Not Close, Do Not Energize, Do Not Operate.]	1910.147(c)(5)(iii)

{Bracket} indicates verbiage included in 1910.269.
[Bracket] indicates verbiage included in 1910.147

TABLE C.8 Comparison of 1910.269 v. 1910.147 Requirements: Energy Isolation/Notification of Employees

29 CFR 1910.269	{Energy Isolation/Notification} [Energy Isolation/ Notification of Employees]	29 CFR 1910.147
1910.269(d)(4)	Lockout and tagout *{device application and removal may only be performed}* [shall be performed only] by the authorized employees who are performing the servicing or maintenance.	1910.147(c)(8)
1910.260(d)(5)	Affected employees shall be notified by the employer or authorized employee of the application and removal of lockout or tagout devices. Notification shall be given before the controls are applied and after they are removed from the machine or equipment.	1910.147(c)(9)

{Bracket} indicates verbiage included in 1910.269.
[Bracket] indicates verbiage included in 1910.147

TABLE C.9 Comparison 1910.269 v. 1910.147 Requirements: Lockout/Tagout Application/Application of Control

29 CFR 1910.269	{Lockout/Tagout Application} [Application of Control]	29 CFR 1910.147
1910.269(d)(6)	The established procedures for the application of energy control (the lockout or tagout procedures) shall include the following elements and actions, and *{these procedures}* shall be performed in the following sequence:	1910.147(d)
1910.269(d)(6)(i)	Before an authorized or affected employee turns off a machine or equipment, the authorized employee shall have knowledge of the type and magnitude of the energy, the hazards of the energy to be controlled, and the method or means to control the energy.	1910.147(d)(1)

(Continued)

TABLE C.9 Comparison 1910.269 v. 1910.147 Requirements: Lockout/Tagout Application/Application of Control—cont'd

29 CFR 1910.269	{Lockout/Tagout Application} [Application of Control]	29 CFR 1910.147
1910.269(d)(6)(ii)	The machine or equipment shall be turned off or shut down using the procedures established for the machine or equipment. An orderly shutdown *[must]* {shall} be *[utilized]* {used} to avoid any additional or increased hazard{s} *[(s)]* to employees as a result of the equipment stoppage.	1910.147(d)(2)
1910.269(d)(6)(iii)	All energy isolating devices that are needed to control the energy to the machine or equipment shall be physically located and operated in such a manner as to isolate the machine or equipment from energy source{s} *[(s)]*.	1910.147(d)(3)
1910.269(d)(6)(iv)	Lockout or tagout devices where used shall be affixed to each energy isolating device by authorized employees.	1910.147(d)(4)(i)
1910.269(d)(6)(iv)(A)	Lockout devices*[, where used,]* shall be *[affixed]* attached in a manner that will hold the energy isolating devices in a "safe" or "off" position.	1910.147(d)(4)(ii)
1910.269(d)(iv)(B)	Tagout devices*[, where used,]* shall be affixed in such a manner as will clearly indicate that the operation or movement of energy isolating devices from the "safe" or "off" position is prohibited.	1910.147(d)(4)(iii)
1910.269(d)(iv)(B)(1)	Where tagout devices are used with energy isolating devices designed with the capability of being locked {out}, the tag attachment shall be fastened at the same point at which the lock would have been attached.	1910.147(d)(4)(iii)(A)
1910.269(d) (6)(iv)(B)(2)	Where a tag cannot be affixed directly to the energy isolating device, the tag shall be located as close as safely possible to the device, in a position that will be immediately obvious to anyone attempting to operate the device.	1910.147(d)(4)(iii)(B)
1910.269(d)(6)(v)	Following the application of lockout or tagout devices to energy isolating devices, all potentially hazardous stored or residual energy shall be relieved, disconnected, restrained, or otherwise rendered safe.	1910.147(d)(5)(i)
1910.269(d)(6)(vi)	If there is a possibility of reaccumulation of stored energy to a hazardous level, verification of isolation shall be continued until the servicing or maintenance is completed or until the possibility of such accumulation no longer exists.	1910.147(d)(5)(ii)
1910.269(d)(6)(vii)	{Before} *[Prior to]* starting work on machines or equipment that have been locked out or tagged out, the authorized employee shall verify that isolation and {deenergizing} *[deenergization]* of the machine or equipment have been accomplished. {If normally energized parts will be exposed to contact by an employee while the machine or equipment is deenergized, a test shall be performed to ensure that these parts are deenergized.}	1910.147(d)(6)

[Brackets] indicate verbiage included in 1910.147
{Brackets} indicate verbiage included in 1910.269.

TABLE C.10   Comparison 1910.269 v. 1910.147 Requirements: Release from Lockout/Tagout

29 CFR 1910.269	{Release from lockout/tagout} [Release from lockout or tagout]	29 CFR 1910.147
1910.269(d)(7)	Before lockout or tagout devices are removed and energy is restored to the machine or equipment, procedures shall be followed and actions taken by the authorized employee{s} [(s)] to ensure the following:	1910.147(e)
1910.269(d)(7)(i)	The work area shall be inspected to ensure that nonessential items have been removed and [to ensure] that machine or equipment components are operationally intact.	1910.147(e)(1)
1910.269(d)(7)(ii)	The work area shall be checked to ensure that all employees have been safely positioned or removed.	1910.147(e)(2)(i)
1910.269(d)(7)(iii)	After lockout or tagout devices have been removed and before a machine or equipment is started, affected employees shall be notified that the lockout or tagout devices {s} [(s)] have been removed.	1910.147(e)(2)(ii)
1910.269(d)(7)(iv)	Each lockout or tagout device shall be removed from each energy isolating device by the authorized employee who applied the lockout or tagout device. {However, if that employee is not available to remove it, the} [Exception to paragraph (e)(3): When the authorized employee who applied the lockout or tagout device is not available to remove it, that] device may be removed under the direction of the employer, provided that specific procedures and training for such removal have been developed, documented, and incorporated into the employer's energy control program. The employer shall demonstrate that the specific procedure provides {a degree of} [equivalent] safety {equivalent to that provided by} [to] the removal of the device by the authorized employee who applied it. The specific procedure shall include at least the following elements:	1910.147(e)(3)
1910.269(d)(7)(iv)(A)	Verification by the employer that the authorized employee who applied the device is not at the facility;	1910.147(e)(3)(i)
1910.269(d)(7)(iv)(B)	Making all reasonable efforts to contact the authorized employee to inform him or her that his or her lockout or tagout device has been removed; and	1910.147(e)(3)(ii)
1910.269(d)(7)(iv)(C)	Ensuring that the authorized employee has this knowledge before he or she resumes work at that facility.	1910.147(e)(3)(iii)

[Brackets] indicate verbiage included in 1910.147
{Brackets} indicate verbiage included in 1910.269.

TABLE C.11   Comparison 1910.269 v. 1910.147 Requirements: Additional Requirements

29 CFR 1910.269	{Additional Requirements} [Additional Requirements]	29 CFR 1910.147
1910.269(d)(8)(i)	{If the} [In situations in which] lockout or tagout devices must be temporarily removed from energy isolating devices and the machine or equipment must be energized to test or position the machine, equipment, or component thereof, the following sequence of actions shall be followed:	1910.147(f)(1)
1910.269(d)(8)(i)(A)	Clear the machine or equipment of tools and materials in accordance with paragraph {(d)(7)(i)} [(e)(1)] of this section;	1910.147(f)(1)(i)
1910.269(d)(8)(i)(B)	Remove employees from the machine or equipment area in accordance with paragraphs {(d)(7)(ii) and (d)(7)(iii)} [(e)(2)] of this section;	1910.147(f)(1)(ii)
1910.269(d)(8)(i)(C)	Remove the lockout or tagout devices as specified in paragraph {(d)(7)(iv)} [(e)(3)] of this section;	1910.147(f)(1)(iii)
1910.269(d)(8)(i)(D)	Energize and proceed with the testing or positioning; and	1910.147(f)(1)(iv)
1910.269(d)(8)(i)(E)	Deenergize all systems and reapply energy control measures in accordance with paragraph {(d)(6)} [(d)] of this section to continue the servicing {or} [and/or] maintenance.	1910.147(f)(1)(v)

[Brackets] indicate verbiage included in 1910.147
{Brackets} indicate verbiage included in 1910.269

TABLE C.12 Comparison 1910.269 v. 1910.147 Requirements: Group Lockout or Tagout/Outside Personnel (Contractors, etc.)

29 CFR 1910.269	{Group Lockout or Tagout/Outside Personnel} [No Heading]	29 CFR 1910.147
1910.269(d)(8)(ii)	When servicing or maintenance is performed by a crew, craft, department, or other group, they shall {use} [utilize] a procedure which affords the employees a level of protection equivalent to that provided by the implementation of a personal lockout or tagout device.	1910.147(f)(3)(i)
1910.269(d)(8)(ii)	Group lockout or tagout devices shall be used in accordance with the procedures required by {paragraphs (d)(2)(iii) and (d)(2)(iv)} [paragraph (c)(4)] of this section including, but not limited to, the following specific requirements:	1910.147(f)(3)(ii)
1019.269(d)(8)(ii)(A)	Primary responsibility{shall be} [is] vested in an authorized employee for a set number of employees working under the protection of a group lockout or tagout device (such as an operations lock);	1910.147(f)(3)(ii)(A)
1019.269(d)(8)(ii)(B)	Provision shall be made for the authorized employee to ascertain the exposure status of all individual group members with regard to the lockout or tagout of the machine or equipment{;} [and]	1910.147(f)(3)(ii)(B)
1019.269(d)(8)(ii)(C)	When more than one crew, craft, department, or other group is involved, assignment of overall job-associated lockout or tagout control responsibility{shall be given} to an authorized employee designated to coordinate affected work forces and ensure continuity of protection; and	1910.147(f)(3)(ii)(C)
1019.269(d)(8)(ii)(D)	Each authorized employee shall affix a personal lockout or tagout device to the group lockout device, group lockbox, or comparable mechanism when he or she begins work and shall remove those devices when he or she stops working on the machine or equipment being serviced or maintained.	1910.147(f)(3)(ii)(D)
1910.269(8)(iii)	[Specific] Procedures shall be {used} [utilized] during shift or personnel changes to ensure the continuity of lockout or tagout protection, including provision for the orderly transfer of lockout or tagout device protection between off-going and on-coming employees, to minimize {their} exposure to hazards from the unexpected {energizing} [energization] or start-up of the machine or equipment or from the release of stored energy.	1910.147(f)(4)

(*Continued*)

TABLE C.12 Comparison 1910.269 v. 1910.147 Requirements: Group Lockout or Tagout/Outside Personnel (Contractors, etc.)—cont'd

29 CFR 1910.269	{Group Lockout or Tagout/Outside Personnel} [No Heading]	29 CFR 1910.147
1910.269(d)(8)(iv)	Whenever outside servicing personnel are to be engaged in activities covered by {paragraph (d) of this section,} [the scope and application of this standard,] the on-site employer and the outside employer shall inform each other of their respective lockout or tagout procedures{,} [.]	1910.147(f)(2)(i)
1910.269(d)(8)(iv)	{, and each} [The on-site] employer shall ensure that his or her personnel understand and comply with restrictions and prohibitions of the energy control procedures being used.	1910.147(f)(2)(ii)

[Brackets] indicate verbiage included in 1910.147

{Brackets} indicate verbiage included in 1910.269.

Sources: Regulations (Standards – 29 CFR) The Control of Hazardous Energy (Lockout/Tagout) – 1910.147 1: US Department of Labor website:

http://www.osha.gov/pls/oshaweb/owadisp.show_document?p_table=STANDARDS&p_id=9804.

Regulations (Standards – 29 CFR) Electric Power Generation, Transmission, and Distribution – 1910.269: US Department of Labor website:

http://www.osha.gov/pls/oshaweb/owadisp.show_document?p_table=STANDARDS&p_id=9868.

Occupational Safety and Health Administration Standard Interpretations 29 CFR 1910.6; 1910.147; 1910.147(c)(4)(ii)

US Department of Labor
Occupational Safety and Health Administration
200 Constitution Avenue, NW
Washington, DC 20210

Standard Interpretations: 11/10/2004 – Recognition of ANSI/ASSE Z244.1-2003 "Control of Hazardous Energy – Lockout/Tagout and Alternative Methods" consensus standard [1]

Standard Number: 1910.6; 1910.147; 1910.147(c)(4)(ii)

November 10, 2004
Mr. xxxxxxxx
Chairman, Z244 ASC
xxxxxxxx
xxxxxxxx

Dear Mr. xxxx:

This is in response to your March 22, 2004 letter to John Henshaw, Assistant Secretary for the Occupational Safety and Health Administration (OSHA). Your letter seeks recognition of the revised American National Standard on the *Control of Hazardous Energy – Lockout/Tagout And Alternative Methods* (ANSI/ASSE Z244.1-2003) and requests that the Agency hyperlink from the OSHA web site to an American Society of Safety Engineers (ASSE) web page, which contains information about both the Z244 American Standards Committee (ASC) and the revised consensus standard. Your letter was forwarded to our office to reply.

The Z244 American Standards Committee requests that OSHA provide "enhanced recognition" of its latest ANSI standard concerning the control of hazardous energy. Your letter conveys the committee's belief that such recognition may help prevent future injuries and would be consistent with provisions of the *National Technology Transfer and Advancement Act (NTTAA)*, 15 USC §272, and the Office of Management and Budget's Circular A-119, *Federal Participation in the Development and Use of Voluntary Consensus Standards and in Conformity Assessment Activities.*

OSHA shares your interest in preventing workplace injuries and recognizes the valuable contribution that your committee has made in developing the new Z244.1 standard. In addition, OSHA is committed to working cooperatively with entities such as ANSI and ASSE to further workplace safety and health. As provided below, we agree that it is prudent to reference the current American National Standard on the control of hazardous energy on OSHA's web site. Moreover, OSHA welcomes other cooperative initiatives that would assist employers and employees to further occupational safety by effectively controlling hazardous energy.

The OMB Circular (consistent with Section 12(d) of the NTTAA) directs agencies to use voluntary consensus standards in lieu of developing government-unique standards, except when such use would be inconsistent with law or otherwise impractical.[1] Neither the NTTAA nor the OMB Circular mandate the revision of an existing standard, such as the *Control of Hazardous Energy (Lockout/Tagout)* (LOTO), 29 CFR §1910.147, whenever a relevant consensus standard is amended. However, OSHA will consider all relevant consensus standards, including the current ANSI standard, if the Agency determines in the future that it is appropriate to revise §1910.147.

Your paraphrased questions and our replies follow:

Question #1: The ANSI Accredited Standards Committee formally requests that OSHA link to the [Z244 Committee] to OSHA's [*Safety and Health Topics: Control of Hazardous Energy (Lockout/Tagout)* page]. [*Links embedded for OSHA web site posting.*]

Reply: OSHA believes that it is appropriate to add the ANSI Z244 Committee link and this letter to the *Control of Hazardous Energy* page in the *Safety and Health Topics* section of the OSHA web site, and we will do so.

Question #2: OSHA references the Z244.1-1982 (R1993) standard in a number of locations on the OSHA web site. The ASC's view is that the Z244.1-2003 standard provides a level of protection superior to the earlier version and recommends that future references to the Z244.1 standard should refer to the 2003 standard. Does OSHA agree with the contention that future references to the Z244.1 standard need to also include the 2003 standard?

Reply: In the future, OSHA will reference the most recent version of the Z244.1 standard, whenever appropriate. Thus, on the OSHA web site, the Agency will refer to the Z244.1-2003 standard when referencing consensus standards and other recognized resources that provide information and guidance concerning the control of hazardous energy. However, OSHA will continue to reference the Z244.1-1982 standard when citing specific sources that the Agency considered in developing 29 CFR 1910.147, see, e.g., 54 *FR* 36644, 36645 (September 1, 1989) (acknowledging the role that the Z244.1-1982 standard played in the development of OSHA's LOTO standard). Likewise, the Agency will reference only the Z244.1-1982 standard when referring to situations where OSHA explicitly considered the Z244.1-1982 standard, as opposed to the Z244.1-2003 standard, in developing a policy or practice regarding the application of the 1910.147.

Question #3: What is OSHA's enforcement position if an employer is complying with the most recent version of a voluntary consensus standard, which was either previously adopted by reference or cited as a reference document during the rulemaking process?

Reply: As you know, OSHA carefully considered the 1982 ANSI standard in developing the agency's LOTO standard, 1910.147. However, OSHA did not adopt the standard by reference, and in some respects the agency deliberately departed from the ANSI standard in order to provide a higher level of employee protection, see 58 *FR* 16617 (March 30, 1993) (Supplemental Statement of Reasons for the Final Rule).

The OSH Act contemplates a distinction between the national consensus standard process and the process of OSHA rulemaking. While the former often produces information useful in the latter, it is not equivalent. Section 5(a)(2) of the OSH Act requires employers to comply with OSHA standards (29 USC §654(a)(2)). Thus, only national consensus standards that have been adopted as, or incorporated by reference into, an OSHA standard pursuant to Section 6 of the OSH Act provide a means of compliance with Section 5(a)(2) of the OSH Act.[2]

While requiring employers to comply with OSHA standards, the OSH Act also authorizes OSHA to treat certain violations, which have no direct or immediate relationship to safety and health, as *de minimis*, requiring no penalty or abatement, see 29 USC §§ 654(a)(2) and 658(a). OSHA's enforcement policy provides that a violation may be *de minimis*, if an employer complies with a proposed standard or amendment or a consensus standard rather than with the standard in effect at the time of the inspection and the employer's action clearly provides equal or greater employee protection, see OSHA Instruction CPL 2.103, *Field Inspection Reference Manual*, Chapter III, Paragraph C(2)(g), September 26, 1994. In applying this principle, OSHA takes heed of its rulemaking findings.

Question #4: We would sincerely hope that OSHA agrees with the position of the Z244 ASC that the Z244.1 standard is a viable document to use as a guideline when implementing lockout/tagout programs.

Reply: OSHA recognizes the value of national consensus standards, and in many respects, the ANSI Z-244.1-2003 standard offers useful guidance for employers and employees attempting to control hazardous energy. However, OSHA has not determined that, in all cases, compliance with specific provisions of the ANSI Z244.1-2003 standard and its annexes would constitute compliance with relevant OSHA requirements.

To a considerable extent, the OSHA LOTO standard is a performance standard, which establishes general employer obligations, but leaves employers latitude to develop and implement specific methods for meeting those obligations. Where this is the case, the detailed discussion in the ANSI Z244.1-2003 standard often can assist employers in developing specific methods to meet their obligations under the OSHA LOTO standard. For example, the OSHA LOTO standard establishes specific minimum criteria relevant to all energy control procedures, see 1910.147(c)(4)(ii). In Annex C, the ANSI Z244.1-2003 standard details a sample energy control procedure for a blasting cabinet and dust extractor. While OSHA cannot ascertain whether the sample procedure provides the breadth and specificity mandated in 1910.147(c)(4)(ii) without more information about the actual machinery and the manner in which servicing and maintenance would be performed, this sample procedure may provide valuable conceptual assistance to an employer who is developing energy control procedures specific to its machinery/equipment as prescribed by the OSHA LOTO standard. In addition, the sample LOTO placards in Annex D are good examples of supplemental tools that provide critical information specific to particular machines and equipment. An employer who chooses to develop a single, generic energy control procedure can supplement its generic procedure with similar placards to comply with 1910.147(c)(4)(ii).

On the other hand, in several important respects, the ANSI standard appears to sanction practices that may provide less employee protection than that provided by compliance with the relevant OSHA provisions. For example, the consensus standard employs a decision matrix that allows employers to use alternative protective methods in situations where OSHA standards require the implementation of lockout/tagout or machine guarding. In addition, the ANSI standard permits the use of tagout programs if they provide "effective" employee protection, while the OSHA LOTO standard allows the use of a tagout program only where the employer demonstrates it provides Full employee protection – i.e., a level of safety equivalent to that obtained by using a lockout program. Further, the *Hazardous energy control procedures, Communication and training*, and Program Review sections of the ANSI standard, while detailed and conceptually valuable, do not appear to mandate certain discrete practices that are prescribed in parallel sections of OSHA's LOTO standard.

OSHA has not formally compared each provision of the ANSI Z244.1-2003 standard with the parallel provisions in OSHA standards. Given the performance nature of both the ANSI standard and OSHA standards addressing the control of hazardous energy, the scope and purpose of OSHA letters of interpretation, and the fact-specific nature of OSHA enforcement decisions, we are not in a position to definitively state the practical effect of applying each of the various ANSI Z244.1 provisions to specific workplace conditions as you have requested. In determining whether it is appropriate to issue a citation for a violation of the OSH Act, OSHA Area Directors have the discretion to apply the Agency's *de minimis* policy as described above. However, when an OSHA standard prescribes a practice, design, or method that provides a requisite level of employee protection, employers may not adopt an alternative approach that provides a lesser level of employee protection, see 29 USC §§654(a)(2) and 655 (respectively requiring employers to comply with occupational safety and health standards promulgated under the OSH Act and providing the Secretary of Labor with authority to promulgate, modify, or revoke OSH Act occupational safety and health standards).

As you may be aware, OSHA's *Control of Hazardous Energy (Lockout/Tagout)* directive (STD 1-7.3) currently is being revised; the revised directive will address the ANSI Z244.1-2003 standard to assure that OSHA personnel are familiar with the ANSI standard and conversant with the Agency's position with respect to the application of the ANSI standard.

Thank you for your interest and participation in activities that further occupational safety and health. OSHA is committed to working cooperatively with parties like ANSI and ASSE, who share the goal of furthering workplace health and safety. We hope that this letter provides the clarification you were seeking and effectively addresses the issues that you raised. OSHA requirements are set by statute, standards, and regulations. Our interpretation letters explain these requirements and how they apply to particular circumstances, but they cannot create additional employer obligations. This letter constitutes OSHA's interpretation of the requirements discussed. Note that our enforcement guidance may be affected by changes to OSHA rules. Also, from time to time we update our guidance in response to new information. To keep apprised of such developments, you can consult OSHA's website at http://www.osha.gov. If you have any further questions, please feel free to contact the Office of General Industry Enforcement at (202) 693-1850.

Sincerely,

Richard E. Fairfax, Director
Directorate of Enforcement Program

Notes

[1] In fact, Congress placed such a high value on consensus standards that it directed the Agency, through legislation, to utilize them in the rulemaking process. Section 6(b)(8) of the Occupational Safety and Health Act of 1970 (Public Law 91-596) states: "Whenever a rule promulgated by the Secretary differs significantly from an existing national consensus standard, the Secretary shall, at the same time, publish in the Federal Register a statement of reasons why the rule as adopted will better effectuate the purposes of this Act than the national consensus standard."

[2] Specific national consensus standards [e.g., American National Standards (ANSI) standards], which the Secretary of Labor adopted on May 29, 1971, were either used as a source standard and published in Part 1910 as an OSHA standard or explicitly incorporated by reference in an OSHA standard. For further details see Section 6 of the OSH Act and §1910.6 for the specific standards incorporated by reference in this part. **[Corrected on 1/09/2006]**

Source: US Department of Labor website: http://www.osha.gov/pls/oshaweb/owadisp.show_document? p_table=INTERPRETATIONS&p_id=24969.

Index

Printed and bound by CPI Group (UK) Ltd, Croydon, CR0 4YY

08/05/2025

01864832-0003